ROUTLEDGE HANDBOOK OF THE HISTORY OF SUSTAINABILITY

The *Routledge Handbook of the History of Sustainability* is a far-reaching survey of the deep and contemporary history of sustainability. This innovative resource will help to define the history of sustainability as an identifiable field. It provides a unique resource for advanced undergraduates, graduate students, and scholars, and delivers essential context for understanding the current state and future path of the sustainability movement.

The history of sustainability is an increasingly important domain within the discipline of history, which draws on an interdisciplinary set of fields, ranging from energy studies, transportation, and urbanism to environmental history, economics, and philosophy. Key sections in this handbook cover the historiography of sustainability, resilience and collapse in historical societies, the deep roots of sustainability (seventeenth century to nineteenth century), the recent history of sustainability (twentieth century to present), and core issues and key debates in sustainability.

This handbook is an invaluable research and teaching tool for those interested in the history and development of sustainability and an essential resource for the many sustainability studies programs that now exist in the world's universities.

Jeremy L. Caradonna has a PhD in History and is Adjunct Professor of Environmental Studies at the University of Victoria, Canada.

ROUTLEDGE HANDBOOK OF THE HISTORY OF SUSTAINABILITY

Edited by
Jeremy L. Caradonna

Routledge
Taylor & Francis Group

LONDON AND NEW YORK

First published 2018 by Routledge

2 Park Square, Milton Park, Abingdon, Oxfordshire OX14 4RN

52 Vanderbilt Avenue, New York, NY 10017

Routledge is an imprint of the Taylor & Francis Group, an informa business

First issued in paperback 2019

British Library Cataloguing-in-Publication Data
A catalogue record for this book is available from the British Library

Library of Congress Cataloging-in-Publication Data
Names: Caradonna, Jeremy L., 1979- editor.
Title: Routledge handbook of the history of sustainability / edited by Jeremy L. Caradonna.
Other titles: Handbook of the history of sustainability
Description: Abingdon, Oxon ; New York, NY : Routledge, 2018. | Includes bibliographical references and index.
Identifiers: LCCN 2017020061 (print) | LCCN 2017044513 (ebook) | ISBN 9781315543017 (eBook) | ISBN 9781138685796 (harback)
Subjects: LCSH: Sustainability--History. | Sustainability--Case studies.
Classification: LCC GE195 (ebook) | LCC GE195 .R679 2018 (print) | DDC 338.9/27--dc23
LC record available at https://lccn.loc.gov/2017020061

ISBN: 978-1-138-68579-6 (hbk)
ISBN: 978-0-367-85584-0 (pbk)

Typeset in Bembo
by FiSH Books Ltd, Enfield

CONTENTS

ILLUSTRATIONS

Figures

Tables

CONTRIBUTORS

Vandana Baweja is an Associate Professor in the School of Architecture at the University of Florida.

Simon Bell is Senior Lecturer in Information Systems at the Open University.

Iris Borowy is Distinguished Professor in the College of Liberal Arts at Shanghai University.

Jeremy L. Caradonna holds a PhD in History and is Adjunct Professor of Environmental Studies at the University of Victoria.

Gareth Dale is Senior Lecturer in Politics and History at Brunel University London.

Ralf Döring is a researcher at the Thünen Institute for Sea Fisheries in Braunschweig, Germany.

John R. Ehrenfeld is Emeritus Senior Research Associate in Technology, Business and Environment at the Massachusetts Institute of Technology.

Pierce Greenberg is a doctoral candidate in the Department of Sociology at Washington State University.

Emma Griffin is a professor in the School of History at the University of East Anglia.

Ulrich Grober is a journalist and historian who has published widely on the history of sustainability.

Richard Heinberg is Senior Fellow at the Post Carbon Institute and a leading expert on the energy economics and oil depletion.

Erik W. Johnson is Associate Professor of Sociology at Washington State University.

Hervé Le Treut is Professor of Climate Science at the Université Pierre et Marie Curie, a member of the Académie des Sciences at the Institut de France, and the Director of the Institut Pierre Simon Laplace.

David Maggs is a post-doctoral researcher at the University of British Columbia's Centre for Interactive Research on Sustainability.

Stephen Morse is Chair in Systems Analysis for Sustainability at the Centre for Environmental Strategy at the University of Surrey.

Jenia Mukherjee is Assistant Professor of Humanities and Social Sciences at the Indian Institute of Technology Kharagpur.

Barbara Muraca is Assistant Professor in the School of Liberal Arts at Oregon State University.

Ana Maria Peredo is Professor of Environmental Studies at the University of Victoria.

Tarik Masud Quadir is an assistant professor at the Department of Philosophy, Necmettin Erbakan University, Turkey.

John B. Robinson is a Professor in the Munk School of Global Affairs at the University of Toronto.

Preston L. Schiller is Visiting Lecturer in the School of Urban and Regional Planning at Queen's University in Kingston, Ontario.

Matthias Schmelzer is a post-doctoral researcher at the Center for Social and Economic History at the University of Zürich.

Teresa Sabol Spezio is Visiting Assistant Professor of Environmental Analysis at Pitzer College.

Pamela J. Stewart is Senior Research Associate in Anthropology at the University of Pittsburgh.

Andrew Strathern is Andrew Mellon Professor of Anthropology at the University of Pittsburgh.

Joseph A. Tainter is Professor of Environment and Society in the College of Natural Resources at Utah State University.

B. L. Turner II is Emeritus Professor of Geography at Clark University and Regent's Professor at Arizona State University.

Thierry Vrain is the retired Head of Biotechnology at Agriculture Canada's Summerland Research Station.

Claire Weill is climate change senior officer in the French National Institute for Agricultural Research (INRA).

Stephen Zavestoski is Professor of Environmental Sociology at the University of San Francisco.

PART I

Introduction

1

INTRODUCTION

Jeremy L. Caradonna

It is my great pleasure to introduce this captivating and engaging volume to readers. The *Routledge Handbook of the History of Sustainability* will become a foundational work in both the historical study of sustainability and in the many interdisciplinary "sustainability studies" programs that are now found in universities across the globe. It brings together some of the leading scholars who study the history, theory, and practice of sustainability, and introduces students and a general readership to historical trends, core issues, and key debates in the field. It can be considered the "state of the art" on the deep roots and recent history of sustainability, which, as a movement, will continue to have a major impact on twenty-first-century global society.

In some ways, this volume will serve as a reference work that is meant to complement the recent research in the field, including the edited volume *Understanding Sustainable Development* (2014) and the multi-volume *Berkshire Encyclopedia of Sustainability* (2009–12), but also more focused monographs, such as Ulrich Grober's *Sustainability: A Cultural History* (2012) and my own *Sustainability: A History* (2014).[1] The chapters in this volume are argument-driven essays, not neutral-sounding textbook entries, that touch on a wide range of subjects, with the intention of introducing readers to the complex, global heritage of sustainability. In putting together a team of contributors, we have sought broad coverage in terms of subject matter, timeframes, and geographies. Thus, readers will find chapters on subjects as diverse as sustainable design, economics and critiques of growth-based capitalism, climate change, water and irrigation, urbanism, agriculture, transportation, sustainability metrics, theoretical approaches to the concept, and more. In terms of the timeframe, this volume includes chapters on ancient civilizations, pre-contact indigenous societies, and early modern Europe, in addition to many chapters on recent and contemporary history. Geographically, the volume covers aspects of Aboriginal and colonial North America, Western Europe and the Mediterranean, the Highlands of Papua New Guinea, the Ancient Maya, the Islamic Middle East, India, and various aspects of international and world history. Despite the wide coverage of this volume, it should not be considered "complete." We would have liked to include more chapters on Asia, Africa, Aborigine Australia, Amazonia, and the history of sustained-yield forestry in Japan and China, in addition to other subjects. There simply is not enough space to cover it all. But our hope is that readers will use these chapters as entryways into subjects and geographical areas that interest them.

The present volume operates with three assumptions. The first is that multiple societies, especially indigenous and pre-industrial ones, lived sustainably for very long stretches of time, and that societal resilience merits more in-depth analyses. Understanding why and how historical societies absorbed disturbance and thrived for millennia is as important as understanding why some historical societies collapsed. The second assumption is that the modern sustainability movement has multiple points of origin. In a sense, the ideal of "living sustainably" was a standard feature of many world societies that co-existed with neighbors and "earth-others" and who fruitfully shaped ecosystems to meet their own needs. But in a more concrete sense, the sustainability movement has origins in the seventeenth and eighteenth centuries, when societies as diverse as Japan and the Germanic states developed new approaches to "resource management." For instance, Europeans came to understand that the industrially driven economy had wiped out forest resources and that a "sustained yield" of timber was a precondition for a prosperous society. The fact that our modern word for sustainability – the German *Nachhaltigkeit* – was coined in the eighteenth century, in the context of a timber crisis, is really no coincidence. The third assumption is that sustainability cannot be reduced to simply one set of principles, one set of institutions, or one legacy. In this volume, sustainability is treated as a discourse, an objective, a philosophy, a historical process, a cultural movement, a critique of industrial society, and a presumed historical state of being. The historiographical essay that I have written lays out a basic definition of sustainability, but the full breadth and complexity of the concept is evident in these pages.

The volume is divided into six sections. In addition to this Introduction (Part I), the sections are entitled "Historiography of Sustainability" (Part II), "Sustainability, Resilience, and Collapse in Historical Societies" (Part III), "The Roots of Sustainability" (Part IV), "The Recent History of Sustainability" (Part V), and "Core Issues and Key Debates on Sustainability" (Part VI). Part II includes one chapter, an overview of the historiography of sustainability – that is, the way in which sustainability now constitutes an historical field. Part III comprises three chapters, each of which focuses on different pre-industrial or indigenous societies. The chapters in this section address what I call "historical sustainability," which is the attempt to understand long-term resilience and/or demographic collapse (that is, unsustainability) in historical societies. This approach has been developed by Jared Diamond and Jospeh A. Tainter, the latter of whom is a contributor here, and often contains insights about how the demise of certain pre-industrial societies relates to our own unsustainable, industrialized world. Part IV consists of four chapters that address, in historicist terms, the roots of the modern sustainability movement. The chapter by Tarik M. Quadir assesses the role of environmental sustainability in the history of Islamic thought, while the remaining chapters argue that a concern for sustainability derives, in part, from Enlightenment and Industrial Revolution Europe. Part V features nine chapters that bring the analysis up to the present day. These works should be considered "contemporary history" and deal principally with the development of sustainability from the 1970s to the present. They reveal the ways in which sustainability has affected everything from transportation, design, analytical metrics, and urbanism to higher education, water management, economic thought, social movements, and international institutions and policies. Finally, the eight chapters in Part VI are generally focused on concepts, debates, and interpretations surrounding sustainability: problems with fossil-fueled economic growth, the concept of "degrowth," climate change and its human dimensions, rethinking sustainable agriculture, and theoretical reconceptualizations of the concept of sustainability.

This volume is both timely and important. As I write this, the parts per million of carbon dioxide in the atmosphere approaches 410. The momentous Paris climate accord was signed by most of the world's countries, and all of the major emitters, but must be ratified or approved by individual states, and then implemented according to self-determined national plans (to limit global warming to below 2°C). The recent right-wing political backlash in many countries threatens to derail not only the accord but positive climate action more generally. Despite the unprecedented growth in renewable energies, finite and pollutive fossil fuels still supply the vast majority of the world's energy needs. The population of *Homo sapiens* approaches 7.5 billion, and much of the demographic growth is occurring in countries that aspire to Western (high-impact) levels of consumption. Cities are sprawling, species are going extinct, and plastics now choke the world's marine species, which in any case have to cope with increased ocean acidification. Climate refugees are fleeing drought, dwindling resources, and war. Their suffering has sparked a severe humanitarian crisis with no end on the horizon. And so on. The world's socio-ecological problems are severe and overwhelming, which is why it is important that a robust and critical scholarship on sustainability exists. Our approach is historical because we believe that knowledge of the deep and recent history of sustainability provides essential wisdom about how we got to where we are and where we should go from here. Thus, ultimately, the hope is that this volume will do its part to empower students, voters, activists, members of faith communities, and political actors to take positive steps toward creating a more sustainable world.

Note

1   John Blewitt, ed., *Understanding Sustainable Development*, 2nd edition (Abingdon: Routledge, 2014); Berkshire Publishing, *Berkshire Encyclopedia of Sustainability* (Great Barrington, MA: Berkshire Publishing Group, 2009–14); Ulrich Grober, *Sustainability: A Cultural History* (Totnes: Green Books, 2012); Jeremy L. Caradonna, *Sustainability: A History* (Oxford: Oxford University Press, 2014).

References

Berkshire Publishing. *Berkshire Encyclopedia of Sustainability*. Great Barrington, MA: Berkshire Publishing Group, 2009–14.
Blewitt, John, ed. *Understanding Sustainable Development*. 2nd edition. Abingdon: Routledge, 2014.
Caradonna, Jeremy L. *Sustainability: A History*. Oxford: Oxford University Press, 2014.
Grober, Ulrich. *Sustainability: A Cultural History*. Totnes: Green Books, 2012.

PART II

Historiography of sustainability

2

SUSTAINABILITY

A new historiography

Jeremy L. Caradonna

Sustainability has become a ubiquitous buzzword in our society. We now see the concept publicized in grocery stores, on university campuses, in corporate headquarters, in governmental departments of environmental management, in natural resource management, and in numerous other domains. Indeed, sustainability has been a standard feature of public and political discourse ever since the United Nations adopted the concept in a series of conventions and reports in the 1980s. By the 1990s it had become a familiar term in the world of policy wonkery—we might think of President Bill Clinton's Council on Sustainable Development, for example—but sustainability had also garnered its first criticism. In 1996, the environmentalist Bill McKibben called sustainability a "buzzless buzzword" that was "born partly in an effort to obfuscate" and which would never catch on in mainstream society: "[It] has never made the leap to lingo—and never will. It's time to figure out why, and then figure out something else." (McKibben, for his part, preferred the term "maturity.")[1]

Sustainability has certainly been, at times, misused and greenwashed, but it is quite clear that McKibben was incorrect about its decline. Since the year 2000, over 5,000 published books have included either the words "sustainable" or "sustainability" in the title, compared to zero such books before about 1976.[2] A quick Google search for the word "sustainable" returns nearly 150,000,000 hits. Moreover, the sustainability movement, as we should now call it, has gained a level of respect and legitimacy that is difficult to dispute. The scholarly fields associated with sustainability have expanded dramatically, new tools and methods have appeared, such as Ecological Footprint Analysis, the Triple Bottom Line, and the Genuine Progress Indicator, which help define, measure, and assess sustainability, and a broad range of governments, businesses, NGOs, and communities have embraced the principles of sustainable living. Virtually every major institution in the industrialized world has either a department or office of sustainability. In a sense, this environmental discourse has won out over rival conceptions of humanity's relationship to the natural world, even if industrial society remains, by any measure, far from sustainable.[3]

A growing consciousness about the pitfalls of industrialization has stimulated interest in sustainability. The so-called developed world is 250 years into an ecological assault on the planet that was triggered by the Industrial Revolution and which has forced a serious reappraisal of the values of industrialism and growth-based capitalism. According to several influential scholars, we now live in a geological epoch called the Age of the Anthropocene, in which "human activity" has become the "dominant driver of the natural environment."[4] We *are*, or have become, a kind of natural disaster. The Fifth Assessment Report (2014) from the Intergovernmental Panel on Climate Change (IPCC), a team of scientists whose job it is to sort through and summarize the state of climate science, makes it clear that Earth's climate system is warming steadily due to "anthropogenic greenhouse gas concentrations," such as carbon dioxide, methane, and nitrous oxide, all of which trap and radiate heat (at infrared wavelengths) that would otherwise escape from the Earth's atmosphere. "It is extremely likely that human influence has been the dominant cause of the observed warming since the mid–20th century."[5] Deforestation, land-use changes, and the burning of long-buried fossil fuels are the primary culprits. Climate change has already begun to alter natural systems and the environment in troubling ways: Increasingly unpredictable temperatures and weather patterns, changes in the hydrological cycle that generate droughts and larger and more frequent storms, rising sea levels from melting ice caps, the die off of some species, and so on.

Furthermore, the mounting population of homo sapiens on the planet, which surpassed the 7 billion mark in 2012, combined with man-made pollutants and the appropriation of approximately 30 per cent of the net primary production (NPP) of organic material (i.e. we use or alter much of what nature has to offer) has resulted in devastating consequences for the world's life-sustaining ecosystems. Here's Sachs again:

> The Millennium Ecosystem Assessment (MEA), a comprehensive study of the state of the world's ecosystems carried out over several years with the input of more than two thousand scientists, found that during the past fifty years humans have degraded most of the world's ecosystems and driven down the abundance of other species, some to extinction.[6]

Finally, we are now dealing with a moribund economic system that has drained the world of many of its finite resources, including fresh water and crude oil, generated a meltdown in global financial systems, exacerbated social inequality in many parts of the world, and driven human civilization to the brink of catastrophe by unwisely advocating for economic growth at the expense of resources and essential ecosystem services.[7] It is with good reason that Nicholas Stern, in his famed *Stern Review* (2006), which calculated the costs of economic action versus inaction on climate change, referred to climate change as an unprecedented form of "market failure."[8]

The growing interest in sustainability in the *present* and *future* has driven interest in the subject as an *historical field*. Historians now have an abundance of evidence to suggest that present-day cultural concerns dictate the kinds of historical events, discourses, and topics that strike scholars as relevant: Recent interest in gay rights and gay marriage has driven interest in the history of same-sex relationships; the reality of anthropogenic climate change has stimulated a rich exploration of past climate change and its effect on historical societies; in the 1970s, women's history and "history from below" were motivated in large part by contemporary concerns for gender and class equality. Likewise, the history of sustainability parallels, or perhaps grew out of, the explicit formulation of the sustainability *movement*, which took shape in the 1980s and 1990s, even if, as many have argued, the *concept* of sustainability stretches back at least into the early modern period, and traces its lineage to several global cultures.

Around thirty years ago, sustainability became an identifiable and publicly discussed concept, and grew in large part out of the work of ecologists, such as Howard Odum and C. S. Holling, economists, such as E. F. Schumacher, E. J. Mishan, and Herman Daly, systems theorists, such as those in the Club of Rome, energy specialists, such as Amory Lovins, environmentalists, such as Paul Hawken and Lester Brown (and his Worldwatch Institute), biologists and other scientists, such as the International Union for Conservation of Nature (IUCN), and diplomats or appointees within the Organisation for Economic Co-operation and Development (OECD) and the United Nations (UN), the latter of whom transformed the concept of sustainability into "sustainable development," and associated it with a new, more ecologically sensitive approach to development in the Third World.[9] The UN also sponsored a whole series of conferences and committees which brought the cause of sustainability to the forefront of the international community's attention: the 1972 Stockholm Conference (and the Stockholm Declaration) on environment and society, the 1980 report called *World Conservation Strategy*, which spoke of sustainable development, and which was written by the IUCN and backed by the United Nations Environment Programme (UNEP), the 1982 "World Charter for Nature" promulgated by the UN General Assembly, the OECD, which wanted to bridge environmental integrity with economic growth,[10] and perhaps most enduringly, the UN-backed World Commission on Environment and Development (1983–7) that produced the so-called Brundtland Report (actually called *Our Common Future*), which popularized the notion that sustainability is about meeting current needs without jeopardizing the ability of future generations to satisfy their own needs.[11] This growing concern for the fate of humanity sparked, at the same time, an interest in tracing these concepts, practices, and discourses back in time. The sustainability movement thus established a need for a history of sustainability.[12]

The history of sustainability, which has been written in an explicit manner only since the 1990s, has begun to differentiate itself from other, complementary approaches to history, the most important of which is environmental history. According to J. Donald Hughes, environmental history comprises three interlocking lines of historical inquiry: Humankind's impact on the natural world, the natural world's impact on humankind, and cultural values, attitudes, and conceptions of nature and the environment.[13] The history of sustainability borrows most heavily from the last of the three features of environmental history, but rarely incorporates the kind of empirical environmental emphasis that one might find in, say, histories of water management, floods, fire-based ecosystem modification, or soil erosion.[14]

The history of sustainability, as with environmental history, is a broadly interdisciplinary field that draws from numerous disciplines across the arts and sciences, but the former is most concerned with the history of "systems thinking," or the ways in which human societies have conceptualized, dealt with, and responded to the relationship between the natural environment, human wellbeing, and economic systems. This approach mirrors the three Es of sustainability: Environment, economy, and equality (or social justice, or social *injustice*). As such, the history of sustainability draws from ecology, economics (and especially ecological economics), social justice and the study of human rights, population studies, urbanism, environmental and climate science, sociology, engineering, energy studies, archaeology, and several branches of history— political, cultural, intellectual, and environmental. Its methods thus flow from the fields from which it borrows, but discourse analysis, comparative analysis, and historical anthropology seem to be the most common methodological tools for sustainability historians.

In the same way that the current sustainability movement could not have existed without the classic environmental movement, an historical approach to sustainability would not have come into existence without environmental history. But the two subfields are not identical. Historians of sustainability are as interested, and necessarily so, in the history of social justice and economic history as they are in environmental history. Works such as Lynn Hunt's 2008 *Inventing Human Rights: A History* and Anthony Brewer's 2010 *The Making of the Classical Theory of Economic Growth* would be valuable references for a history of sustainability in eighteenth-century Europe, for instance, but neither work has any real relevance to environmental history narrowly defined. The challenge of writing the history of sustainability is to find linkages between environmental thought and practices, economic policy, and social wellbeing, which can incorporate equality, democracy, mental and physical health, life satisfaction, and so on.

To a certain extent, some environmental historians have been writing the history of sustainability for quite some time, even though they have not necessarily been using the explicit language of sustainability or sustainable development. It is obviously not the case that all environmental history focuses narrowly on the natural environment without discussing linkages to social and economic issues, but classic environmental history was often accused of ignoring, in particular, economics. That said, monographic studies such as Andrew Hurley's excellent *Environmental Inequalities: Class, Race, and Industrial Pollution in Gary, Indiana, 1945–1980,* for instance, could be seen as contributing to the history of sustainability (or in this case, *unsustainability*), since it pays such close attention to the interplay between economics, society, and environment.[15] But the history of sustainability differentiates itself from environmental history both in its explicit discussion of the history and concept of sustainability, in its awareness of and attention to systems thinking and the sustainability movement, and its standard interest in balancing social issues, environmental concerns, and economics. As an offspring of contemporary sustainability studies, it also stands out for its emphasis on the *future* wellbeing of human society and for its relative optimism, in contradistinction to what many see as the gloom and doom of environmentalism and even environmental history. But certainly the nuances can be quite subtle between the history of sustainability and some forms of environmental history, and, as noted above, placing sustainability in an historical framework is an exercise that simply could not have come into existence without the resources and model of environmental history.

One could divide the historiography of sustainability into two broad categories. The first category comprises works that analyze the genesis and development of the concept of sustainability, as well as the formation of the actual sustainability movement at the end of the twentieth century. The second category, which we might call "historical sustainability," brings together a range of scholarship that seeks to understand the fate of historical societies—that is, how and why some societies collapsed, such as Ancient Rome and the Maya, whereas other societies, such as the Highlanders of New Guinea, have thrived for thousands of years. Both branches of the historiography focus on the complex relationships between sustainability and social collapse—either the outright collapse of historical societies in the past, or the threat of collapse in the future. Scholars of sustainability are thus always interested in sustainability's binaried other: Unsustainability.

This first approach is clearly an attempt to historicize a set of ideas and a movement that exists in the present day. It addresses the following questions: "Where did sustainability come from?"; "What does the concept mean, necessitate, and imply?"; "When, how, and why did people come to see industrial society as unsustainable?" "How did an economic system based on growth and resource consumption come to dominate (or even create) modern industrial societies?" Defining sustainability, and therefore historicizing it, is no easy task, and theorists such as Richard Heinberg, David Holmgren, Albert A. Bartlett, John Dryzek, and others have formulated somewhat different definitions of sustainability.[16] Based on my own synthesis of the historiography, however, the four main ideas that historians of sustainability tend to analyze and historicize are these:

1   *The idea that human society, the economy, and the natural environment are necessarily interconnected.* This is an ecological idea that coalesced in the mid twentieth century and which considers human society and economy as part of the broader ecosystem.[17]

2   *The contention that human societies must operate within ecological limits if they expect to persist over a long period of time.* Sustainability historians are always on the lookout for historical actors who express—either in words or in actions—an interest in living within the limits dictated by nature.[18]

3   *The notion that human society must engage in wise and sensible future-oriented planning.* The inter-generational component of sustainability has become an important part of the discourse in the present day, but its roots can be found in numerous world cultures, including some aboriginal societies.[19]

4   *The idea that industrial society, above all, needs to adopt the logic of the small and the local and move away from the logic of the big and the centralized if it hopes to survive and thrive in the long term.* The industrialized world has made things big and centralized, but how did this process unfold, who were its critics, and what alternatives does the study of history reveal?[20]

One of the pioneers of this branch of the historiography is the eminent historian Donald Worster, who discussed the history of sustainability in *The Wealth of Nature: Environmental History and the Ecological Imagination* (1994). This collection of essays includes a chapter called "The Shaky Grounds of Sustainable Development," which addresses the conceptual roots of sustainability in forestry and resource economics before criticizing the ambiguity of the idea.[21] Worster also laid some important groundwork for studying historical conceptions of sustainability in his earlier work, *Nature's Economy: A History of Ecological Ideas* (1994), which is an intellectual history of changing conceptions of humanity's relationship to the natural world.[22] He made it clear that both ecological "arcadians" and "imperialists" created the conditions, at least in European society, for seeing human society and economy as constituent parts of nature.

Worster is an historian of the Western world, and thus his work focuses almost exclusively on Europe and European settler societies in North America. The same goes for more recent historians of sustainability, who have generally been Europeanists, Americanists, or historians of the Atlantic world. Far less has been written on the discourse of sustainability in the non-Western world (at least by Anglophone scholars), although Richard H. Grove's *Green Imperialism: Colonial Expansion, Tropical Island Edens and the Origins of Environmentalism, 1600–1860* (1995) represents an important contribution to a more globalized historiography. He argues that European-controlled islands in the Caribbean, the Atlantic Ocean, and the Indian Ocean had a major impact on the development of modern environmental consciousness and the idea of "sustainable development," which he sees as a blend of East Asian, South Asian, and European ideas about managing the natural world.[23] But in terms of sources, most of what has been written about the history of sustainability has been based on printed sources written by social, intellectual, and political elites in Europe and North America.

John Robinson added to the historiography with an influential article called "Squaring the Circle? Some Thoughts on the Idea of Sustainable Development" (2004), which deals with the differences between "sustainability" and "sustainable development." For Robinson, sustainability traces its roots back to John Muir's eco-centric "preservationist" movement, whereas sustainable development is an elaboration of Gifford Pinchot's pro-business and pro-growth conception of "conservationism." He then goes on to criticize sustainable development as little more than business-as-usual economic development that does not value the idea of living within biophysical limits (a common critique of the UN's approach to sustainable development).[24] Robinson has historicized in a very helpful way the ongoing debate over whether sustainable development is merely a greenwashed approach to economics and resource exploitation in the developing world. William Cronon has also weighed in on the growing interest in the history of sustainability, and did so in an important plenary address that he gave at the 2011 conference of the American Society for Environmental History. Although not a published study, Cronon's insightful address argued that the concept of sustainability stretches back long before the word began to buzz. He also discussed the hopeful optimism of the concept and its shortcomings in the political arena.[25]

Only in the past few years have historians begun to craft overarching narratives of sustainability in the European Atlantic world. Simon Dresner perhaps set this trend with his concise book of 2002, updated in 2008, called *The Principles of Sustainability*. It deals with the period from the late nineteenth century to the present, emphasizes the political aspects of sustainability, and argues, in part, that the collapse of Communism opened up new opportunities (and challenges) for green values.[26] Ulrich Grober, the journalist and scholar, has written extensively on the origins of sustainability, which he traces back even earlier than the nineteenth century, to new forms of forest management in England, Germany, and France around 1700. In works such as "Deep Roots: A Conceptual History of Sustainability" (2007) and *Sustainability: A Cultural History* (2012), Grober argues that the history of sustainability begins with the forestry treatises produced by John Evelyn in England, Jean-Baptiste Colbert in France, and especially Hans Carl von Carlowitz in Saxony (Holy Roman Empire). It was the latter, in fact, who invented the word sustainability (*Nachhaltigkeit*) in his 1713 treatise on forestry called *Sylvicultura Oeconomica*.[27] Grober has argued persuasively that, in Europe, deforestation and subsequent timber shortages drove interest in creating what was later called sustainable yield forestry.

Grober shows that trees were to early modern European society what fossil fuels are to industrial society: Utterly foundational. The decline of available forest resources spelt disaster both for the poor, who faced higher wood prices, and social elites who managed wood-reliant industries, including mining and metallurgy, which happened to be Carlowitz's domain in Saxony. Although Grober's book moves forward to the present day, his real emphasis, and contribution is situating the origins of sustainability in forestry. Sustainability in the eighteenth century was not yet a blanket critique of a particular mode of existence so much as it was a technical recalibration of governmental policy by a social elite with the training and influence to make that determination. But Carlowitz and others nonetheless laid the conceptual foundation for a more explicit sustainability movement, especially after sustainable yield forestry began to dominate forestry schools in Germany, France, and elsewhere.

My own overview of sustainability in Europe and North America, *Sustainability: A History* (Caradonna 2014), covers the period from the late seventeenth century to the present day, and also discusses the future challenges of the sustainability movement. It draws on the work of historians, such as Worster and Grober, but also makes significant use of economic history and ecological economics. It analyzes the development of sustainable yield forestry and early "systems thinking" in the eighteenth century, but focuses primarily on the period from the nineteenth century to the present. It shows that there were widespread critiques of environmental destruction, resource overconsumption, population growth, and growth-based economics throughout the Industrial Revolution. It makes significant use of the writings of Thomas Malthus, William Stanley Jevons, John Stuart Mill, David Ricardo, Friedrich Engels, and others on the right, left, and center who criticized the myth of industrial progress, or aspects of it, which became *the* metanarrative of Western society in the modern era. It then moves on the environmental movement and the growth of ecological economics in the 1960s and 1970s, and shows the extent to which the modern sustainability movement grew out of activism and steady-state economics.

The culmination of *Sustainability: A History* is three chapters that deal with recent history, the present day, and the future. The explicit objective is to untangle the complex strands of thought that created the conditions for the emergence of the modern sustainability movement.

Chapter 5 investigates the formation of an explicit sustainability movement in the 1980s and 1990s, with particular attention paid to the politics, treaties, and reports of the United Nations. Chapter 6 profiles the different ways in which sustainability has become integrated into contemporary society: Sustainable design and green building; methods and measurement tools; energy; transportation; housing; higher education; business and finance; economics; urbanism; food systems and localism; and government planning and policymaking.[28] The final chapter discusses ten challenges for the future of the sustainability. The goal for both myself and Grober is to demonstrate the extent to which modern sustainability traces back its lineage at least to the eighteenth century. The Enlightenment is, at once, the origin of unsustainable industrialism *as well as* ideas and practices that shaped sustainability.

Aside from the overarching studies, there have been many monographic articles and books since the 1990s that have added to our understanding of sustainability, sustainable development, and the formation of the sustainability movement. Studies by Carl Mitcham (1995), Desta Mebratu (1998), Anne Dale (2001, 2012), and Jacobus A. Du Pisani (2006) have added immensely to the historiography of sustainability by focusing on the institutional adoption and cultural normalization of sustainability concepts.[29] Equally important has been the burgeoning body of work on economic "Degrowth," led by Serge Latouche, Giorgos Kallis, François Schneider, Christian Kerschner, and Joan Martinez-Alier, which offers a speculative framework for pursuing economic sustainability.[30] The final works to cite are Stephen Macekura's *Of Limits and Growth* (2015) and Iris Borowy's *Defining Sustainable Development for Our Common Future* (2014), both of which have made crucial contributions to understanding the birth, growth, and character of sustainable development.[31]

Although the historiography on sustainability is fairly new, and dates only to the 1990s, it is important to note that much of the recent historical work draws on the historical forays of first-wave ecological economics (late 1960s and 1970s).[32] That is, it is not just environmental history but also economic history—and especially the work of ecological economists—that has served as a crucial source base for historians of sustainability, who have made extensive use of E. J. Mishan's 1967 *The Cost of Economic Growth*, the Club of Rome's 1972 *The Limits to Growth* (Meadows et al.), E. F. Schumacher's 1973 *Small is Beautiful*, Herman Daly's 1973 *Toward a Steady-State Economy*, Daly's 1977 *Steady-State Economics*, E. J. Mishan's 1977 *The Economic Growth Debate*, Amory Lovins's 1977 *Soft Energy Paths: Toward a Durable Peace*, in addition to works by Kenneth Boulding, Howard T. Odum, and Nicholas Georgescu-Roegen. These economic thinkers are important not only because they created ecological economics in the 1970s, but because all of these economists and systems thinkers incorporated historical analyses into their respective economic arguments.[33] In short, they wrote their own economic histories, which challenged the hegemonic economic discourse of the twentieth century. They were all aware that mainstream and neoclassical economists had crafted a narrative that made a certain mode of capitalist economics seem "natural," "normal," and "inevitable." This ubiquitous narrative of economic progress begins with Adam Smith, A.-R.-J. Turgot, Jean-Baptiste Say, and William Huskisson in the Industrial Revolution, passes through Friedrich Hayek, Milton Friedman, and, to a lesser extent, John Maynard Keynes, in the middle twentieth century, and on to the post-war period. Although these economic thinkers represented different strands of economic thought, they also shared much in common.

By contrast, the ecological economists not only rejected the fundamental tenets of neoclassical and growth-oriented economics, with its apathy for the natural world, its adoration of limitless growth, and its ignorance about biophysical limits, but they also revived historical interest in past economic thinkers who had challenged endless economic and population growth, privatization, and/or industrial pollution: Rousseau, Malthus, Ricardo, Jevons, Mill,

Engels, and so on.[34] It becomes clear in reading these economists (and systems thinkers) that there are at least two economic traditions in the Western world: the pro-growth and often (but not always) laissez-faire tradition, on the one hand, and the steady-state, ecological tradition on the other (with some figures, such as Mill and Ricardo, playing a role in both). As diverse as these thinkers were, they have been appropriated by ecological economists since the 1970s as part of an alternative genealogy, in which forgotten economic thinkers have been rehabilitated and/or well-known ones, such as Mill, have been reconsidered and reconceptualized. Thus the economics of sustainability, which is today practiced by William E. Rees, Mathis Wackernagel, Peter Victor, Tim Jackson, Daniel O'Neill, Richard Heinberg, and many others, exists as part of a long economic tradition that runs from Rousseau, Malthus, and Mill, through to the ecological economics and systems thinking of the 1970s, and up to the present day.[35] It's clear from the historiography that ecological economics provides an indispensable set of sources for historians of sustainability, not to mention those in the present working on building a green economy.

The second branch of the historiography is what we might refer to as "historical sustainability," and is, in a sense, an older, more geographically, and more temporally diverse approach to history. The most emblematic books in this branch are Joseph A. Tainter's *The Collapse of Complex Societies* (1988) and Jared Diamond's best-selling *Collapse: How Societies Choose to Fail or Succeed* (2005). Diamond's earlier best-seller, *Guns, Germs, and Steel: The Fates of Human Societies* (1997) also finds a place in the historiography, as does Daron Acemoglu's and James Robinson's *Why Nations Fail: The Origins of Power, Prosperity, and Poverty* (2012), Ian Morris's *Why the West Rules—For Now: The Patterns of History, and What They Reveal about the Future* (2010), and Ian Morris's *The Measure of Civilization: How Social Development Decides the Fate of Nations* (2013).[36] These books, and the many others like them, are not interested in tracing the origins and development of the sustainability movement in the modern world, although most of these authors, and certainly Tainter and Diamond, are concerned for the fate of modern industrial society. It seems clear, however, that both this branch of the historiography and the one discussed above reflect deep-seated anxieties about the world's current ecological crisis.

But whereas Worster, Grober, Robinson, Caradonna, and Borowy are *historicists* who focus on context, concepts, and culture, the historical sustainability of Tainter, Diamond, Morris, and others tends to employ either structuralist techniques, overarching theories or typologies of collapse, and/or arguments by analogy. (Diamond, in fact, responds to accusations that he's a determinist and a structuralist in the opening pages of *Collapse*.[37]) Both Tainter, who is an anthropologist and historian, and Diamond, who is a scientist and historian, undertook their respective studies of failed societies because they fear that similar forms of social collapse could occur in the twenty-first century—indeed, for Diamond, collapses have already occurred. Tainter and Diamond are, in a sense, more interested in unsustainability and collapse than they are in identifying the secrets and strategies of successful long-standing societies. The idea in these works is that we (in the present) should not make the same mistakes made by those in the past. These scholars tend to make these arguments by incorporating diverse historical data sets, extrapolating inductive conclusions, and drawing explicit or implicit parallels with modern society. In a sense, historical events are treated as *exempla* to be followed or avoided.

What is most salient about this branch of the historiography, for the purposes of this essay, is the assumption that modern society is "vulnerable" (Tainter's term) to Roman- or Mayan-style social collapse. The implication is that we should learn from the shipwrecks of history because our own world is structurally similar to—or at least subject to the same problems as—the failed experiments in civilization that were Norse Greenland, pre-Colombian Maya, the

Native American Anasazi, Easter Island, and so on. Here, Tainter discusses his methods and his concern for the present:

> The concern [with collapse] crosses the social and intellectual spectrum, from the responsible scientists and business leaders who make up the Club of Rome, to the more extreme fringes of the "survivalist" movement. In between one finds a variety of serious, well-meaning persons: environmentalists, no-growth advocates, nuclear-freeze proponents, and others. All fear, for one reason or another, that industrial civilization is in danger. Such fears are frequently based on historical analogy with past civilizations that have disappeared (and indeed it is sometimes suggested that we are about to go the way of the dinosaurs).[38]

Tainter makes it clear that he is creating a "general explanation of collapse, applicable in a variety of contexts."[39] Even though his book is ostensibly about the decline of the Roman Empire, the Western Chou Empire, the Egyptian Old Kingdom, the Hittite Empire, and so on, including many "simpler societies,"[40] the book is *really* meant as a warning about the vulnerabilities and perils of our own industrial order.

Similarly, Diamond is interested in collapse because he fears that industrial society is headed for the same fate as the Greenland Vikings and the Anasazi. His strategy is to generalize lessons from a wide variety of contexts, and then use these lessons, inductively, as the basis for a universal theory of failed societies. The idea here is that we should not assume that industrial society invented unsustainable living. Indeed, many societies before the nineteenth century dealt with deforestation, desertification, soil erosion, silted rivers, urban air pollution, drought, crop failure, resource shortages, and population pressures. In *Collapse*, Diamond formulates a five-point framework to understand the collapse of such historical societies as those living on Easter Island, Pitcairn Island, and Henderson Island (all located in the South Pacific), the Anasazi Native Americans who lived in present-day New Mexico, the Maya Civilization of the Yucatán and surrounding areas, and the Vikings who once lived in Southern Greenland. The five factors are as follows:

1 environmental damage;
2 climate change;
3 hostile neighbors;
4 friendly trade partners (or lack thereof); and
5 social responses to environmental problems.[41]

Diamond argues that modern industrial societies face these problems, too, and that an inability to prevent them—or cope resiliently—will lead to population decline and disintegration of the social order.

Of course, not all of the work on societal collapse is comparative, typological, or overarching, and many more localized studies have appeared in recent years. Examples include the archaeologist Arthur Demarest's *Ancient Maya: The Rise and Fall of a Rainforest Civilization* (2005), the historian Charles C. Mann's *1491: New Revelations of the Americas Before Columbus* (2006) and *1493: Uncovering the New World Columbus Created* (2011), and the article "Climate Change During and After the Roman Empire: Reconstructing the Past from Scientific and Historical Evidence," published by Michael McCormick et al. (2012), which draws largely from scientific data to understand the climatic context during the demise of the Roman Empire in the West.[42] The latter article uses a range of data sets to show that Rome enjoyed surprising climatic stability during the rise of the Empire and that Egypt, which became Rome's bread-basket in this period, benefitted from favorable growing conditions. But then the climate became more erratic toward the end of the Empire—it became cooler and drier in the 200s AD, possibly as a result of several volcanic eruptions in the period, before eventually returning to a period of sustained warming. "Such rapid short-term changes," the authors argue, "would have had a great capacity to disrupt food production during the most difficult decades that the Roman Empire had faced so far; the political, military, and monetary crisis peaked between c. 250 and 290."[43]

Although significant differences exist between the two branches of the historiography—the first is historicist, conceptual, and cultural; the second is often structuralist, comparative, typological, and empiricist—each approach has added to our understanding of the past, and perhaps more importantly, our relationship to it. The concern for the present that characterizes the history of sustainability has created a knowledge base that is practical, relevant, and informative, and which can empower the citizens, leaders, and decision-makers who confront the ecological challenges of the present day. Above all the usefulness of a history of sustainability is that it provides essential context for understanding and addressing our current ecological predicament. The body of knowledge that is in the process of forming provides helpful answers to the following questions:

- "How did the sustainability movement come about and what does it criticize and counteract?"
- "How did industrial society become so unsustainable—why are we living in 'global overshoot'"?
- "What kinds of alternative economic models does history offer us?"
- "How can our own society avoid the fate of collapsed societies?"
- "How do social, economic, and environmental factors interrelate?"

By addressing these and similar questions, the history of sustainability has become a culturally and politically charged subfield of the historical discipline, akin to gender history, labor history, race history, and other approaches that eschew a pretense of detached neutrality. The raving success of Jared Diamond's books is just one indicator that the public is deeply interested in this growing body of knowledge.

How can this field improve? Where will it go from here? The history of sustainability could develop in a range of ways in the coming years. First, it needs to establish its own professional and academic identity separate from both environmental history and economic history. Even though sustainability (and sustainable development) is in the process of becoming a set of identifiable academic disciplines, replete with scholarly experts, university courses, degree programs (the College of Sustainability at Dalhousie University, the BA in Environment and Sustainability at the University of British Columbia, the PhD program in Sustainable Development at Columbia University, etc.), journals (too many to list), and so on, the development within the *history* of sustainability has been rather slow going. As of 2016, there is no academic journal that is dedicated uniquely to the subject, and as a result, works that fit within this body of knowledge often end up in journals such as *Environmental History*, *Environment and History*, and even *Ecological Economics*. Clearly, the history of sustainability needs its own journal—and probably its own conferences and/or panels—if it expects to develop its own academic and professional identity. There also needs to be more university courses on the subject. My own seminars on the history of sustainability are still something of an anomaly.

Second, the field needs greater specialization. The relative lack of work on the subject has meant that pioneering historians have had little in the way of historiographical baggage to weigh them down. As a result, studies such as Grober's and my own have ranged over time and space. But the broader narratives need to rely on microhistories and monographical analyses, which will hopefully emerge in the next decade. Greater attention to local conditions and histories will help nuance the broader understanding that we have about sustainability, its past, its present, and maybe even its future. To a certain extent, though, academic publishing houses have begun to take note of the history of sustainability. Michael Egan, for instance, is editing a book series for MIT Press called "History for a Sustainable Future" that has already begun to publish monographs.[44] Also, it seems clear that at least some of the work on the origins and structures of sustainability is being done outside of traditional history departments—in peace studies, environmental studies, ecology, resource economics, food studies, environmental sociology, and, indeed, "sustainability studies" (a new interdisciplinary academic field), meaning that historians should collaborate with and learn from colleagues outside of history.[45]

Third, with specialization will come a diversification of sources. Although the historical sustainability of Tainter, Diamond, and others has dealt with non-Western societies and archaeological sources, the bulk of sustainability histories have relied on fairly "traditional," printed sources produced by intellectuals, politicians, bureaucrats, economists, ecologists, and so on. Moreover, relatively little attention has been paid to how such countries as China, India, and Japan contributed to the global sustainability movement, and if such studies are being written in those respective countries, then there has not been enough in the way of cross-cultural exchange. Many of the world's indigenous societies appear only sporadically, or not at all, in the current literature (except when they suffer collapse). I hope to see the history of sustainability become more globalized, less Eurocentric, and more nuanced in the years to come.

Fourth, historians of sustainability need to identify and refine their methods. Currently, as noted, there's a split between the more culturally oriented historians and those that rely on comparative, typological history. But how will the field develop from here? Will this division remain or could it be collapsed? It seems clear that whatever happens, the history of sustainability needs to build an interdisciplinary framework that accommodates economic, social, and environmental perspectives.

Fifth, there needs to be a comparative history of resilience. Thus far, both branches of the historiography have focused on either collapse or the rise (and critique) of an unsustainable industrial society. But why is there so much more of an emphasis on failure, decline, and collapse than there is on success, resilience, and survival? This bias is most recognizable in the examination of indigenous societies in historical sustainability, in which societies such as the Maya and the Anasazi enter the story only when they undergo precipitous decline. Why not, at the same time, attempt to understand why some pre-industrial societies *survived* for so long and under such difficult circumstances? We shouldn't only fear collapse; we should also admire resilience. Indeed, virtually nothing has been written about the dynamic endurance of many indigenous societies.[46]

Sixth and finally, the role of economic and social history must remain central to the study of sustainability, just as economics and social justice constitute two of the three Es of sustainability. There has been a tendency, understandable to a certain extent, to cast sustainability as an "environmental" discourse appropriate only for environmental historians. But systems thinking and its history—along with the history of collapse and resilience—requires a dynamic understanding of how society, environment, and economics interrelate and contribute to the successes and failures of human societies.

19

As the sustainability movement continues to grow, and as our world sinks deeper into ecological crisis, the history of sustainability will continue to expand and develop. It seems like only a matter of time until the field gets its own journals, its own experts, its own PhD students, its own identity. It's exciting, and relatively rare these days, to be involved in the formation of a new academic arena. Just as those who work in the world of sustainability have an opportunity to impact the future of industrial society, so too do historians of sustainability have the opportunity to influence how we in the present view our relationship to the past, and where we want our society to go in the future.

Acknowledgements

This chapter is adapted from an earlier version published as "The Historiography of Sustainability: An Emergent Subfield," *Economic and Ecohistory* 11 (2015), 7–18.

My thanks to Iris Borowy for her helpful feedback on revising this chapter.

Notes

1 Bill McKibben, "Buzzless Buzzword," *New York Times* (April 10, 1996).
2 See the Hollis Catalog at Harvard University.
3 See, for instance, the Ecological Footprint Analysis in Mathis Wackernagel et al., "Tracking the Ecological Over-Shoot of the Human Economy," *Proceedings of the National Academy of Sciences* (July 9, 2002), 9269.
4 Jeffrey D. Sachs, *Common Wealth: Economics for a Crowded Planet* (London: Penguin, 2008); see also, Paul Crutzen, "Geology of Mankind," *Nature* 415 (2002), 23.
5 IPCC, *Climate Change 2013: Synthesis Report* (Cambridge: Cambridge University Press, 2013). It strengthens the language and findings of IPCC, *Climate Change 2007: Synthesis Report* (Cambridge: Cambridge University Press, 2007). For instance, the 2014 report found that there is a 95–100 percent chance that human actions are the primary cause of the warming of the past few decades, whereas the 2007 put the figure at 90–100 percent.
6 Sachs, *Common Wealth*, 139. See also MA, *Millennium Ecosystem Assessment* (Washington, DC: Island Press, 2005), online: http://millenniumassessment.org/en/Condition.html; for another take on human appropriation of NPP, see Fridolin Krausmann et al., "Global Human Appropriation of Net Primary Production Doubled in the 20th Century," *PNAS* 110, no. 25 (2013), 10, 324–9.
7 For instance, Paul Mason, *Meltdown: The End of the Age of Greed* (London: Verso, 2010).
8 Nicholas Stern, *Stern Review on the Economics of Climate Change* (London: HM Treasury, 2006).
9 Most of these authors and institutions are cited below in the notes. For more on the origins of sustainable development, see Michael Redclift, *Sustainable Development: Exploring the Contradictions* (London: Routledge, 1987). Note, also, that my own book discusses the conceptual differences between sustainability and sustainable development, but I will not dwell on the subject too much in this essay. See Jeremy L. Caradonna, *Sustainability: A History* (Oxford: Oxford University Press, 2014). John A. Robinson, whose article is cited below, also contrasts sustainability and sustainable development. Finally, this paragraph merely evokes some of the major names and organizations associated with sustainability, but there is obviously much more to the story. See also John Blewitt, ed., *Sustainable Development*, 3 vols. (London: Routledge, 2014).
10 OECD, *The State of the Environment in OECD Countries* (Paris: OECD, 1979).
11 World Commission on Environment and Development (WCED), *Our Common Future* (Oxford: Oxford University Press, 1987).
12 There simply is not enough time or space to summarize the formation of and current practices of the sustainability movement. This history is discussed by myself, in *Sustainability: A History*, and in Simon Dresner, *The Principles of Sustainability* (London: Earthscan, 2008). For the purposes of this essay, I will focus on the historiography.
13 J. Donald Hughes, *What is Environmental History?* (Cambridge: Polity Press, 2006). See also J. R. McNeill and A. Roe, eds., *Global Environmental History: An Introductory Reader* (Abingdon: Routledge, 2012); John R. McNeill, *Something New Under the Sun: An Environmental History of the Twentieth-Century World* (New York: W. W. Norton & Company, 2001).

14 There are many examples to cite, but see, for instance: David Soll, *Empire of Water: An Environmental and Political History of the New York City Water Supply* (Ithaca, NY: Cornell University Press, 2013); Emily O'Gorman, *Flood Country: An Environmental History of the Murray-Darling Basin* (Collingwood: CSIRO Publishing, 2012); Steven J. Pyne, *Fire: A Brief History* (Seattle, WA: University of Washington Press, 2001); J. R. McNeil and V. Winiwarter, eds., *Soils and Societies: Perspectives from Environmental History*, 2nd revised edition (Winwick: White Horse Press, 2010); David R. Montgomery, *Dirt: The Erosion of Civilizations* (Berkeley, CA: University of California Press, 2007).

15 Andrew Hurley, *Environmental Inequalities: Class, Race, and Industrial Pollution in Gary, Indiana, 1945–1980* (Chapel Hill, NC: University of North Carolina Press, 1995).

16 See Richard Heinberg and D. Lerch, eds., *The Post-Carbon Reader: Managing the 21st Century's Sustainability Crisis* (Heraldsburg, CA: Watershed Media, 2010); David Holmgren, *Permaculture: Principles and Pathways Beyond Sustainability* (Hepburn, Australia: Holmgren Design Services, 2002); Albert A. Bartlett, "Reflections on Sustainability, Population Growth, and the Environment— Revisited," *Renewable Resources Journal* 15, no. 4 (Winter 1997–8), 6–23; John Dryzek, *Politics of the Earth: Environmental Discourses*, 2nd edition (Oxford: Oxford University Press, 2005).

17 This idea is seen, for instance, in the work of ecological economists from the late 1960s and 1970s, discussed below in the text, from the Club of Rome and their approach to systems, and in the historical work of Donald Worster, also cited later in the text.

18 This idea is found in the Brundtland Report (WCED, *Our Common Future*), in the work of the Club of Rome, and in Herman Daly's principal works, cited below.

19 This idea is most commonly associated with the Brundtland Report, in addition to other UN and UN-backed documents, including the *World Conservation Strategy: Living Resource Conservation for Sustainable Development* (Gland: IUCN, 1980).

20 This idea is forever associated with E. F. Schumacher, whose writings remain a crucial inspiration to the contemporary sustainability movement. His work is also cited below.

21 Donald Worster, *The Wealth of Nature: Environmental History and the Ecological Imagination* (Oxford: Oxford University Press, 1994).

22 Donald Worster, *Nature's Economy: A History of Ecological Ideas*, 2nd edition (Cambridge: Cambridge University Press, 1994).

23 Richard H. Grove, *Green Imperialism: Colonial Expansion, Tropical Island Edens and the Origins of Environmentalism, 1600–1860* (Cambridge: Cambridge University Press, 1995).

24 John Robinson, "Squaring the Circle? Some Thoughts on the Idea of Sustainable Development," *Ecological Economics* 48 (2004), 369–84.

25 William Cronon, "Sustainability: A Short History for the Future," ASEH plenary talk, Phoenix, Arizona, April 14, 2011.

26 Dresner, *Principles of Sustainability*, ch. 9.

27 Ulrich Grober, *Deep Roots: A Conceptual History of "Sustainability"*, (Berlin: Wissenschaftszentrum Berlin für Sozialforschung, February 2007); Ulrich Grober, *Sustainability: A Cultural History*. Trans. Ray Cunningham (Totnes: Green Books, 2012).

28 Caradonna, *Sustainability*.

29 Carl Mitcham, "The Concept of Sustainable Development: Its Origins and Ambivalence," *Technological Society* 17 (1995), 311–26; Desta Mebratu, "Sustainability and Sustainable Development: Historical and Conceptual Review," *Environmental Impact Assessment Review* 18, no. 6 (November 1998), 493–520; Ann Dale, *At The Edge: Sustainable Development in the 21st Century* (Vancouver: UBC Press, 2001); Ann Dale, "Introduction," in *Urban Sustainability: Reconnecting Space and Place*, A. Dale, W. T. Dushenko, P. Robinson, eds. (Toronto: University of Toronto Press, 2012); Jacobus A. Du Pisani, "Sustainable Development—Historical Roots of the Concept," *Environmental Science* 3, no. 2 (2006), 83–96.

30 See, for instance, Serge Latouche, *Farewell to Growth* (Cambridge: Polity, 2010) and F. Schneider, G. Kallis, and J. Martinez-Alier, "Crisis or Opportunity? Economic Degrowth for Social Equity and Ecological Sustainability. Introduction to this Special Issue," *Journal of Cleaner Production* 18 (2010), 511–18.

31 Stephen Macekura, *Of Limits and Growth: The Rise of Global Sustainable Development in the Twentieth Century* (Cambridge: Cambridge University Press, 2015); Iris Borowy, *Defining Sustainable Development for Our Common Future: A History of the World Commission on Environment and Development (Brundtland Report)* (New York: Earthscan, 2014).

32 See, for instance, Caradonna, *Sustainability* and Grober, *Sustainability*.

33 See, for instance, the following works: Donella H. Meadows, Dennis L. Meadows, Jørgen Randers, and William W. Behrens III (The Club of Rome), *The Limits to Growth* (New York: Universe Books, 1972); E. F. Schumacher, *Small is Beautiful: Economics as If People Mattered* (London: HarperCollins, 1973); Herman E. Daly, ed., *Toward a Steady-State Economy* (New York: W. H. Freeman and Company, 1973); Herman E. Daly, *Steady-State Economics* (New York: W. H. Freeman and Company, 1977); E. J. Mishan, *The Cost of Economic Growth* (Staples, 1967); E. J. Mishan, *Economic Growth Debate: An Assessment* (London: Allen & Unwin, 1977); Amory Lovins, *Soft Energy Paths: Toward a Durable Peace* (London: Penguin, 1977); Kenneth Boulding, "The Economics of the Coming Spaceship Earth," reproduced in Daly, *Toward a Steady-State Economy*. Originally published in *Environmental Quality in a Growing Economy* (Baltimore, MD: JHU Press, 1966); Howard T. Odum and Elisabeth C. Odum, *Energy Basis for Man and Nature* (New York: McGraw-Hill Book Company, 1976); Nicholas Georgescu-Roegen, *The Entropy Law and the Economic Process* (Cambridge, MA: Harvard University Press, 1971).

34 All of these thinkers remain important figures in the sustainability movement. Historians such as Nick Cullather, Matthew J. Connelly, Thomas Robertson, and Jared Diamond, in various writings, continue to engage directly with Malthus and his concerns about resource consumption and over-population. Mill has been recast as economist of sustainability ever since Herman Daly revived interest in Mill's work in the 1970s. Rousseau, Jevons, Engels, and Ricardo are also common referents for ecological economists and other defenders of the green economy. See Caradonna, *Sustainability*.

35 See, for instance: William E. Rees, "Ecological Footprints and Appropriated Carrying Capacity: What Urban Economics Leaves Out," *Environment and Urbanization* 4, no. 2 (October 1992), 121–30; Mathis Wackernagel and William E. Rees, *Our Ecological Footprint: Reducing Human Impact on the Earth* (Gabriola Island: New Society, 1996); Peter Victor, *Managing Without Growth: Slower By Design, not Disaster* (Cheltenham: Edward Elgar Publishing, 2008); Tim Jackson, *Prosperity Without Growth: Economics For a Finite Planet* (London: Earthscan, 2009); Richard Heinberg, *The End of Growth: Adapting to Our New Economic Reality* (Gabriola Island: New Society, 2011).

36 Jared Diamond, *Guns, Germs, and Steel: The Fates of Human Societies* (New York: W. W. Norton, 1997); Daron Acemoglu and James Robinson, *Why Nations Fail: The Origins of Power, Prosperity, and Poverty* (New York: Crown Business, 2012); Ian Morris, *Why the West Rules—For Now: The Patterns of History, and What they Reveal about the Future* (London: Picador, 2010); Ian Morris, *The Measure of Civilization: How Social Development Decides the Fate of Nations* (Princeton, NJ: Princeton University Press, 2013).

37 See the opening chapter in Jared Diamond, *Collapse: How Societies Choose to Fail or Succeed* (New York: Viking Press, 2005). Moreover, there has been a fair bit of testy exchange between Diamond and his historical critics. In 2003 Diamond spoke at the American Society for Environmental History, and faced major criticisms from environmental historians for his methods and conclusions. See also William H. McNeill's earlier critique of *Guns, Germs, and Steel*: "History Upside Down," *New York Review of Books* (May 15, 1997). And see also J. R. McNeill's "The World According to Jared Diamond," *The History Teacher* 34, no. 2 (2001), 1–8, in addition to Diamond's various responses to McNeill and other critics.

38 Joseph Tainter, *The Collapse of Complex Societies* (Cambridge: Cambridge University Press, 1988), 2–3.

39 Ibid., 3.

40 Ibid., 24.

41 Diamond, *Collapse*, 11.

42 Arthur Demarest, *Ancient Maya: The Rise and Fall of a Rainforest Civilization* (Cambridge: Cambridge University Press, 2005); Charles C. Mann, *1491: New Revelations of the Americas Before Columbus* (New York: Vintage, 2006); Charles C. Mann, *1493: Uncovering the New World Columbus Created* (New York: Vintage, 2011); Michael McCormick et al., "Climate Change During and After the Roman Empire: Reconstructing the Past from Scientific and Historical Evidence," *Journal of Interdisciplinary History* 43, no. 2 (August 2012), 169–220.

43 McCormick et al., "Climate Change During and After the Roman Empire," 186.

44 See Michael Egan's webpage on the series: http://eganhistory.com/book-series/.

45 The literature on sustainability and sustainable development is vast and I do not intend to invoke all of it here. My book, *Sustainability: A History*, discusses the current state of the literature and current practices associated with the movement. For the purposes of this essay, I will mention but a few titles. Bill Adams, *Green Development: Environment and Sustainability in a Developing World* (New York: Routledge, 2008); S. Sorlin and P. Warde, eds., *The Future of Nature* (New Haven, CT: Yale University Press, 2013); Wackernagel, "Tracking the Ecological Over-Shoot of the Human Economy"; Richard

Heinberg, *The End of Growth: Adapting to Our New Economic Reality* (Gabriola Island: New Society, 2011); Jan Gehl, *Cities for People* (Washington, DC: Island Press, 2010); John Ehrenfeld, *Sustainability by Design: A Subversive Strategy for Transforming Our Consumer Culture* (New Haven, CT: Yale University Press, 2009).

46 Resilience is a domain of ecology that looks at the ability of ecosystems and species to respond to disturbance and change. C. S. Holling developed the approach in the 1970s, and since then it has become a growing component of ecology. However, virtually nothing has been written on the history of socio-ecological resilience. For the important contemporary studies, see C. S. Holling, "Resilience and Stability of Ecological Systems," *Annual Review of Ecology and Systematics* 4 (November 1973), 1–23; Michael Lewis and Pat Conaty, *The Resilience Imperative: Cooperative Transitions to a Steady State Economy* (Gabriola Island: New Society, 2012); Brian Walker, C. S. Holling, Stephen R. Carpenter, and Ann Kinzig, "Resilience, Adaptability and Transformability in Social-Ecological Systems," *Ecology and Society* 9, no. 2 (2004), article 5; Andrew Zolli and Ann Marie Healy, *Resilience: Why Things Bounce Back* (New York: Free Press, 2012).

References

Acemoglu, Daron and James Robinson. *Why Nations Fail: The Origins of Power, Prosperity, and Poverty*. New York: Crown Business, 2012.

Adams, Bill. *Green Development: Environment and Sustainability in a Developing World*. New York: Routledge, 2008.

Bartlett, Albert A. "Reflections on Sustainability, Population Growth, and the Environment—Revisited." *Renewable Resources Journal* 15, no. 4 (Winter 1997–8), 6–23.

Blewitt, John, ed., *Sustainable Development*, 3 vols. London: Routledge, 2014.

Borowy, Iris. *Defining Sustainable Development for Our Common Future: A History of the World Commission on Environment and Development (Brundtland Report)*. New York: Earthscan, 2014.

Brewer, Anthony. *The Making of the Classical Theory of Economic Growth*. London: Routledge, 2010.

Caradonna, Jeremy. *Sustainability: A History*. Oxford: Oxford University Press, 2014.

Cronon, William. "Sustainability: A Short History for the Future." ASEH plenary talk, Phoenix, Arizona, April 14, 2011.

Crutzen, Paul. "Geology of Mankind." *Nature* 415 (2002), 23.

Dale, Ann. *At The Edge: Sustainable Development in the 21st Century*. Vancouver: UBC Press, 2001.

Dale, Ann. "Introduction." In *Urban Sustainability: Reconnecting Space and Place*, A. Dale, W. T. Dushenko, P. Robinson, eds. Toronto: University of Toronto Press, 2012.

Daly, Herman E., ed. *Toward a Steady-State Economy*. New York: W. H. Freeman and Company, 1973.

Daly, Herman E. *Steady-State Economics*. New York: W. H. Freeman and Company, 1977.

Demarest, Arthur. *Ancient Maya: The Rise and Fall of a Rainforest Civilization*. Cambridge: Cambridge University Press, 2005.

Diamond, Jared. *Collapse: How Societies Choose to Fail or Succeed*. New York: Viking Press, 2005.

Diamond, Jared. *Guns, Germs, and Steel: The Fates of Human Societies*. New York: W. W. Norton, 1997.

Dresner, Simon. *The Principles of Sustainability*. London: Earthscan, 2008.

Dryzek, John. *Politics of the Earth: Environmental Discourses*, 2nd edition. Oxford: Oxford University Press, 2005.

Du Pisani, Jacobus A. "Sustainable Development—Historical Roots of the Concept." *Environmental Science* 3, no. 2 (2006), 83–96.

Ehrenfeld, John. *Sustainability by Design: A Subversive Strategy for Transforming Our Consumer Culture*. New Haven, CT: Yale University Press, 2009.

Gehl, Jan. *Cities for People*. Washington, DC: Island Press, 2010.

Georgescu-Roegen, Nicholas. *The Entropy Law and the Economic Process*. Cambridge, MA: Harvard University Press, 1971.

Grober, Ulrich. *Deep Roots: A Conceptual History of "Sustainability"*. Berlin: Wissenschaftszentrum Berlin für Sozialforschung, February 2007.

Grober, Ulrich. *Sustainability: A Cultural History*. Trans. Ray Cunningham. Totnes: Green Books, 2012.

Grove, Richard H. *Green Imperialism: Colonial Expansion, Tropical Island Edens and the Origins of Environmentalism, 1600–1860*. Cambridge: Cambridge University Press, 1995.

Heinberg, Richard. *The End of Growth: Adapting to Our New Economic Reality*. Gabriola Island: New Society, 2011.

Heinberg, Richard and D. Lerch, eds. *The Post-Carbon Reader: Managing the 21st Century's Sustainability Crisis*. Heraldsburg, CA: Watershed Media, 2010.

Holling, C.S. "Resilience and Stability of Ecological Systems." *Annual Review of Ecology and Systematics* 4 (November 1973), 1–23.

Holmgren, David. *Permaculture: Principles and Pathways Beyond Sustainability*. Hepburn, Australia: Holmgren Design Services, 2002.

Hughes, J. Donald. *What is Environmental History?* Cambridge: Polity Press, 2006.

Hunt, Lynn. *Inventing Human Rights: A History*. New York: W. W. Norton, 2008.

Hurley, Andrew. *Environmental Inequalities: Class, Race, and Industrial Pollution in Gary, Indiana, 1945–1980*. Chapel Hill, NC: University of North Carolina Press, 1995.

IPCC. *Climate Change 2007: Synthesis Report*. Cambridge: Cambridge University Press, 2007.

IPCC. *Climate Change 2013: Synthesis Report*. Cambridge: Cambridge University Press, 2013.

IUCN. *World Conservation Strategy: Living Resource Conservation for Sustainable Development*. Gland: IUCN, 1980.

Jackson, Tim. *Prosperity Without Growth: Economics For a Finite Planet*. London: Earthscan, 2009.

Krausmann, Fridolin, et al., "Global Human Appropriation of Net Primary Production Doubled in the 20th Century." *PNAS* 110, no. 25 (2013), 10, 324–9.

Latouche, Serge. *Farewell to Growth*. Cambridge: Polity, 2010.

Lewis, Michael and Pat Conaty. *The Resilience Imperative: Cooperative Transitions to a Steady State Economy*. Gabriola Island: New Society, 2012.

Lovins, Amory. *Soft Energy Paths: Toward a Durable Peace*. London: Penguin, 1977.

MA. *Millennium Ecosystem Assessment*. Washington, DC: Island Press, 2005. online: http://millenniu-massessment.org/en/Condition.html.

Mann, Charles C. *1491: New Revelations of the Americas Before Columbus*. New York: Vintage, 2006.

Mann, Charles C. *1493: Uncovering the New World Columbus Created*. New York: Vintage, 2011.

Mason, Paul. *Meltdown: The End of the Age of Greed*. London: Verso, 2010.

McCormick, Michael, et al. "Climate Change During and After the Roman Empire: Reconstructing the Past from Scientific and Historical Evidence." *Journal of Interdisciplinary History*. 43, no. 2 (August 2012), 169–220.

McKibben, Bill. "Buzzless Buzzword." *New York Times* (April 10, 1996).

McNeill, J. R. *Something New Under the Sun: An Environmental History of the Twentieth-Century World*. New York: W. W. Norton & Company, 2001.

McNeill, J. R. "The World According to Jared Diamond." *The History Teacher* 34, no. 2 (2001), 1–8.

McNeil, J. R. and V. Winiwarter, eds. *Soils and Societies: Perspectives from Environmental History*. 2nd revised edition. Winwick: White Horse Press, 2010.

McNeill, J. R. and A. Roe, eds. *Global Environmental History: An Introductory Reader*. Abingdon: Routledge, 2012.

McNeill, William H. "History Upside Down." *New York Review of Books* (May 15, 1997).

Macekura, Stephen. *Of Limits and Growth: The Rise of Global Sustainable Development in the Twentieth Century*. Cambridge: Cambridge University Press, 2015.

Meadows, Donella H., Dennis L. Meadows, Jørgen Randers and William W. Behrens III (The Club of Rome). *The Limits to Growth*. New York: Universe Books, 1972.

Mebratu, Desta. "Sustainability and Sustainable Development: Historical and Conceptual Review." *Environmental Impact Assessment Review* 18, no. 6 (November 1998), 493–520.

Mishan, E. J. *The Cost of Economic Growth*. Staples, 1967.

Mishan, E. J. *Economic Growth Debate: An Assessment*. London: Allen & Unwin, 1977.

Mitcham, Carl. "The Concept of Sustainable Development: Its Origins and Ambivalence." *Technological Society* 17 (1995), 311–26.

Montgomery, David R. *Dirt: The Erosion of Civilizations*. Berkeley, CA: University of California Press: 2007.

Morris, Ian. *The Measure of Civilization: How Social Development Decides the Fate of Nations*. Princeton, NJ: Princeton University Press, 2013.

Morris, Ian. *Why the West Rules—For Now: The Patterns of History, and What they Reveal about the Future*. London: Picador, 2010.

Odum, Howard T. and Elisabeth C. Odum. *Energy Basis for Man and Nature*. New York: McGraw-Hill Book Company, 1976.

OECD, *The State of the Environment in OECD Countries*. Paris: OECD, 1979.

O'Gorman, Emily. *Flood Country: An Environmental History of the Murray-Darling Basin*. Collingwood:

CSIRO Publishing, 2012.

Pyne, Steven J. *Fire: A Brief History*. Seattle, WA: University of Washington Press, 2001.

Redclift, Michael. *Sustainable Development: Exploring the Contradictions*. London: Routledge, 1987.

Rees, William E. "Ecological Footprints and Appropriated Carrying Capacity: What Urban Economics Leaves Out." *Environment and Urbanization* 4, no. 2 (October 1992), 121–30.

Robinson, John. "Squaring the Circle? Some Thoughts on the Idea of Sustainable Development." *Ecological Economics* 48 (2004), 369–84.

Sachs, Jeffrey D. *Common Wealth: Economics for a Crowded Planet*. London: Penguin, 2008.

Schneider, F., Kallis, G., and Martinez-Alier, J. "Crisis or Opportunity? Economic Degrowth for Social Equity and Ecological Sustainability. Introduction to this Special Issue." *Journal of Cleaner Production* 18 (2010), 511–18.

Schumacher, E. F. *Small is Beautiful: Economics as If People Mattered*. London: HarperCollins, 1973.

Soll, David. *Empire of Water: An Environmental and Political History of the New York City Water Supply*. Ithaca, NY: Cornell University Press, 2013.

Sorlin, S. and P. Warde, eds. *The Future of Nature*. New Haven, CT: Yale University Press, 2013.

Stern, Nicholas. *Stern Review on the Economics of Climate Change*. London: HM Treasury, 2006.

Tainter, Joseph. *The Collapse of Complex Societies*. Cambridge: Cambridge University Press, 1988.

Victor, Peter. *Managing Without Growth: Slower By Design, Not Disaster*. Cheltenham: Edward Elgar Publishing, 2008.

Wackernagel, Mathis and William E. Rees. *Our Ecological Footprint: Reducing Human Impact on the Earth*. Gabriola Island: New Society, 1996.

Wackernagel, Mathis, et al. "Tracking the Ecological Over-Shoot of the Human Economy." *Proceedings of the National Academy of Sciences* (July 9, 2002), 9269.

Walker, Brian, C.S. Holling, Stephen R. Carpenter, and Ann Kinzig. "Resilience, Adaptability and Transformability in Social-Ecological Systems." *Ecology and Society* 9, no. 2 (2004), article 5.

World Commission on Environment and Development (WCED). *Our Common Future*. Oxford: Oxford University Press, 1987.

Worster, Donald. *Nature's Economy: A History of Ecological Ideas*. 2nd edition. Cambridge: Cambridge University Press, 1994.

Worster, Donald. *The Wealth of Nature: Environmental History and the Ecological Imagination*. Oxford: Oxford University Press, 1994.

Zolli, Andrew and Ann Marie Healy. *Resilience: Why Things Bounce Back*. New York: Free Press, 2012.

PART III

Sustainability, resilience, and collapse in historical societies

3

WHAT IS SUSTAINABLE?

Some views from Highlands Papua New Guinea

Andrew Strathern and Pamela J. Stewart

The answer to the question posed in our title here must be that this depends on two issues. One is the framework of time and space that is in focus; the other is whether we are regarding sustainability as a state of simple continuity or as a state that adjusts to, or incorporates change into, itself. A framework of time and space needs to be specified, or at least sketched-out, bearing in mind the aphorism that nothing lasts forever. Correlatively, the concept of sustainability needs to include some recognition of adaptation and change because in order to maintain some variables, or patterns of behavior and practice, as well as the values of the social actors involved, change has to come into play in order to sustain these values. For example, if an agricultural regimen that depends on a certain crop is rendered at risk by the entry of a disease that attacks this crop, people may seek to find a comparable crop that can replace it as a major source of subsistence.

In this chapter, we will address the question of sustainability using some perspectives on Highlands Papua New Guinea. Given the fact that Highlanders have continuously inhabited the area for approximately 50,000 years, the set of cultures in the Highlands provide a useful basis from which to consider successful approaches to long-term resilience.

In one sense, the history of sustainability must be coterminous with (at least) the history of human occupation of parts of the Earth, since all such histories of populating places depend on the availability of ecological niches for people to practice their ways of life in, and those niches vary in their capacity to sustain such ways of life. The crucial factor is whether there is an overall growth in population over time or whether there is a stable state. Where the idea of economic development takes hold and is coupled with a rise in population levels, the issue of sustainability becomes crucial. This is obviously why sustainability has come to the fore as a key global issue nowadays. Huge cities demand vast amounts of energy. Large populations require to be fed. All this in turn is driven by the mere fact of biological reproduction and the values associated with it in human kinship systems that continue to underpin inter-generational growth and change.[1] In state-based structures these biologically underpinned processes are also greatly affected by government policies and programs that look to the continuity of an available work force in order to sustain the economic system and to provide producers and consumers of goods. Sustainability as such is lost or mired in such secondary imperatives of adaptation, and the growth of population levels is taken as axiomatic, rather than contingent on choice and planning.

It is useful, in such a context, to look at the practices, and the results of practices, among pre-industrial peoples studied by anthropologists around the world. Do, or did, such people practice sustainable ways of living, and if so by what mechanisms? There have been many important works on the history of societal collapse, but less has been written on what might be called the long-term history of sustainable resilience. Hunter-gatherer populations generally consisted of groups, spread over large ranges of land, with clear divisions of labor and only limited capacity to increase either their population sizes or their sustainable resources. The cultivation of crops in garden patches or larger fields at once modified this pattern, bringing people together in a more sedentary pattern and increasing the capacity to sustain levels of production over long periods of time.

Practices related to reproduction are involved here. Contraception or induced abortions can help to reduce population growth, as can of course illness conditions or killings resulting from conflict. Reduced fertility can be a result of poor nutrition and disease loads. Among settled agriculturalists, such as are found in the Central Highlands of the island of New Guinea, we find the spacing of children customarily occurs as a result of lengthy post-partum taboos on intercourse between partners and long periods of lactation and breast-feeding on the part of the mothers. Such taboos obviously conduce towards preservation of the bodily health of mothers as well as their children. However, they tend to be supported not by any ideology of conservation as such but by ideas of a ritual kind. People explain that a man's sperm, which is instrumental in the production of children, can be lethally harmful to an unweaned child, acting as a kind of toxin that spoils the mother's breast milk. This is an automatically self-reinforcing taboo, predicated on the generally correct presumption that the genitor will not wish to inflict bodily harm on his young child, and the mother likewise will not wish to engage in sexual relations that would have such an effect. This long-established, traditional taboo contributes to the sustainability of the population as a whole by limiting the rate of population growth. The interpretive argument here is functional in character but does not imply that there is any overt or conscious effort by the persons involved to justify their practices in functionalist terms.

Similar considerations apply when we move to the related issues having to do with settlement patterns, warfare, and the putative pressure on resources of land that would result from population growth.

A lively debate concerning the effects of land pressure on the incidence of warfare between clan groups in the Highlands emerged during the 1960s when relatively early detailed studies were undertaken by a wave of researchers during a period of colonial pacification by Australian government forces in what were then the two separate administrative areas of Papua and New Guinea. Government officers had patrolled the central belt of populations straddling these two administrative divisions and everywhere had sought to eliminate armed aggression between groups and gradually to introduce cash cropping (mostly through coffee) and institutions of elected local government councils that could preside over road-building and inform people about political and economic change in general, linked to the collection of council taxes and their disbursement on public projects.

Central Highlands groups were fairly intensive gardeners and at the time they first came into contact with explorers and government officers from the outside world they had developed flourishing systems of wealth exchanges linked in one way and another to both warfare and peacemaking. The tools of both gardening and fighting were relatively simple, consisting of wooden spades and digging sticks used by men and women respectively, and bows, arrows, spears, and shields deployed by men in concerted aggressive encounters. Production relied largely on the mounding and draining of fertile garden areas. The staple crop was (and is) sweet potato, which had replaced an earlier suite of ancient crops within the last thousand years and perhaps for only a few hundred years.

A clear gendered division of labor operated. Men turned the soil and made drains with these long wooden spades, and women employed digging sticks to till, plant, and harvest sweet potatoes and taro (the older and much longer established crop.) The land in many places is fertile, especially where the soil is volcanic, and the sweet potato is a hardy tuber that can be grown at altitudes higher than the older staple (*Colocasia* taro).

Sweet potato can also be fed to pigs and its advent in the Highlands led to the development of the complex of activities recorded by early observers from the outside world in the 1930s onward; high density populations of pigs and people, the pigs being used for life-cycle and ceremonial exchange events in the sphere of local politics.

Origin stories of groups in the Mount Hagen area stress the power of fertility channeled through male ancestors and granted to them by transcendent spirit powers of the cosmos. An explicit aim of groups was to increase their numbers and correspondingly to migrate out from their original place and colonize new areas. With such an underpinning ideology it is clear that conflict between groups would be likely to emerge, with competition for land resources. In that case the groups would run into problems of sustainability, because warfare can be harmful and destructive to crops. Indeed, the Highlanders gained a reputation as warriors keen to maintain their standing and hold on to their resources. So how was conflict between groups controlled or contained?

The answer is that it was limited by a number of social mechanisms, but these mechanisms were by no means perfect. Narratives of the fortunes of groups often contain episodes of the group being routed and dispersed, seeking to live with extra-clan kinsfolk with whom there were ties resulting from inter-marriage.[2] An elaborate segmentary structure of degrees of enmity and alliance between groups operated to define the fields of social relations. Among allies and minor enemies, transfers of pigs served to repair conflicts and maintain or renew friendly ties. With major enemies, marriages were infrequent and compensation for deaths by planned violence or imputed as a result of covert sorcery was not paid. This outer structure defined the limits of solidarity. Within the circle of solidarity, especially close relations held between groups named as paired alliance sets, and disputes over resources within these same ambits of sociality could be settled. Inside such ambits, there was no driving out of people from their land and access to it could thus be accommodated, keeping sustainability within the security circle. If a severe conflict with major enemies arose, and people were driven out, as happened historically to the Kawelka people, they were able to find refuge with an extended network of kin and to consolidate their recovery by entering into a new pairing with a host group that took them in, within a territory that was perhaps sparsely occupied prior to their arrival as refugees.

These fluid arrangements for sustainability in the face of conflict contrast somewhat with the classic picture painted by a stream of notable figures in the history of anthropology, centered on the Highlands fringe population of the Maring living north of the area occupied by the Melpa speakers of Mount Hagen in the Western Highlands of Papua New Guinea. Prominent among anthropologists who studied the Maring were Roy Rappaport and Andrew Vayda, both engaged in ecological work and in seeking functionalist accounts of how periods of war and peace among groups, strictly calibrated in terms of ritualized sequences, played out over time. Rappaport worked with a very small group, the Tsembaga, numbering some 200 people.[3] He showed how political and ecological time was divided between a period of peace (or suspended hostilities) and war (open declaration of hostilities) and that the division was regulated by a complex series of rituals. Warfare was pursued in order to take revenge for deaths inflicted in previous bouts of conflict, and was inaugurated by the collective uprooting of red cordyline plants described as the spirits of men (*yu min rumbim*). Peace was marked when hostilities were suspended, pigs were sacrificed, and the sacred cordyline plants were rooted again in the soil of the group's territory. During the ritual time of peace people bent their energies to making gardens and building up numbers of domesticated pigs, destined to be sacrificed, after some years, at the next pig festival (*kaiko*). The ecological part of Rappaport's argument was that as the numbers of pigs increased, caring for them became more arduous, and the decision to hold the *kaiko* was triggered when the burden became too great for the workers, especially women who fed the pigs, to bear. The "system" of pressure on resources was thus brought back to a sustainable level, in the same way as a thermostat regulates temperatures and maintains them within a range.

This argument, as far as it went, was obviously correct. However, after the *kaiko* was over, the group would go to war with whoever had killed one of their own in a previous bout of fighting. The effects of war were also regulated, but perhaps not so predictably or clearly. Rappaport's analysis most convincingly portrayed the ritual alternation between war and peace, ensuring longish periods of peace and build-up of resources and shorter periods of fighting with loss of life and occasional temporary displacement from land. Andrew Vayda's work carried this mode of analysis more widely into a general argument that warfare itself could be seen as an ecological mechanism by which, over time, land was redistributed among groups in response to population pressure. In a later recantation of this thesis, Vayda noted that his previous assumption of population pressure as a condition affecting or potentially affecting all Maring groups was unwarranted, and indeed rested on just two particular case histories, centered on the Kauwatyi and the Kundagai groups. These two groups had large areas of grassland and secondary forest in their territories, and were thus short of the fertile primary forest they needed for new gardens and for pigs to root around in.[4] Correspondingly, they had population densities greater than any others of the many Maring groups.

Tension over gardening land was clearly exhibited in the fighting histories of the 1950s that Vayda collected. The pattern of fighting in which a group might be driven off their land usually resulted in their later return to their territory, but in a few cases led to annexation. The conflict between the Kauwatyi and the Kundagai also broke custom in its fierceness and disregard for formalities, and appears not to have been marked by the holding of any *kaiko* in advance of it or by an uprooting of the *rumbim* plants. Only the external force of the colonial Australian patrol officers, who reinstated the original landowners on their land and forbade warfare, halted the fighting and annexation of land. As an aside, but a significant one, Vayda here notes that one of the groups attacked, the Tyenda, gave a considerable portion of their territory to their allies, the Kundagai, in return for their giving them help. The Kundagai, in turn, had been the other group that had aggressively intruded on the land of their own neighbors, the Ambrakui, who had been weakened by population decline. In other words, warfare was in fact correlated in these two cases with a complex pattern of transactions in land. Vayda, therefore, is right to point out that territorial conquest was an intentional aim in the actions of the Kauwatyi and the Kundagai; but, in systemic or processual terms (*pace* Vayda's rejection of process as a concept) it is equally interesting that a voluntary gift of land to allies was a part of the broader picture, and indeed falls into line with the importance of allies both in the Maring case and in Hagen. From the point of view of ritual analysis it is also important to note that the case of the Kauwatyi shows that a fine-tuned ritual regulation of war and peace and of controlled modes of fighting may be abandoned if circumstances dictate this. Ritually managed homeostasis, then, works well until it does not; and when it does not, the way lies open for more and more lethal encounters. This is indeed what happened in Hagen with the introduction of guns into warfare.[5]

In his life work, Andrew Vayda moved from "ecological functionalism" (as in his first take on Maring warfare) to what he calls "event ecology," that is, the study of empirically identifiable actions and events that have ecological results—and therefore, we can add, also bear upon sustainability.

In promoting event ecology, Vayda, working with Bradley Walters, also took a stand against a popular spin-off from activist ecological studies, or political ecology, which in turn is deeply informed by critical political economy theorizing.[6] Basically, these theorists adduce political factors and influences from outside as major explanatory factors in ecological studies. Clearly, there is some truth in this perspective, which is akin to the perspectives found in critical medical anthropology. However, from his own research-focused perspective, Vayda wishes to examine local details, and he proposes event ecology as his term for this endeavor. In other words, instead of backing supposed "laws" of processes, he is scrutinizing the contingences of local histories as sources of ecological results and causes of results. Such contingencies can provide needed clues to the sustainability or unsustainability of practices.

In terms of general theories and discussions about sustainability, a number of sites or venues of investigation emerge. One is the theme of global and regional climate change and its relationship to the incidence of adverse physical events that produce conditions which people experience as disasters. This theme feeds into the study of such disasters and how people are able or not able to recover from them. A second theme lies in the importance of maintaining an ecological viewpoint on the question of sustainability. Ecology is concerned with the total mix of environmental relationships among different biota. From the perspective of human ecology this focus narrows down to the effects of human interactions with the environment. In turn, such effects have largely to do with the creation, maintenance, and exhaustion of resources for sustaining life, and the major message here is that there is an intricate web of relationships among species, leading to a balance or imbalance between them based on symbiosis. Imbalance leads to the destruction or disappearance of certain species as they are outcompeted by others or are deliberately targeted by others.

With the development of sophisticated and powerful tools of technology that both generate and require huge amounts of energy to operate, it is obvious that the question of the finite character of natural resources swings into view. It is urban civilization and its massive artificial built environments that causes this problem to be severe, along with another dynamic factor— the steady growth in population numbers throughout the world. Birth control and limitations on reproduction thus become crucial factors of ecology and sustainability, but a concern to limit population growth runs counter to prevailing capitalist values that envision persons as consumers and producers of commodities that generate monetary income. It might seem inevitable, then, that there is a contradiction between sustainability and the ideology of economic growth. The technological answer to this contradiction is that with new, more efficient technology, sustainability of food production can be achieved. An example would be urban, indoor hydroponic gardens that can produce crops without the cultivation of large areas of land. Such a technology, however, itself depends on a high consumption of energy.

Global climate change further complicates this picture. Such contemporary change is due to the expansion of industry and the release of toxic elements into the atmosphere. The climate changes that are currently being observed do not have a benign appearance but rather seem connected with increasingly fierce and destructive storms, hurricanes, tornadoes, tropical cyclones, flooding, and the like. Meanwhile, tsunamis resulting from earthquakes in regions of the world that are in zones of turbulence between tectonic plates continue to cause havoc, along with volcanic eruptions in seismically active arenas. All of these processes cause large numbers of deaths, material destruction, and the displacement of people into supposedly safer areas of occupation.

Intensive capitalist-style use of resources is usually signaled as instrumental in disrupting sustainability, although here we must note the problem of scale. For example, in many parts of the world, large-scale dams have been built to supply water for cities and/or to provide hydro-electric sources of power for the same purpose. This process occurred earlier in the US, and has been happening currently in mainland China. The Aswan dam in Egypt is another well-known example, and one that destroyed one of the most stable agricultural regions on Earth. In almost every case, such large-scale developments have had unfavorable effects for some animal species and/or for some indigenous peoples on whose traditional lands these large-scale installations have been imposed. Operations of this kind provide temporal sustainability for the cities they serve. By the same token, they cause unsustainability for the peoples whose riverine and riparian resources are destroyed. So our general question in this survey of "what is sustainable?" can be parsed as "what is sustainable for whom?"

An alternative vision of sustainability is provided by a set of studies by Anne Ross and collaborators, with the title of *Indigenous Peoples and the Collaborative Stewardship of Nature*.[7] The basis for this volume is the idea that an indigenous stewardship model provides the foundation for the sustainable use of nature. In turn, the hypothesis here is that indigenous peoples, living in close proximity to and symbiosis with nature, have enshrined in their cultures a respect for these resources that precludes them from harmfully overusing them. This interpretation certainly could be applicable in Highlands Papua New Guinea, where resources have been used sustainably for tens of thousands of years That said, there is no reason to suppose that this model applies universally. Maori people in Aotearoa (New Zealand), for example, seem to have hunted the large indigenous birds known as Moas into extinction prior to the European colonization and the more powerful ecological changes that it brought in its wake. Further, several indigenous societies experienced precipitous collapse prior to contact with Europeans. The model works best, however, in areas where colonial influences have had deleterious effects, and strategies of conservation and regeneration are built on the premises of a cosmic world-view of respect for living species. Summarizing their overall argument after a survey of complex data from many different indigenous peoples in Australia, America, and Asia, the authors conclude that indigenous stewardship models represent the agency and efforts of indigenous peoples themselves to revitalize their traditional knowledge systems in the service of caring for the environment.[8] In our view, it is valid for industrial society to look to indigenous societies for wisdom and inspiration, but this must be done cautiously and in a non-exploitative manner.

A related approach is found in the literature on eco-tourism. The volume *Ecotourism and Sustainable Development: Who Owns Paradise?* explores this theme in depth, using studies from the Galapagos Islands, Costa Rica, Cuba, Tanzania, Zanzibar, Kenya, and South Africa.[9] The overall aim of ecotourism is to sponsor tourist visits to natural areas where the habitat is special and may be at risk, fostering respect for the environment and helping to provide resources to protect and conserve it. Whereas indigenous stewardship is often centered on areas that have already suffered considerable adverse change, ecotourism, in the best of cases, focuses on bringing money into protected areas or areas that may need protection, in order to conserve some aspect of a pristine or intact ecosystem.

Costa Rica is one area, Honey writes, that has a relatively long-established track-record of ecotourism operators. One of the founding figures there was Michael Kaye, who shifted from nature and adventure tours into the ecotourism sphere by trying to prevent the building of a particular mega-resort in Costa Rica. Failing to do so, he began to build eco-lodges himself in competition with bigger operators. Such efforts, Honey indicates, have received much money from the US Agency for International Development (USAID), contributing to an international "enterprise-based" approach to conservation.[10] The World Wildlife Fund and the Nature

Conservancy are also involved. Such efforts at combining tourism with conservation of the local environment and cultural practices are not without their contradictions. Local people are not always happy with backpackers who somewhat intrusively try to learn from their hosts and also live in cheap accommodation. Honey perceptively notes that achieving the most acceptable balance in relationships and "authentic cultural exchange" between hosts and visitors is a difficult task.[11] In the majority of cases, the eco-tourists are paying money for their experience, and the hosts may or may not be ploughing this back into the development of their business and/or the environment in which it operates. A beneficial factor generally provided by government is the provision of official park areas within which tourists can both seek adventure and learn about the landscape, history, and contemporary concerns of the park itself, as a managed environment accessible to visitors. Since Honey's book was published in 1999 many more ecotourism projects have been launched. Honey provides figures that suggest why this may be so, since eco-tourism generates a considerable amount of money while doing less damage to the natural environment than some other economic activities in many of the areas studied.[12] The reason for this would seem to lie in the fact that eco-tourists are able to pay more for their experience than their hosts can make from other sources of income. In other words, the money comes from outside and if visitors do not come there is no income at all, since eco-tourists like to visit areas free of other kinds of development, areas with putatively pristine or unaltered environments (although such an assumption may not be grounded in historical fact). The sustainability of eco-tourism thus rests on the effectiveness of marketing it to the outside world and on attracting visitors with enough resources to pay for it. There must also often be a fine line between adventure tourism that incorporates a modicum of learning about the environment and eco-tourism that incorporates a certain degree of adventure, if only because of the difficulties of reaching the areas visited.

Eco-tourism is likely to be pursued in areas where aspects of the environment are traditionally regarded as sacred and therefore not to be disturbed because of the offense that such disturbance would give to guardian spirits who might in turn enact punishment or revenge on those who violate their taboos. The whole concept of "taboo" as such stems from Polynesian practices whereby a chief could forbid the use of a certain natural resource for a period of time, lifting the taboo perhaps when the resource was needed for acts of ritual feasting. A parallel to such impositions by chiefly authority is to be found in the famous ritual complex of the *kaiko* among the Maring of Papua New Guinea, in which peace held in a group as long as the red cordyline plants that contained male life force remained planted in the ground and war was initiated by uprooting these plants and triggering a sequence of rituals leading to warfare.[13] While the sacred cordylines remained planted, pigs were intensively reared over a period of years, with the ultimate aim of killing these as sacrifices to spirits and unleashing warfare in order to revenge the deaths of kinsfolk in previous bouts of fighting. The ritually configured alternation between peace and war obviously contributed to limiting the damages of war as well as ensuring that pig herds were reduced before they became seriously parasitic on human occupation of the land.[14] Of course, while the sacrifices of pigs that constituted central ritual acts in the *kaiko* cycle generally took place before the environment was placed at risk, there is assuredly a likelihood that the Maring aimed at rearing and killing an impressive number of these animals in order to earn the approval of both the spirits of their dead and the minds of their living allies in warfare. Practical and cosmic considerations flow together in such processes. But the salient point is that the *kaiko* cycle, intentionally or not, helped maintain social and ecological balance.

This theme of the relationship of ritual practices and ecology finds resonance with a much larger, global theme, on the importance of sacred sites that are often loci of valuable biodiversity because the plants in them are protected from being destroyed. Gloria Pungetti and collaborators have extensively documented the importance of this point across the globe, and interestingly enough include English churchyards in their purview.[15] Another chapter in this same volume recounts a situation extremely common in the history of struggles between indigenous peoples and settlers. Joseph S. Te Rito, a member of the Ngati Hinemana kin group of Maori in the province of Hawkes Bay in Aotearoa (New Zealand), tells of the struggle of his people to regain control over Puketapu, a sacred hill that had been renamed Fernhill by settlers in 1879. A hundred years later, in 1989, the Maori revived the ancient name, meaning "sacred hill," and opposed the local Council's attempts to build a housing section on it or to sell it to vineyard operators.[16] In this way, cultural sacrality acted as a rampart against urban sprawl or sudden land-use changes.

Puketapu gained its name as a result of a battle between Maori groups that entailed bloodshed and deaths. After that the Hinemanu banned residence on the hill, considering that their warriors' bones were buried in its soil. In spite of the historic provisions of the Waitangi Treaty in 1840, incoming settlers obtained control over Maori lands by pressuring the Crown's representatives to release it for division among settler families. From 1989 onward, the Maori leaders organized a movement to oppose the sale of land on the hill to outside interests and to prevent the Council from dividing it into residential lots. Interestingly, the Maori won over numbers of non-Maori local people and they started the Puketapu/Fernhill Reserve Trust together, with the aim of keeping the land as a conservation area and free of dwelling houses like those occupied by the descendants of settlers who had obtained parts of the hill much earlier.

The author records that little of the original vegetation remains on the hill, and it has been turned into sheep pasture.[17] However, he notes that the Trust members hope to replant it with indigenous species and to concentrate on "ecological restoration" turning Puketapu into a nature reserve for the general public, following the award of a certificate of title to the Trust on February 22, 2011.[18]

Te Rito's narrative is one of hundreds around the world where indigenous people turn to promoting reserves and parks, either for themselves, or for themselves and a wider public, as a way of keeping indigenous land out of capitalist development projects. Rights to the land are central in the struggles to reclaim heritage areas, and partnerships with benevolent aims are a way to mediate the struggles on behalf of shared wider values. In large part this pathway has opened up because of the institutions of creating parks and reserves that have come to be a part of government policies. If we turn to the US or the UK for examples, we find that national or general ordinances have been instrumental, as well as state authorities in the US. The philosophical background to the provision of parks feeds to some extent off the radical traditions embracing solitude and reflection generated by Henry David Thoreau in the mid-nineteenth century and expounded in his book on Walden pond near Concord, Massachusetts, where he lived from July 1845 to September 1847.[19] Thoreau wanted to get completely away from the way of life of people in the developed urban world and to immerse himself totally in a remote rural locality, living and studying the environment around him, and writing about it in step with the seasons. Thoreau consciously sought to emulate a simpler, pre-industrial mode of existence, and took inspiration from indigenous Native American cultures. The microcosm of Walden stood for him as an icon of the wider world of nature untrammeled by industrial influence. He observed everything closely and wrote about it on a regular basis. His book on his experience was published in 1854, and shows the influence on his thinking of the philosopher Ralph Waldo Emerson. Thoreau's sensibilities were in harmony with Emerson's theory of "transcendental revelation," a mystical experience stemming from deep contemplation of the world and the spiritual meanings that are immanent in it.

Thoreau's ideas influenced a figure who became very influential in founding and promoting wilderness parks areas in the USA, for example Yosemite. This was John Muir, an extraordinary explorer with a remarkable, if not uncanny, ability to survive in difficult places without any elaborate food supplies, weapons, or camping gear.[20] Brought up strictly, first in Scotland in Dunbar not far from the Lammermuir Hills, and later, from the age of ten onwards, in America where his father took the family in search of religious freedom, Muir struck out from the family farm in Wisconsin, executing a number of strenuous traverses on foot over mountains and maintaining a prolific record of his experiences in a series of journals. He spent time at Yosemite from 1871 onward, exploring all its canyons and peaks. Much later, in 1889, he mapped out, with Robert Underwood Johnson, a plan to preserve Yosemite as a natural park, and in 1903 he went on a trip with President Theodore Roosevelt and presented him with a successful plan for preserving several wilderness areas.

Muir's life work represents one salient side of conservation efforts, to preserve wilderness for its own sake. Another side is the aim of introducing conservation efforts into the world of agriculture itself, as shown in the environmental regulations of the European Union today via its Common Agricultural Policy, with provisions for setting aside areas for wildlife on field edges and its designation of Sites of Special Scientific Interest for conservation.[21] John Muir would also have been interested to know about the creation of the Cairngorms National Park in the Grampians area in Central Scotland, as reported by Kathy Rettie in the volume we have just mentioned.[22]

Conservation efforts around the world have taken on a new urgency with the threat to the global environment implicated in climate change. Sustainability has often most effectively been both studied and promoted on local and regional scales of time and space. However, global climate change raises the stakes to the widest level of life on Earth, so that the question of what is sustainable is swallowed up in the question "Is life on Earth sustainable for humans and the species whose lives are interwoven with those of humans?"[23] Of course, the long-term geological history of the emergence and disappearance of land areas outruns the current concern with human-induced climate change, but there is little doubt that the dangers of rising sea levels affecting low-lying islands in parts of the Pacific are very real and are likely to make necessary the displacement and migration elsewhere of some of the islanders whose intrepid seafaring ancestors first discovered and colonized these specks of land in the ocean.[24]

In the context of ecological crisis, it is instructive to consider the Highlands of Papua New Guinea and other areas where indigenous societies have survived and thrived over tens of thousands of years—especially when one considers that industrialism is a mere 250 years old. The reverence for the natural world, the concept of the sacred, and a whole host of social and ecological practices helped the Highlanders live comfortably within their ecological limits. Given how little has been written on long-term social resilience, it would be useful to flip around Jared Diamond's focus on collapse and ask an equally important question: How have some societies been able to live resiliently and sustainably for such long periods of time?[25] What Western industrial society can learn from this question remains to be seen, but learning from sustainable indigenous societies would be a potential means of reversing global power dynamics that impose unsustainable practices on the indigenous societies of the world.

Notes

1   See Andrew Strathern and Pamela J. Stewart, *Kinship in Action* (Upper Saddle River, NJ: Prentice Hall, 2011).

2   See, for example, the history of settlement among the Kawelka people of Mount Hagen detailed in Andrew Strathern, *One Father, One Blood. Descent and Group Structure Among the Melpa People* (Canberra: Australian National University Press, 1972).

3   See Roy Rappaport, *Pigs for the Ancestors* (New Haven, CT: Yale University Press, 1968).

4   Andrew P. Vayda, *Explaining Human Actions and Environmental Changes* (Lanham, MD: Alta Mira Press, 2009), 214.

5   See, for example, Andrew Strathern and Pamela J. Stewart, *Peace-Making and the Imagination: Papua New Guinea Perspectives* (St Lucia: University of Queensland Press, 2011).

6   Vayda, *Explaining*, 130.

7   See Anne Ross et al., *Indigenous Peoples and the Collaborative Stewardship of Nature* (Walnut Creek, CA: Left Coast Press, 2011).

8   Ibid., 260.

9   See Martha Honey, *Ecotourism and Sustainable Development. Who Owns Paradise?* (Washington, DC: Island Press, 1999).

10 Ibid., 76.

11 Ibid., 90.

12 Ibid., 391.

13 Rappaport, *Pigs for the Ancestors*.

14 Compare, for further details and discussion, Susan Lees, "Kicking Off the *Kaiko*: Instability, Opportunism, and Crisis in Ecological Anthropology" (pp. 49–63) with Andrew Strathern and Pamela J. Stewart, "Rappaport's Maring: The Challenge of Ethnography" (277–90), both in *Ecology and the Sacred: Engaging the Anthropology of Roy A. Rappaport*, E. Messer and M. Lambek, eds. (Ann Arbor, MI: University of Michigan Press, 2001).

15 Gloria Pungetti, Gonzalo Oviedo, and Della Hooke, eds., *Sacred Species and Sites: Advances in Biocultural Conservation* (Cambridge: Cambridge University Press, 2012), esp. ch. 7 by Nigel Cooper on Rivenhall in Essex.

16 Joseph S. Te Rito, "Struggles to Protect Puketapu, a Sacred Hill in Aotearoa," in Pungetti et al., *Sacred Species and Sites*, 165.

17 Ibid., 176.

18 Ibid., 177.

19 Henry David Thoreau, *Walden; or, Life in the Woods* (Boston, MA: Ticknor and Fields, 1854).

20 Edwin Way Teale, ed. *The Wildernesss World of John Muir* (Boston, MA: Houghton Mifflin Company, 1954), xii, an introduction to Muir's own autobiographical accounts.

21 See discussion in ch. 1 of Pamela J. Stewart and Andrew Strathern, eds., *Landscape, Heritage, and Conservation* (Durham, NC: Carolina Academic Press, 2010), which reports on our own field studies in Scotland and Ireland since 1996.

22 Kathy Rettie, "Place, Politics and Power in the Cairngorms National Park: Landscape, Heritage, and Conservation," in *Landscape, Heritage, and Conservation: Farming Issues in the European Union*, P. J. Stewart and A. Strathern, eds. (Durham, NC: Carolina Academic Press, 2010), 107–39.

23 See Susan A. Crate and Mark Nuttall, eds., *Anthropology and Climate Change From Encounters to Actions*, 2nd edition (New York: Routledge, 2016).

24 For an overview, see Patrick D. Nunn, *Vanished Islands and Hidden Continents of the Pacific* (Honolulu, HI: University of Hawai'i Press, 2009).

25 Jared Diamond, *Collapse: How Societies Choose to Fail or Succeed* (Penguin, 2005).

References

Crate, Susan A., and Mark Nuttall, eds. *Anthropology and Climate Change from Encounters to Actions*, 2nd edition. New York: Routledge, 2016.

Diamond, Jared. *Collapse: How Societies Choose to Fail or Succeed*. Penguin, 2005.

Honey, Martha. *Ecotourism and Sustainable Development. Who Owns Paradise?* Washington, DC: Island Press, 1999.

Lees, Susan. "Kicking Off the Kaiko: Instability, Opportunism, and Crisis in Ecological Anthropology." In *Ecology and the Sacred: Engaging the Anthropology of Roy A. Rappaport*, E. Messer and M. Lambek, eds., 49–63. Ann Arbor, MI: University of Michigan Press, 2001.

Messer, E. and M. Lambek, eds. *Ecology and the Sacred: Engaging the Anthropology of Roy A. Rappaport*. Ann Arbor, MI: University of Michigan Press, 2001.

Nunn, Patrick D. *Vanished Islands and Hidden Continents of the Pacific.* Honolulu, HI: University of Hawai'i Press, 2009.

Pungetti, Gloria, Gonzalo Oviedo, and Della Hooke, eds. *Sacred Species and Sites: Advances in Biocultural Conservation.* Cambridge: Cambridge University Press, 2012.

Rappaport, Roy. *Pigs for the Ancestors.* New Haven, CT: Yale University Press, 1968.

Rettie, Kathy "Place, Politics and Power in the Cairngorms National Park: Landscape, Heritage, and Conservation." In *Landscape, Heritage, and Conservation: Farming Issues in the European Union*, P. J. Stewart and A. Strathern, eds., 107–39. Durham, NC: Carolina Academic Press, 2010.

Ross, Anne, et al. *Indigenous Peoples and the Collaborative Stewardship of Nature.* Walnut Creek, CA: Left Coast Press, 2011.

Stewart, Pamela J., and Andrew Strathern, eds. *Landscape, Heritage, and Conservation.* Durham, NC: Carolina Academic Press, 2010.

Strathern, Andrew. *One Father, One Blood. Descent and Group Structure Among the Melpa People.* Canberra: Australian National University Press, 1972.

Strathern, Andrew, and Pamela J. Stewart. "Rappaport's Maring: The Challenge of Ethnography." In *Ecology and the Sacred: Engaging the Anthropology of Roy A. Rappaport, E. Messer and M. Lambek*, eds., 277–90. Ann Arbor, MI: University of Michigan Press, 2001.

Strathern, Andrew, and Pamela J. Stewart. *Kinship in Action.* Upper Saddle River, NJ: Prentice Hall, 2011.

Strathern, Andrew, and Pamela J. Stewart. *Peace-Making and the Imagination: Papua New Guinea Perspectives.* St Lucia: University of Queensland Press, 2011.

Te Rito, Joseph S. "Struggles to Protect Puketapu, a Sacred Hill in Aotearoa." In *Sacred Species and Sites: Advances in Biocultural Conservation*, Gloria Pungetti, Gonzalo Oviedo, and Della Hooke, eds., 165. Cambridge: Cambridge University Press, 2012.

Teale, Edwin Way, ed. *The Wilderness World of John Muir.* Boston, MA: Houghton Mifflin Company, 1954.

Thoreau, Henry David. *Walden; or, Life in the Woods.* Boston, MA: Ticknor and Fields, 1854.

Vayda, Andrew P. *Explaining Human Actions and Environmental Changes.* Lanham, MD: Alta Mira Press, 2009.

4

UNDERSTANDING SUSTAINABILITY THROUGH HISTORY

Resources and complexity

Joseph A. Tainter

In understanding the meaning of sustainability, it is useful to resort to basics. The term "sustain" comes originally from the Latin *sustinere*, and into English through the Old French *soustenir*. Both terms mean literally "to hold underneath"—in other words, to uphold or support. The *Shorter Oxford English Dictionary*, sixth edition, lists nine definitions of "sustain." Two of these are particularly useful. Number three reads "cause to continue in a certain state; maintain at the proper level or standard." Number five, which is consistent with biophysical concepts of sustainability, reads "support life in; provide for the life or needs of."[1] Both definitions are consistent with the original Latin and French terms: to sustain something is to support its continuation. Sustainability is the science of continuity. Sustainability emerges therefore from success in addressing existential problems, that is, problems of continuity. As a historical endeavor, sustainability concerns the long-term success of problem-solving efforts.

If sustainability is the attempt to maintain continuity, unsustainability manifests itself in transformation or collapse. Both transformation and collapse may mean that an existing way of life is not sustained. The difference between transformation and collapse involves simplification and the rate of changes in complexity. As defined elsewhere, a collapse is the rapid loss of an established level of complexity. When a society collapses, it rapidly simplifies. Elements of social and political structure, hierarchy, technology, and information are lost. The classic example is the emergence of the Dark Ages in Western Europe after the collapse of the Western Roman Empire. Transformation is cultural change that may not involve simplification.

In addition to often-vague conceptualization, the study of sustainability suffers from a short-term outlook. The processes that make a society sustainable or vulnerable to collapse develop over long time periods, typically stretching to generations or centuries.[2,3] It is difficult to discern, let alone comprehend, these processes within a single lifespan. Statements about sustainability must be based on long-term observations, and those that are not should be considered suspect.

Complexity is central to long-term trends in sustainability. Human societies over the past 12,000 years have increased greatly in complexity. The hunter–gatherer societies of our ancestors, for example, consisted of no more than a few dozen types of social roles and personalities, while modern censuses recognize 10,000 to 20,000 distinct occupations.[4] This is just one contrast in complexity. Modern societies also differ greatly from earlier ones in technology, organization, institutions, economic interconnections, amounts and kinds of information processed, and so forth. But what do we mean by complexity? It has many colloquial meanings (usually as a synonym for "complicated") and, today, several technical ones. The term is used here in the anthropological sense of differentiation in structure (more parts to a system and, in particular, more kinds of parts) and increase in organization. Organization is defined as constraints on behavior.[5] A complex society, then, is one that has a highly differentiated structure, with corresponding organization to bind the parts into a functioning system.

One of the most important points to understand is that complexity costs. In the realm of complex systems there is, to use a colloquial expression, no free lunch. This observation derives from the second law of thermodynamics. It takes energy to maintain a system away from equilibrium (entropy), and more complex systems are farther from equilibrium than are simpler ones. Thus, in any living system, increased complexity carries a metabolic cost. In non-human species this cost is a straightforward matter of additional calories that must be found and consumed. Among humans the cost is calculated in such currencies as resources, effort, time, or money, or by more subtle matters such as annoyance. While humans find complexity appealing in spheres such as art, music, or architecture, we usually prefer that someone else pay the cost. We are averse to complexity when it unalterably increases the cost of daily life without a clear benefit to the individual or household. Before the development of fossil fuels, increasing the complexity and costliness of a society meant that *people* worked harder.

The development of complexity is thus a paradox of human history. Over the past 12,000 years, we have developed technologies, economies, and social institutions that cost more labor, time, money, energy, and annoyance, and that go against our aversion to such costs. We have progressively adopted ways of life that impose increasing costs on both societies and individuals, and that contravene some of our deepest inclinations. Why, then, did human societies ever become more complex?

At least part of the answer is that complexity is a basic problem-solving tool. Confronted with problems, we often respond by developing more complex technologies, establishing new institutions, adding more specialists or bureaucratic levels to an institution, increasing organization or regulation, or gathering and processing more information. Such increases in complexity work in part because they can be implemented rapidly, and typically build on what was developed before. While we usually prefer not to bear the cost of complexity, our problem-solving efforts are powerful complexity generators. All that is needed for growth of complexity is a problem that requires it. Since problems continually arise, there is persistent pressure for complexity to increase.[6]

Growth of complexity to address problems is well illustrated in the response to the attacks on the US of September 11, 2001. In the aftermath, steps taken to prevent future similar attacks focused on creating new government agencies, such as the Transportation Security Administration and the Department of Homeland Security, consolidating existing functions into some of the new agencies, and increasing control over realms of behavior from which a threat might arise. In other words, the US's first response was to complexify—to diversify structure and function, and to increase organization or control. The report of the government commission convened to investigate the attacks (colloquially called the 9/11 commission) recommended steps to prevent future attacks. The recommended actions amount, in effect, to more complexity requiring more costs in the form of resources, time, or annoyance.[7]

The costliness of complexity is not a mere annoyance or inconvenience. It conditions the long-term success or failure of problem-solving efforts, and thus of sustainability. Complexity can be viewed as an economic function. Societies and institutions invest in problem solving, undertaking costs and expecting benefits in return. In any system of problem solving, early efforts tend to be simple and cost-effective. That is, they work and give high returns per unit of effort. This is a normal economic process: humans always tend to pluck the lowest fruit, going to higher branches only when those lower no longer hold fruit. This is known as the best-first principle.[8] In problem-solving systems, inexpensive solutions are adopted before more complex and expensive ones. In the history of human food-gathering and production, for example, labor-sparing hunting and gathering gave way to more labor-intensive agriculture, which in some places has been replaced by industrial agriculture that consumes more energy than it produces.[9] We produce minerals and energy whenever possible from the most economic sources. Our societies have changed from egalitarian relations, economic reciprocity, *ad hoc* leadership, and generalized roles to social and economic differentiation, specialization, inequality, and full-time leadership. These characteristics are the essence of complexity, and they increase the costliness of any society.

As best-first (high-return) solutions are progressively implemented, only more costly solutions remain. After the highest-return ways to produce resources, process information, and organize society are applied, continuing problems must be addressed in ways that are more costly and less cost-effective. As the costs of solving problems grow, the point is reached where further investments in complexity do not give a proportionate return. Increments of investment in complexity begin to yield smaller and smaller increments of return. The *marginal* return (that is, the return per extra unit of investment) starts to decline (Figure 4.1). This is the long-term challenge faced by problem-solving institutions: diminishing returns to complexity. If allowed to proceed unchecked, eventually it brings ineffective problem solving and even economic stagnation. A prolonged period of diminishing returns to complexity is a major part of what makes problem solving ineffective and societies or institutions unsustainable.[10]

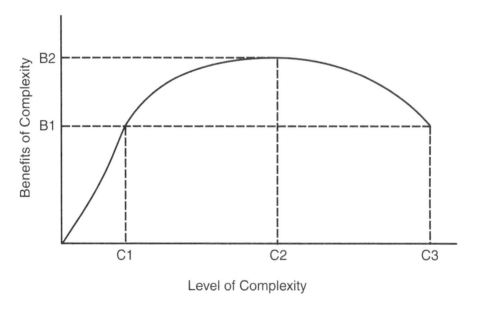

Figure 4.1   The marginal productivity of increasing complexity. At a point such as B1, C3, the costs of complexity exceed the benefits, and complexity is a disadvantageous approach to problem solving

Most of the time, cultural complexity increases in a purely mundane manner, from day-to-day exercises in solving problems. Complexity that emerges in this way will usually appear *before* there are additional resources to support it. Complexity thus compels increases in resource production. This understanding of the temporal relationship between complexity and resources has implications for sustainability that diverge from what is commonly assumed. These implications will be explored at the end of this essay. It is useful first to present historical case studies that illustrate the points made in this section.

Case studies in resources and complexity

Two historical cases illustrate the relationship of resources to problem solving, complexity, and sustainability. These are the collapse of the Western Roman Empire in the fifth century AD and the collapse of the Byzantine Empire in the seventh century AD, followed by Byzantine recovery. These cases are chosen for the lessons they impart about sustainability today.

Collapse of the Western Roman Empire

The economics of an empire such as the Romans assembled are seductive but illusory. The returns to any campaign of conquest are highest initially, when the accumulated surpluses of the conquered peoples are appropriated. Thereafter the conqueror assumes the cost of administering and defending the province. These responsibilities may last centuries, and are paid for from yearly agricultural surpluses.

The Roman government was financed by agricultural taxes that barely sufficed for ordinary administration. When extraordinary expenses arose, typically during wars, the precious metals on hand frequently were insufficient. Facing the costs of war with Parthia and rebuilding Rome after the Great Fire, Nero began in 64 AD a policy that later emperors found irresistible. He debased the primary silver coin, the denarius, reducing the alloy from 98 to 93 percent silver. It was the first step down a slope that resulted two centuries later in a currency that was worthless and a government that was insolvent (Figure 4.2).

In the half-century from 235 to 284, the empire nearly came to an end. There were foreign and civil wars almost without interruption. The period witnessed 26 legitimate emperors and perhaps 50 usurpers. Cities were sacked and frontier provinces devastated. The empire shrank in the 260s to Italy, the Balkans, and North Africa. By prodigious effort the empire survived the crisis, but it emerged at the turn of the fourth century AD as a very different organization.

In response to the crises, Diocletian and Constantine, in the late third and early fourth centuries, designed a government that was larger, more complex, and more highly organized. They doubled the size of the army. To pay for all this, the government taxed its citizens more heavily, conscripted their labor, and dictated their occupations. Villages were responsible for the taxes on their members, and one village could even be held liable for another. Despite several monetary reforms a stable currency could not be found (Figure 4.3). As masses of worthless coins were produced, prices rose higher and higher. Moneychangers in the east would not convert imperial currency, and the government refused to accept its own coins for taxes.

With the rise in taxes, population could not recover from plagues in the second and third centuries. There were chronic shortages of labor. Marginal lands went out of cultivation. Faced with taxes, peasants would abandon their lands and flee to the protection of a wealthy landowner. By 400 AD most of Gaul and Italy were owned by fewer than twenty senatorial families.

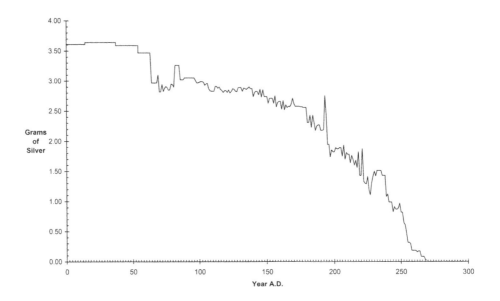

Figure 4.2 Debasement of the denarius to 269 AD

Source: J. A. Tainter, "La Fine dell'Amministrazione Centrale: il Collaso dell'Impero Romano in Occidente," in *Storia d'Europa, Volume Secondo: Preistoria e Antichità*, eds. J. Guilaine and S. Settis (Turin: Einaudi, 1994): 1217.

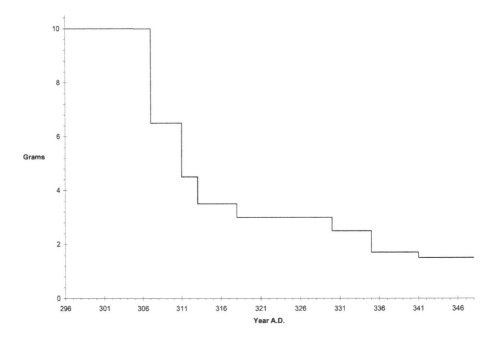

Figure 4.3 Reductions in the weight of the Roman follis, 296–348 AD

Source: data from D. Van Meter, *The Handbook of Roman Imperial Coins* (Nashua, NH: Laurion Numismatics, 1991): 47.

From the late fourth century the peoples of central Europe could no longer be kept out. They forced their way into Roman lands in Western Europe and North Africa. The government came to rely almost exclusively on troops from Germanic tribes. When finally they could not be paid, they overthrew the last emperor in Italy in 476.[11]

The strategy of the later Roman Empire was to respond to a near-fatal challenge in the third century by increasing the size, complexity, power, and costliness of the primary problem-solving system—the government and its army. The higher costs were undertaken not to expand the empire or to acquire new wealth, but to maintain the status quo. The benefit/cost ratio of imperial government declined. In the end the Western Roman Empire could no longer afford the problem of its own existence.[12]

Collapse and recovery of the Byzantine Empire

The Eastern Roman Empire (usually known as the Byzantine Empire) survived the fifth-century debacle. Efforts to develop the economic base, and to improve the effectiveness of the army, were so successful that by the mid sixth century Justinian (527–65) could engage in a massive building program and attempt to recover the western provinces.

By 541, the Byzantines had conquered North Africa and most of Italy. Then that year bubonic plague swept over the Mediterranean for the first time. Just as in the fourteenth century, the plague of the sixth century killed from one-fourth to one-third of the population. The loss of taxpayers caused immediate financial and military problems. By the early seventh century the Slavs and Avars had overrun the Balkans. Italy was largely lost to the Lombards. The Persians conquered Syria, Palestine, and Egypt.

The emperor Heraclius cut soldiers' pay by half in 616, and proceeded to debase the currency (Figure 4.4). These economic measures facilitated his military strategy. In 626 the siege of Constantinople was broken. The Byzantines destroyed the Persian army and occupied the Persian king's favorite residence. The Persians had no choice but to surrender all the territory they had seized. The Persian war lasted 26 years, and resulted only in restoration of the status quo of a generation earlier.

The empire was exhausted by the struggle. Arab forces, newly converted to Islam, defeated the Byzantine army decisively in 636. Syria, Palestine, and Egypt, the wealthiest provinces, were lost permanently. The Arabs raided Asia Minor nearly every year for two centuries, forcing thousands to hide in underground cities. Constantinople was besieged each year from 674 to 678. The Bulgars broke into the empire from the north. The Arabs took Carthage in 697. From 717 to 718 an Arab force besieged Constantinople continuously for over a year. It seemed that the empire could not survive. The city was saved in the summer of 718, when the Byzantines ambushed reinforcements sent through Asia Minor, but the empire was now merely a shadow of its former size.

Third- and fourth-century emperors had managed a similar crisis by increasing the complexity of administration, the regimentation of the population, and the size of the army. This was paid for by such levels of taxation that lands were abandoned and peasants could not replenish the population. Byzantine emperors could hardly impose more of the same exploitation on the depleted population of the shrunken empire. Instead they adopted a strategy that is truly rare in the history of complex societies: systematic simplification.

Around 659 military pay was cut in half again. The government had lost so much revenue that even at one-fourth the previous rate it could not pay its troops. The solution was for the army to support itself. Soldiers were given grants of land on condition of hereditary military service. The Byzantine fiscal administration was correspondingly simplified.

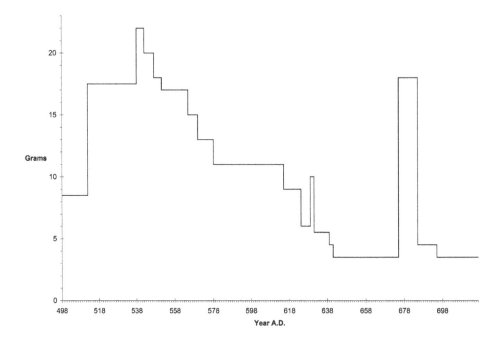

Figure 4.4   Weight of the Byzantine follis, 498–717 AD

Source: data from K. W. Harl, *Coinage in the Roman Economy, 300 BC to AD 700* (Baltimore, MD: Johns Hopkins University Press, 1996): 197.

The transformation ramified throughout Byzantine society. Both central and provincial government were simplified, and the costs of government were reduced. Provincial civil administration was merged into the military. Cities across Anatolia contracted to fortified hilltops. The economy developed into its medieval form, organized around self-sufficient manors. There was little education beyond basic literacy and numeracy, and literature itself consisted of little more than lives of saints. The period is sometimes called the Byzantine Dark Age.

The simplification rejuvenated Byzantium. The peasant-soldiers became producers rather than consumers of the empire's wealth. The solar energy that had formerly sustained the empire had flowed from the sun to agricultural fields, to farmers, to currency, to taxes, and finally to soldiers as their pay. Now energy flowed from the sun directly to farmers (Figure 4.5). Transaction costs were reduced, and energy was no longer lost at each transformation, as it is in ecological systems. By lowering the cost of military defense, the Byzantines secured a better return on their most important investment. Fighting as they were for their own lands and families, soldiers performed better.

During the next century, campaigns against the Bulgars and Slavs gradually extended the empire in the Balkans. Greece was recaptured. Pay was increased after 840, yet gold became so plentiful that in 867 Michael III met an army payroll by melting down 20,000 pounds of ornaments from the throne room. When marines were added to the imperial fleet it became more effective against Arab pirates. In the tenth century the Byzantines reconquered parts of coastal Syria. Overall after 840 the size of the empire was nearly doubled. The process culminated when Basil II (963–1025) conquered the Bulgars and extended the empire's boundaries again

Energetics of the Byzantine Army

600 A.D.
Sun→crops→farmers→tax collectors→central government→army

700 A.D
Sun→crops→army

Figure 4.5   Energetics of the Byzantine Army, 600–700 AD

to the Danube.[13] In two centuries the Byzantines had gone from near disintegration to being the premier power in Europe and the Near East, an accomplishment won by decreasing the complexity and costliness of problem solving.

Discussion

The Roman and Byzantine case studies illustrate different outcomes to complexification, and offer different lessons for understanding sustainability. The Roman collapse exemplifies the thesis of this essay, that increasing complexity precedes the availability of resources and subsequently compels increases in their production. The Byzantine collapse and recovery illustrate a different but also important point, which will be discussed shortly.

The Roman Empire is a single case study in complexity and problem solving,[14] but it is an important and representative one. It illustrates one of the basic processes by which societies increase in complexity. Societies adopt increasing complexity to solve problems, including problems of sustainability, becoming at the same time more costly. In the normal course of economic evolution, this process at some point will produce diminishing returns. This is due to the best-first principle, in which the most cost-effective solutions are adopted first, and subsequent solutions are more costly and yield diminishing returns. Once diminishing returns develop, a problem-solving institution must either find new resources to continue the activity, or fund the activity by reducing the share of resources available to other economic sectors. The latter is likely to produce economic contraction, popular discontent, and eventual collapse. This was the fate of the Western Roman Empire.

This understanding of complexity and resources has implications for our contemporary discussions of resources and sustainability. Both popular and academic discourse on sustainability commonly make the following assumptions: that (a) future sustainability requires that industrial societies consume a lower quantity of resources than is now the case,[15] and (b) sustainability will result automatically if we do so. Sustainability emerges, in this view, as a passive consequence of consuming less. Thus sustainability efforts are commonly focused on reducing consumption through voluntary or enforced conservation, perhaps involving simplification, and/or through improvements in technical efficiencies.

This perspective on sustainability follows logically from the common view that societies grow in complexity solely due to innovation.[16] Complexity, in this view, is a voluntary matter. Human societies became more complex by choice rather than necessity. By this reasoning, we should be able to choose to forego complexity and the resource consumption that it entails. It is widely thought in sustainability studies that societies can deliberately reduce their consumption of resources and thus achieve sustainability.

The fact that complexity and costliness increase through mundane problem solving suggests a different conclusion with a startling implication: Contrary to what is typically advocated as the route to sustainability, *it is usually not possible for a society to reduce its consumption of resources voluntarily over the long term*. To the contrary, as problems great and small inevitably arise, addressing these problems requires complexity and resource consumption to increase. Historically, as illustrated by the Roman Empire and other cases, this has commonly been the case.[17]

The Byzantine collapse becomes important at this point. It is the only case of which I am aware in which a large, complex society systematically simplified, and reduced thereby its consumption of resources. While this case shows that societies can reduce resource consumption and thrive, it offers no hope that this can be done commonly. In the Byzantine case simplification was forced, made necessary by a gross insufficiency of revenues. The Byzantines undertook simplification and conservation because, to use a colloquial expression, their backs were to the wall. The empire had no choice. The Byzantine simplification was also temporary. As Byzantine finances recovered, emperors again expanded the size and complexity of their armed forces.[18] The Byzantine chronicler Anna Comnena, daughter of emperor Alexius I (1081–1118), described her father's marching army as like a moving city.[19]

Many students of sustainability will find it a disturbing conclusion that long-term conservation is not possible, contravening as it does so many assumptions about future sustainability. Naturally we must ask: Are there alternatives to this process? Can we find a way out of this dilemma? Regrettably, no simple solutions are evident. Consider some of the approaches commonly advocated:

1   *Voluntarily reduce resource consumption.* While this may work for a time, its longevity as a strategy is constrained by the factors discussed in this essay: Societies increase in complexity to solve problems, becoming more costly in the process. Resource production must subsequently increase to fund the increased complexity. To implement voluntary conservation long term would require that a society be either uniquely lucky in not being challenged by problems, or that it avoids addressing the problems that confront it. The latter strategy would at best reduce the legitimacy of the problem-solving institution, and at worst lead to its demise.

  I will not address in depth the question of whether long-term voluntary conservation is possible at the level of individuals and households. I am confident that usually it is not, that humans will not ordinarily forego affordable consumption of things they desire on the basis of abstract projections about the future. I raise the possibility of voluntary conservation only because of its perennial popularity.

There are societies that seem to incorporate an ethic of conservation. Japan, as described by Caldararo, may be such a society. Caldararo argues that Japan participates in the system of industrial nations in its own way: low fertility, comparatively low consumption, high savings, acceptance of high prices, and tolerance of institutions that are economically inefficient but socially rational. "Japan," Caldararo believes, "is building a sustainable economy for the 21st century."[20] Yet even if Caldararo's assessment is accurate, such a case does not contravene the arguments presented here. Even in societies that do voluntarily consume less than they could, problem solving must in time cause complexity, costliness, and resource consumption to grow. These things may grow from a smaller base, but the fundamental process of increasing complexity remains unaltered.

2   *Employ the price mechanism to control resource consumption.* This is currently the *laissez-faire* strategy of industrialized nations. Since humans do not commonly forego affordable consumption of desired goods and services, economists consider it more effective than voluntary conservation. Both approaches, however, lead eventually to the same outcome: As problems arise, resource consumption must increase at the societal level even if consumers as individuals purchase less.

3   *Ration resources.* Because of its unpopularity, rationing is possible in democracies only for clear, short-term emergencies. This is illustrated by the reactions to rationing in England and the United States during World War II. Moreover, rationed resources may become needed to solve societal problems, belying any attempt to conserve through rationing. Something like this can be seen in the fiscal stimulus programs enacted in late 2008 and early 2009.

4   *Reduce population.* While this would reduce aggregate resource consumption temporarily, as a long-term strategy it has the same fatal flaw as the first two: Problems will emerge that require solutions, and those solutions will compel resource production to grow.

5   *Hope for technological solutions.* I sometimes call this a faith-based approach to our future. We members of industrialized societies are socialized to believe that we can always find a technological solution to resource problems. Technology, within the framework of this belief, will presumably allow us continually to reduce our resource consumption per unit of material wellbeing. Conventional economics teaches that to bring this about we need only the price mechanism and unfettered markets. Consider, for example, the following statements:

> "No society can escape the general limits of its resources, but no innovative society need accept Malthusian diminishing returns."[21]

> "All observers of energy seem to agree that various energy alternatives are virtually inexhaustible."[22]

> "By allocation of resources to R&D, we may deny the Malthusian hypothesis and prevent the conclusion of the doomsday models."[23]

Our society's belief in technical solutions is deeply ingrained.

One flaw in this reasoning was pointed out by Jevons in his work of 1866: As technological improvements reduce the cost of using a resource, total consumption will eventually increase.

The Jevons paradox (also known as the rebound effect) is widely in effect,[24] among economic levels ranging from nations to households and individuals, including in many sectors of daily life.[25]

A second flaw in the approach known as technological optimism is that it is based on the implicit assumption that investments in innovation yield at least constant returns. Yet in innovation, as in other activities, we are constrained by variations of the best-first principle. That is, we make the easiest and most productive innovations first. Thereafter research problems become more difficult and costly to resolve. Where scientific breakthroughs were once made by lone-wolf naturalists, such as Charles Darwin and Gregor Mendel, research now requires interdisciplinary teams, large institutions, and supporting staffs.[26] The costliness of innovation grows, and productivity declines. Innovation investments reach diminishing returns. Figure 4.6 shows that the productivity of innovation, measured as patents per inventor, declined from 1974 to 2005 by over 20 percent.[27] As the productivity of innovation declines, innovation itself becomes less suitable for solving problems. The fatal flaw in technological optimism is that continued investments in research and development yield declining marginal returns. In time this will curtail the usefulness of innovation in producing sustainability.

Thus, conventional solutions to problems of resource consumption can be effective only for short periods of time. Over the long term, problem solving compels societies to grow in complexity and increase consumption. Because of this it is useful to think of sustainability in the metaphor of an athletic game: It is possible to "lose"—that is, to become unsustainable, as happened to the Western Roman Empire. But the converse does not hold. Because we continually confront challenges, there is no point at which a society has "won"—that is, become sustainable in perpetuity, or at least for a very long time. Success, rather, consists of remaining in the game.

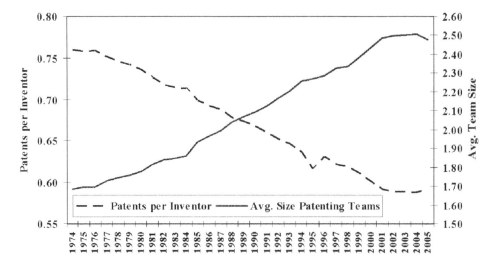

Figure 4.6 The productivity of innovation, measured as the average size of patenting teams and patents per inventor, 1974–2005, in US patents. More than half of US patents come from entities outside the US, so the data measure global innovation trends

Source: after D. Strumsky, J. Lobo, and J. A. Tainter, "Complexity and the Productivity of Innovation," *Systems Research and Behavioral Science* 5 (2010): 502.

What can societies do when faced with increasing complexity, increasing costs, and diminishing returns in problem solving? There appear to be seven possible strategies, *all of which are effective only for a time.*[28] These are not sequential steps, nor are they mutually exclusive. They are simply ideas that can work alone or in combination. Some of these strategies would clearly have only short-term effects, while others may be effective for longer. The first strategy, however, is essential in all long-term efforts toward sustainability.

1   *Be aware.* Complexity is most insidious when the participants in an institution are unaware of what causes it. Managers of problem-solving institutions gain an advantage by understanding how complexity develops, and its long-term consequences. It is important to understand that unsustainable complexity may emerge over periods of time stretching from years to millennia, and that cumulative costs bring the greatest problems. It is important to point out that complexity is not intrinsically good or bad. Complexity does provide solutions to problems. Rather, complexity is either useful and affordable or it is not.

2   *Don't solve the problem.* This option is deceptively simple. The worldview of Western industrial societies is that ingenuity and incentives can solve all problems. Ignorance of complexity, combined with the fact that the cost of solving problems is often deferred or spread thinly, reinforces our problem-solving inclination. Yet often we do choose not to solve problems, either because of their cost or because of competing priorities. Appropriators and managers do this routinely. Yet as noted above, not solving problems may reduce the legitimacy of an institution, and bring about its eventual demise.

3   *Accept and pay the cost of complexity.* This is a common strategy, perhaps the most common in coping with complexity. It, too, is deceptively simple. Governments are often tempted to pay the cost of problem solving by increasing taxes, which reduces the share of national income available to other economic sectors. Businesses may do the same by increasing prices. The problem comes when taxpayers and consumers rebel, or when a firm's competitors offer a similar product at a lower cost.

4   *Find subsidies to pay costs.* This has been the strategy of modern industrial economies, which have employed the subsidies of fossil and nuclear energy to support our unprecedented levels of complexity. As seen since the adoption of coal,[29] the right subsidies can sustain complex problem solving for centuries. Anxiety over future energy is not just about maintaining a standard of living. It also concerns our future problem-solving abilities. Fossil fuels, after all, are finite.

5   *Shift or defer costs.* This is one of the most common ways to pay for complexity, and it can work in the short term. Budget deficits, currency devaluation, and externalizing costs exemplify this principle in practice. This was the strategy of the Roman Empire in debasing its currency, which shifted to the future the costs of containing current crises. Governments before the Roman Empire also practiced this subterfuge, as have many since. Governments today defer costs through borrowing. As seen in the case of the Romans, this is a strategy that can work only for a time. When it is no longer feasible, the economic repercussions may be far worse than if costs had never been deferred.

6   *Connect costs and benefits.* If one adopts the explicit goal of controlling complexity, costs and benefits must be connected so explicitly that the tendency for complexity to grow can be constrained by its costs. In an institution this means that information about the cost of complexity must flow accurately and effectively. Yet in a hierarchical institution, the flow of information from the bottom to the top is frequently inaccurate and ineffective.[30] Thus the managers of an institution are often poorly informed about the cost of complexity and feel free to deploy more.

7   *Recalibrate or revolutionize the activity.* This involves a fundamental change in how costs and benefits are connected, and is potentially the most far-reaching technique for coping with complexity. The strategy may involve both new resources and new types of complexity that lower costs, combined with positive feedback among new elements that amplifies benefits and produces growth. True revolutions of this sort are rare, so much so that we recognize them in retrospect with a term signifying a new era: the Agricultural Revolution and the Industrial Revolution, for example. Today's Information Revolution may be another such case. Fundamental changes of this sort depend on opportunities for positive feedback, where elements reinforce each other. For example, Watt's steam engine facilitated the mining of coal by improving pumping water from mines. Cheaper coal meant more steam engines could be built and put to use, facilitating even cheaper coal.[31] Put a steam engine on rails and both coal and other products can be distributed better to consumers. Combine coal, steam engines, and railroads, and one had most of the components of the Industrial Revolution, all mutually reinforcing each other. The economic system became more complex, but the complexity involved new elements, connections, and subsidies that produced increasing returns.

The transformation of the US military since the 1970s provides a more recent example. So profound is this transformation that it is recognized by its own acronym: RMA, the Revolution in Military Affairs. The revolution involves extensive reliance on information technology, as well as the integration of hardware, software, and personnel. Weapons platforms are just part of this revolution, since weapons now depend on integration with sensors, satellites, software, and command systems.[32] This is a military that is vastly more complex than ever before. That complexity is of course costly, but the benefits include both greater effectiveness and significant cost savings. Being able to pinpoint targets means less waste of ordnance, less need for large numbers of weapons platforms, and a need for fewer people.

The fact that such revolutions do occur gives hope that a way out of our current dilemma may be found. Yet complex systems at the societal level cannot be designed. They emerge on their own or they do not. To rely on some hoped-for revolution involving innovation, energy, and positive feedback is, like relying on technological innovation, a faith-based approach to our future.

Concluding remarks

Sustainability is not the achievement of an idealized steady state. It is not a passive consequence of having fewer humans who consume more limited resources. One must work at being sustainable. The challenges to sustainability that any society (or other institution) might confront are, for practical purposes, endless in number and infinite in variety. This being so, sustainability is a matter of problem solving, an activity so commonplace that we perform it with little thought to its long-term implications.

The notion of progress is ingrained in industrial societies, so much so that it is part of our cosmology, a fundamental element of our ancestor myth. Just as our ancestors, we believe, "pulled themselves up" through ingenuity, so today we continue this tradition. In the conventional framework, all that past societies required for progress was innovation. Complexity, it is believed, emerged voluntarily, and if this was so then we should be able to forego complexity voluntarily and reduce our consumption of the resources that it requires. This is the conventional approach to sustainability, which implicitly sees the future as a condition of stasis with no challenges.

In actuality, complexity commonly increases in response to problems, problems that are sometimes large-scale, urgent, and even existential. Increased complexity requires increased resources, although when a problem is addressed long-term costs are typically not considered fully. Complexity emerging through problem solving typically precedes the availability of resources, and compels increases in their production. Complexity is not voluntary, nor is it something that we can ordinarily choose to forego. Complexity is required to solve problems, as are the resources that support it. Since sustainability emerges from success in solving societal problems, it usually requires its own allocation of resources. As noted above, sustainability emerging from lower resource consumption is so rare in human history that the Byzantine recovery is the only known case among large-scale, complex societies.

Applying this understanding to the problem of sustainability leads to two conclusions that are not presently recognized in the sustainability movement. The first is that the solutions commonly recommended to promote sustainability—conservation, simplification, pricing, and innovation—can do so only in the short term. Secondly, long-term sustainability depends on solving major societal problems that will converge in coming decades, and this will require increasing complexity and resource production. Sustainability is demonstrably not a condition of stasis. It is, rather, a process of continuous adaptation, of perpetually addressing new or ongoing problems and securing the resources to do so.

Acknowledgements

This is a substantially revised version of the author's previously published work, "Resources and Cultural Complexity: Implications for Sustainability," *Critical Reviews in Plant Sciences* 30 (2011): 24–34. Every reasonable attempt was made to obtain permission for reproduction.

I am pleased to express my appreciation to Jeremy Caradonna for the invitation to submit this paper, and to Temis Taylor for comments.

Notes

1   A. Stevenson, ed. *Shorter Oxford English Dictionary on Historical Principles*, 6th ed. (Oxford: Oxford University Press, 2007), 3126.
2   J. A. Tainter, *The Collapse of Complex Societies* (Cambridge: Cambridge University Press, 1988).
3   Ibid.; Timothy F. H. Allen, Joseph A. Tainter, and Thomas W. Hoekstra, *Supply-Side Sustainability* (New York: Columbia University Press, 2003).
4   R. H. McGuire, "Breaking Down Cultural Complexity: Inequality and Heterogeneity," in *Advances in Archaeological Method and Theory*, Vol. 6, ed. M. B. Schiffer (New York: Academic Press, 1983), 91–142.
5   L. L. Gatlin, *Information Theory and the Living System* (New York: Columbia University Press, 1972).
6   Tainter, *Collapse of Complex Societies*; J. A. Tainter, "Complexity, Problem Solving, and Sustainable Societies," in *Getting Down to Earth: Practical Applications of Ecological Economics*, eds. R. Costanza, O. Segura, and J. Martinez-Alier. (Washington, DC: Island Press, 1996), 61–76; J. A. Tainter, "Problem Solving: Complexity, History, Sustainability," *Population and Environment* 22 (2000): 3–41; J. A. Tainter, "Social Complexity and Sustainability," *Ecological Complexity* 3 (2006): 91–103.
7   9/11 Commission, *The 9/11 Commission Report: Final Report of the National Commission on Terrorist Attacks Upon the United States.* (New York: Norton, 2004), 367–428.
8   Cutler J. Cleveland, "Natural Resource Quality," in *The Encyclopedia of Earth*, ed. C. J. Cleveland (Washington, DC: Environmental Information Coalition, 2008).
9   E. Boserup, *The Conditions of Agricultural Growth: The Economics of Agrarian Change Under Population Pressure* (Chicago, IL: Aldine, 1965); C. Clark and M. Haswell, *The Economics of Subsistence Agriculture* (London: Macmillan, 1966); M. N. Cohen, *The Food Crisis in Prehistory: Overpopulation and the Origins of Agriculture* (New Haven, CT: Yale University Press, 1977).

10 J. A. Tainter, "Post-Collapse Societies," in *Companion Encyclopedia of Archaeology*, ed. G. Barker, 988–1039 (London: Routledge, 1999); Tainter, *Collapse of Complex Societies*; Tainter, "Problem Solving: Complexity, History, Sustainability"; Tainter, "Social Complexity and Sustainability."

11 A. E. R. Boak, *Manpower Shortage and the Fall of the Roman Empire in the West* (Ann Arbor, MI: University of Michigan Press, 1955); J. C. Russell, "Late Ancient and Medieval Population," *Transactions of the American Philosophical Society*, 48, 3 (1958); A. H. M. Jones, *The Later Roman Empire, 284-602: A Social, Economic and Administrative Survey* (Norman, OK: University of Oklahoma Press, 1964); A. H. M. Jones, *The Roman Economy: Studies in Ancient Economic and Administrative History* (Oxford: Basil Blackwell, 1974); G. A. J. Hodgett, *A Social and Economic History of Medieval Europe* (London: Methuen, 1972); R. MacMullen, *Roman Government's Response to Crisis, A.D. 235–337* (New Haven, CT: Yale University Press, 1976); Charles Wickham, "The Other Transition: From the Ancient World to Feudalism," *Past and Present* 103, 1 (1984): 3–36; S. Williams, *Diocletian and the Roman Recovery* (New York: Methuen, 1985); Tainter, *Collapse of Complex Societies*; J. A. Tainter, "La Fine dell'Amministrazione Centrale: il Collaso dell'Impero Romano in Occidente," in *Storia d'Europa, Volume Secondo: Preistoria e Antichità*, eds. J. Guilaine and S. Settis, 1207–55 (Turin: Einaudi, 1994); R. Duncan-Jones, *Structure and Scale in the Roman Economy.* (Cambridge: Cambridge University Press, 1990); K. W. Harl, *Coinage in the Roman Economy, 300 B.C. to A.D. 700* (Baltimore, MD: Johns Hopkins University Press, 1996).

12 Tainter, *Collapse of Complex Societies*; Tainter, "La Fine dell'Amministrazione Centrale"; Tainter, "Problem Solving: Complexity, History, Sustainability"; Tainter, "Social Complexity and Sustainability"; Allen, Tainter, and Hoekstra, *Supply-Side Sustainability*; J. A. Tainter and C. L. Crumley, "Climate, Complexity, and Problem Solving in the Roman Empire," in *Sustainability or Collapse? An Integrated History and Future of People on Earth*, Eds. R. Costanza, L. J. Graumlich, and W. Steffen, 61–75, Dahlem Workshop Report 96 (Cambridge, MA: MIT Press, 2007).

13 W. Treadgold, *The Byzantine Revival, 780–842* (Stanford, CA: Stanford University Press, 1988); W. Treadgold, *Byzantium and its Army, 284–1081* (Stanford, CA: Stanford University Press, 1995); W. Treadgold, *A History of the Byzantine State and Society* (Stanford, CA: Stanford University Press, 1997); J. F. Haldon, *Byzantium in the Seventh Century: The Transformation of a Culture* (Cambridge: Cambridge University Press, 1990); Harl, *Coinage in the Roman Economy*.

14 For other examples, see J. A. Tainter, "A Framework for Archaeology and Sustainability," in *Encyclopedia of Life Support Systems* (Ocford: EOLSS Publishers, 2002), www.eolss.net; Tainter, *Collapse of Complex Societies*; Tainter, "Problem Solving: Complexity, History, Sustainability"; Tainter, "Social Complexity and Sustainability"; Allen, Tainter, and Hoekstra, *Supply-Side Sustainability*.

15 See, for instance, L. R. Brown, *Plan B 4.0: Mobilizing to Save Civilization* (New York: W. W. Norton, 2009); N. Caldararo, *Sustainability, Human Ecology, and the Collapse of Complex Societies: Economic Anthropology and a 21st Century Adaptation*, Mellen Studies in Anthropology 15 (Lewiston, New York: Edward Mellen Press, 2004); and Richard Heinberg, *Power Down: Options and Actions for a Post-Carbon World* (Gabriola Island, BC: New Society Publishers, 2004).

16 D. Strumsky, J. Lobo, and J. A. Tainter, "Complexity and the Productivity of Innovation," *Systems Research and Behavioral Science* 5 (2010): 496–509.

17 See Tainter, *Collapse of Complex Societies*; Tainter, "Problem Solving: Complexity, History, Sustainability"; Tainter, "A Framework for Archaeology and Sustainability"; Tainter, "Social Complexity and Sustainability"; Allen, Tainter, and Hoekstra, *Supply-Side Sustainability*.

18 McGeer, E. *Sowing the Dragon's Teeth: Byzantine Warfare in the Tenth Century*, Dumbarton Oaks Studies XXXIII (Washington, DC: Dumbarton Oaks, 1995); Treadgold, *Byzantium and Its Army, 284–1081*.

19 See J. F. Haldon, *Warfare, State and Society in the Byzantine World, 565–1204* (London: UCL Press, 1999), 156.

20 N. Caldararo, *Sustainability, Human Ecology, and the Collapse of Complex Societies: Economic Anthropology and a 21st Century Adaptation*, Mellen Studies in Anthropology 15 (Lewiston, NY: Edward Mellen Press, 2004), 256.

21 H. J. Barnett and C. Morse, *Scarcity and Growth: The Economics of Natural Resource Availability* (Baltimore, MD: Johns Hopkins Press, 1963), 139.

22 R. L. Gordon, *An Economic Analysis of World Energy Problems* (Cambridge, MA: MIT Press, 1981), 109.

23 R. Sato and G. S. Suzawa, *Research and Productivity: Endogenous Technical Change* (Boston, MA: Auburn House, 1983), 81.

24 J. M. Polimeni, K, Mayumi, M. Giampietro, and B. Alcott, *The Jevons Paradox and the Myth of Resource Efficiency Improvements* (London: Earthscan, 2008).

25 J. A. Tainter, "Foreword," in *The Jevons Paradox and the Myth of Resource Efficiency Improvements*, Eds. J. M. Polimeni, K. Mayumi, M. Giampietro, and B. Alcott, ix–xvi (London: Earthscan, 2008).

26 See, for instance, S. Wuchty, B. F. Jones, and B. Uzzi, "The Increasing Dominance of Teams in Production of Knowledge," *Science* 316 (2007): 1036–9; B. F. Jones, S. Wuchty, and B. Uzzi, "Multi-University Research Teams: Shifting Impact, Geography, and Stratification in Science," *Science* 322 (2008): 1259–62; Strumsky, Lobo, and Tainter, "Complexity and the Productivity of Innovation."

27 Strumsky, Lobo, and Tainter, "Complexity and the Productivity of Innovation."

28 Tainter, "Social Complexity and Sustainability."

29 R. G. Wilkinson, *Poverty and Progress: An Ecological Model of Economic Development* (London: Methuen, 1973).

30 R. J. McIntosh, J. A. Tainter, and S. K. McIntosh, "Climate, History, and Human Action," in *The Way the Wind Blows: Climate, History, and Human Action*, Eds R. J. McIntosh, J. A. Tainter, and S. K. McIntosh (New York: Columbia University Press, 2000), 29–30.

31 Wilkinson, *Poverty and Progress*.

32 R. L. Paarlberg, "Knowledge as Power: Science, Military Dominance, and US Security," *International Security* 29 (2004): 122–51.

References

9/11 Commission. *The 9/11 Commission Report: Final Report of the National Commission on Terrorist Attacks Upon the United States*. New York: Norton, 2004.

Allen, T. F. H., J. A. Tainter, and T. W. Hoekstra. *Supply-Side Sustainability*. New York: Columbia University Press, 2003.

Barnett, H. J. and C. Morse. *Scarcity and Growth: The Economics of Natural Resource Availability*. Baltimore, MD: Johns Hopkins Press, 1963.

Boak, A. E. R. *Manpower Shortage and the Fall of the Roman Empire in the West*. Ann Arbor, MI: University of Michigan Press, 1955.

Boserup, E. *The Conditions of Agricultural Growth: The Economics of Agrarian Change Under Population Pressure*. Chicago, IL: Aldine, 1965.

Brown, L. R. *Plan B 4.0: Mobilizing to Save Civilization*. New York: W. W. Norton, 2009.

Caldararo, N. *Sustainability, Human Ecology, and the Collapse of Complex Societies: Economic Anthropology and a 21st Century Adaptation*. Mellen Studies in Anthropology 15. Lewiston, NY: Edward Mellen Press, 2004.

Clark, C. and M. Haswell, *The Economics of Subsistence Agriculture*. London: Macmillan, 1966.

Cleveland, C. J. "Natural Resource Quality." in *The Encyclopedia of Earth*, ed. C. J. Cleveland. Washington, DC: Environmental Information Coalition, 2008. Retrieved on November 29, 2016 from http://editors.eol.org/eoearth/wiki/Natural_resource_quality.

Cohen, M. N. *The Food Crisis in Prehistory: Overpopulation and the Origins of Agriculture*. New Haven, CT: Yale University Press, 1977.

Duncan-Jones, R. *Structure and Scale in the Roman Economy*. Cambridge: Cambridge University Press, 1990.

Gatlin, L. L. *Information Theory and the Living System*. New York: Columbia University Press, 1972.

Gordon, R. L. *An Economic Analysis of World Energy Problems*. Cambridge, MA: MIT Press, 1981.

Haldon, J. F. *Byzantium in the Seventh Century: The Transformation of a Culture*. Cambridge: Cambridge University Press, 1990.

Haldon, J. F. *Warfare, State and Society in the Byzantine World, 565–1204*. London: UCL Press, 1999.

Harl, K. W. *Coinage in the Roman Economy, 300 B.C. to A.D. 700*. Baltimore, MD: Johns Hopkins University Press, 1996.

Heinberg, R. *Power Down: Options and Actions for a Post-Carbon World*. Gabriola Island, BC: New Society Publishers, 2004.

Hodgett, G. A. J. *A Social and Economic History of Medieval Europe*. London: Methuen, 1972.

Jevons, W. S. *The Coal Question: An Inquiry Concerning the Progress of the Nation and the Probably Exhaustion of our Coal-Mines*, 2nd ed. London: Macmillan, 1866.

Jones, A. H. M. *The Later Roman Empire, 284–602: A Social, Economic and Administrative Survey*. Norman: University of Oklahoma Press, 1964.

Jones, A. H. M. *The Roman Economy: Studies in Ancient Economic and Administrative History*. Oxford: Basil Blackwell, 1974.

Jones, B. F., S. Wuchty, and B. Uzzi. "Multi-University Research Teams: Shifting Impact, Geography, and Stratification in Science." *Science* 322 (2008): 1259–62.

MacMullen, R. *Roman Government's Response to Crisis, A.D. 235–337.* New Haven, CT: Yale University Press, 1976.

McGeer, E. *Sowing the Dragon's Teeth: Byzantine Warfare in the Tenth Century.* Dumbarton Oaks Studies XXXIII. Washington, DC: Dumbarton Oaks, 1995.

McGuire, R. H. "Breaking Down Cultural Complexity: Inequality and Heterogeneity." In *Advances in Archaeological Method and Thoery*, Vol. 6, ed. M. B. Schiffer, 91–142. New York: Academic Press, 1983.

McIntosh, R. J., J. A. Tainter, and S. K. McIntosh. "Climate, History, and Human Action." In *The Way the Wind Blows: Climate, History, and Human Action*, eds R. J. McIntosh, J. A. Tainter, and S. K. McIntosh, 1–42. New York: Columbia University Press, 2000.

Paarlberg, R. L. "Knowledge as Power: Science, Military Dominance, and US Security." *International Security* 29 (2004): 122–51.

Polimeni, J. M., K, Mayumi, M. Giampietro, and B. Alcott. *The Jevons Paradox and the Myth of Resource Efficiency Improvements.* London: Earthscan, 2008.

Russell, J. C. "Late Ancient and Medieval Population." *Transactions of the American Philosophical Society* 48, 3 (1958).

Sato, R. and G. S. Suzawa. *Research and Productivity: Endogenous Technical Change.* Boston, MA: Auburn House, 1983.

Stevenson, A., ed. *Shorter Oxford English Dictionary on Historical Principles*, 6th ed. Oxford: Oxford University Press, 2007.

Strumsky, D., J. Lobo, and J. A. Tainter. "Complexity and the Productivity of Innovation." *Systems Research and Behavioral Science* 5 (2010): 496–509.

Tainter, J. A. *The Collapse of Complex Societies.* Cambridge: Cambridge University Press, 1988.

Tainter, J. A. "La Fine dell'Amministrazione Centrale: il Collaso dell'Impero Romano in Occidente." In *Storia d'Europa, Volume Secondo: Preistoria e Antichità*, eds. J. Guilaine and S. Settis, 1207–55. Turin: Einaudi, 1994.

Tainter, J. A. "Complexity, Problem Solving, and Sustainable Societies." in *Getting Down to Earth: Practical Applications of Ecological Economics*, eds. R. Costanza, O. Segura, and J. Martinez-Alier, 61–76. Washington, DC: Island Press, 1996.

Tainter, J. A. "Post-Collapse Societies." In *Companion Encyclopedia of Archaeology*, ed. G. Barker, 988–1039. London: Routledge, 1999.

Tainter, J. A. "Problem Solving: Complexity, History, Sustainability." *Population and Environment* 22 (2000): 3–41.

Tainter, J. A. "A Framework for Archaeology and Sustainability." In *Encyclopedia of Life Support Systems.* Ocford: EOLSS Publishers, 2002, www.eolss.net.

Tainter, J. A. "Social Complexity and Sustainability." *Ecological Complexity* 3 (2006): 91–103.

Tainter, J. A. "Foreword." In *The Jevons Paradox and the Myth of Resource Efficiency Improvements*, Eds. J. M. Polimeni, K. Mayumi, M. Giampietro, and B. Alcott, ix–xvi. London: Earthscan, 2008.

Tainter, J. A. and C. L. Crumley. "Climate, Complexity, and Problem Solving in the Roman Empire." In *Sustainability or Collapse? An Integrated History and Future of People on Earth*, Eds. R. Costanza, L. J. Graumlich, and W. Steffen, 61–75. Dahlem Workshop Report 96. Cambridge, MA: MIT Press, 2007.

Treadgold, W. *The Byzantine Revival, 780–842.* Stanford, CA: Stanford University Press, 1988.

Treadgold, W. *Byzantium and its Army, 284–1081.* Stanford, CA: Stanford University Press, 1995.

Treadgold, W. *A History of the Byzantine State and Society.* Stanford, CA: Stanford University Press, 1997.

Van Meter, D. *The Handbook of Roman Imperial Coins.* Nashua, NH: Laurion Numismatics, 1991.

Wickham, Cs. "The Other Transition: From the Ancient World to Feudalism." *Past and Present* 103, 1 (1984): 3–36.

Wilkinson, R. G. *Poverty and Progress: An Ecological Model of Economic Development.* London: Methuen, 1973.

Williams, S. *Diocletian and the Roman Recovery.* New York: Methuen, 1985.

Wuchty, S., B. F. Jones, and B. Uzzi. "The Increasing Dominance of Teams in Production of Knowledge." *Science* 316 (2007): 1036–9.

5

THE ANCIENT MAYA

Sustainability and collapse?

B. L. Turner II

The problem

Arguably no civilization collapse has received more attention, either in the public or scholarly literature, than that of the Classic Period lowland Maya. The "mystery" of this collapse is amplified by the massive depopulation of the Classic Period heartlands, creating a rather unique occupational history. Why? Holding places constant, large populations within them typically surge and retract in size over the millennia.[1] Few large-scale depopulations have failed to recover on such time scales.[2] This occupational history surrounding the Maya collapse, make the heartland demise a post-child case for unsustainable occupation.[3] The various evidence and arguments surrounding this collapse and depopulation are complex, including the role of climate and economic change. Full consideration of these issues casts doubt on simple population "overshoot" theses. The role of high stress, human–environment conditions as a factor in the Maya case grows increasingly strong, however.[4]

Defining elements

To understand the Maya collapse and depopulation a number of factors that define the case require clarification.

- *Collapse*. The Classic Period lowland Maya collapse refers to cessation of the construction of monumental architecture, the demise of city-state polities and alliances, and the abandonment of most major city-states and their hinterlands in the "affected area" (below), especially during the Terminal Classic Period, about 800 to 1000 CE. For these reasons, the term collapse is used in this assessment, although Maya culture persists to this day.[5]
- *Affected area*. Most of the lowlands beyond the plains of northern Yucatán and the littoral of the peninsula witnessed collapse and depopulation (Figure 5.1). Especially hard hit was the elevated interior area of the Maya lowlands,[6] extending from the Puuc Hills of northern Yucatán to mountains arising just beyond the contemporary Petén, Guatemala, border (Figure 5.1). That portion of the elevated interior most densely occupied and replete with large city-states stretched from central Quintana Roo and Campeche, Mexico, to the central lakes of the Petén. The area has been referred to as the central Maya lowlands or heartlands,[7] the latter term applied throughout this assessment.[8]

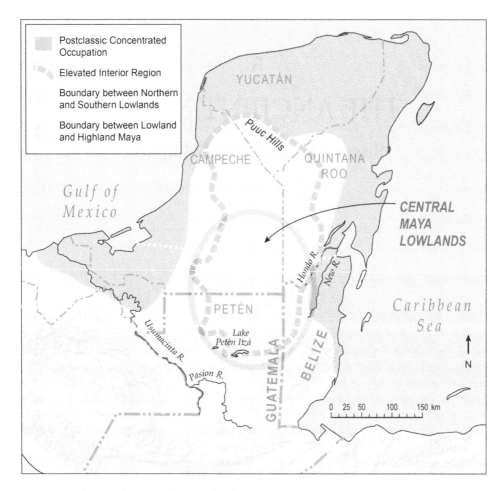

Figure 5.1 The central Maya lowlands or heartlands

Source: Barbara Trapido-Laurie.

- *Occupation.* The maximum population in the heartlands before the collapse was large, with numerous cities housing 20,000–50,000 people and several approaching 100,000,[9] often with large rural areas of dense occupation. Overall population numbers and density estimates are debated, owing in part to boundaries imposed on the calculations and to assumptions about the number of house sites occupied simultaneously and the number of occupants within them. Nevertheless, some parts of the heartlands likely approached 100 people per km^2. Regardless of the precise number, the density of population was multiple times larger than it has been subsequent to the collapse.[10]
- *Sustained depopulation.* The abandonment of the heartlands left so few people that the Cortez-led expedition of 1524–6 from Mexico to Honduras almost perished in an attempt to cross them, owing to the density of the forest and paucity of trails and settlements, the latter a requisite for food to supply the expedition. Re-occupation has subsequently never approached the size of Classic Period Maya.

Causes of the collapse and depopulation of the Maya heartlands

The causes of the Classic Period lowland Maya collapse in and the depopulation of the elevated interior area, especially the heartlands, range from fanciful and simple to complex.[11] The simple or single cause explanations include class conflict and peasant revolt, foreign invasion, city-state warfare, inadequate agriculture, soil exhaustion or sedimentation of shallow lakes (water sources), general constraints of a tropical environment, epidemic crop or human diseases, trade route losses, climate change (wetter or drier), marginal returns from or loss of elite control due to increasing social complexity, among others.[12] For example, one of the earliest explanations linked an increasingly wetter environment to increasing disease vectors, such as malaria.[13] None of the simple explanations for the Maya case withstand scrutiny, however.[14] Neither do they provide a strong, logical rationale for the subsequent millennial-long paucity of occupation of the heartlands.

Interestingly, human–environment interactions have long been proposed as the collapse-depopulation cause.[15] This kind of explanation has become progressively sophisticated as the array of research expertise (paleo-environmental to complex system modeling) has expanded the range of evidence brought to bear on the problem. In tandem, the various research communities dealing with the Maya case appear to have coalesced around the pivotal role of human–environment stresses in the context of climate change. Past interpretations have implicitly or explicitly addressed population "overshoot"[16] relative to the natural resources of the lowlands. The apparent emerging consensus asserts that during the Late and Terminal Classic periods the lowland Maya, foremost the occupants of the heartlands, reached a tipping point in their socioeconomic, technological, environmental relationships, complete with "bitebacks" from nature, generating decisions to terminate their occupation of the interior elevated lands.[17] This decision, however, was likely influenced by the loss of trade flowing across the heartlands.

(Un)sustainability and the Maya case

No society can reach the size and density of occupation as that witnessed in the Maya heartlands without large demands on resources, profound landscape changes, and drawdowns in environmental services, in some cases with significant environmental consequences.[18] These changes and their consequences were experienced at different times in the occupational development up to the collapse. For the most part, however, they reached a crescendo during the last several centuries before the collapse and depopulation took place.

The large number and density of Maya sites (people and structures) generated environmental pressures, from big city-states to rural compounds, and especially in the heartlands. While disagreement exists over how to transform the physical evidence of occupation into population estimates, the number of Maya was large by most any comparison to other Mesoamerican areas or societies over such an extensive area (see above). To provision this population, in many cases at rather high levels of consumption for the time in question, the Maya constructed and maintained an expansive "built environment," especially in regard to water infrastructure and landscape engineering.[19] This built environment reflected the opportunities and constraints of the base environmental conditions across the elevated interior lands.[20]

The heartlands reside in a seasonal tropical environment that entertains an extended winter dry season increasing in length and intensity from south to north. The topography everywhere is karst (limestone in which surface water percolates to subsurface aquifers), with rolling uplands (*c.* 150–300 m above sea level) punctuated by large, seasonal wetlands. Permanent surface water is found in only a few dispersed shallow lakes and springs. Rivers are limited to the eastern and western flanks of the elevated interior, especially towards the south (Figure 5.1). Aquifers in the heartlands are typically deep (*c.* 100 m or more). The better drained upland soils are shallow (typically < 1m) and subject to erosion when cleared of vegetation. Phosphorus is the limiting plant nutrient. Wetland soils of thick clays are seasonally flooded. A major provisioning constraint confronting long-term, large occupation is water. Food production is hampered by variable seasonal precipitation and common droughts in the uplands and inundation in the wetlands. Potable water is constrained by the paucity of surface water throughout much of the area and difficulties in accessing aquifers.

Transformed landscape

The Maya confronted the challenges of the heartlands through various techno-managerial strategies. Initial upland forest clearing triggered significant erosion[21] that was subsequently countered with terracing, which also helped to retain soil moisture on slopes.[22] Terraces were often connected by stone walls that may have served as field boundaries, walkways or as impediments to crop damage from storm winds.[23] House sites, both rural and peri-urban, maintained orchard-gardens, leading to at least one assessment crediting the sustainability of Maya cities to this land use.[24] More extensive patches of managed forests, often with demarcated boundaries, were transformed into large orchard-gardens or forest gardens,[25] the relics of which are observable today.[26] The huge demand for wood fuel for cooking and mortar preparation as well as timber for construction, required extensive patches of "native" forests as well. Later in their occupation, as desiccation increased, wetlands along the edges of the elevated interior were cultivated by diverting water to crisscrossing channels, or canals, and cropping the drier "island-parcels" between them.[27] By the latter stages of the Classic Period, the heartlands were a significantly managed landscape composed of a mosaic of land-uses with corresponding land-covers, all of which had implications for human–environment interactions.[28]

Part of this management involved the sheer number of settlements—cities to hamlets—that required potable water, the access to which was not abundant (above). Collection and storage of water was tackled in a number of ways.[29] City-states constructed large reservoirs, often involving dams, adjacent to their centers, along with drainage systems that steered rainwater to collection points throughout the urban area.[30] Smaller reservoirs (or *aguadas*), often with deep catchment wells within them, dotted the peri-urban and rural landscape;[31] small, natural sinks (or sinkholes) may have been used to store water;[32] and wells in context of fault springs or adjacent perched water tables[33] were employed. Maya-made *chultunes*, or small, near-surface, subterranean cavities, often replete among individual house sites, also may have served to store water,[34] although alternative uses for these features have been proposed, such as food storage. Clearly, the management of potable water was a serious matter.

The environmental consequences

The amount of Maya deforestation in the lowlands has long been debated and surely varied by location and history of occupation.[35] For example, one study at Copan (Honduras), a site environmentally atypical of the heartlands, suggests a resurgence of forest in the later stages of Classic Period.[36] The paleo-environmental research from the heartlands, however, consistently finds evidence of large-scale forest decline, registered by major fall-offs in arboreal (tree) pollen and major increases in disturbance species pollen (e.g., grasses and ferns) in lake cores, indicating substantial "open" lands long before the collapse.[37] This evidence is consistent with that which documents the various landscape infrastructures employed widely across the lowlands, but especially in the heartlands (above).

Other indicators of pressures on forests and loss of timber stocks substantiate these base findings. For instance, timber use at Tikal declined over time from large, upland tree species (e.g., mahogany) to the use of smaller, wetland species, suggesting an exhaustion of available upland older growth.[38] It is likely that the substantial and sustained demands for wood fuel for cooking and mortar at such sites also stressed the wood reserves of the forests,[39] land for which cultivation and forest gardens competed. Other environmental stresses include the evidence from the southern lowlands of decreased Maya consumption of large animal species, especially white tailed deer, suggesting overhunting in the Late Classic Period.[40]

Major deforestation of the heartlands had a number of environmental consequences. A major source of the limiting plant nutrient in soils, phosphorus, was the capture of atmospheric dust by the tree canopy. Opening land for cultivation reduced this important source of phosphorus replenishment.[41] Canopy loss also increased soil temperature and decreased soil moisture, affecting crop production or its upkeep. Importantly, large-scale deforestation lowered evapotranspiration, likely reducing precipitation and amplifying climatic drought.[42]

Climate change

The greater Maya lowlands witnessed a prolonged period of decreasing precipitation from about 2500 to 1000 BCE, a period of Maya development in the lowlands, reducing overall forest cover.[43] Subsequently, two prolonged drought episodes punctuated the climate of the Peninsula, the first from which the Maya not only recovered but moved into their "classic" phase development, before collapsing toward the end of the second prolonged drought period. This precipitation record has been well established from a variety of paleo-environment evidence and analyses,[44] giving rise to a number of drought-driven collapse themes.[45] Modest disagreement exists over the intensity of drought in the collapse period, with at least one study concluding that it comprised a modest decline in precipitation, perhaps by about 40 percent.[46] Other studies conclude that this drought was the most intense in the heartland area and southward.[47]

The paleo-environmental data, largely derived over the past two decades, constitutes a game-changer regarding the collapse. It cannot be dismissed. That said, to reduce the collapse and depopulation of the lowlands to this factor alone errs in several ways. Despite the general intensity of the drought in the heartlands, a distinctive spatio-temporal pattern of city-state abandonment consistent with drought intensity or access to river and lake water does not appear to exist.[48] In addition, the fact that the Maya had combatted previous long-term drought and emerged from it even more numerous and materially advanced begs the question concerning the conditions that made the second episode special. This question, more so than any other, raises the issue of sustainability.

Human–environment interactions: a tipping point?

Had millennia of occupation of the lowlands, foremost the intensity of that occupation in the heartlands, generated human–environment conditions especially vulnerable or non-resilient to sustained drought? Had these conditions become unsustainable, reaching a tipping point? An increasing number of interpretations of the evidence lead to such a conclusion, or at least to the assessment that such conditions played a role in the collapse and depopulation.[49]

One of the most explicit assessments links the various land-use and management practices in the heartland to their environmental consequences.[50] In this assessment, the variety and scale of land-cover changes wrought in heartlands to enhance food and fuel production and access to potable water generated a set of environmental feedbacks that became increasingly difficult to counteract during a long-term drought. Sustained problems of invasive species on croplands, increased soil temperature, decreased soil moisture and phosphorus, and decreased evapotranspiration that amplified climate desiccation surely placed hardships on food production. Increasing aridity apparently disrupted even cultivation in perennial wetlands.[51] Additionally, extended drought played havoc with potable water collection and storage.

Were these the conditions of a tipping point precipitating the collapse? The sheer pressures of occupation were indeed high and rising. Were there no measures to be taken to provision demand, such as the intensive cultivation of the higher elevated, seasonal wetlands adjacent to major city-states? Had these wetlands become too dry? Or were the pre-collapse human–environment conditions simply too costly to maintain in the face of extended drought?

The shifting economic impacts: aiding the tipping point?

Maya city-states controlled land, water, and people to provision food, fuel, and fiber, although the evidence mounts regarding inter-site trade of everyday items.[52] Various necessities, such as salt and obsidian, and luxury items, however, involved longer distance trading. It appears that during the ascendency of the heartland city-states, substantial trade flowed from the south and the Caribbean across the heartlands to the Usumacinta River and northward to central Mexico and vice versa.[53] These trade routes surely benefitted the coffers of the heartland city-states. During the period leading up to the collapse, however, this routing of trade declined significantly, shifting to a sea-based route that circumnavigated the peninsula.[54] The synergy between this economic loss and the high land, labor, infrastructure investments required to confront aridity surely made the provisioning of the heartlands problematic, challenging the authority of the power elite and triggering inter-city-state warfare and, perhaps, class conflict.[55] In the end, large-scale abandonment of the heartlands, not a reconfiguring of the political economy, proved to be the dominant decision.

Collapse and complex system dynamics?

The narrative that the heartland Maya succumbed to high-stress human-environment conditions confronting climate change, is not so much new as it is a more data-rich, informed one than that of its predecessors.[56] It may also be placed within contemporary conceptual models and their accompanying vocabulary. The collapse itself fits within the dynamics of complex systems in which "legacy effects" and tipping points, among other system attributes, may trigger a regime shift, in this case from pre- to post-collapse conditions in the heartlands. At least one argument holds that the response of Maya during the collapse phase may have been hastened by "sunk-cost" effects, that is, their long-term agricultural infrastructure investments created a legacy that impeded constructive response options and decisions to the tipping point conditions.[57] Complex systems reorganize, however. That reorganization for the heartlands presents a conundrum because it involved the elimination of most of the social subsystem.

Confounding elements

The demise of civilizations is the norm, or so history informs us. Massive depopulation and long-term failure of occupational recovery is not. This failure makes explanations of the Maya collapse problematic. Why was the depopulation so dramatic? Why was there no substantial post-collapse reoccupation? Does the answer provide any clues about the environmental sustainability of the region? Interestingly, little attention has been given to these issues.[58]

With the cessation of the drought period, it took no more than 80 to 260 years for old-growth forest to return, followed by soil stabilization in 120 to 280 years.[59] Yet few Maya re-established a sustained occupation of the heartlands—as the Cortez expedition (above) discovered—save for a few settlements along the central lakes of the Petén. This occupational history has little to do with the environmental conditions for resettlement in the elevated interior, which were not too dissimilar from those encountered by the pre-collapse Maya. Rather, it would appear that reoccupation of the interior elevated lands was stymied by one or both of the following conditions:

1   Land and resource pressures along the coasts and north of the Puuc Hills in Yucatán were not sufficient to trigger the use of interior lands.
2   The cost of reconstructing the relict infrastructure of the interior to permit a lifestyle commensurate with the coastal areas was deemed to be too large to undertake.

Either way, diseases and subjugation with the arrival of the Spaniards in the 1500s truncated the subsequent potential land and resource pressures of the occupied Maya lowlands. During this time period, the abandoned elevated interior lands served as a refuge for Maya escaping Spanish and later Mexican domination.[60]

Concluding observations

Assessments of the sustainability of the ancient lowland Maya must begin with the recognition that the civilization flourished for 3000 years through innovative environmental engineering and resource management. This time frame matches or exceeds that of other major civilizations worldwide, all of which collapsed. The sustainability question for the lowland Maya, therefore, is raised not by the Classic Period collapse per se, even in the face of extended aridity, but by the sheer scale of the depopulation of the interior elevated lands and subsequent sparse reoccupation of them. Multiple lines of evidence point to major environmental stresses and feedbacks in context of long-term drought and the loss of trade flows through lowlands as conditions in which the collapse and depopulation took place. The Maya were either unable or unwilling to maintain their city-states in these conditions and moved elsewhere. Complex questions of human–environment sustainability are raised in this storyline. Why such a large portion of the population departed and why substantial reoccupation subsequently failed to take place, at least in regard to the evidence, appears to have little to do with the post-collapse sustainability of the vacated heartland, but with the subsequent socio-demographic history of the Maya region.

Notes

1   See Whitmore et al. (1990). Due to the large number of sources in this chapter, abbreviated citations will be given for all general references. For complete citations, see the references.
2   See, for example, Schwartz and Nichols (2010).
3   See, for examples, Diamond (2005); Ponting (1991).
4   The more recent literature is favored throughout this assessment, with reference to the older literature (i.e., prior to 2000) only where needed. For the most part, the later literature provides insights and references to the older.

5   Given the non-depopulation of Maya along the littoral of the Yucatán Peninsula and among the lowly
     elevated and climatically drier northern interior, and the persistence of the Maya in much of the
     lowlands today, some experts favor referring to abandonment and depopulation of the heartland
     rather than collapse. See Aimers (2007); McAnany and Yoffee (2009). This reframing does not chal-
     lenge the notion of the sociopolitical downfall that took place in the heartlands by the 11th century,
     but seeks to recognize the persistence of Maya culture to this day. The term collapse remains abun-
     dant in the literature, however, and is used here in reference to the sociopolitical consequences in the
     heartlands of the abandonment.
6   See Dunning et al. (2011).
7   See Turner and Sabloff (2012).
8   The Maya lowlands are most commonly divided into the northern and southern areas (Figure 5.1)
     with the division extending east–west across the peninsula just above the contemporary border of
     Mexico and Guatemala. This division is too simplistic to capture the human–environment relation-
     ships confronting the ancient Maya. The far north is dominated by a semi-arid climate and extreme
     karst topography, and the far south by a wet tropical climate that gives rise to rivers and small lakes,
     although limestone bedrock dominates much of the area. The central Maya lowlands or heartlands
     constitutes a transition zone between the north–south extremes. See Dunning et al. (2011); Turner
     and Sabloff (2012). See the text for details of the heartland environment.
9   See Culbert (1973); Scarborough and Burnside (2010).
10   Turner and Sabloff (2012). Skeletal examination of Classic Period Maya suggests large-scale, interac-
     tion among the populations of city-states. See Scherer (2007). Strontium analysis confirms that the
     Maya traded bush meat over considerable distances. See Thornton (2011). Both sets of evidence are
     indicative of densely settled, interactive communities, and not that of spatially isolated city-states with
     minimal contact that relied solely on their individual hinterlands for food provisions.
11   Fanciful explanations abound in "ancient alien" books and television series, the most noted of which
     is the claim in E. Von Daniken's *The Chariots of the Gods* that the Maya voluntarily went into space
     or were abducted (Von Daniken 1999).
12   The number of books and articles reviewing or advancing various causes of the Maya collapse and
     depopulation are too numerous to reference in mass. A selection includes: Acuna-Soto et al. (2005);
     Cooke (1931); Culbert (1973); Gill (2001); Hamblin and Pitcher (1980); Lowe (1985); Lucero (2002);
     Santley, Killion, and Lycett (1986); Tainter (1988); Webster (2002).
13   See Huntington (1917).
14   For examples, Webster (2002); Robichaux (2002).
15   For instance, Cooke (1931).
16   See Tainter (2006).
17   See, for instance, Diamond (2005); Kennett and Beach (2013); Scarborough and Burnside (2010);
     Turner and Sabloff (2012).
18   Environmental or ecosystem services refers to more than natural resources or provisioning services,
     such as food production. The term refers to the full range of outputs from nature that support our
     species, from pollination to climate or flood regulation to soil regeneration and water purification.
     See MEA (2005).
19   See, for instance, Beach et al. (2009); Dunning et al. (2011); Scarborough and Burnside (2010).
20   For details on the environments of the elevated interior, including references, see Turner and Sabloff
     (2012); Dunning et al. (2011); Beach et al. (2009).
21   Anselmetti et al. (2007).
22   See Beach et al. (2006, 2010).
23   Turner (1974).
24   Isendahl and Smith (2013).
25   Ford and Nigh (2009).
26   See Gómez-Pompa et al. (1987); Ross (2011).
27   Beach et al. (2009); Luzzader-Beach et al. (2012); Turner and Harrison (1981). To date, the docu-
     mentation of ancient Maya wetland cultivation has been limited to the lower-lying eastern and
     western flanks of the interior elevated lands, from sea level to about 150 m above. See Luzzader-Beach
     et al. (2012). The higher elevated wetlands, those associated with such super-sites as Calakmul and
     Tikal, have yet to reveal evidence of this activity. As yet, understanding of this distribution of wetland
     cultivation is inadequate, but likely has something to do with the hydrology of the elevated wetlands.
28   See Turner and Sabloff (2012).

29 Wyatt (2014).
30 See Davis-Salazar (2006); Lucero (2002); Scarborough and Gallopin (1991); Scarborough et al. (2012).
31 Akpinar-Ferr et al. (2012).
32 Weiss-Krejci and Sabbas (2002).
33 Johnston (2004).
34 Licea and Ruiz (2010).
35 For instance, Wahl et al. (2006).
36 See McNeil et al. (2012); but see Paine and Freter (1996).
37 For a review, see Turner and Sabloff (2012).
38 Lentz and Hockaday (2009).
39 Abrams et al. (1996).
40 See Emery (2007). This evidence does not appear to be related to significant changes in landscape conditions, such as major increases in agricultural lands in the Late Classic period. See Emery et al. (2000). The Petexbatún region where the study was undertaken resides to the southwest of the heartlands.
41 Lawrence et al. (2007).
42 See Oglesby et al. (2010); Shaw (2003).
43 See Mueller et al. (2009).
44 Aimers and Hodell (2011); Curtis et al. (1996); Hodell et al. (1995, 2001); Webster et al. (2007).
45 Haug et al. (2003); Gill (2001); Gill et al. (2007).
46 Medina-Elizalde and Rohling (2012).
47 Douglas et al. (2015).
48 Aimers and Hodell (2011); Douglas et al. (2015).
49 Dunning et al. (2011); Kennett and Beach (2013); Robichaux (2002); Scarborough and Burnside (2010).
50 Turner and Sabloff (2012); also Kennett and Beach (2013).
51 Luzzader-Beach et al. (2012).
52 For example, Thornton (2011).
53 Barrett and Guderjan (2006); Golitko et al. (2012); McKellop (2005a).
54 See McKellop (2005b); Sabloff (2007); Sosa et al. (2014). Evidence for the shift from land- to sea-based trade continues to mount. As yet, however, it is not clear if the abandonment of the heartlands triggered this shift or if the shift helped to precipitate the abandonment, as proposed here.
55 Barrett and Scherer (2005); Foias and Bishop (1997); Hamblin and Pitcher (1980); Webster (2000).
56 For example, Cooke (1931); Culbert (1973); Lowe (1985); Tainter (1988).
57 For instance, Downey et al. (2016); Janssen et al. (2002).
58 See Turner (2010).
59 Mueller et al. (2010).
60 Jones (1989).

References

Abrams, E. M., A. Freter, D. J. Rue, and J. D. Wingard. "The Role of Deforestation in the Collapse of the Late Classic Copán Maya." In *Tropical Deforestation: The Human Dimension*, edited by L. E. Sponsel, T. N. Headland, and R. C. Bailey, 555–75. New York: Columbia University Press, 1996.

Acuna-Soto, R., D. W. Stahle, M. D. Therrell, S. G. Chavez, and M. K. Cleaveland. "Drought, Epidemic Disease, and the Fall of Classic Period Cultures in Mesoamerica (AD 750–950). Hemorrhagic Fevers as a Cause of Massive Population Loss." *Medical Hypotheses* 62 (2005): 405–9.

Aimers, J. J. "What Maya Collapse? Terminal Classic Cariation in the Maya Lowlands." *Journal of Archeological Research* 15 (2007): 329–77.

Aimers, J. J. and D. A. Hodell. "Societal Collapse: Drought and the Maya." *Nature* 479 (2011): 44–5.

Akpinar-Ferr, E., N. Dunning, D. L. Lentz, and J. G. Jones. "Use of Aguadas as Aater Management Sources in Two Southern Maya Lowland Sites." *Ancient Mesoamerica* 23 (2012): 85–101.

Anselmetti, F. S., D. A. Hodell, D. Ariztegui, M. Brenner, and M. F. Rosenmeier. "Quantification of Soil Erosion Rates Related to Ancient Maya Deforestation." *Geology* 35 (2007): 915–18.

Barrett, J. W. and T. H. Guderjan. "An Ancient Maya Dock and Dam at Blue Creek, Rio Hondo, Belize." *Latin American Antiquity* 17 (2006): 227–39.

Barrett, J. W., and A. K. Scherer. "Stone, Bones, and Crowded Plazas: Evidence for Terminal Classic Maya

Warfare at Colha, Belize." *Ancient Mesoamerica* 1 (2005): 101–16.

Beach, T., N. Dunning, S. Luzzadder-Beach, D. E. Cook, and J. Lohse. "Impacts of the Ancient Maya on Soils and Soil Erosion in the Central Maya Lowlands." *Catena* 65 (2006): 166–78.

Beach, T. P., S. Luzzadder-Beach, and N. Dunning. "A Review of Human and Natural Changes in Maya Lowland Wetlands Over the Holocene." *Quaternary Science Reviews* 28 (2009): 1710–24.

Beach, T., S. Luzzadder-Beach, N. Dunning, J. Hageman, and J. Lohse. "Upland Agriculture in the Maya Lowlands: Ancient Maya Soil Conservation in Northwestern Belize." *Geographical Review* 92 (2010): 372–97.

Cooke, C. W. "Why the Mayan Cities of the Peten District, Guatemala, Were Abandoned." *Journal of the Washington Academy of Sciences* 21 (1931): 281–7.

Culbert, T. P., ed. *The Classic Maya Collapse.* Albuquerque, NM: University of New Mexico Press, 1973.

Curtis, J. H., D. A. Hodell, and M. Brenner. "Climate Variability on the Yucatan Peninsula (Mexico) During the Past 3500 Years, and Implications for Maya Cultural Evolution." *Quaternary Research* 46 (1996): 37–47.

Davis-Salazar, K. L. "Late Classic Maya Drainage and Flood Control at Copan, Honduras." *Ancient Mesoamerica* 17 (2006): 125–38.

Diamond, J., *Collapse: How Societies Choose Life, and Death.* New York: Viking, 2005.

Douglas, P. M., M. Pagani, M. A. Canuto, M. Brenner, D. A. Hodell, T. I. Eglinton, and J. H. Curtis. "Drought, Agricultural Adaptation, and Sociopolitical Collapse in the Maya Lowlands." *Proceedings of the National Academy of Sciences* 112 (2015): 5607–12.

Downey, S. S., W. R. Haas, Jr., and S. Shennan. "European Neolithic Societies Showed Early Warning Signals of Population Collapse." *Proceedings of the National Academy of Sciences* 113 (2016): 9751–6.

Dunning, N. P., T. P. Beach, and S. Luzzadder-Beach. "Kax and Kol: Collapse and Resilience in Lowland Maya Civilization." *Proceedings of the National Academy of Sciences* 109 (2011): 3652–7.

Emery, K. F. "Assessing the Impact of Ancient Maya Animal Use." *Journal for Nature Conservation* 15 (2007): 184–95.

Emery, K. F., L. E. Wright, and H. Schwarcz. "Isotopic Analysis of Ancient Deer Bone: Biotic Stability in Collapse Period Maya Land-Use." *Journal of Archaeological Science* 6 (2000): 537–50.

Foias, A. E. and R. L. Bishop. "Changing Ceramic Production and Exchange in the Petexbatun Region, Guatemala: Reconsidering the Classic Maya Collapse." *Ancient Mesoamerica* 8 (1997): 275–91.

Ford, A. and R. Nigh. "Origins of the Maya Forest Garden: Maya Resource Management." *Journal of Ethnobiology* 29 (2009): 213–36.

Gill, R. B. *The Great Maya Droughts: Water, Life, and Death.* Albuquerque, NM: University of New Mexico Press, 2001.

Gill, R. B., P. A. Mayewski, J. Nyberg, G. H. Haug, and L. C. Peterson. "Drought and the Maya Collapse." *Ancient Mesoamerica* 18 (2007): 283–302.

Golitko, M., J. Meierhoff, G. M., Feinman, and P. R. William. "Complexities of Collapse: The Evidence of Maya Obsidian as Revealed by Social Network Graphical Analysis." *Antiquity* 86 (2012): 507–23.

Gómez-Pompa, A., J. Salvador Flores, and V. Sosa. "The 'Pet Kot': A Man-Made Tropical Forest of the Maya." *Interciencia* 12 (1987): 10–15.

Hamblin, R. L. and B. L. Pitcher. "The Classic Maya Collapse: Testing Class Conflict Hypotheses." *American Antiquity* 45 (1980): 246–67.

Haug, G. H., D. Günther, L. C. Peterson, D. M. Sigman, K. A. Hughen, and B. Aeschlimann. "Climate and the Collapse of Maya Civilization." *Science* 299 (2003): 1731–5.

Hodell, D. A., J. H. Curtis, and M. Brenner. "Possible Role of Climate in the Collapse of Classic Maya Civilization." *Nature* 375 (1995): 391–4.

Hodell, D. A., M. Brenner, J. H. Curtis, and T. Guilderson. "Solar Forcing of Drought Frequency in the Maya Lowlands." *Science* 292 (2001): 1367–70.

Huntington, E. "Maya Civilization and Climate Change." *Proceedings of the 19th Congress of Americanists, 5–10 October 1914, Washington, DC* (1917): 150–64.

Isendahl, C. and M. E. Smith. "Sustainable Agrarian Urbanism: The Low-Density Cities of the Mayas and Aztecs." *Cities* 31 (2013): 132–43.

Janssen, M. A., T. A. Kohler, and M. Scheffer. "Sunk-Cost Effects and Vulnerability to Collapse in Ancient Societies." *Current Anthropology* 101 (2002): 98–112.

Johnston, K. J. "Lowland Maya Water Management Practices: The Household Exploitation of Rural Wells." *Geoarchaeology* 19 (2004): 265–92.

Jones, G. D. *Maya Resistance to Spanish Rule: Time and History on a Colonial Frontier.* Albuquerque, NM:

University of New Mexico Press, 1989.

Kennett, D. J., and T. P. Beach. "Archeological and Environmental Lessons for the Anthropocene from the Classic Maya Collapse." *Anthropocene* 4 (2013): 88–100.

Lawrence, D., P. D'Odorico, L. Diekmann, M. DeLonge, R. Das, and J. Eaton. "Ecological Feedbacks Following Deforestation Create the Potential for a Catastrophic Ecosystem Shift in Tropical Dry Forest." *Proceedings of the National Academy of Sciences* 104 (2007): 20, 696–701.

Lentz, D. L., and B. Hockaday. "Tikal Timbers and Temples: Ancient Maya Agroforestry and the End of Time." *Journal of Archaeological Science* 36 (2009): 1342–53.

Licea, D. M., and J. L. M. Ruiz. "From Pre-Hispanic Technologies to Appropriate Technologies." *Plurimondi* 4 (2010): 105–23.

Lowe, J. W. G. *The Dynamics of Apocalypse: A Systems Simulation of the Classic Maya Collapse.* Albuquerque, NM: University of New Mexico Press, 1985.

Lucero, L. J. "The Collapse of the Classic Maya: A Case for the Role of Water Control." *American Anthropologist* 104 (2002): 814–26.

Luzzadder-Beach, S., T. P. Beach, and N. Dunning. "Wetland Fields as Mirrors of Drought and the Maya Abandonment." *Proceedings of the National Academy of Sciences* 109 (2012): 3646–51.

McAnany, P. A. and N. Yoffee, eds. *Questioning Collapse: Human Resilience, Ecological Vulnerability, and the Aftermath of Empire.* Cambridge: Cambridge University Press, 2009.

McKellop, H. "Finds in Belize Document Late Classic Maya Salt Making and Canoe Transport." *Proceedings of the National Academy of Sciences* 102 (2005a): 5630.

McKellop, H. *In Search of Maya Sea Traders.* College Station, TX: Texas A&M Press, 2005b.

McNeil, C. L., D. A. Burney, and L. P. Burney. "Deforestation, Agroforestry, and Sustainable Land Management Practices Among the Classic Period Maya." *Quaternary International* 249 (2012): 19–30.

MEA (Millennium Ecosystem Assessment). *Ecosystem and Human Well-Being: Synthesis.* Washington, DC: Island Press, 2005.

Medina-Elizalde, M. and E. J. Rohling. "Collapse of Classic Maya Civilization Related to Modest Reduction in Precipitation." *Science* 335 (2012): 956–9.

Mueller, A. D., G. A. Islebe, M. B. Hillesheim, D. A. Grzeski, F. S. Anselmetti, D. Ariztegui, M. Brenner, J. H. Curtis, D. A. Hodell, and K. A. Venz. "Climate Drying and Associated Forest Decline in the Lowlands of Northern Guatemala During the Late Holocene." *Quaternary Research* 71 (2009): 133–41.

Mueller, A. D., G. A. Islebe, F. S. Anselmetti, D. Ariztegui, M. Brenner, D. A. Hodell, I. Hajdas, Y. Hamann, G. H. Huag, and D. J. Kennett. "Recovery of the Forest Ecosystem in the Tropical Lowlands of Northern Guatemala after Disintegration of Classic Maya Polities." *Geology* 38 (2010): 523–26.

Oglesby, R. J., T. L. Sever, W. Saturno, D. J. Erickson, III, and J. Sirkishen. "Collapse of the Maya: Could Deforestation have Contributed?" *Journal of Geophysical Research* 115 (2010). doi:10.1029/2009JD011942.

Paine, R. R., and A. Freter. "Environmental Degradation and the Classic Maya Collapse at Copan, Honduras (AD 600–1250): Evidence from Studies of Household Survival." *Ancient Mesoamerica* 7 (1996): 37–47.

Ponting, C. *Green History of the World. The Environment and Collapse of Great Civilizations.* New York: Penguin, 1991.

Robichaux, H. R. "On the Compatibility of Epigraphic, Geographic, and Archaeological Data, with a Drought-Based Explanation for the Classic Maya Collapse." *Ancient Mesoamerica* 13 (2002): 341–5.

Ross, N. J. "Modern Tree Species Composition Reflects Ancient Maya 'Forest Gardens' in Northwest Belize." *Ecological Applications* 21 (2011): 75–84.

Sabloff, J. A. "It Depends on How We Look at Things: New Perspectives on the Postclassic Period in the Northern Maya Lowlands." *Proceedings of the American Philosophical Society* 108 (2007): 11–25.

Santley, R. S., T. W. Killion, and M. T. Lycett. "On the Maya Collapse." *Journal of Anthropological Research* 42 (1986): 123–59.

Scarborough, V. L. and D. A. Burnside. "Complexity and Sustainability: Perspectives from the Ancient Maya and the Modern Balinese." *American Antiquity* 75 (2010): 327–63.

Scarborough, V. L. and G. G. Gallopin. "A Water Storage Adaptation in the Maya Lowlands." *Science* 251 (1991): 658–2.

Scarborough, V. L., N. Dunning, K. Tankersley, C. Carr, E. Weaver, L. Grazioso, B. Lane, J. Jones, P. Buttles, F. Valdez, and D. Lentz. "Water and Sustainable Land Use at the Ancient Tropical City of Tikal, Guatemala." *Proceedings of the National Academy of Sciences* 109 (2012): 12, 408–13.

Scherer, A. K. "Population Structure of the Classic Period Maya." *American Journal of Physical Anthropology*

132 (2007): 367–80.

Schwartz, G. M. and J. J. Nichols, eds. *After Collapse: The Regeneration of Complex Societies.* Tucson, AZ: University of Arizona Press, 2010.

Shaw, J. M. "Climate Change and Deforestation: Implications for the Maya Collapse." *Ancient Mesoamerica* 14 (2003): 157–67.

Sosa, T. S., A. Cucina, T. D. Price, J. H. Burton, and V. Tiesler. "Maya Coastal Production, Exchange, Life Style, and Population Mobility: A View from the Port of Xcambo, Yucatan, Mexico." *Ancient Mesoamerica* 25 (2014): 221–38.

Tainter, J. A. *The Collapse of Complex Societies.* Cambridge: Cambridge University Press, 1988.

Tainter, J. A. "Archaeology of Overshoot and Collapse." *Annual Review of Anthropology* 35 (2006) 59–74.

Thornton, E. K. "Reconstructing Ancient Maya Animal Trade through Strontium Isotope (87Sr/86Sr) Analysis." *Journal of Archaeological Science* 38 (2011): 3254–63.

Turner, B. L., II. "Prehistoric Intensive Agriculture in the Mayan Lowlands." *Science* 185 (1974): 118–24.

Turner, B. L., II. "Unlocking the Ancient Maya and Their Environment: Paleo-Evidence and Dating Resolution." *Geology* 38 (2010): 523–6.

Turner, B. L., II and P. D. Harrison. "Prehistoric Raised-Field Agriculture in the Maya Lowlands." *Science* 213 (1981): 399–405.

Turner, B. L., II and J. Sabloff. "Classic Period Collapse of the Central Maya Lowlands: Insights about Human–Environment Relationships for Sustainability." *Proceedings of the National Academy of Sciences* 109 (2012): 13, 908–14.

Von Daniken, E. *The Chariots of the Gods.* London: Penguin, 1999.

Wahl, D., R. Byrne, T. Schreiner, and R. Hansen. "Holocene Vegetation Change in the Northern Peten and its Implications for Maya Prehistory." *Quaternary Research* 65 (2006): 380–89.

Webster, D. L. "The Not So Peaceful Civilization: A Review of Maya War." *Journal of World Prehistory* 14 (2000): 65–119.

Webster, D. L. *The Fall of the Ancient Maya: Solving the Mystery of the Maya Collapse.* New York: Thames & Hudson, 2002.

Webster, J. W., G. A. Brook, L. B. Railsback, H. Cheng, R. L. Edwards, C. Alexander, and P. P. Reeder. "Stalagmite Evidence from Belize Indicating Significant Droughts at the Time of Preclassic Abandonment, the Maya Hiatus, and the Classic Maya Collapse." *Paleogeography, Paleoclimatology, Paleoecology* 250 (2007): 1–17.

Weiss-Krejci, E. and T. Sabbas. "The Potential Role of Small Depressions as Water Storage Features in the Central Maya Lowlands." *Latin American Antiquity* 13 (2002): 343–57.

Whitmore, T., D. Johnson, B. L. Turner II, and R. W. Kates. "Long-Term Population Change." In *The Earth as Transformed by Human Action*, edited by B. L. Turner II, W. C. Clark, R. W. Kates, J. F. Richards, J. T. Mathews, and W. B. Meyer, 25–39. Cambridge: Cambridge University Press, 1990.

Wyatt, A. R. "The Scale and Organization of Ancient Maya Water Management." *Wiley Interdisciplinary Reviews: Water* 1 (2014): 449–67.

PART IV

The roots of sustainability

6

SUSTAINING WHAT?

Scarcity, growth, and the natural order in the discourse on sustainability, 1650–1900

Gareth Dale

Introduction

Questions of sustainability—the interaction between human economic practices, on one hand, and natural resources and the web of life, on the other—are ancient. Concerns over deforestation and soil erosion are documented in literatures from ancient Mesopotamia, Greece and Rome, and Mauryan India, to name a few.[1] But in the early modern era a distinctive discourse of sustainability arose, even if it initially lacked a standardized vocabulary.[2] The capitalist heartlands of Europe and their ex-colonial appendages deployed economic and military advantage to seize control of the world's land surface—from 10 percent in 1700 to 30 percent in 1800 and 85 percent in 1900. The transformations associated with capital and colonialism sparked, and propelled, a revolution in nature-society relations, with the construction of what Jason Moore terms regimes of abstract social nature: spatio-temporal practices of quantification, mapping, and standardization that facilitated the "quantitative expansion of abstract labor."[3] With new regimes of economic order and abstract social nature came new ideas concerning the purposes and dynamics of economic activity and the management of nature.

It was above all at the frontiers of colonial expansion that new environmental discourses and conservation practices evolved. The expansion of European maritime travel and settlement, Richard Grove has shown, stimulated new and more complex ways of apprehending nature–society relations, and a more vivid sense of humanity's ability to radically alter its physical environment. It was fed by information flowing from all continents, including the systematic observation, recording and classification of the natural world, experiments in forest, soil and animal conservation and water-pollution control by scientists in the employ of the colonial companies, and the acquisition of local knowledge of the natural world and its symbolism. European encounters with new territories, societies, fauna and flora promoted the attachment of "a new kind of social significance to nature, reflected especially in the philosophies underlying the transfer to, and development of the Middle Eastern idea of the botanical garden in Europe and in the emergence of the tropical and oceanic island as an important new social metaphor and image of nature."[4]

If the emergence of ideas and techniques of conservation and sustainability were most evident at the colonial periphery, they were also assimilated into the research of scholars in the imperial core. And here, too, new ideas concerning society's interaction with nature were deeply marked by the imperial framework within which they grew. This is the central concern of this chapter. Its subject is British environmental and environmental-economic theory between the seventeenth and nineteenth centuries, as well as the interconnections among discourses of scarcity, sustainability and economic growth.

Empire of biomass

Would it be hyperbolic to suggest that the concepts of sustained economic growth, scarcity and sustainability were triplets—the more or less simultaneous progeny of the transformations mentioned above? Consider sustainability. The evolution of the concept, according to Ulrich Grober, is best understood with reference to the Enlightenment-era crumbling of the belief in divine Providence. Earlier, in medieval times, the preservation of Creation had been God's responsibility: His providential hand sustained the world meticulously, continuously and perpetually. That belief disintegrated in the seventeenth and eighteenth centuries but it left its imprint on subsequent social and economic thought, notably Adam Smith's invisible hand of the market, and, Grober shows, "the modern discourse of sustainability."[5]

Grober's magisterial survey locates a number of originary impulses in the story of sustainability in eighteenth-century Germany. They include the first use of "sustainable" ("nachhaltig") in its modern sense, yoking together commitments to the preservation and continuous utilization of natural resources as well as respect for nature. This was in *Sylvicultura oeconomica* (1713) by the tax accountant and mining administrator Hans Carl von Carlowitz, as illustrated in his advice that the conservation and cultivation of timber should be practiced in order to ensure its "continuous, steady and *sustaining [nachhaltende]* use."[6] Of equal importance in the history of sustainability is its Romantic modification, exemplified in the work of Alexander von Humboldt and Johann Wolfgang von Goethe. Humboldt's ecological writings drew on his travels in the New World as well as the holistic thought of Hindu philosophy.[7] Goethe challenged the utilitarian and Cartesian conception of sustainability by establishing the "'economy of nature' as a foundation for *all* economic activity," thereby planting ecology "at the heart of sustainability."[8]

The concept of sustainability today contains both these strands of thought, the utilitarian or instrumental and the ecological or aesthetic. They were vividly present, too, in a precursory work that Grober also explores—John Evelyn's *Sylva, or a Discourse of Forest Trees and the Propagation of Timber in His Majesties Dominions*.[9] Evelyn was a proto-green thinker. He railed against urban pollution, originated the idea of the garden city, and maintained that it is possible to live on "wholesome vegetables, both long and happily."[10] His *Sylva* is a remarkable document in its aspiration to marry utility to aesthetics, bankability to beauty, the preservation of the natural world to the imperative of sustained profitability. Landlords dedicated to making a quick buck—condemned by Evelyn for their "avarice," with its associated sins of "*Pride, Effeminacy* and *Luxurie*"[11]—erode nature's substance, which, he advises, would undermine profitability in the long run, too. Tree planting, by contrast, is manly and heroic, driven in large part by self-interested landlords and championed by Evelyn out of concern for national power and autonomy. (No trees, no navy or trade.) Over and again, Evelyn enjoins his landowning readers that if they follow his silvicultural guidance their woodlands "will both prove more profitable, and more delightful"; the trees thus preserved and managed will augment "both profit and pleasure," for they will "increase the beauty of forests, and value of timber, more in ten or twelve years, than all other imaginable plantations can do in forty or fifty." In preserving woodland for judicious exploitation, "Persons who are Owners of Land" will reap "infinite delight, as well as profit," in addition to supporting national sovereignty. "What can be more delightful," Evelyn rhapsodises, "than for noble persons to adorn their goodly mansions and demesnes with trees of venerable shade and profitable timber."[12] Moreover, for Evelyn trees are not simply a feast for the eyes and a revenue stream, they also form the basis of (what we would call) ecosystems and make their presence felt in local climates. In "the Indies" in particular, trees contribute to cloud formation, such that "if their woods were once destroyed, they might perish for want of rains; upon which account Barbadoes grows every year more torrid, and has not near the rain it formerly enjoyed when it was better furnished with trees; and so in Jamaica at Gunaboa, the rains are observed to diminish, as their [sugar] plantations extend."[13]

Little wonder, then, that Evelyn heads the pantheon of theorists of sustainable development. In Henry David Thoreau's eyes, "Evelyn is as good as several old druids, and his 'Silva' is a new kind of prayer book, a glorifying of the trees and enjoying them forever, which was the chief end of his life."[14] "There would not have been a sustainability movement," according to Jeremy Caradonna, "without John Evelyn, Jean-Baptiste Colbert, Hans Carl von Carlowitz, Jean-Jacques Rousseau, Thomas Malthus … and countless other men and women who developed ecological thinking."[15] John Bellamy Foster concurs, adding that Evelyn's works serve to remind us that Enlightenment conceptions of nature were not as vulgar as sometimes supposed.[16] The seventeenth-century scientific revolution of which he was a pre-eminent figure was not only associated "with a new conception of the domination of nature but also with a new materialist understanding of nature—one in which human beings were not simply the center of God's universe and could not simply dominate nature at whim, but rather were compelled to develop a sustainable relation with the natural world." Grober, similarly, hails *Sylva* as a seminal text in the modern sustainability discourse, in that it is about preserving resources to enable human consumption but transcends a purely managerial approach to nature—the Cartesian view of nature as an object to be subjugated, as advocated in the same era, for example, by Colbert.[17]

Grober reads *Sylva* as a "plea for responsible management of renewable resources," rooted in "an ethics for a provident and responsible society," as contrasted with two rival strategies for the achievement of continuous prosperity: "to import natural resources that were becoming scarce at home from the entire globe," and "to substitute other raw materials for any threatened by shortages." For many of Evelyn's contemporaries, then, "colonialism, plus technical innovation, became the key to a future at whose end we have arrived today."[18] Such contrast as there may have been, however, was more blurred than this formulation would suggest.

Who was Evelyn, and what social forces did he represent? He was heir to a fortune that "his grandfather had accumulated under James I and Charles I through his royal monopoly on saltpeter," an essential ingredient (with sulfur and charcoal) of gunpowder.[19] Evelyn senior had "forcibly ransacked stables, barns, dovecots, pigeon houses in search of potassium nitrate," and his son's sensitivity to silvicultural over-exploitation was doubtless informed by the problems encountered by his father in sourcing sufficient quantities of wood for charcoal production. Evelyn's career included positions in public service, too, as an armaments administrator and a naval functionary. He spoke for the modernizing Anglican nobility who sought to uphold the traditional structure of English hierarchy despite—or indeed through—their embrace of the thrusting new spirit of enterprise and "improvement." Thus, although Evelyn was a courtier and a devoted Royalist and Cavalier who fought for Charles I in the English civil war and dedicated *Sylva* to Charles II (and was indeed "on terms of intimacy with all the Stuart kings"[20]), he would on occasion pay visits to the "honest and learned" Anglo-Prussian agronomist Samuel Hartlib, "a public spirited and ingenious person, who had propagated many useful things and arts,"[21] despite Hartlib's close connections with Cromwell and the Commonwealth. Together with other Hartlib associates, Evelyn founded the Royal Society—he coined its name, he was its Secretary in the early 1670s, and *Sylva* was the first book to bear its imprimatur.

A utilitarian streak runs throughout *Sylva*. Its author was a modernist: he sought to transform the world through the application of knowledge—to "inlarge the Empire of Operative Philosophy," in his words.[22] But *Sylva* is not simply a manual for silvicultural improvement and for addressing an immediate crisis of resource depletion. It is also and above all a handbook for the English landed gentry. In providing silvicultural advice it simultaneously guides them in the arts and artistry of governance and affirms their role as arbiters of taste and discernment, confirming their natural right to rule. That the text is adorned with quotations from the poets, botanists and improving landowners of the glittering empires of antiquity—the likes of Theophrastus, Virgil, Cato, Pliny and Columella—is suggestive of its ambit and ambition. One of Evelyn's goals is to encourage his fellow aristocrats to raise their game and become "industrious planters."[23] Instead of frittering their time and money on hunting, or racing dogs and horses, they should turn to silviculture, an altogether more rewarding enterprise, aesthetically and financially.

If *Sylva*'s target audience was the modernizing gentry, it also carried a message for the state. Specifically, it addresses the monarch as himself a landowner, propounding "proposals for the planting and improvement of his Majesty's forests."[24] More generally, it notes, where trees are "cherished and orderly dressed, what a commodity [arises] to the owner, *and the Commonwealth*."[25] *Sylva*'s modernity, however, is apparent above all in its appeals to the state to manage the nation's forests. In order to conserve timber, Evelyn proposes the creation of an office of state, "accountable to the Lord Treasurer, and to the principal officers of his Majesty's Navy." Its remit would be to inspect and manage "all the woods and forests in his Majesty's dominions," ensuring for example "that such proportions of timber, &c. were planted and set out upon every hundred, or more of acres," surveying "the growth and decay of woods, and of their fitness for publick uses and sale," and reporting on these findings in order that any "defect in their ill governing may be speedily remedied."[26]

What is noteworthy here is that sustainability is articulated with broader social projects; it is linked to a particular matrix of forces. In this, Evelyn's work reveals two points. First, it shows that, even in the seventeenth century, European elites were aware that the natural environment, especially natural resources, is subject to deterioration and despoliation. Evelyn was concerned in particular about deforestation, the loss of soils, and air pollution.[27] Second, although environmental concerns are evident, sustainability was not really "about nature" so much as it was about managing natural resources in a way that benefits resource owners (landowners) and the monarchical state (of which the gentry functioned as its main pillar). Evelyn's sustainability agenda was designed to empower the state, and to serve the interests of the state and ruling elites; it was associated with money-making, self-interest and power.

For Evelyn, a Tory modernizer, the ethos of sustainability required the delegation of responsibility for the management of nature–society relations to the state. He was a pioneer of the principle that the state's role is to ensure sustainability, in the sense of regulating the conditions of production in order that capital can appropriate them, while limiting its excesses.[28] But states are not neutral. They serve some groups at the expense of others. In Evelyn's day, the state was wedded to the interests of the great landowners and merchants—his own milieu. Accordingly, while *Sylva* delegates overarching responsibility for forest management to the state, immediate responsibility—and reward—is accorded the class of great landowners. Evelyn's project is saturated with a perspective on land and natural resources that apprehends them through the lens of ownership and profitability.[29] His aim, in Peter Linebaugh's summary, "was to make an inventory of English trees in terms of their use values, and to convey this knowledge from commoners to commercial, scientific, and military markets."[30]

Evelyn's agenda of top-down sustainability existed in antagonistic relation to common-use rights and local customs of resource management.[31] He favored enclosures (that is, the privatization of land), and sharply criticized the bias of England's laws "in favour of *Custom*" as being "indulgent," a regrettable sop aimed at "the satisfying of a few clamorous, and rude Commoners."[32] In order to "gratifie the quaeries of the honourable the principal officers and commissioners of the Navy" (i.e. to supply more wood for shipbuilding), Evelyn advised that enclosure of woodland "would be an excellent way." But, he warned, "the people, viz. foresters, and borderers, are not generally so civil and reasonable, as might be wished; and therefore to design a solid improvement in such places, his Majesty must assert his power, with a firm and high resolution to reduce these men to their due obedience."[33]

Sustainability, for Evelyn, was above all an imperial mission, and *Sylva* was born of militarism. In the mid-seventeenth century England's admiralty was troubled by a looming shortage of timber. This concern was increasing, but not new—a full century earlier, the government had ordered "an inquiry into timber wastage and deforestation."[34] But from the 1640s onward several developments spurred an unprecedented demand for iron and timber: trade growth and the agricultural revolution (with soaring demand for iron tools), as well as the Cromwellian revolution that precipitated a rapid naval build-up.[35] By the 1660s, some reports warned that England did not have enough wood to produce the iron required for armaments.[36]

It was in this context that Henry Oldenburg, secretary of the Royal Society, commissioned four experts to look into the timber crisis.[37] One was Evelyn; his response was *Sylva*. It begins with a warning: "There is nothing which seems more fatally to threaten a weakening of this famous and flourishing nation [than the] decay of her wooden walls"—the depletion of the navy's material substance.[38] "Forest culture," accordingly, should be "vigorously encouraged and promoted" in order to ensure "a competent advance of the most useful timber for the building of ships."[39] Landowners must roll up their sleeves, for modern war is total war, fought with ordnance made of wood, and from charcoal-forged iron. "Our forests," Evelyn intones, "are undoubtedly the greatest magazines of the wealth and glory of this nation; and our oaks the truest oracles of its perpetuity and happiness, as being the only support of that navigation which makes us fear'd abroad, and flourish at home."[40]

Evelyn's manifesto for English naval expansion was made at greater length some years later, in *Navigation and Commerce* (1674). It credits England's "flourishing" and its ascendancy "over the rest of the World" to its "glorious, and formidable Navy." Penned as the preface to a projected history of the Dutch wars, the juncture at which England usurped the Netherlands as Europe and the Atlantic's hegemonic power, *Navigation and Commerce* urges England to consolidate its supremacy over the oceans and thereby over the planet as a whole. "A Spirit of Commerce, and strength at Sea to protect it," it argues, "are the most certain marks of the Greatness of Empire," for "whoever Commands the Ocean, Commands the Trade of the World, and whoever Commands the Trade of the World, Commands the Riches of the World, and whoever is Master of That, Commands the World it self." The book is a paean to the conquerors of the Old World—Roman imperialists, and "the Exploits, and glorious Atchievments" of the Crusaders—and of the New World: the Spanish conquistadores, Francis Drake (who "terrif[ied] the whole Ocean, sack'd St. Iago, Domingo, and Cartagena and carried away with him incredible Booty"), and all those hearty European settlers busy with the "Planting of Colonies."[41] Evelyn himself, as an appointed member of the Council of Plantations and Trade, played a prominent role in England's colonial and commercial expansion.[42] He was keen to ramp up the profitability of England's slave plantations by expanding the area of the Caribbean to be given over to non-native species such as nutmeg, and he cast longing silvicultural eyes toward the forests of New England, where, thanks to abundant biomass energy (and the absence of the undergrowth of common rights, custom and political interest that were proving so refractory in England[43]), numerous iron mills had been constructed. Felicitously, Evelyn's ecological beliefs—notably that deforestation in North America (and in England's other temperate colonies, such as Ireland) was beneficial, making previously "gloomy tracts" healthy and habitable[44]—chimed with his imperialist agenda. To manage England's forests sustainably, given the necessity of naval expansion, required land grabs abroad, and the appropriation of what Jason Moore calls "Cheap Nature"—here, in the shape of New England timber.[45] Evelyn enjoined his countrymen to emigrate to North America, for with their "surfeit of the Woods *which we want*" (for our own iron mills), England's American colonists were well positioned to supply iron to the mother country. This would allow its forests to recover even as its navy expanded, and would, in turn and above all, enable Charles II to become "the great sovereign of the ocean [and] free commerce."[46] The consequences for the sustainability of the colonial forestry barely entered Evelyn's field of vision.[47]

In several ways, this brief survey has shown Evelyn's sustainability doctrine was lashed to an immeasurably larger social project: English imperialism. Its scientific wing centered on the Royal Society. Together with his good friend (and fellow Royal Society founder) William Petty, Evelyn promoted the idea that England's colonies represented vital sources of raw materials, to be exploited in order to enhance the competitiveness of England's industries and navy.[48] He channeled what one historian has described as the "aggressive will to possess and alter land" which flourished first in England and its settlement colonies and culminated in the globe-transforming land rush of the nineteenth century.[49] In his writings and in his practice as an imperial functionary, Evelyn fashioned links between the conservation of timber in England, its cultivation in colonial settlements ("plantations," in the parlance of the time), the manifest destiny of the English nation, and its monarch as ruler of the waves and champion of commerce.[50]

Thoreau was mistaken. Trees were not the chief end of Evelyn's life. They were a means to an end. And all of this would have profound implications for how the concept of sustainability would take shape in Europe and North America. Since the seventeenth century, sustainability has been tied up with the interests of capital ("sustained profit") and the needs of the state—institutions that continue to stamp their agendas upon environmental theory and practice up to the present day.

Harmony through tooth and claw

In the temperate and forested lands of England and New England, Mike Davis points out, "energy flows through the environment in a seasoned pattern that varies little from year to year. Geology is generally quiescent, and it's easy to perceive natural powers as orderly and incremental, rarely catastrophic. Frequent rainfall is the principal geomorphic agency, and the landscape seems generally in equilibrium with the vector of forces acting upon it." A conception of nature as essentially harmonious characterized the emerging "imperial" ecological viewpoint—the hub of which in England was the Royal Society—as well as the "Arcadian" view, which was more concerned with the wellbeing of nature. The "canonical evocations" of rural England and New England, the Reverend Gilbert White's *Natural History of Selborne* (1788) and Thoreau's *Walden* (1854), Davis continues, "were microcosmic celebrations of nature's gentle balance (even as Thoreau sounded the tocsin against the potentially catastrophic environmental threat of the industrial revolution)."[51] This gradualist view of the natural process was molded by Charles Lyell "into one of the great dogmas of Victorian science." The earth, in Lyell's conception, appeared and worked much the same in the past as it does in the present. It is a conservative, moderate and orderly system characterized by "slow processes that unfold over time at even, predictable rates."[52] His was a vision, in short, of a world "in constant motion, but always the same in substance and state."[53]

The conception of nature as balanced and self-regulating, Fredrik Albritton Jonsson proposes in *Enlightenment's Frontier,* found itself mirrored in an image of the market economy. Taking inspiration from Isaac Newton's model of a (mostly) self-regulated physical order, and foreshadowed by Evelyn's friends (William Petty) and acquaintances (Dudley North, John Locke) and, a generation later, by Bernard Mandeville and Richard Cantillon,[54] the vision of the market economy as *by nature* balanced and self-regulating reached its apotheosis in the thinkers of the Scottish Enlightenment, and in Joseph Townsend's *Dissertation on the Poor Laws* (1786). For David Hume and Adam Smith and their liberal followers, "nature served as a hand-maiden for exchange in a double sense. They looked to the natural world for a model of self-regulating balance that justified their own faith in market exchange. At the same time, they championed the market as the best means of managing the balance of nature." They did not hold up conservation as a strategic necessity and assumed that nature was a cornucopia that could be extensively exploited, without real ramifications, in the progressive state of capital.[55]

The parallels are striking. Conceptions of markets, the sustainability of the natural resources, and the Newtonian idea of the law-bound physical universe were conceived alike as more or less self-regulating and mutually reinforcing "natural" systems that only periodically required some kind of external intervention to reconstitute the proper order—whether that interven-tion came from God, the state, or an enlightened expert class. Moreover, these ideas were used to reinforce, legitimate and explain one another.

In this regard, Townsend's work was seminal. He explained the workings of the market economy in terms of natural law, invoking the allegory of goats and dogs on a deserted island to argue that, the suffering of individual goats notwithstanding, when considered from the perspective of the available resources nature's laws of predatory competition result in harmo-nious equilibrium. In human society, by extension, economic behavior rests upon a biological basis and should be modelled as such: market competition establishes "balance"—notwith-standing the suffering of the poor—in exactly the manner of competition between goats and dogs in an island ecosystem.[56] "It is with the human species," as he put it, "as with all other arti-cles of trade without a premium; the demand will regulate the market."[57] Townsend's *Dissertation,* Philipp Lepenies has remarked, was tremendously influential in "fostering a belief in the superiority of self-regulating markets and in warning against outside interference in the market mechanism." Its originality consisted in its argument that it is not human beings but nature that makes markets, with nature governed by atomistic competition and markets behav-ing according to "natural laws" that "should not be tampered with."[58] In other words, Townsend was a pioneer of *social naturalism,* an approach that assumes the workings of the social order can be boiled down to natural laws.[59]

With Townsend, and still more with Robert Malthus and David Ricardo, a sense emerged of the economy as an institutionally distinct sphere which, though supported by the state, could and should be generally left to its own devices. These authors, Karl Polanyi argues, "established the modern concept of a separate autonomous economic system, governed by economic motives, and subject to the economic principle of formal rationality." In their work the "substantive" aspects of economic life were reduced to "the postulates of population (Malthus) and diminishing returns of the soil (Ricardo)."[60] Malthus, in positing a tendency of population growth to outstrip the available food supply, a conflict that emerges directly from the basic drives of hunger and sex, was attempting to set political economy upon a naturalistic basis. The new conception was of economic behavior as resulting "from the boundlessness of man's wants and needs, or, as it is phrased today, from the fact of scarcity."[61] Ricardo shared this Malthusian perspective and applied it to the theory of differential rent: rent is determined by the differ-ential fertility of units of land and not, as Smith had supposed, by their absolute fertility.[62] For

Ricardo, the declining natural fertility of land as the margin of cultivation approaches represented the ultimate barrier to progress. For this reason, one can regard him as a theorist of ecological limits to production.[63] Yet Ricardo's marginalist theory, applied to land valuation, was to be generalized later by neoclassical economists, as I show below, in a process that elided nature as a subject of economic inquiry.

The conception of market exchange founded on a benign natural order, both regarded as models of spontaneous stability and self-regulation, enabled Smith and his successors in the tradition of classical political economy, in the words of Jonsson, to develop "a universal model of growth that could be extended across the British Empire from New Jersey to Bengal." On this assumption, the basic problem of economic development "consisted in mastering the state rather than the natural order."[64] This is one of many examples of Enlightenment thinkers carrying the concept of nature and natural law into other domains of life, subtly making social and economic changes appear as naturally given. It is no coincidence, argues Jonsson, that this universalist vision of nature–economy harmony flourished in a particular agro-climatic regime, of intensive cereal production and animal husbandry in England and Lowland Scotland. That said, this strategic use of nature was controversial and even in these regions it did not go unchallenged. For a "loose constellation of natural historians, agricultural improvers, and conservative landowners … the natural order was too complex or fragile to be left unregulated." In Germany, cameralists did not take the stability of natural economies for granted, and focused on vulnerabilities of supply.[65] It was here, in the eighteenth century, that the origins of the term sustainability may be traced, in the concept of sustained-yield forest management, the calibrating of timber harvest to the rate of new growth. In this conception, science, backed by benign regulation, would ensure a judicious balance between biological and economic growth, providing the basis for rising revenues for landowners and a stable order for society.[66]

The invention of scarcity

It is perhaps no accident that the same period that saw the coining of "sustainability" [*Nachhaltigkeit*[67]] also saw the advent of sustained economic growth. In Europe, the "long nineteenth century" was the first to see rapid economic and population growth become the norm. In previous centuries, per capita output growth had crawled along at roughly 0.1 percent per annum; that figure soared in the nineteenth by an order of magnitude. In mid-eighteenth-century Britain, when Adam Smith published *Theory of Moral Sentiments*, the notion that continuous "material progress" was possible and desirable was beginning to take hold, and the first treatise that set as its explicit goal the theorization of economic growth was the same author's *Inquiry into the Nature and Causes of the Wealth of Nations*. For the classical economists, growth was perceived as ultimately self-limiting. Capitalist economies would follow a parabolic growth rate—it would eventually level out and decline. At the same time, however, they were the first to take growth as a core concern and to theorize its causes. Against older traditions that associated luxury with excess and greed, philosophers such as Hume and Smith reconceived need and desire as "conceptually indistinguishable"; they deemed it impossible "to separate, morally or conceptually, needs and luxuries."[68] Desire was, in a sense, naturalized, and was seen to stimulate demand, which in turn spurs trade and the creation of wealth, further exciting the proliferation of desires. In parallel with this circle of desire ran another (and more potent) one, according to which trade encourages the development of the division of labor, enabling specialization that yields productivity gains and market expansion. This was the engine of self-sustaining growth discussed by Smith in *Wealth of Nations*. That treatise takes growth for granted (in Smith's lexicon, the "progress of opulence"). It does not elaborate a conception of growth as it is understood today—a sustained and potentially infinite increase in per capita income and living standards.[69] Rather, its originality consists in its account of the mutual influence of technical, commercial and moral progress, and their integration into a comprehensive theory of growth. For Smith, economic growth, although not endless, is self-reinforcing in its dynamics, and it is good. In these assumptions, he was followed by Ricardo and Malthus.

There exists a widespread belief that the classical economists concerned with scarcity were growth sceptics, that they pioneered the modern idea of sustainability. Malthus is the classic case. In *Sustainability*, Robert Goodland proposes that "a notion of economic sustainability was firmly embodied in the writings of T. R. Malthus."[70] In *Sustainability*, Kent Portney cites a voluminous literature that traces "the essential seeds of sustainability to ideas put forth by Malthus."[71] In *Sustainability*, Peter Jacques refers to Malthus as "sustainability's Godfather,"[72] while in *Sustainability: A History*, Caradonna credits Malthus with having developed an "ecological" approach to the question of economic growth.[73]

The reason Malthus takes center stage in the history of sustainability is his emphasis on scarcity and the risk of overshoot. Given that the contradictory outcomes of the drives for food and sex ensure that "absolute scarcity" is an inherent feature of human society, there will always be a mass of poor people. By absolute scarcity I refer, following Herman Daly, to a quantitative relationship between the requirements of human beings and available resources, as contrasted with relative scarcity, denoting situations of choice among desired alternatives. Taking the limited availability of land and population pressure as givens, Malthus theorized his version of the law of diminishing returns. As population expands, food production per agricultural laborer tends to decline, bringing starvation to the poor and lower population growth, tendencies that combine to decrease the rate of economic growth. Starvation, alongside war and disease, figure as "positive" checks on population growth, but "preventive" checks (e.g. sexual abstinence) can produce a similar outcome with less suffering.

Malthus's admirers express surprise that his views encounter hostility, and muse that this must be due to the critics' "anthropocentrism" and their subscription to an "ideology of Progress [which] must be indefinite or not at all."[74] Yet, when reading Malthus, one searches in vain for any challenge to anthropocentrism or to the ideology of progress or, for that matter, to the exploitation of the earth's resources. Although he made his name by castigating the utopian perfectionism of radicals, for him the future did not preclude a "gradual and progressive improvement in human society,"[75] and he could "easily conceive" of substantial growth of population and of its welfare.[76] He partook of the consensus of his time and milieu, that humanity was on a path of progress from savagery to civilization; he shared not only Smith and Ricardo's enthusiasm for economic growth but also their assumption that land is nature's "free gift" to capital and their belief that nature's principal function is to furnish raw materials for human consumption (at least by the rich).[77] Although the production of food for a growing population faces constraints in the supply of available land, manufacturing, in Malthus's view, faces no such limitations. The raw materials it requires exist "in great plenty," and as demand grows, the necessary supply is inevitably called forth. Underlying Malthus's economic theory (and Ricardo's too), John Weaver points out, was "an expansionary vision" in which the modernization of land tenure and land taxation would "coax landlords and tenant farmers to make improvements that would produce more," thereby raising living standards and propelling the British state to martial success.[78] Malthus held that a country could cheerfully "go on increasing in riches and population for hundreds, nay, almost thousands of years,"[79] and that the happiness of a country's population depends not on its poverty or wealth, or it sparseness or density, "but upon the rapidity with which it is increasing, upon the degree in which the yearly increase of food approaches to the yearly increase of an unrestricted population."[80] For Malthus, in the judgment of Giorgios Kallis, population growth "was not a problem, but a goal: a happy nation for Malthus is one that its numbers are increasing as much as possible." Happiness "is the degree to which the population of a country approximates a geometric rate of growth, which for Malthus is the natural rate of growth."[81] Limits were a challenge, to be sure, but one that could be overcome or greatly forestalled.

Like his fellow Anglican (and Surreyite) John Evelyn before him, Malthus justified his political-economic hypotheses with reference to God and natural science. But whereas Evelyn struggled to navigate a path between "religious orthodoxy, which assumes the world to be unique, and progress in science, which suggested to him that there was a multiplicity,"[82] for Malthus the relationship was simple. In his theology, the Supreme Being's cardinal command is that mankind fully cultivate the earth. It was his "fervent belief," observes Don Worster, "that man must obey the command of Jehova to multiply and replenish the land and to achieve mastery over it."[83] The deity-decreed laws of nature operate in fulfilment of this plan. Indeed, the contradiction between population growth and food supply itself betokens the cunning of Providence, for it is ordained by divine-natural law precisely to spur progress. Malthus's view of working people places extraordinary emphasis upon their sloth, which he saw as a regrettable drag on the accumulation of capital, and he railed against measures, such as the poor laws, that risked alleviating poverty thereby lessening the motivation to work among the lower orders. At bottom, Malthus's law of population is a divinely fashioned whip to spur human beings, naturally indolent as they are, to develop civilization.

His preoccupation with scarcity notwithstanding, Malthus was also troubled by abundance, in the restricted sense of the overproduction of goods relative to effective demand. Alongside Simonde de Sismondi, he was an early theorist of underconsumption as a major cause of economic crisis, although, unlike his Swiss contemporary, his proposed remedy consisted in augmenting the spending power of the classes of unproductive consumers—aristocrats, bureaucrats, parsons and their ilk.[84] The power to consume should be withheld from the lower classes in the name of controlling the pressure of population on resources, while the upper classes should consume as heartily as they can in order to sustain aggregate demand. The thrust of Malthus's political economy was unambiguous: the market should function as a disciplinary mechanism to enforce scarcity upon the poor and channel abundance to the rich.

Malthus was writing during the final act of the centuries-long historical drama in Britain that saw a regime of custom, with large tracts of common land, replaced by a regime based on absolute private property. In the countryside around his parsonage, ordinary people were being "literally excluded by fences enclosing the common lands that had sustained them for centuries."[85] Scarcity was being *produced,* in the form of their welfare needs, and it is this that formed the basis of the moral panic whose flames Malthus was so eager to fan. Rather than theorize the construction and distribution of scarcity, he sought instead to naturalize it by locating its cause solely and exclusively in purportedly "natural" facts: the sloth and fertility of working people and the constrained supply of cultivable land. In this way, scarcity came to be seen as the inevitable condition of economic life on which the self-regulating market was based. Remove it and you remove the fear of hunger that spurs the laborers to work and to adhere to Christian prescriptions of thrift and sexual abstinence. Malthus's genius was to conscript the new discovery, scarcity, to his normative and deeply reactionary political program. He did so by postulating it as a force of nature.

Although he chopped and changed his theoretical framework, famously making swingeing revisions to his *Essay,* Malthus stuck doggedly to his core manifesto: inequality is eternal; the distribution of resources should be left to the market; consumption of the poor must never be subsidized by the public purse. Through his position at the East India Company he taught a generation of future administrators of the British Empire about the menace of overpopulation and the futility of charity; they implemented his principles around the empire—in India on a genocidal scale.[86] At home, Malthus provided an economic-liberal apology for free-market capitalism as it was consolidating in early-nineteenth-century England, and a neo-conservative defense of private property and social inequality as these were facing challenges from 1789-inspired radicalism and nascent socialist and anarchist currents. In opposition to these movements, his arguments served to bolster and reinvent "the elite fashion of identifying the poor's plight as their fate by using mathematics to explain why the killings troubling his society were unavoidable. The oppressed were recast as sacrifices, and those who scapegoated them as upholders of society."[87]

Of particular salience to my argument is that Malthus's approach melded discourses of scarcity, natural law, and the use, distribution and conservation of natural resources. By appealing to the natural "facts" of scarcity and human egotism he sought to yoke the authority of natural science to an otherwise eminently contestable chain of reasoning: that the contradiction between the infinitude of human wants and the limits of the land gives rise to economic life; that individuals invariably seek as large a share of nature's feast as possible, necessitating private property; and that relative valuations differ, leading individuals to exchange on markets, ergo economic activity is to be understood as the striving of individuals to maximize utility by minimizing labor expended and maximizing profit. Markets, private property, and utilitarianism, according to this logic, are ordained by natural law.[88] These arguments translated into policy prescriptions—notably the 1834 Poor Law Amendment—and simultaneously steered the nascent discipline of political economy onto a positivist and utilitarian track. Malthus played a seminal role in elevating scarcity to the axiomatic, naturalized status that it thenceforth enjoyed, in successively modified forms, in the works of John Stuart Mill and, later, in the neoclassical orthodoxy. As the founding father, with Ricardo, of a successful "social movement from above," one that, following its 1834 breakthrough, achieved hegemony and provided the script for a new mode of governance, Malthusian ideas acquired "common sense" status.

With the advent of neoclassical economics, scarcity divided into its absolute and relative forms. As discussed above, the former denotes the actual use or existence of a resource in relation to requirements, the latter refers to the alternative uses of resources in relation to competing wants.[89] This distinction underpins the philosophical and policy differences between the Malthusian and neoclassical traditions. Malthusians emphasize the natural limits of resources, set great store by population control, and tend to a pronounced skepticism toward technological remedies. This is the reading of Malthus that has most influenced the economics of sustainability. For the neoclassical tradition, scarcity plays a role not so much in its methodology and substance but in the definition of the subject matter of economics (the science of resource allocation among competing goals in a situation of scarcity) and as a "legitimizing device for the general application of the technical apparatus and formal deductive methods of mainstream economics."[90] Nonetheless, its underlying tone carries an unmistakably Malthusian ring. Common to Malthus and his neoclassical successors are the unquestioned assumptions that scarcity, being the outcome of the contradiction between the infinity of society's wants and the finitude of productive potential at its disposal, is natural, essential to the human condition, and the driver of the economic process. Common to both is the assumption that needs and wants are not socially constructed but "givens," rooted in the hedonism of individual consumers. Malthus held that needs are fixed in human nature, while neoclassical economics, in reducing consumption to a relationship between the abstract "household" and the world of objects, implicitly denies their sociality.[91] By framing scarcity as an inherent characteristic of resources, Adel Daoud and Steve Rayner have each argued, Malthus and neoclassical economics alike ignore the social construction of scarcity and the possibility of "states of abundance and sufficiency," and, in so doing, they naturalize the failure of societies to provide for the needs of their populations.[92] Scarcity comes to function as a rationale for inequitable allocation, for the justification of skewed access to control over finite and limited resources, and for the diversion of attention away from causes of poverty and inequality that may implicate powerful groups.[93]

The Jevons paradox and the paradox of Jevons

For Evelyn and Malthus, as supporters and functionaries of the British Empire, the concern for the sustainability of natural resources was bound up with questions of the provision of land and labor for the imperial machine. In the nineteenth century its supply of resources was amply assured. Thanks to empire and the "imperialism of free trade," Britain's economy was liberated from reliance on the produce of its own fields and forests.[94] As the economist William Jevons put it in 1866, "the plains of North America and Russia are our cornfields; Chicago and Odessa are our granaries; Canada and the Baltic are our timber forests, Australia contains our sheep farms, and in Argentina and on the western prairies of North America are our herds of oxen; Peru sends her silver, and the gold of South Africa and Australia flows to London; the Hindus and the Chinese grow tea for us, and our coffee, sugar and spice plantations are all in the Indies. Spain and France are our vineyards, and the Mediterranean our fruit garden."[95] A century before the term "ghost acres" was coined, Jevons had its sense clearly in mind.

Worried by the prospect of British imperial decline, Jevons undertook an investigation of its energy basis: coal. His *The Coal Question* has been characterized as "an extended exercise in Malthusian projection."[96] He adapted Malthus's method vis-à-vis population to resource use in general. Jevons's *Coal Question* represented the culmination of what Albritton Jonsson describes as a debate between "catastrophists" and "cornucopians," inspired by the new realities of sustained economic growth and fossil-fueled machine production.[97] Was industrialism a solid and lasting phase of history in which technological ingenuity would overcome all obstacles, or a precarious moment, heavily reliant on a diminishing resource? The latter perspective was articulated, already in 1789, by a Welsh mining engineer, John Williams. He announced the beginning of peak coal—a forecast, notes Jonsson, that was remarkable "for its precocious recognition of the centrality of mineral energy to the British economy," and for its catastrophic predictions.[98] Coal's exhaustion would bring an end to Britain's "prosperity and glory." Britain's towns would become "ruinous heaps for want of fuel," its factories would fail and its trade flows would dry up. Its inhabitants would be forced to live, "like its first inhabitants, by fishing and hunting."[99] Government action to sustain the nation's coal reserves, Williams, concluded, was imperative. Jevons expanded upon this idea and used extensive economic data to show that the fossil-fueled good times could not last forever.

On the "cornucopian" side of the debate there was for example the physician and inventor Erasmus Darwin. His *Economy of Vegetation* (1791) saluted "the revolutionary potential of coal" while *The Temple of Nature* (1803) celebrated humanity's capacity to create a new artificial world and imagined Britain's landscapes transformed, "with hydraulically engineered rivers and teeming cities filled with vast skyscrapers."[100] But the high priests of machine fetishism were Charles Babbage and Andrew Ure, as well as the Scottish economist John Ramsay McCulloch. Machinery has overwhelmingly beneficial effects, argued McCulloch in 1821, for its introduction in one sector "necessarily occasions an equal or greater demand for the disengaged labourers" in another.[101] His *Principles of Political Economy* (1825) declared that "there are no limits to the bounty of nature in manufactures," and that the nation's coal stocks were "inexhaustible."[102] Another significant figure was a government commissioner, John Leifchild. His *Our Coal and Our Coal-Pits; the People in Them, and the Scenes Around Them* begins with a comparison of coal with gold. The author relishes the assonance of the two words, their contrasting symbolism and their shared connotation with wealth. Yet if one is "the apparent representative of the country's wealth, the other [is] its real representative," and whereas hordes of prospectors are lured by gold all over the world, "the true gold diggings are at home"—in Durham, Fife and Yorkshire. "The true source of wealth," in short, "is coal."[103] It is indispensable for "civilized communities." It supplies the steam power that spins the wheels and moves locomotives. It gives Britain "incalculable commercial advantages," underpinning "our prosperity as a nation, and possibly our supremacy." If it were ever to run out, future historians of empire "would date the decline and fall of the vast dominion of Britain from the period when her supplies of mineral fuel were exhausted and her last coal-field worked out!" But that dreadful vision, Leifchild (1853) hazards—against the catastrophists—will surely not arrive for a millennium or more.

This was the debate into which Jevons intervened. In 1863 he penned a tract on gold that presents the element as "one of the last things which can be considered wealth," and two years later published his anxiety-wracked paean to coal. *The Coal Question* (1865) hails the black stuff as "the mainspring of modern material civilization, … a Promethean gift" that his fellow Britons had bestowed upon the world. He voices concern that the depletion of its coal supplies will lead to Britain's industrial decline and consequent loss of imperial dominance, and deploys Malthusian metaphors in proposing that "geometric" increases in population and the constraints on output growth would result from "arithmetic" increases in food supply and diminishing returns on investment.[104] The equation on which the book rests is that the geometric growth of population and industry will run into the buffers of "a fixed amount of material resources."[105] Of course, writing during an epoch of rapid growth, Jevons (1865) recognized the degree to which the Malthusian equation could be stretched. In contrast to the agricultural produce that was Malthus's subject, "the new applications of coal are of an unlimited character," and this had enabled the "unchecked course of discovery and growth" in the use of coal.[106] Jevons conceded that although "continuous multiplication"—he is referring to population, manufacturing output and trade—"is seldom long possible, owing to the material limits of subsistence … up to the present time our growth is unchecked by any such limits, and is proceeding at uniform or rising rates of multiplication." Indeed, "the rate of multiplication is in recent years many times as great as during preceding centuries." However, this growth will inevitably hit the buffers of resource limits. "We cannot long maintain our present rate of increase of consumption," Jevons concluded. Within a century, "the check to our progress" will become perceptible.[107] This argument about the limits of fossil fuels is one of the main reasons that Jevons is still cited approvingly by sustainability economists. He prefigured theories of energy security and resource exhaustion, including the question of peak oil. In extending Malthus's "geometric versus arithmetic" model to energy sources and raw materials, Jevons developed an approach that was to re-surface in the 1970s under the rubric "limits to growth."

The other reason he is still remembered is for the paradox that bears his name. One seemingly obvious solution for coping with the finitude of resources would be to effect improvements in the efficiency of steam engines and furnaces—surely this would lead to a reduction in the rate of depletion. Paradoxically, however, the opposite occurs. After James Watt introduced his coal-fired steam engine, an improvement upon Newcomen's earlier design, Britain's coal consumption soared. Watt's innovation made coal a more cost-effective energy source, and this helped pave the way to steam.[108] As the steam engine was adopted in a broader range of industries, this increased total coal consumption even as the quantity of coal required for any particular application decreased. Efficiency improvements, Jevons observed, tend to increase overall fuel use. It was "the very economy" of the use of coal that led to its more rapid consumption. "To see how this paradox arises" was simple.

> If the quantity of coal used in a blast-furnace be diminished in comparison with the yield, the profits of the trade will increase, new capital will be attracted, the price of pig-iron will fall, but the demand for it increase; and eventually the greater number of furnaces will more than make up for the diminished consumption of each. And if such is not always the result within a single branch, it must be remembered that the progress of any branch of manufacture excites a new activity in most other branches, and leads indirectly, if not directly, to increased inroads upon our seams of coal.[109]

Rapid growth in the use of coal, Jevons warned, would lead to the need to dig deeper mines and extract poorer quality coal. And unlike land, which, under pressure to feed a growing population, may "continue to yield for ever a constant crop," a coal mine, "once pushed to the utmost will soon begin to fail and sink towards zero."[110] This, then, was the Jevons paradox: it is a fallacy to believe "that the economical use of fuel is equivalent to diminished consumption. The very contrary is the truth … No one must suppose that coal thus saved is spared—it is only saved from one use to be employed in others."[111]

What practical conclusions did Jevons draw? He initially considers alternatives to coal, but dismisses them as equally subject to depletion (such as oil) or relatively inefficient, such as solar (the use of which he predicted), wind, water and tidal power. A conversion to renewable sources would leave Britain unable "to compete with nations enjoying yet undiminished stores" of fossil fuels. This left the inevitability of industrial and imperial decline, for "the cost of fuel must rise, perhaps within a lifetime, to a rate injurious to our commercial and manufacturing supremacy; and the conclusion is inevitable, that our present happy progressive condition is a thing of limited duration."[112] Realistically, therefore, the only alternatives were to allow industrial growth to proceed apace, only to see it decline rapidly at a later date, or to reduce its growth in the present in order to string out the inevitable demise. Jevons preferred the former, for, "in fearlessly following our instincts of rapid growth we may rear a fabric of varied civilization, we may develop talents and virtues, and propagate influences which could not have resulted from slow restricted growth however prolonged."[113]

For ecological economics, the Jevons paradox has proved important in furthering an understanding of the parallel trends toward greater energy efficiency and energy consumption. The paradox highlights the absurdity relying on technological efficiency improvements alone as a pathway to a sustainable society—in a capitalist framework, technological improvements tend to increase the rate of overall throughput, because relative unit costs are lowered, freeing up capital for alternative uses. The real paradox, however, is that Jevons sought to lock society into structures that are susceptible to his paradox. Not only did he contribute to the fetishism of coal, with his claim that it could not be replaced as an energy source by oil or renewables, but his concern with resources was purely with their extraction and not at all with ecological sustainability. He showed no concern for

> the environmental problems associated with the exhaustion of energy reserves. He even failed to address the air, land, and water pollution that accompanied coal production. … Indeed, there was in Jevons no concern for nature as such. He simply assumed that the mass disruption and degradation of the earth was a natural process.[114]

Nowhere did he seek to explain the drive for accumulation, focusing only on its promethean consequences. His framework was thus ill-equipped, argue Foster, Clark and York, "to deal concretely with issues of accumulation and economic growth."[115] Increases in population, manufacturing industry and the demand for coal were all, in his view, simply the product of a Malthus-style "Natural Law of Social Growth." Viewing capitalism as a natural phenomenon, "he could find no explanation for continuously increasing economic demand, other than to point to individual behavior, Malthusian demographics, and the price mechanism."[116] Hence he was unable to situate his paradox as the product of a specific economic system, one that tends systemically to increase the throughput of energy and natural resources on the macro scale, even as technological innovation improves efficiency at the micro scale.

Jevons's paradox remains a valuable insight. But the other Jevons paradox is equally significant for the history of sustainability. His concern for resource depletion, and even his awareness of the limitations of efficiency improvements, meant little or nothing, from an ecological vantage point, given the system of competitive accumulation in which he invested his undying faith. This same paradox remains central to the institutionally most significant variant of the sustainability discourse, sustainable development. Ever since the 1980s, when it gained support from the United Nations, sustainable development policies have sought to limit pollution and the exploitation and degradation of nature, but on the basis of underlying socio-economic relations and mechanisms that ensure their continuation at unsustainable levels.[117]

The elision of nature and the naturalization of growth

In the late nineteenth century, economic theory underwent a transformation and a division. The transformation was in value theory, and in a redefinition of the scope of economics as the hypothetical behavior of rational individuals in conditions of scarcity. The new value theory held "marginal utility," rather than cost of production (as Smith had it), to be the determinant of exchange value. The utility theory of value had been adumbrated by earlier theorists, such as Jeremy Bentham and his friend Joseph Townsend (as a matter of record, the first individual to be called a utilitarian), and also by Malthus and Jean-Baptiste Say. To this, Jevons and his contemporaries added the notion of diminishing marginal utility. According to this concept, as Ismael Hossein-Zadeh explains, "the utility derived from the use or consumption of a commodity diminishes with every additional unit consumed." Although Jevons's addition of the idea of marginal utility to the received utility theory of value was simple and straightforward, "it nonetheless proved to be instrumentally a very important notion in neoclassical economics":

> For, the term 'marginal' was soon extended to other economic categories such as marginal cost, marginal revenue, marginal propensity to consume, and the like; thereby paving the way for the application of differential calculus to economics. By introducing the notion of marginalism into utilitarian economics, Jevons had found a way in which the utilitarian view of human beings as rational, calculating maximizers could be put into mathematical terms.[118]

The new understanding of the scope of economics jettisoned the classical economists' concern with social structure, and generalized Malthus's concern with scarcity into a methodological principle. The division was in method, such that two aspects of the method of the classical economists were taken in different directions, one toward "hard science," the other toward inductive social science. The classical economists had been ambivalent in their disciplinary self-definition. Adam Smith and John Stuart Mill, for example, attempted to put political economy on a scientific footing, yet regarded it as inextricable from the other social sciences and from ethics. In the late nineteenth-century divide in economics, one branch, institutionalist economics, followed the "morals" and "politics" path. The other direction was taken by marginalism, which branched into the neoclassical and Austrian traditions. The neoclassical tradition in particular modelled the newly reconstituted discipline upon natural science. Whereas earlier economists, notably the physiocrats, had regarded economics as the physiology of society, neoclassical economists tended to see the world "as a smoothly functioning machine," one that could be adequately modelled with mathematical equations.[119] It is no coincidence that Jevons and the French economist Léon Walras, both of whom developed marginal value, were originally trained as natural scientists. Jevons likened the notion of value in economics to that of energy in mechanics, and depicted marginal utility as analogous to gravitational force. He envisaged the mind of the economic agent as balancing the forces of pleasure and pain, so that exchange could be depicted as a balance, "using the analogy of a lever in equilibrium."[120] Walras maintained that "the pure theory of economics is a science which resembles the physic-mathematical sciences in every respect."[121] Carl Menger, although averse to the mathematicization of his discipline, nonetheless conceived of economics as dealing in definite laws, those that condition the economic activity of human beings and which are "entirely independent of human will."[122] The inspiration from natural science lay not with its subject matter (the material world), but with its methods, and especially the relationship between the presumed rational agent and the objective natural world. As Phillip Mirowski has shown, the neoclassical economists invoked physical conservation principles (of matter, of energy) "to argue that physical science dictated that production did not really exist in a physical sense, thus absolving the new economics from having to deal with the questions of material production processes that had so occupied the classical economists."[123] Thus Jevons, in *Principles of Economics,* proposed that although "we speak familiarly of creating wealth" this expression means only that we create "utility."[124] The same point is echoed in Alfred Marshall's influential textbook of the same name:

> Man cannot create material things. In the mental and moral world indeed he may produce new ideas; but when he is said to produce material things, he really only produces utilities; or in other words, his efforts and sacrifices result in changing the form or arrangement of matter to adapt it better for the satisfaction of wants. All that he can do in the physical world is either to readjust matter so as to make it more useful, as when he makes a log of wood into a table; or to put it in the way of being made more useful by nature, as when he puts seed where the forces of nature will make it burst out into life.[125]

Unlike the classical economists, who brought the production of wealth as well as its distribution (and, correspondingly, social structure) into the frame, the subject for marginalism is exchange alone. The natural world does not have any meaning or value in this system and, in a sense, market capitalism is merely one of many natural phenomena in which humans participate. It is in markets that human beings express their desires, and markets, in the absence of deliberate human intervention, automatically produce the most efficient outcomes. Marginalist economics in this sense is the disciplinary avatar of pure exchange value. Its practitioners abstract the analysis of exchange relations from the concrete character and conditions of production of the commodities whose price movements they model, from the historically specific relations of production and from the institutional matrix that they entail, revealing an interplay of economic forces that are homogenous, reversible and quantitatively unlimited. Indeed, they saw this as their scientific accomplishment.

The vantage point of marginalism is the aggregate decisions of individual consumers, each making rational economizing decisions in a situation of scarcity. Whereas Smith, whose concern was with the creation of wealth in terms of the production of goods, conceived of labor as the source of value, Menger, by contrast,

> is concerned with the issue of maximizing satisfaction, so he looks for the natural foundation of exchange not in terms of labor or of natural propensities, but rather in the calculation people make, in a context of scarcity and individual property, that they would rather have something that someone else has than all or part of something in their own possession.[126]

In these ways, Nicholas Xenos argues, neoclassical economics systematized the scarcity postulate—it "discovered what the eighteenth century had invented: a universal condition of scarcity."[127] The tendency to what has recently become known as "economics imperialism" is inscribed within this. Neoclassical economics appears to restrict its scope by limiting it to scarcity situations, but in reality "it expects, in Menger's version, that more and more goods will enter the realm of scarcity (and hence become economic) through a constant growth in human needs. As civilization advances, therefore, scarcity situations will become generalized, and along with it the applicability of economic science."[128]

With the partial exception of Jevons, whose theory of value was pedigree marginalism but whose theory of production was classical (centered on dynamics of population and natural resources, their limits, and the eventual decline of growth), the marginalists shifted the focus of economics further away from the causes and consequences of growth and from society's interaction with nature. Following its subjectivist re-theorization, value found itself abstracted from labor and from nature. Surplus value disappeared from view, through the attribution of the entire revenue to productive factors, according to their marginal productivity. Natural resources, likewise, were elided, by means of new categories such as "decreasing marginal utility and productivity, and by denying 'that they are a different factor from capital' at all."[129] Once elided, nature was then reintroduced into economics, not as its basis but merely as an issue area, with its appropriate sub-discipline, environmental economics.[130]

The relevance of marginalism to questions of sustainability lies not only in the elision of nature but also in the naturalization of economic growth. Relative to their classical forebears, marginalist economists paid scant attention to economic change. Insofar as they did, it was assumed to be gradual, non-disruptive and equilibrating. Following Say, they conflated wealth with exchange value and took the clearing of markets as a given, obliterating at a stroke both the role of nature in the process of wealth production and the problem of economic crisis. The categories they introduced, such as decreasing marginal utility, served to assimilate natural resources to the category of capital.[131] Rather than theorize the causes and conditions of economic growth, they developed a static general equilibrium analysis of the allocation of factors of production—how a given quantity of resources can be allocated most efficiently among individual consumers and firms.[132] With regard to policy, the assumption was that economic efficiency, and by extension economic growth, would be promoted by leaving business to its own devices. Hardly a line, Heinz Arndt has suggested, "is to be found in the writings of any professional economists between 1870 and 1940 in support of economic growth as a policy objective."[133]

Yet if mainstream economists in the late nineteenth century paid little overt attention to explaining economic growth, and many were inclined to criticize growth-oriented industrial policy, it would be erroneous to assume that they did not contribute to the development of the growth paradigm. Indeed, crucially for our purposes, some of the economists who have had the biggest impact on the history of sustainability economics, notably Jevons, were as much part of this paradigm-formation as anyone. In doing so, these economists made three assumptions. First, they took growth as a given, as the natural end of a market economy, and championed the proposal that it is not merely wealth but, in Jevons's phrase, "*growing* wealth that makes a happy and prosperous country."[134] Second, they redefined economic activity as an essentially quantitative field. In this, Jevons was the pioneer. It would not be a travesty of his economic theory to summarize it as "Bentham + mathematics." Essentially, Jevons holds that *all* economic decisions can be reduced to expressions of pleasure seeking and pain aversion, i.e. utility and disutility (with labor assumed to be pain, and consumption assumed to be pleasure). Thus reduced to a utility spectrum, economic behavior is rendered quantifiable and subject to treatment as mathematical formulae. "Being concerned with quantities," as Jevons put it, economics is necessarily "mathematical in its subject."[135] His *Theory of Political Economy* was the first English text that "discussed all prices in terms of 'the laws of supply and demand' and explained all economic actions in terms of marginal utility, using the calculus and geometry."[136] Third, they contributed to the ideological affirmation of capitalism, the social system of which the growth paradigm is an organic ideological accompaniment.[137] They "naturalized" capitalism, conceiving it as the telos of human nature: capitalist norms and institutions manifest the imputed behavior of the rational individual, and therefore represent the "end of history." Their insistence that the market be left to its own devices provided sanction for the growth imperative. In their understanding, economic growth, once initiated, becomes "automatic and all-pervasive, spreading among nations and trickling down among classes."[138] Paradoxically therefore, some of the most influential thinkers whose work has been appropriated by the modern sustainability movement were central figures in justifying the very growth-based, unfettered, industrial capitalism that seemingly elicited the sustainability movement in the first place.

Conclusion

The centuries under discussion in this chapter witnessed a transformation in society's understanding of nature. God was gently elbowed aside by natural science, and the natural world was re-imagined as a law-governed realm. Stupendous intellectual advances were achieved—from Newton's physics and Linnaeus's taxonomy to Watt's steam engine and a thousand other technological inventions. These same centuries also saw the rise of capitalism, centered in northwest Europe and its colonial and post-colonial appendages, accompanied by a set of inter-related ideological moves: the theorization of "the economy" as a distinct, law-governed sphere, the justification of "self-interest" and the burial of earlier suspicions of avarice, the conceptualization of both nature and the market economy as essentially balanced and benign, the justification of capital accumulation and the delegation to the state of responsibility for its management and for the suppression of its excesses, and the justification of colonial expansion by reference to the colonizers' superior ability to "improve" the soil (with improvement often understood as short-term yield and monetary profit).

The period 1650 to 1900 also witnessed the development of a discourse on sustainability, and it is clear that this concept is more in line with the intellectual and imperial developments of the period than is often supposed. Although the world of sustainability today is complex and contested, the roots of the concept and the movement lie with thinkers who were concerned principally with industrial and colonial expansion, the wellbeing of the state and social elites, and "the economy" as a law-governed mechanism. "As odd and unsavory as it may seem," in Caradonna's words, "sustainability traces its roots primarily to imperialists who cared *very little* about nature or social justice and *very much* about state power, industrialization, and profit."[139] Indeed, sustainability emerged from the same cultural milieu that viewed markets, akin to the physical universe, as a law-bound sphere. In a sense, the concern for nature has increased in recent decades, but "dealing with nature" to ensure growth and economic might is altogether consistent with the concerns of Evelyn, Carlowitz, Malthus, Jevons, and many others in the pantheon of sustainability.

The naturalism of such political economists as Malthus or Ricardo has led some to suppose that they were ecologically minded. But, as I have argued in this chapter, nothing could be further from the truth. They used scarcity, limits and diminishing returns to naturalize the prevailing economic system, and to edge nature out of the subject matter of economics, a process that reached its apotheosis with the marginalist revolution of the late nineteenth century. Resources such as fuels or soils were a concern, but the economic system they favored was about not ecological sustainability but sustained wealth creation.

Thanks in part to fossil fuels and the ghost acres of the colonies, and to the application of science to processes of production, a sharp increase in output per worker was achieved. One consequence was a rapid increase in pollution—both visible (soot, etc.) and invisible (the first significant and "sustained" signs of anthropogenic global warming can be dated to the mid nineteenth century[140]), which spurred a variety of forms of nineteenth-century environmentalism (above all, aristocratic anti-industrialism). Another consequence was concern for resource depletion. Paradoxically, as Jonsson puts it, "the fear of limits emerged precisely at the moment when Enlightenment ideology and industrialization began to make sustained economic growth imaginable."[141] This was particularly so in the nineteenth century. It saw an increasing concern with resource scarcity, and with scarcity per se. Yet, in the same moment, the natural world was evacuated from the discipline of economics. The apotheosis of this process was neoclassical economics, with its myth of the infinite substitutability of resource inputs. And yet, as this chapter has demonstrated, this myth was not born in the late nineteenth century. It had a pre-history. It was foreshadowed in the work of earlier economic theorists, notably Malthus and Ricardo, and indeed in the utilitarianism of that seventeenth-century so-called theorist of sustainability, John Evelyn. He envisaged nature as a utilitarian commodity, to be managed for the common good, understood as landowners' power and English empire.

Sustainable development and green growth, the attempts to manage the natural realm in a manner compatible with an economic system governed by the imperative of capital accumulation, have a centuries-long pre-history. "All that is new in the Brundtland Report" and many other similar documents, as Don Worster puts it, "is that they have extended the idea *to the entire globe*. Now it is Planet Earth, not merely a beech forest, that is to be managed by trained minds, an eco-technocratic elite."[142]

Acknowledgements

The research for sections of this paper was funded by a British Academy Mid-Career Fellowship (2013–14, project title "Economic Growth as Ideology: Origins, Evolution and Dilemmas of the 'Growth Paradigm'"). I am grateful to Jeremy Caradonna for his assiduous and insightful editorial suggestions.

Notes

1   Caradonna 2014, p. 40; Grove 1995.
2   Caradonna 2014, p. 53.

3   Moore 2014, p. 204.
4   Grove 1995, p. 23.
5   Grober 2012, p. 39.
6   Ibid., p. 83.
7   Grove 1995, p. 11.
8   Grober 2010, p. 95.
9   Evelyn 1664b.
10   Thomas 1991.
11   Evelyn, in Wood 1984, p. 99.
12   Evelyn 1664b.
13   Ibid. See also Grove 1995, 1997.
14   Thoreau, in Worster 1994, p. 87.
15   Caradonna 2014, p. 138.
16   Foster 1999, p. 186.
17   Grober 2012, p. 68.
18   Ibid., pp. 69–70.
19   Linebaugh 2008, p. 90.
20   Carter 1953, p. 228.
21   In Evelyn's words: Evelyn 1640–1706.
22   Evelyn 1664b.
23   Darley 2006, p. 182.
24   Evelyn 1664b.
25   Ibid.
26   Ibid.
27   In addition to his work on air pollution and deforestation, Evelyn 1675.
28   On this in general, see Keucheyan 2016.
29   Wood 2003; di Muzio 2015.
30   Linebaugh 2008, p. 90.
31   Cf. Albritton and Albritton Jonsson 2016, p. 11.
32   Evelyn, quoted in Wood 1984, p. 61.
33   Evelyn 1664b.
34   Cipolla 1977, p. 180.
35   Brinley Thomas, quoted in di Muzio 2015, p. 72.
36   For example the Marquis de Seignelay, cited in di Muzio 2015, p. 66.
37   Grober 2012.
38   Evelyn 1664a.
39   Evelyn 1664b.
40   Ibid.
41   Evelyn 1674.
42   At the council, "Cold exchanges about slaves, human 'commodities,' as they were termed, traded by the Africa Company like cocoa beans, were minuted dispassionately alongside mundane discussions of premises, duties and appointments and the reports from Jamaica, New England or Surinam" (Darley 2006, p. 247).
43   Grove 1995, p. 57.
44   Ibid., p. 58.
45   Moore 2014; see also Lessenich 2016.
46   Evelyn 1664b.
47   In the two centuries after the colonists arrived, North America "lost more woodland than Europe had lost in 1000 years" (David Nye quoted by di Muzio 2015, p. 88).
48   Wood 2003, p. 85.
49   Weaver 2003, p. 43.
50   Rogers 2005, p. 299.
51   Davis 1998, p. 15.
52   Gould 1988, p. 105; Ghosh 2016, p. 26. A related sensibility arose in literature too. The novel facilitated a new kind of "narrative pleasure compatible with the new regularity of bourgeois life," in which the imagined cosmos is rationalized, converted into "a world of few surprises … and no miracles at all" (Franco Moretti, quoted in Ghosh 2016, p. 25). See also Eagleton 1990.

53   Gould 1988, p. 105
54   I develop this point in Dale 2015, 2017.
55   Jonsson 2013, p. 3.
56   Townsend 1971.
57   Townsend, in Lepenies 2014, p. 451.
58   Lepenies 2014, p. 447.
59   Block and Somers 2014, p. 228.
60   Karl Polanyi, in Dale 2016, p. 230.
61   Ibid.
62   Clarke, 1982.
63   Caradonna 2014, p. 75.
64   Albritton Jonsson 2013, p. 49.
65   Jonsson 2013, p. 3.
66   Worster 2009, p. 137.
67   The English words "sustainability" and "sustainable" only entered the language in the twentieth century. Prior to that, the German term was generally used (or other descriptors of the same concept). Of note, however, is that the first recorded use of "sustainable" was in an economics dictionary of 1965 that used the term "sustainable growth." See Caradonna 2014, p. 7.
68   Lasch 1991, p. 52.
69   Friedman 2006, p. 47.
70   Goodland 2009, p. 213.
71   Portney 2015, p. 5.
72   Jacques 2015, p. 67.
73   Caradonna 2014, p. 73.
74   Curry 2011, p. 201.
75   Levin 1966, p.104.
76   McNally 1993, p. 82.
77   Worster 1994, p. 53; Foster 2000, p. 177.
78   Weaver 2003, p. 26.
79   Malthus 1989, pp. 43, 437.
80   Malthus 1798, quoted in Kallis 2016.
81   Kallis 2016.
82   Darley 2006, p. 141.
83   Worster 1994, p. 151.
84   Malthus 1836, pp. 413–14.
85   Boal 2010.
86   Pearce 2010; Davis 2001.
87   Lohmann 2005, p. 95.
88   Matthaei 1984.
89   Daoud 2010.
90   Fine 2010, p. 81.
91   Matthaei 1984.
92   Daoud 2010; Rayner 2010.
93   Mehta 2010.
94   Gallagher and Robinson 1953.
95   Jevons, in Hardin 1995, p. 134.
96   Mayhew 2014, p. 135.
97   Jonsson 2013.
98   Ibid, p. 167. Compare Adam Smith's *Wealth of Nations*, published only a decade earlier, which was blind to the importance of fossil fuels.
99   Ibid, p. 174.
100 Jonsson 2014, p. 161.
101 McCulloch, in Tribe 1981, p. 118.
102 McCulloch, in Jonsson 2014, p. 163.
103 Leifchild 1853, pp. 11–24.
104 Peart 1996, p. 33.
105 Jevons 1865.

106 Ibid.
107 Ibid.
108 For other factors, see Malm 2016.
109 Jevons 1865.
110 Ibid.
111 Ibid.
112 Ibid.
113 Ibid.
114 Foster, Clark, and York 2010.
115 Ibid.
116 Ibid.
117 See the essays in Borowy and Schmelzer 2017.
118 Hossein-zadeh 2014.
119 Ormerod 1994, p. 45.
120 White 1994, p. 205.
121 In Mirowski 1989, pp. 217–18.
122 Menger 1950, p. 48.
123 Mirowski 1989, p. 289.
124 Jevons 1905.
125 Marshall 1890.
126 Xenos 1989, p. 73.
127 Ibid., p. 69.
128 Ibid., p. 71.
129 Koch 2011, p. 18.
130 Lane 2014; Barry 2012, p. 143.
131 Koch 2011.
132 Ormerod 1994, p. 42.
133 Arndt 1978, p. 13.
134 Jevons 1865.
135 Jevons 1866.
136 White 1994, p. 197; see Jevons 1871.
137 Dale 2012.
138 Nugent and Yotopoulos, quoted in Hettne 1983, p. 248.
139 Caradonna 2014, p. 45. For an alternative slant, see Grove 1995.
140 Worland 2016.
141 Jonsson 2013, p. 4.
142 Worster 2009, p. 139.

References

Albritton Jonsson, Fredrik (2013) *Enlightenment's Frontier: The Scottish Highlands and the Origins of Environmentalism*. New Haven, CT: Yale University Press.

Arndt, Heinz (1978) *The Rise and Fall of Economic Growth*. Harlow: Longman.

Block, Fred and Margaret Somers (2014) *The Power of Market Fundamentalism: Karl Polanyi's Critique*. Cambridge, MA: Harvard University Press.

Boal, Iain (2010) "Specters of Malthus: Scarcity, Poverty, Apocalypse." Retrieved from www.economics.arawakcity.org/node/519.

Borowy, Iris and Matthias Schmelzer, eds. (2017) *History of the Future of Economic Growth: Historical Roots of Current Debates on Sustainable Degrowth*. Abingdon: Routledge.

Caradonna, Jeremy (2014) *Sustainability: A History*. Oxford: Oxford University Press.

Carter, Sydney (1953) "John Evelyn: A Study in Royalist Piety." *The Churchman* 67(4).

Clarke, Simon (1982) *Marx, Marginalism and Modern Sociology: From Adam Smith to Max Weber*. Basingstoke: Macmillan.

Curry, Patrick (2011) *Ecological Ethics: An Introduction* (2nd edition). Cambridge: Polity.

Dale, Gareth (2012) "The Growth Paradigm: A Critique." *International Socialism* 134: 55–88.

Dale, Gareth (2015) "Delusions of Green Growth." *International Socialist Review* 97.

Dale, Gareth (2016) *Karl Polanyi: A Life on the Left*. New York: Columbia University Press.

Dale, Gareth (2017) "Seventeenth-century Origins of the Growth Paradigm." In Iris Borowy and Matthias Schmelzer, eds, *History of the Future of Economic Growth: Historical Roots of Current Debates on Sustainable Degrowth.* Abingdon: Routledge.

Daoud, Adel (2010) "Robbins and Malthus on Scarcity, Abundance, and Sufficiency: The Missing Sociocultural Element." *The American Journal of Economics and Sociology* 69(4): 1206–29.

Darley, Gillian (2006) *John Evelyn: Living for Ingenuity.* New Haven, CT: Yale University Press.

Davis, Mike (1998) *Ecology of Fear: Los Angeles and the Imagination of Disaster.* London: Picador.

Davis, Mike (2001) *Late Victorian Holocausts: El Niño Famines and the Making of the Third World.* London: Verso.

di Muzio, Tim (2015) *Carbon Capitalism: Energy, Social Reproduction and World Order.* Lanham, MD: Rowman & Littlefield.

Eagleton, Terry (1990) *The Ideology of the Aesthetic.* Oxford: Blackwell.

Evelyn, John (1640–1706) *The Diary of John Evelyn.* Retrieved from https://archive.org/stream/diaryofjohnevely01eveliala/diaryofjohnevely01eveliala_djvu.txt.

Evelyn, John (1664a) *Sylva,* volume I. Retrieved from www.gutenberg.org/ebooks/20778.

Evelyn, John (1664b) *Sylva,* volume II. Retrieved from https://archive.org/stream/ sylvaordiscourse 02evelrich/sylvaordiscourse02evelrich_djvu.txt.

Evelyn, John (1674) *Navigation and Commerce.* Retrieved from http://quod.lib.umich.edu/e/eebo/A38802.0001.001?rgn=main;view=fulltext.

Evelyn, John (1675) *Discourse on Earth, Mould and Soil.*

Foster, John Bellamy (1999) "Introduction to John Evelyn's Fumifugium." *Organization and Environment* 12(2).

Foster, John Bellamy (2000) *Marx's Ecology: Materialism and Nature.* New York: Monthly Review Press.

Foster, John Bellamy, Brett Clark, and Richard York (2010) "Capitalism and the Curse of Energy Efficiency: The Return of the Jevons Paradox." Retrieved from http://monthlyreview.org/2010/11/01/capitalism-and-the-curse-of-energy-efficiency.

Friedman, Benjamin (2006) *The Moral Consequences of Economic Growth.* London: Vintage.

Gallagher, John and Ronald Robinson (1953) "The Imperialism of Free Trade." *Economic History Review.* 6(1): 1–15.

Ghosh, Amitav (2016) *The Great Derangement: Climate Change and the Unthinkable.* London: Penguin.

Goodland, Robert (2009) "The Concept of Environmental Sustainability." In Tom Campbell and David Mollica, eds., *Sustainability.* Farnham: Ashgate.

Gould, Stephen Jay (1988) *Time's Arrow, Time's Cycle: Myth and Metaphor in the Discovery of Geological Time.* London: Penguin.

Grober, Ulrich (2010) *Die Entdeckung der Nachhaltigkeit: Kulturgeschichte eines Begriffs.* Munich: Kunstmann.

Grober, Ulrich (2012) *Sustainability: A Cultural History.* Cambridge: Green Books.

Grove, Richard (1995) *Green Imperialism: Colonial Expansion, Tropical Island Edens and the Origins of Environmentalism, 1600–1860.* Cambridge: Cambridge University Press.

Hardin, Garrett (1995) *Living within Limits: Ecology, Economics, and Population Taboos.* Oxford: Oxford University Press.

Hettne, Björn (1983) "The Development of Development Theory." *Acta Sociologica* 26(3–4).

Hirschman, Albert (1991) *The Rhetoric of Reaction: Perversity, Futility, Jeopardy.* Cambridge, MA: The Belknap Press of Harvard University Press.

Hossein-zadeh, Ismael (2014) "Ideological Foundations of Neoclassical Economics." Retrieved from http://philosophersforchange.org/2014/11/25/taking-notes-41-ideological-foundations-of-neoclassical-economics-class-interests-as-economic-theory.

Jacques, Peter (2015) *Sustainability.* Abingdon: Routledge.

Jevons, William (1865) *The Coal Question: An Inquiry Concerning the Progress of the Nation, and the Probable Exhaustion of Our Coal-Mines.* Retrieved from www.eoearth.org/article/The_Coal_ Question:_Of_our_Consumption_of_Coal.

Jevons, William (1866) *Brief Account of a General Mathematical Theory of Political Economy.* Retrieved from www.marxists.org/reference/subject/economics/jevons/mathem.htm.

Jevons, William (1871) *The Theory of Political Economy.* London: Macmillan and Co.

Jevons, William (1905) *The Principles of Economics and Other Papers.* Retrieved from https://archive.org/stream/principlesofecon00jevouoft/principlesofecon00jevouoft_djvu.txt.

Jonsson, Fredrik Albritton (2013) *Enlightenment's Frontier: The Scottish Highlands and the Origins of Environmentalism*. New Haven, CT: Yale University Press.

Jonsson, Fredrik Albritton (2014) "The Origins of Cornucopianism: A Preliminary Genealogy." *Critical Historical Studies* 1(1): 151–68.

Kallis, Giorgos (2016) "Limits Without Scarcity: Why Malthus Was Wrong." Unpublished manuscript.

Keucheyan, Razmig (2016) *Nature is a Battlefield: Towards a Political Ecology*. Cambridge: Polity.

Koch, Max (2011) *Capitalism and Climate Change: Theoretical Discussion, Historical Development and Policy Responses*. Basingstoke: Palgrave.

Lasch, Christopher (1991) *The True and Only Heaven*. New York: Norton.

Leifchild, John (1853) *Our Coal and Our Coal-Pits; the People in them, and the Scenes Around Them*. London: Longman, Brown, Green and Longmans.

Lepenies, Philipp (2014) "Of Goats and Dogs: Joseph Townsend and the Idealisation of Markets—a Decisive Episode in the History of Economics." *Cambridge Journal of Economics* 38: 447–57.

Lessenich, Stephan (2016) *Neben uns die Sintflut: Die Externalisierungsgesellschaft und ihr Preis*. Hanser Berlin.

Levin, Samuel. (1966) "Malthus and the Idea of Progress." *Journal of the History of Ideas* 27(1): 92–108.

Linebaugh, Peter (2008) *The Magna Carta Manifesto: Liberties and Commons for All*. Berkeley, CA: University of California Press.

Lohmann, Larry (2005) "Malthusianism and the Terror of Scarcity." In Betsy Hartmann, ed., *Making Threats: Biofears and Environmental Anxieties*, 81–98. Lanham, MD: Rowman & Littlefield.

Malm, Andreas (2016) *Fossil Capital: The Rise of Steam Power and the Roots of Global Warming*. London: Verso.

Malthus, Thomas Robert (1836) *Principles of Political Economy* (2nd edition). London: William Pickering.

Marshall, Alfred (1890) *Principles of Economics*. Retrieved from www.econlib.org/library/Marshall/marP.html.

Matthaei, Julie (1984) "Rethinking Scarcity: Neoclassicism, NeoMalthusianism, and NeoMarxism." *Review of Radical Political Economics* 16(2–3): 81–94.

Mayhew, Robert (2014) *Malthus: The Life and Legacies of an Untimely Prophet*. Cambridge, MA: Harvard University Press.

McNally, David (1993) *Against the Market: Political Economy, Market Socialism, and the Marxist Critique*. London: Verso.

Mehta, Lyla (2010) "Introduction." In Lyla Mehta, ed., *The Limits to Scarcity: Contesting the Politics of Allocation*, 1–10. London: Earthscan.

Menger, Carl (1950) *Principles of Economics*. Chicago, IL: The Free Press.

Mirowski, Philip (1989) *More Heat Than Light. Economics as Social Physics: Physics as Nature's Economics*. Cambridge: Cambridge University Press.

Moore, Jason (2014) *Capitalism in the Web of Life: Ecology and the Accumulation of Capital*. London: Verso.

Ormerod, Paul (1994) *The Death of Economics*. London: Faber & Faber.

Pearce, Fred (2010) *Peoplequake: Mass Migration, Ageing Nations and the Coming Population Crash*. London: Eden Project Books.

Peart, Sandra (1996) *The Economics of W. S. Jevons*. Abingdon: Routledge.

Portney, Kent (2015) *Sustainability*. Cambridge, MA: MIT Press.

Rayner, Steve (2010) "Foreword." In Lyla Mehta, ed., *The Limits to Scarcity: Contesting the Politics of Allocation*, xvii–xxi. London: Earthscan.

Rogers, Pat (2005) *Pope and the Destiny of the Stuarts: History, Politics, and Mythology in the Age of Queen Anne*. Oxford: Oxford University Press.

Thomas, Keith (1991) *Man and the Natural World: Changing Attitudes in England 1500–1800* (2nd edition). London: Penguin.

Townsend, Joseph (1971) *A Dissertation on the Poor Laws, by a Well-wisher to Mankind*. Berkeley, CA: University of California Press.

Tribe, Keith (1981) *Genealogies of Capitalism*. Basingstoke: Macmillan.

Weaver, John (2003) *The Great Land Rush and the Making of the Modern World, 1650–1900*. Montreal: McGill-Queen's University Press.

White, Michael (1994) "The Moment of Richard Jennings: The Production of Jevons's Marginalist Economic Agent." In Philip Mirowski, ed., *Natural Images in Economic Thought: 'Markets Read in Tooth and Claw'*. Cambridge: Cambridge University Press.

Wood, Ellen (2003) *Empire of Capital*. London: Verso.

Wood, Neal (1984) *John Locke and Agrarian Capitalism*. Berkeley, CA: University of California Press.

Worland, Justin (2016) "Humans Have Caused Global Warming for Longer Than We Thought." *Time* (24 August). Retrieved from http://time.com/4461719/global-warming-climate-change-humans.

Worster, Donald (1994) *Nature's Economy: A History of Ecological Ideas*. Cambridge: Cambridge University Press.

Worster, Don (2009) "The Shaky Ground of Sustainable Development." In Tom Campbell and David Mollica, eds., *Sustainability*. Farnham: Ashgate.

Xenos, Nicholas (1989) *Scarcity and Modernity*. London: Routledge.

7

ETERNAL FOREST, SUSTAINABLE USE

The making of the term "*Nachhaltig*" in seventeenth- and eighteenth-century German forestry

Ulrich Grober

The idea of sustainability is not a brainchild of the modern environmental movement. It has deep roots in many cultures of the world and is, in fact, a world cultural heritage. As far as the modern vocabulary, in which we conceptualize this idea, is concerned, a distinct blueprint can be found in the professional language of German forestry. After the first appearance of the term "*nachhaltende Nutzung*" (sustainable use[1]) in print in 1713, "*Nachhaltigkeit*" gradually became the guideline—and indeed the holy grail—of German forestry. Translated as "sustained forest yield" or "*rendement soutenu des forêts*" it turned into a household word of international forestry. This paper traces the roots of the concept back to the age of early European Enlightenment. In that era, German cameralists, driven by the fear of an imminent timber shortage and inspired by new approaches and best practices in England, France, other European and non-European countries, coined the term and began to manage their territories' woodlands "*nachhaltig*" in order to hand them along undiminished and as intact as possible to future generations.

"Perpetual, continual and eternal"

"God created the forests for the salt spring so that they—like it—may continue eternally (ewig). Therefore humans should keep it that way: before the old forest is exhausted the young should have regrown for cutting."[2]

The time is 1661; the place the alpine Bavarian town of Reichenhall. The notion of "eternal forest" appears in a letter written by the alderman, "*Rathskanzler*" Schmidt, to the "*Salzmayster*"Aufhammer, the representative of the salt works, which had formed the backbone of the local economy since the times of the Roman Empire. The notion is still embedded in the concept of divine Providence, the belief that God not only created the world, but continues to be present within it, cares for its "*conservatio*," its preservation and leads all of his creatures towards their destiny, which is salvation. Thus the notion radiates a spirit of basic trust in creation, of collective hope and optimism. On the other hand, it makes clear that the entire community has to assume responsibility for the continual regeneration of the forest in order to ensure the long-term steadiness of wood supply and the continuation of salt boiling in

Reichenhall. In fact, the concern for a "perpetual, continual and eternal" (*perpetuierlich, continuierlich und ewig*[3]) forest was more and more driven by a collective anxiety: The fear of the exhaustion of the woods in the vicinity, an imminent firewood-shortage and the subsequent collapse of the Reichenhall salt-boiling industry had been steadily growing since the end of the sixteenth century.

The Bavarian councilors would have found ample inspiration in a contemporary standard reference work of German cameralism: "*Teutscher Fürsten-Staat*" (German Princely State) had been published in 1656. Its author, Veit Ludwig von Seckendorff, was at the time head of the "*Cammer*," the Chamber or Treasury, in the Thuringian Duchy of Saxe-Gotha. Following the collapse of the state's infrastructure during the Thirty Years' War, Duke Ernest the Pious (incidentally, a direct predecessor of the current English queen) attempted to build a modern model state in this small, heavily wooded territory. He saw himself in the role of a "*Landes- und Haus-Vater*," good house-keeper and father of the state. His program aimed at a "*reformatio vitae*," a reform of daily life on the basis of Luther's catechism. Seckendorff's book provided a detailed description of the model. Several chapters were devoted to the right use of natural resources. Seckendorff set up a rule against the over-exploitation of the forests by demanding that they should be used "*pfleglich*" (with care), so that they would provide a "steady yield" (*beständige Nutzung*). To use the forests "with care" means "to handle them in such a way that they give a steady revenue over an infinite number of years.[4] The harvest should "not exceed the capacity of the woods. Instead the forests should provide an everlasting and steady yield of wood for the use of the Prince and a continual supply of firewood and wood for other uses for the inhabitants, year after year, presently and in the future, for posterity."[5]

The fear of a severe wood shortage was rampant not only in the saline towns of the Northern Alps or the Thuringian forests, but also in many other regions of Europe. There seems to have been no general shortage of wood at that time. But there were regional supply shortfalls, especially in the centers of wood-consuming industries such as ore melting, glass-making and ship-building. These shortages were extremely worrying, because wood-transport over long distances was expensive and complicated, as it depended on suitable water-courses and rafting. Multiple regional scarcities of wood occurring simultaneously combined to create a general perception of crisis. What concerned Europeans profoundly was the prediction of a severe wood shortage, the prospect of a widespread resource crisis within the next generations, in case the devastation of the forests were to continue at a similar rate. Around the middle of the seventeenth century this anxiety had reached the capitals and the intellectual hotspots of western Europe. The problem warranted high political priority. It was widely regarded as a question of survival. The vision of a "perpetual forest" providing a "steady yield" gained sharper contours. "Conservation" or its synonym at the time, "preservation," were key terms in that quest.

A European dream

"May such Woods as do yet remain intire be carefully Preserv'd and such as are destroy'd sedulously Repair'd," wrote John Evelyn in his famous book *Sylva*, published in 1664 by the British Royal Society.[6] The *conservatio* concept of enlightened philosophy appears here in a new context. The *conservatio* of the old doctrine of Providence returns here secularized and in a new and limited context—as the duty of the whole society not to overexploit natural resources.

Evelyn describes "the furious devastation of so many goodly Woods and Forests," and points at the economic and social drivers: "The late increase of shipping, the multiplication of Glass-Works, Iron-Furnaces and the like," above all "the disproportionate spreading of Tillage" and "the destructive razing and converting of Woods to Pasture"[7] have brought about the annihilation of "the greatest magazines of the wealth and glory of this nation."[8] "This devastation," he continues, "is now become … Epidemical."[9] The root causes? Evelyn attacks head on those of his contemporaries "so miserably lost in their speculations" that they seek only "to satisfie an impious and unworthy Avarice."[10]

Can a natural resource be exploited over a long time and at the same time be substantially preserved? This question, still central to every sustainability strategy today, is Evelyn's central question. He indicates a possible answer by telling the story of his own father and the family's iron foundry in the woods at Wotton Place in Surrey. "I have heard my own Father … affirm, that a Forge, and some other Mills, to which he furnish'd much Fuel, were a means of maintaining, and increasing his Woods."[11] This "Paradox," as he called it, is resolved a few lines later. Evelyn writes that forests are "inexhaustible magazines" when they are handled "with care." Evelyn's formulation for this is "to manage Woods discreetly." This means to differentiate between individual forests, treat them in accordance with their respective ecological characteristics and "by increasing the Industry of planting." Evelyn is convinced that the human mind can impose a new order on wild nature, and that must be done, not least in the interests of the coming generations. "Let us arise then and plant, and not give it over until we have repaired the havoc" is this passionate plea illustrated by numerous examples of good practice drawn from all over Europe.

The leitmotif of his book concerns provision for the future, for "Posterity." Evelyn quotes a Latin saying: each generation is "*non sibi solus natus*"—not born for itself alone, but rather "born for Posterity." But, he adds accusingly, his own contemporaries were apparently "*fruges consumeri nati*"—born to consume the fruits of the earth.[12]

At this point, Evelyn develops an ethic for a provident and responsible society:

> I do not pretend that a man … is obliged to attend so many ages ere he fell his Trees; but I do by this infer, how highly necessary it were, that men should perpetually be planting; that so posterity might have Trees for their service … which it is impossible they should have, if we thus continue to destroy our Woods, without this providential planting in their stead, and felling what we do cut down with great discretion, and regard of the future.[13]

"Good husbandry"[14] is the central concept of Evelyn's underlying economic theory. This traditional term describes the careful, efficient, and truly "economical" landowner's use of available resources. It transcends a purely managerial approach to nature. For what is the secret of good husbandry? "To obey Nature."[15] And what is its goal? "To render the countries habitable." We are very close to the vocabulary of our modern discourse on ecology and sustainability. Evelyn's vision was an "*Elysium Britannicum*," turning his country into the "isles of the blessed." *Sylva* became a bestseller in Britain. Its enormous influence as a handbook on forestry as well as a cultural history of land use and a guide for ecological living spread to the continent.

"*La France perira faute de bois*"[16] (France will perish for lack of wood) became the alarm-call, which defined Jean-Baptiste Colbert's long and fierce campaign for a forestry reform in France. As Louis XIV's minister of finance from 1665 to 1683, Colbert oversaw the creation of the famous "*Ordonnances sur fait des Eaux et Forets*" of 1669, which summarized centuries worth of forestry policies and added new ones, often with state needs in mind. It was focused on the "conservation des bois"—the conservation of the woods—"*à perpetuité*." Again the vision of the "eternal forest" appears in the discourse on forestry.

A short look at contemporary European philosophy opens another fascinating perspective: During those years, a new concept occupied the discourse: "*Conservatio sui*"—self-preservation, survival—became the major concern of the early stage of the Enlightenment. One of the most advanced theories was put forward by the Dutch philosopher and lens-grinder Baruch Spinoza. From his works, a highly sophisticated theory of sustainability can be distilled. In 1661, the same year the council of Reichenhall debated the conservation of the woods, he moved from Amsterdam to the village of Rijnsburg near the university town of Leiden and started working on his opus magnum, *Ethica*, published posthumously in 1678.

Spinoza rejected the notion of Providence. In his philosophy "*suum esse conservare*,"[17] to persist in one's own being, was the main human drive and the starting point of all economic activities. Spinoza identified Nature with God. His fundamental formula was "*deus sive natura*": God or—if you wish—nature. There is only one universal substance that is in everlasting pulsing creation. Human beings are but reflexes of this creativity and thus part of nature. The project of human self-preservation is therefore embedded in a wider ecological context: the web of life. What does this mean for the structuring of a sustainable society? Reason requires us, says Spinoza, to see self-preservation as linked not only to the preservation of ecological integrity but also to the welfare of others. It is apparent "that men can provide for their wants much more easily by mutual help and that only by uniting their forces can they escape from the dangers that on every side beset them."[18]

At that point Spinoza brings into play the concept of "infinite duration," the endless continuation of existence. "It is in the nature of reason to perceive things under the aspect of eternity (*sub aeternitatis specie*)."[19] Here we are at the heart of sustainability. When we perceive things under the aspect of eternity, future things become equally true and relevant as the things of our present and past. In fact, we even regard as present many things that will only fully emerge in the future. Assuming responsibility for the future and future generations thus becomes a constant of our thinking.

The expertise of Hans Carl von Carlowitz

A semantic innovation with the potential to substitute "conservation" as a guiding concept for the discourse on resources and future generations first appeared in a book with the title *Sylvicultura oeconomica*, published in Leipzig in 1713.[20] Its author was Hans Carl von Carlowitz, a high-ranking senior official in Electoral Saxony's mining administration. He had just been appointed head of the "Chief Inspectorate of Mining" (*Oberbergamt*). Located in the old silver-mining town of Freiberg in the foothills of the Ore Mountains, he ran hundreds of ore mines, smelting works and hammer mills with a workforce of about 10,000 employees.

What qualified Carlowitz for his achievement? The portrait medallion featured in his book depicts him as a baroque nobleman. His head is revealed in a three-quarter profile. The furrows in his brows are deep and vertical. The mouth with its narrow lips, appears energetic, the expression serious and studious. The dark ringlets of his long French wig fall upon the iron of a decorative suit of armor, over which he has thrown a velvet cape. Wrapped around his neck he wears a light scarf. The family crest rounds off the portrait of an aristocratic personality, a cosmopolitan, a virtuoso.

Carlowitz was born in 1645 at Burg Rabenstein, a small but bulky medieval castle near Chemnitz. The Thirty Years' War, which had ravaged Electorate Saxony like no other German territory, slowly ground to a halt. His family were of ancient Saxon nobility. Over several generations the management of the forests in Saxony's Ore Mountains had been almost their exclusive domain. Hunting, forestry, and timber rafting were still closely linked. The secure supply of the mines and smelting works with wood and charcoal were of strategic importance for the economy of Electoral Saxony. Alongside water power and, of course, the power of human and animals' muscles, this resource was the principle source of energy for the extraction and smelting of metals. A veteran of the war, Carlowitz's father was promoted to the office of the Elector's "*Land-Jägermeister*"—Master of the Hunt—forest superintendent and overseer of timber rafting in one.

In spite of the "then meager times afflicting our beloved fatherland,"[21] Carlowitz's parents invested "all diligence and expense" in their son's upbringing. His education was carefully mapped out in advance as to prepare him systematically and purposefully to seek ways out of the impending resource crisis. He received a sterling humanistic education at one of the country's most eminent grammar schools, then enrolled for a year at the University of Jena where he dedicated himself to "the learning of the laws and matters of state and the study of old and recent history." But as "foreign lands are the best schools of wise performance," in 1665, at the age of 20, Carlowitz set off on an extended "*peregrinatio academica*," that over the course of five years took him across Europe. His itinerary stretched from Sweden all the way to Malta and included longer stays—"good times and diligent studies"—in Leiden, London, Paris, Rome, and Venice. Such a "grand tour" was obligatory for the sons of princes and nobles in order to broaden their general intellectual horizon and their savoir vivre, but equally to deepen their specific expert knowledge. Role models were the "*uomo universale*" and the "virtuoso," a personality combining a broad horizon and deep knowledge in special fields.

Carlowitz was in Leiden, the Dutch stronghold of early European Enlightenment, when Spinoza's pantheistic philosophy was first discussed in small circles. His stay in London coincided with the *annus mirabilis* 1666. He witnessed the aftermath of the great plague, watched the Great Fire destroy the City of London, and the Dutch navy attack the British fleet on the River Thames. Although Carlowitz neither mentioned the author's name nor quoted from the book, it is highly probable that he came in touch with circles around John Evelyn and, given his interest in forestry, became acquainted with *Sylva*. The following year Carlowitz spent in France. It was the time when Louis XIV's forestry reforms were well under way.[22] The edicts of King Louis XIV, Carlowitz wrote in his *Sylvicultura oeconomica*, already implied "*das gantze summarium*" (the complete summary) of his own book. Later on he spent some time in Venice and probably paid a visit to "Il Montello," the famous oak forest in the foothills of the Alps which for more than 200 years had supplied the "Arsenale," the shipyard of Venice, with timber. After five years of travelling, Carlowitz returned to his native Saxony and entered the state service.

In 1678, Elector Johann Georg II appointed the 33-year-old Hans Carl von Carlowitz to the position of deputy chief of the Mining Inspectorate in Freiberg. His superior was Abraham von Schönberg, a superb mining expert who ran the district with an iron fist and an innovative mind until his death in 1711. In that era the annual yield of silver and other metals rose slowly but surely, making the Ore Mountains once again one of the leading mining regions in Europe, competing with "*cerro rico*" of Potosì, the rich mountain, in the Spanish colony of Alto Perú. The silver from the Ore Mountains helped finance the blossoming of Dresden into its baroque glory, as well as August the Strong's takeover of the Polish crown in 1697, and Saxony's engagement in the Northern War between 1700 and 1706. The office in Freiberg became a nerve center of a scientific-technical revolution. Influenced by the advance of Enlightenment, various forms of esoteric knowledge were banned and gave way to "knowledge, hard-won from experience" becoming "the mistress of all activity."[23] Step by step the divining rod was substituted by techniques of prospecting based on the emerging science of geology. Alchemy was developing into chemistry. Several technical and administrative innovations were introduced. The Mines Inspectorate was deeply involved in the successful efforts being made around 1700 by Saxony to discover the secrets of porcelain production. The method of mining the ore by roasting the rock underground by fire was abandoned in favor of blasting the rock from the mountain by black powder. The smelteries and hammer mills were centralized and brought under the complete control of the state. The latter innovations were basically wood-saving measures. The specter of a resource crisis was haunting Saxony.

Apparently Carlowitz was little occupied with the operative management of the mining activities. Clearly he had free rein to focus on the forecast crisis, on practical and local measures, as well as on conceptual solutions. He was a member of a wood commission, installed by the government in Dresden during the 1680s, which surveyed the problems of the forests and of wood supply for the mines and smelteries.

Coining the term "*Nachhaltig*" in 1713

In 1713, Hans Carl von Carlowitz summed up his experiences in a long life of study, travels, research, and administrative practice in a book with the full title, in translation, of *Sylvicultura oeconomica or Economic Report and Instruction on the Cultivation of Wild Trees according to Nature.* Carlowitz combines an urgent warning against an impending timber shortage and an outline for long-term solutions with lively descriptions of the most common forest tree species. He embellishes his account with numerous quotations from the humanists, classic Latin writers, and the Bible.

The parallels to John Evelyn's *Sylva* are obvious, not only in the broad outline of the book. The illustrations—oak-felling, deforestation, charcoal kiln—even resemble Evelyn's iconography. The direct influence of Evelyn's work is clearly proven by looking at some details. Whole passages are fairly close translations of the corresponding descriptions in Evelyn's *Sylva*. Carlowitz does, however, refer to and quote from the French edicts of 1669 as a central source, where "the whole summary of our project may be found."[24]

But then Carlowitz goes beyond his British predecessor. The text gradually moves ever closer to the idea and the term of sustainability. His starting point is the key role of the resource wood, which is essential for the "conservation" of humans, as "no economy can dispense with fire and with wood."[25] He outlines the European perspective: "In Europe", he writes, "within but a few years more wood was felled than grew over many ages."[26] The outcome of this development had already been foreseen by Philip Melanchthon, the Lutheran reformer, who had prophesied that, "at the end of the world, people will suffer from great lack of wood."[27]

The impending crisis is attributed to the growing population, the impact of industrialization and the increasing greed evident in society. He criticizes that much of the thinking of his time is oriented towards short-term monetary gain—to making money. A cornfield may be harvested annually and will bring an annual return, whereas one has to wait for decades in order to obtain wood—and money—from a forest. However the relentless conversion of forest to tillage and pasture is folly. The common man is not inclined to spare the young trees because he senses that he himself will not benefit from their wood. He uses them wastefully, believing them to be inexhaustible. Thus through the sale of wood, "considerable sums of money can be earned in the short term. Yet when the woods are ruined, so the income thereof is postponed for countless years, the treasury is exhausted so that beneath the same apparent profit lays irreparable damage."[28]

The book advocates a bundle of practical measures, an efficiency revolution by means of "wood saving techniques," including improved heat insulation in house construction and the use of energy-saving smelting ovens, tiled stoves, and kitchen ranges. Carlowitz proposes planned afforestation through the "sowing and planting of native trees" and the search for "*surrogata*" for wood such as peat.

Carlowitz develops his overarching idea—that the "consumption of wood" must remain within the limits of what "*der Wald-Raum / zu zeugen und zu tragen vermag*" (the woodland is able to generate and to carry or sustain).[29] Here Carlowitz is very close to the modern English term "sustainable."

The traditional vocabulary is still dominant when he calls for a "careful" use of wood, "which is as important as our daily bread," so that a balance between re-growing and harvesting is achieved and its benefits can be enjoyed "continuously" and "perpetually." He continues: "For this reason, we must manage our economy such that we suffer from no lack of it, and where it has been clear-cut, that we apply our best efforts to ensure that new growth will replace it."[30] Proverbial wisdom is used to underline the message: "Old clothes should not be discarded until one has new ones. Just so, the stock of grown timber must not be felled until it is seen that enough has grown back to replace it."[31] Resorting to the key term of the European discourse on wood-shortage and to the central category of contemporary enlightened philosophy, Carlowitz employs "Conservation" at numerous points of his book. This Germanized form of the Latin *conservatio* happens to be identical with the English and French words. Carlowitz uses the word in connection with the conservation of timber, of the forests, of humans and even more generally of life as a whole ("*Conservation des Lebens*"). He even speaks of the "sustentation and conservation" ("*Sustentation und Conservation*") of a country, thereby employing the Latin root-word of the modern term sustainability. So far he is moving within the limits of the well-known terminology. But then he ventures into uncharted territory.

Apparently the author finds neither the word "careful" nor the Latin term *conservatio* precise and illustrative enough to express the long-term continuity of the use of natural resources. In a key moment of his book a new term appears. The sentence is long and complex. It reads: "But as the earth's underground has through labor and expenses revealed its ores, we are confronted with a scarcity of wood and charcoal, that needs to be remedied, therefore the greatest technical skills, science, diligence and management of this country must address how such a conservation and cultivation of wood can be achieved so as to make possible a continual, steady and *sustainable use*, as this is an indispensable matter, without which the country cannot maintain its Being."[32]

The first thought here is directed to the dependence of metallurgy on wood, and the impending shortage of this source of energy. Is a scenario conceivable that allows wood to be harvested but the forest to be preserved? The objective of conservation is usage, but a long-term enduring usage. To emphasize this aspect, and to further refine it, Carlowitz strings together three closely related temporal attributes: "*continuierlich*" (continual) in order to signify the regularity and permanence of the process, "*beständig*" (steady) which combines the implication of unlimited duration with locally defined stability, and finally: "*nachhaltend*" (sustaining or sustainable). The German verb "*nachhalten*" literally means: to hold (*halten*) and continue to hold after (*nach*) a certain point in time, which is to retain something for future usage. The present participle form "*nachhaltend*" signals the simultaneity of the two actions "using" and "sustaining." The specific manner of using something intends and effects through its specific course that something is preserved. The verbal expression compels action and systematic thinking. Furthermore the notion of trusteeship resonates in the word. "*To trower hand naholden*" (retaining at faithful hand) is an expression from the medieval language of law, meaning to steward in trust something for someone else for a later time.

Carlowitz's sentence includes the Latin words *conservatio* and *esse*. The message here is that without the resource of wood and its "sustaining use," "the country cannot maintain its Being," that is its very existence and its well-being. Implied is Spinoza's description of humanity's most vital task: "*Suum esse conservare*," preserving one's own self—survival. At the very beginning, sustainability is perceived as an antidote to "collapse," and although its roots are in forestry, the concept, from the beginning, had a wider resonance.

The *Sylvicultura oeconomica* presents not only the historical blueprint of the modern word "*nachhaltig*" (sustainable), but also—embryonically—the triple bottom line of sustainability, a combined consideration of ecology, economy, and social justice.

How does Carlowitz speak of nature? She is "bounteous" and "benevolent"—Mater Natura—Mother Nature. He speaks of the "wonder of vegetation," of the "life-giving power of the sun," of the "awe-worthy nourishing life-spirit" that the earthly realm contains and the "wonderous and nourishing life force" within the soil. The plant is "*corpus animatum* … a living body (*belebter Cörper*) … that grows out of the soil, draws from it its nourishment, grows bigger and multiplies."[33] The outer appearance of trees is intrinsically linked to the "inner form, signature, constellation of the sky" under which they green. Nature is "unspeakably beautiful. She is never to be comprehended" and "keeps much hidden from humans." Yet, even so, we may read in the book of nature and, through experiment, uncover "how nature plays … and contemplate her extraordinary marvels."[34]

Anticipating notions of deep ecology, it is obvious to the author that (what we would call) renewable resources consist of living beings. They are not just biomass. They require the sun for their photosynthesis. For growth and propagation they rely on the fertility of the soil and on clean water. They do not blossom and ripen everywhere and at any time. Their availability is not unlimited. They depend on the ecological integrity of their environment. They have an aesthetic and even spiritual value. They demand care and respect.

How is Carlowitz's economic thinking conceived? The starting point is the simple observation that humans no longer find themselves in the Garden of Eden. They cannot rely on nature to provide an eternal bounty. Like John Evelyn before him, Carlowitz quotes from Genesis 2:15 the divine command to "dress and keep the soil"—in Luther's translation, which Carlowitz was well acquainted with—"*bebauen und bewahren*." In following this rule humans should refrain from "acting against nature." Instead they have to "follow" her, "come to her aid" and be true housekeepers with her offerings. "Humans should replicate nature, as she best knows what is useful, necessary and profitable." The role model is the wise "*pater familiae*" (*Haus-Vater*). Economy is not about making money but about husbandry, the economical usage of the resources available. Its goal is to obtain from a minimum of resources a maximum of well-being for all involved. The notion of an economy embedded in the natural order of creation is still deeply rooted in the *Sylvicultura oeconomica*. The ideal of a morally based economy in harmony with the natural environment is clearly to be seen.

Accordingly, the book outlines a social ethics. Fundamental is the idea that everybody is entitled to "sufficient nourishment and sustenance." Included here are the "poor subjects" (*arme Unterthanen*) and the "dear posterity" (*liebe Posterität*). This refers in the first instance to basic needs (*Nothdurfft*). But Carlowitz also envisages development. In his dedication of the book to August the Strong he speaks of the "*Auffnehmen*," that is, the advancement of the country. This is specified by the terms "*Wohlfahrt, Flor und Glorie*" (welfare, flourishing, and glory).[35] Manifest in Carlowitz's thinking is the ethical principle that permeates the discourse on sustainability from the beginning to this very day: assume responsibility for future generations. Despite his noble heritage and emphasis on state needs, Carlowitz demonstrates clear concern for both nature and general social wellbeing.

Translating "Nachhaltigkeit"

Carlowitz's semantic innovation "sustaining (or sustainable) use," introduced in 1713, gradually established itself as a clearly defined concept for scientific methods of managing forests.

"*Nachhaltigkeit*" appears as a guiding principle in a decree on forestry by Duchess Anna Amalia of Saxe-Weimar in 1760, and as a fully developed and defined term in the writings of German foresters around the year 1800. To quote just one example, in 1804 Georg Ludwig Hartig, head of the Prussian Forestry Department offered a much-cited definition:

> No long-term forestry can be imagined or expected, if the yield of timber is not calculated with regard to sustainability [*Nachhaltigkeit*] … Every wise forest manager must therefore immediately evaluate the public forests for their maximum yield, but must seek to utilize them in such a way that posterity derives at least as much benefit from them as does the present generation.[36]

This is clearly an echo of Carlowitz's writing, as well as a stunning anticipation of the Brundtland Report's famous definition of sustainability from 1987.

During the nineteenth century the principle was adopted in other European countries, and the need to translate "*Nachhaltigkeit*" became urgent.[37] It was in this context that derivations of the Latin word *sustinere* were introduced. In the bilingual forestry literature of Switzerland "*Nachhaltigkeit*" was, as early as 1802 translated into French as "*produit soutenu et égal*" or "equal and sustained product." The term "*rendement soutenu*" was adopted a few years later by the founders of the French Forest Academy in Nancy, most of whom had studied in Germany. In the English-speaking countries, the term "sustained yield forestry" established itself due to the influence of German forestry experts in the service of the British colonial administration, such as Dietrich Brandis and Wilhelm Schlich.

Almost a hundred years later, in 1951, the Food and Agriculture Organization (FAO), the global nutrition organization of the United Nations, adopted its "Principles of Forest Policy." Each country, it stated, should seek "to derive in perpetuity, for the greatest number of its people, the maximum benefits available from the protective, productive and accessory values of its forests." Therefore forests should be managed in such a way as to obtain "a sustained yield." The principle—and the term—of sustainability had arrived at the United Nations.[38] Although the concept of sustainability now extends well beyond the domain of forestry, it very clearly has roots, at least in part, in the milieu of forest conservation in Britain, France, and the Germanic states.

Notes

1   This phrase is challenging to translate. It might be rendered more accurately as "sustaining use," although this phrase is less familiar in English.
2   Götz von Bülow, *Die Sudwälder von Reichenhall* (Munich: Institut der Forstlichen Forschungsanstalt, 1962), 159.
3   Ibid., 155.
4   Veit Ludwig von Seckendorff, *Teutscher Fürsten-Staat* (Aalen: Scientia Verlag, 1972; reprinted from Jena edition of 1737), 474.
5   Ibid., 471.
6   John Evelyn, *Sylva, or a Discourse on Forest-Trees and the Propagation of Timber*, in *The Writings of John Evelyn*, ed. Guy de la Bédoyère (Woodbridge: Boydell Press, 1995), 190.
7   Ibid., 319.
8   Ibid., 322.
9   Ibid., 198.
10   Ibid., 189.
11   Ibid., 320.
12   Ibid., 324.

13 Ibid., 298.
14 Ibid., 329.
15 Ibid., 327.
16 See Ulrich Grober, *Sustainability: A Cultural History*, trans. Ray Cunningham (Totnes: Green Books, 2012), 71–5.
17 Benedict de Spinoza, *The Ethics of Spinoza*, trans. R. H. M. Elwes (New York: Dover, 1988), part 4, proposition 18.
18 Ibid., part 2, proposition 44.
19 Ibid., part 4, proposition 35.
20 See Grober, *Sustainability*, 76–85.
21 This and the following quotes from: Hieronymus Joachim Wäger, *Leichenpredigt* [*Funeral Sermon*] *für Hans Carl von Carlowitz* (Dresden: Sächsische Landesbibliothek, 1714).
22 See John Croumbie Brown, *The French Forest Ordinance of 1669* (Edinburgh: Oliver & Boyd, 1883).
23 Ibid., n.p.
24 Hans Carl von Carlowitz, *Sylvicultura oeconomica, oder Haußwirthliche Nachricht und Naturmäßige Anweisung zur Wilden Baum-Zucht*, ed. J Hamberger (Munich: Oekom Verlag, 2013), 194. Unless otherwise noted, translations from the German are by Ray Cunningham.
25 Ibid., 152.
26 Ibid., 154.
27 Ibid., 160.
28 Ibid., 197 (translation modified slightly by the editor).
29 Ibid., 99.
30 Ibid., 209.
31 Ibid., 198.
32 Ibid., 216; italics added.
33 Ibid., 131.
34 Ibid., 148.
35 Ibid., 158.
36 Cited in Grober, *Sustainability*, 117.
37 See ibid., 140–54. See also, Ulrich Grober, "The Discovery of Sustainability: The Geneology of a Term," in *Theories of Sustainable Development*, eds. J. C. Enders and M. Remig (London: Routledge, 2015), 6–15.
38 *Unasylva*, vol. 5, no. 1, retrieved on December 30, 2016 from www.fao.org/docrep/x5358e/x5358 e00.htm.

References

Brown, John Croumbie. *The French Forest Ordinance of 1669*. Edinburgh: Oliver & Boyd, 1883.

Evelyn, John. *Sylva, or a Discourse on Forest-Trees and the Propagation of Timber*. In *The Writings of John Evelyn*, ed. Guy de la Bédoyère. Woodbridge: Boydell Press, 1995.

Grober, Ulrich. "The Discovery of Sustainability: The Geneology of a Term." In *Theories of Sustainable Development*, eds. J. C. Enders and M. Remig, 6–15. London: Routledge, 2015.

Grober, Ulrich. *Sustainability: A Cultural History*, trans. Ray Cunningham. Totnes: Green Books, 2012.

Spinoza, Benedict de. *The Ethics of Spinoza*, trans. R. H. M. Elwes. New York: Dover, 1988.

Unasylva, vol. 5, no. 1, retrieved on December 30, 2016 from www.fao.org/docrep/x5358e/x5358 e00.htm.

Von Bülow, Götz. *Die Sudwälder von Reichenhall*. Munich: Institut der Forstlichen Forschungsanstalt, 1962.

Von Carlowitz, Hans Carl. *Sylvicultura oeconomica, oder Haußwirthliche Nachricht und Naturmäßige Anweisung zur Wilden Baum-Zucht*, ed. J Hamberger. Munich: Oekom Verlag, 2013.

Von Seckendorff, Veit Ludwig. *Teutscher Fürsten-Staat*. Aalen: Scientia Verlag, 1972; reprinted from Jena edition of 1737.

Wäger, Hieronymus Joachim. *Leichenpredigt* [*Funeral Sermon*] *für Hans Carl von Carlowitz*. Dresden: Sächsische Landesbibliothek, 1714.

8

THE INDUSTRIAL REVOLUTION
Social costs and social change

Emma Griffin

Most historians agree that the Industrial Revolution was a defining moment in British history, but what were the social consequences of this landmark historical event? What benefits – if any – did industrialisation and mechanisation bring to the nameless men, women and children who worked in the factories and made it all happen? These are questions charged with political overtones, and they have polarised historians who have inevitably interpreted the disappearance of traditional society and the advent of capitalism in very different ways. In this chapter, we shall consider these debates and ask what happened to both working people's incomes and their quality of life when Britain transitioned from an agrarian to an industrial economy.

Let us start, however, with an account of what we mean by the "Industrial Revolution". There are many different pathways to industrialisation: different nations will follow their own route according to their own unique resource base and historical development. In Britain, however, we must look primarily to coal. Coal provided the energy needed for cooking, heating, and powering industry, and its significance can be most easily grasped by thinking about the alternatives and their limitations. If energy is not derived from coal, it might instead be harnessed from the wind or from rivers by means of windmills and waterwheels. Alternatively, energy may be obtained from the soil, either directly through the burning of wood, or more indirectly, by the muscles of humans or animals – in this instance, the energy is provided by consuming food or fodder, which is largely derived from the soil. But the landmass, rivers, and wind energy of any given country is fixed. Though scientific and technological improvement might enable us to eke a little more power from our soil, rivers, and wind, the energy that can be obtained from these sources cannot be increased indefinitely. These energy constraints effectively placed a cap on the extent of economic growth that could be achieved, and explain why economic growth in pre-industrial societies always occurred at a very slow pace.

When industry began to burn coal rather than wood, the age-old constraints that had placed a ceiling on the development of industry in previous centuries were at a stroke removed, and new and previously unimaginable rates of growth were now attainable. By digging under the soil, vast new expanses of economic possibility were opened; an apparently limitless source of fuel permitted a long-term rise in manufacturing on a previously unimaginable scale. No longer were the needs of industry and human sustenance in competition for the same set of resources: the coalmines could provide fuel for industrial processes, cooking and heating, and the land could be used to grow crops to feed the growing population that an expanding industrial sector required. Though coal is of course itself a finite resource, in the context of the period, and in comparison with the fuel sources that had preceded it, the opportunities it presented for economic growth were seemingly endless.[1]

The Industrial Revolution, therefore, was a watermark in the history of Britain. It refers to the moment when the nation stopped trying to make all its goods by hand and by organic energy forms, and started to use coal and machinery to do the work instead. While different nations have expanded their energy base in different ways, the trend towards greater energy use is universal. And as we have been learning in recent decades, our increasing energy consumption has placed considerable stress upon our planet and contributed to a worrying rise in global temperatures.[2] There is now clear evidence that economic growth goes hand in hand with environmental costs, but what is the human cost of modern economic development? Did industrialisation and city growth undermine our quality of life? Did it make the rich richer while the poor simply grew more impoverished? Or did economic modernisation trigger improvements in living standards for the bulk of ordinary people as well as for the social elites?

This question is one that has preoccupied commentators and historians for almost two hundred years. In 1839, the historian Thomas Carlyle coined the expression the "Condition of England",[3] and Victorian writers and thinkers Charles Dickens, Elizabeth Gaskell, Benjamin Disraeli, Friedrich Engels to Karl Marx continued to debate the Condition of England throughout the 1840s and beyond.[4] These writers expressed a complex web of concerns about the ways in which rapid industrial and urban growth were changing the fabric of traditional society and tended to believe that the condition of the labouring poor was worsening as part of this process. Carlyle, for example, thought that the "working body of this rich English Nation has sunk or is fast sinking into a state, to which, all sides of it considered, there was literally never any parallel".[5] Friedrich Engels in his influential survey of the working class in Manchester argued that not only were the workers' living and working conditions worse than ever before, but that their wages were lower too: pre-industrial workers, he wrote, had lived a "passably comfortable existence … and their material position was far better than that of their successors".[6]

It perhaps goes without saying that much of the vitality of the Condition of England debate derived from the few dissenting voices that pictured the working class enjoying new levels of prosperity and contentment. The defenders of industrialisation constituted an equally diverse alliance of writers and thinkers – Harriet Martineau, Andrew Ure, Edward Baines, and John Rickman, to name a few. Edwin Chadwick, whose pioneering research into the living conditions of urban slums formed the basis of the nation's earliest public health measures, nonetheless believed that in strictly monetary terms, the new town dwellers were sharing in the nation's newfound wealth: "wages, or the means of obtaining the necessaries of life for the whole mass of the labouring community", he wrote, "have advanced, and the comforts within the reach of the labouring classes have increased with the late increase of population".[7] Yet for the most part his contemporaries were not convinced. Just beneath the veneer of Victorian economic and cultural progress, they feared, laid a mass of unnecessary human suffering. Worse still, their comforts were bought at the expense of the workers' well-being.

And just as contemporaries were deeply divided over the relative gains of industrialisation for working people, so did historians continue to disagree throughout the twentieth century. It is perhaps significant that Arnold Toynbee, who first popularised the term "Industrial Revolution" in the 1880s, was also the first historian to present a decisive argument about the consequences of this revolution for working-class living standards. Toynbee was emphatic that the effects of industrialisation had been entirely deleterious for the men, women, and children whose labour had underpinned it: "The steam-engine, the spinning-jenny, the power-loom", he wrote, "had torn up the population by the roots ... The effects of the Industrial Revolution prove that free competition may produce wealth without producing well-being."[8] In this way, Toynbee produced a hugely influential interpretation of the Industrial Revolution as social catastrophe that informed writing about working-class living standards in Britain for several years.

In the decades following the publication of Toynbee's *Lectures*, historical opinion divided into two opposing camps, the "optimists" and the "pessimists", and this dichotomy continued to frame historical debate throughout the twentieth century. In the early twentieth century, the popular historians Sidney and Beatrice Webb and J. L and Barbara Hammond echoed Toynbee's view of the Industrial Revolution as social disaster, bringing no immediate gains for the labouring poor.[9] Against this view, however, the "optimists" argued that by raising incomes, the Industrial Revolution improved the lot of the labouring poor. For example, Clapham's *Economic History of Modern Britain*, published in three volumes between 1926 and 1928, argued that the Industrial Revolution was responsible for steady rises in the standard of living.[10] This optimistic view was strongly restated by Thomas Ashton in the 1950s, by R. M. Hartwell, Phyllis Deane, and W. A. Cole in the 1960s, and by Lindert and Williamson in the early 1980s.[11] Yet the "pessimists" continued to dispute that the working classes shared in the rising wealth of the nation, and to argue that whatever the paltry financial gains realised by the working classes, they were more than outweighed by the disadvantages – the urban slums, the relentless discipline of the factory, the breaking of traditional family and community bonds – that the Industrial Revolution also brought in its wake. In the 1960s, two highly influential Marxist historians, Eric Hobsbawm and E. P. Thompson, both took a bleak view of the impact of the Industrial Revolution on working-class living standards, and breathed powerful new life into the pessimistic interpretation.[12]

In the past two decades, analyses of the social effects of industrialisation have become considerably more sophisticated, and historians have moved away from measurements of real wages and explored new and innovative ways in which the people's living standards may be measured. Death rates;[13] infant mortality;[14] life expectancy;[15] the consumption of "luxuries", such as tea, coffee and tobacco;[16] body heights;[17] and complex measures of "well-being" incorporating mortality, heights, as well as political rights and literacy levels[18] have all been exploited to produce a detailed picture of the ways in which working people's lives changed during the Industrial Revolution.

Yet though today's historians attempt to answer old questions in new ways, there is rather less that is new about their conclusions. With the exception of Gregory Clark, most writers tend to see the industrialising world in the same way.[19] However living standards are measured, historians report stagnation or decline. Evidence of modest improvement is gloomily dismissed as a paltry recompense for the labouring families that had done the most to create the substantial economic growth that occurred over the period. And even these modest improvements are reduced to nothing when the deteriorating industrial environment is brought into the balance. In the final analysis, today's intellectuals understand the Industrial Revolution in much the same way as the educated elites who lived through it. Historians, like contemporaries, evince a degree of guilt over the price that others paid for the comforts and privileges that we enjoy in a mature industrial society. It has been a very long time since the critics of industrialisation could plausibly deny the long-term benefits of industrial growth. Nonetheless, even the most circumspect continue to claim that these long-term gains must have been bought with the blood, sweat, and tears of the workers who experienced at first hand its grinding effects.

Yet there is a problem with this account of deteriorating living standards. Implicit in all these interpretations is the suggestion that something was lost when Britain transitioned from an agrarian to an urban and industrial economy. Yet when we take a longer-term view and consider the achievements of the pre-industrial economy, there is less cause to celebrate the older ways of life. Certainly, there is a large literature indicating that in the settled agricultural economies that preceded industrialisation real wages and calorie availability were consistently lower than in industrialised economies. Across the ancient world, even the wealthiest societies struggled to produce enough food to ensure that all members were adequately fed.[20] Despite some economic growth by the medieval period, there had been no sustained improvement in living standards, with both peasant economies and market economies still struggling to produce enough wealth to ensure adequate living standards for all.[21] Long-term analyses of real wage growth suggest that on the eve of industrialisation in Britain, wages were at their highest point, yet even here living standards and calorie availability were considerably lower than those established a century later once the Industrial Revolution had taken place.[22]

Not only were material living standards low in agrarian societies, other measurable indicators of quality of life such as morbidity and mortality testify to a depressed standard of living in pre-industrial societies. In Sweden, life expectancy at birth rose no higher than thirty-seven years throughout the eighteenth and nineteenth centuries.[23] In Britain, it hovered around forty years of age.[24] Such low levels of life expectancy were in large part driven by very high levels of infant and child mortality. Estimates of child and infant mortality for those societies for which robust demographic exist suggest that somewhere between 30 and 40 per cent of the population died before the age of fifteen.[25] Such patterns were replicated throughout the ancient and early modern world and are also evident in non-industrialised sectors of today's world. Low standards of living have ever characterised pre-modern societies. Low wages, poor diets, and high mortality rates have always prevailed in agricultural economies, and it is important to recognise that there are no known pre-industrial societies that have bucked this trend.

Establishing levels of welfare in hunter–gatherer and semi-agricultural societies is more difficult. There is some evidence to suggest that such societies were more equal, and therefore may have been more successful at protecting their members from hunger,[26] though these claims can be pressed only so far. There is considerable diversity between different hunter–gatherer societies, rendering general claims suspect, and gathering robust evidence about patterns of diet and mortality in societies untouched by market economies is extremely difficult. That said, the more equal nature of hunter–gatherers and semi-agriculturalists enabled at least some societies to sustain better diets at the population level. For instance, the coastal First Peoples of British Columbia enjoyed an abundance of food and a high population density pre-contact (1780s), although this relative plenty did not improve the quality of health care.[27] Infant and child mortality was always high in pre-industrial communities, which meant higher mortality levels than are recorded in modern, industrial societies.[28] Some pre-industrial societies persisted for long periods of time and could be said to have achieved a high degree of sustainability, but life expectancy and the modern indicators of living standard were comparatively low.

In pointing out that living standards were low, it should not be assumed that agrarian economies were not capable of growth or of change. In fact, there is good evidence that pre-modern market economies did grow. It is rather that the growth in pre-industrial economies was never sufficient to effect a permanent and meaningful uplift in the population's living standards, and tended to occur over such a long time-frame that commentators were not generally aware of its existence.[29] Historians of medieval Britain have sketched out a period of sustained expansion between the eleventh and the mid-thirteenth century; further gains were made over the sixteenth and seventeenth centuries as well.[30] But no matter what advances were made during these earlier periods of prosperity, they were never sufficient to bring about significant improvements in living standards. The limited nature of economic progress prior to industrialisation can be grasped by considering Gregory Clark's estimates of the craftsman's real wage over the six centuries between 1200 and 1800.[31] His figures indicate sizeable fluctuations in the real wage over the period yet no long term increase. At the end of the period, a builder's wage bought him little more in the way of food, housing, and clothing than had been enjoyed by his medieval ancestors, leaving large numbers eking out a precarious living on the margins of a decent existence. Why had economic growth not resulted in any substantial gains in living standards? The difficulty was that earlier periods of economic growth had always given rise to demographic growth; so although the economy was considerably larger in 1800 than it had been in 1200, so too was the population.[32] The outcome was an economy capable of feeding larger numbers, but not able to feed them very much better. And this was the pattern that had prevailed across the globe prior to the onset of industrialisation. *Homo sapiens* emerged as a sub-species around 200,000 years ago. Yet throughout this very long period no human society succeeded in decisively and permanently protecting every one of its members from the threat of an empty belly.[33]

With the coming of the Industrial Revolution, Britain was the first market economy to break out of this older cycle of limited growth and sustained improvements in living standards were established. By the end of the nineteenth century the value of the real wage was three times above its value in 1800 and wages were higher than at any previous point in history. Not only did the real wage steadily increase over the nineteenth century, but these increases were maintained and extended in the years that followed: by the end of the twentieth century real wages had risen a further fourfold.[34] And in this respect, the nineteenth-century economy was behaving in a way that was fundamentally different to the way it had ever behaved previously. Although the population was continuing to expand, economic growth was now outstripping population growth by a comfortable margin, giving rise to a prolonged upward trend in living standards. The past two centuries have seen material gains for every man, woman and child that were simply unimaginable for earlier generations, and in this respect the achievements of industrialisation stand out as an event of great historical significance.

Yet the question still remains: when did ordinary people start to share in the gains? Much of the debate about working-class living standards has turned upon the matter of timing. All agree that real wages began to rise after 1850,[35] but the onset of industrialisation is usually dated to the late eighteenth century, and this, it is argued, left at least two generations of workers toiling away in the industrial heartlands without taking home an equitable share of the nation's gains. One way into these questions is by thinking about food availability in the pre-1850 period.

While we inevitably lack detailed information about food production and distribution in the early nineteenth century, we do possess a valuable collection of family budgets collected in the 1790s and the 1840s and 1850s by gentlemen investigators interested in the lives of their poorer neighbours. Their research followed the same basic formula: information about income, the composition of the household, and the sum of money spent weekly on rent and food was gathered and tabulated. Most, though not all, researchers also collected information about weekly income, and some tried to include items such as clothes and shoes, which were bought on an intermittent rather than weekly basis.

Using this data, we are able to consider when and where working people's diets began to improve. We shall ask simply: what proportion of income did these families spend on food? And what proportion of income did they spend on bread (or flour)? According to Engel's law, poor families spend the greater share of their income on the cheapest, energy-dense foods, and as income rises, the proportion of income spent on food starts to decline. Examining expenditure by looking at the proportion spent on food and at the way in which spending was allocated across the food groups is a relatively simple procedure which provides a crude measure of the economic well-being of the family and permits us to compare and contrast the fortunes of the different workers during the first phase of industrialisation.

Let us turn first to the agricultural workforce. The situation for agricultural labourers before 1850 was grim. In the late eighteenth century, farm labourers were on average spending 75 per cent of their income on food. Nearly two-thirds of all food spending (62 per cent) was devoted to bread, meaning that over half of all income (52. 5 per cent) was spent on bread alone. This bread was almost always made by hand and rarely bought as a pre-baked loaf.[36]

This picture changed relatively little in the following half-century. A further 84 budgets were collected during the 1830s and 1840s, and these show a broadly similar picture. The overall proportion of household income devoted to food remained high; 75 per cent of family income was devoted to food. Once again, the proportion of income spent on bread was high: 71 per cent of food spending was on bread, with the result that 55.5 per cent of all income was spent on this one item alone.[37] The overall picture, then, is clear: the diets of agricultural workers were both highly restricted and remained largely unchanged down to 1850.

Turning to the miners and industrial workers, matters look dramatically different. Let us consider the miners first. Whereas farm-workers in the 1840s were spending 75 per cent of their income on food, miners were spending 58 per cent of their income on food, and only 40 per cent of their food expenditure was on bread. The total proportion of their income devoted to this staple was about 25 per cent − less than half the proportion paid by agricultural labourers.[38] The pattern among factory workers was broadly similar: 60 per cent of their income was spent on food, and of this 36 per cent was spent on bread. The overall proportion of family income that was spent on bread was 23 per cent.[39]

The family budgets demonstrate a clear difference between agricultural and industrial families in the proportion of income devoted to food and bread. Before proceeding further, it is necessary to explore the possible reasons for this difference. Because food is a necessary good with a low price elasticity of demand, the proportion of income spent on food will decrease as income rises. However, the proportion of income spent on food is also determined to some degree by family size. So in order to establish that the evidence presented here does capture differences in income across our three occupational groups, we need to control for family size.

In fact, average family was broadly similar across the occupational groups, always oscillating around four children per household. By contrast, incomes did vary significantly across the three occupational groups. In agricultural families in the 1830s and 1840s, the weekly household income was a meagre 11s 6d. The average family income of a mining family was nearly three times this − 29 shillings. In the manufacturing families, it was higher again − over 33 shillings. Factoring for family size, this provided a weekly income of 5s 6d per person in both mining and manufacturing families, more than double the 2s 3d per person that was available in the agricultural families.

Higher family incomes in the mining and manufacturing families had two components. In the first instance, male wages were significantly higher. The mean male wage in agriculture was 10 shillings a week, and there was very little variance in the wage − the median was 10 shillings too. In mining it was 16 shillings and in the factories it was 18s 6d. Industrial wages were also more widely dispersed around the mean. The standard deviation for male wages was 2.47 in agriculture; in mining it was 18. The rest of the difference in family incomes was owing to the contribution made by child workers. In the industrial districts, children started work as young as six or seven and stayed at home until they married: their earnings made up around 40 per cent of family earnings. In agriculture, children started working later and they left home earlier, so their contribution to the family coffers was less: they contributed between 10 and 15 per cent of the total. As with male earnings, there was much greater variance in child earnings in the industrial sectors than in the agricultural. Where children were very young, they made no contribution to the family income, but large numbers of children of working age had the potential to push family earnings considerably higher than the breadwinner's wage – a situation that never obtained among the rural families. Variance in the earnings of both male breadwinners and their children in the industrial sector gave rise to a much more diverse range of experiences. Although the average manufacturing family earned 33 shillings a week, the sample contains several who were earning nearly three times that amount: in each case there were seven or eight children living at home. These outlying, large and high-earning families inevitably serve to push up the average family income for the sector, though the median family income – 23 shillings for both mining and manufacturing families – was still double the median of the agricultural families.

The information collected by our nineteenth-century researchers on working–class budgets permits us to identify some clear and important differences between the industrial and rural regions. In agricultural areas, family incomes were low and provided for a very limited diet based largely on bread, with occasional scraps of meat, butter, and cheese, and very small quantities of tea and sugar. There was no discernible improvement between the 1790s and 1850, and no evidence that allotments or self-provisioning made good these deficiencies. Furthermore, male wages, children's earnings, and household income were all very tightly clustered around the mean, with the consequence that experiences among the rural poor were remarkably homogeneous. The information about incomes and diets among workers in the industrial sector paints a very different picture. In the 1830s and 1840s, families in mining and factory districts earned higher incomes than their rural counterparts, owing to a combination of higher male wages and higher children's earnings. Families invested much of that extra income in food and were therefore able to consume a significantly better diet. Experiences among this sector were also much more diverse with a few families living on a per capita income that was on a par with the agricultural workers, a few earning four or five times as much, and the rest scattered widely between these two extremes.

The budget data provide a fairly clear account of the different diets consumed by agricultural workers on the one hand and industrial workers on the other, but do these differences reflect genuine differences in diet, or could they be a reflection of the ways in which information about diets was collected? Numerous individuals collected the budgets at different times and for different reasons, and we do not know why particular surveys were conducted at that particular point in time. If investigators interrogated farm-workers during the bad years and the industrial workers during the boom times, we would expect their evidence to show that the industrial workers ate the better diet. This would be a reflection of the original data-gathering exercise rather than of an underlying social reality. It is thus important to contrast the information contained within the budgets with alternative sources of evidence about diets and welfare during the early period of industrialisation.

In addition to the budget data, we have a wealth of autobiographical material covering the period 1750–1850 in which working people reflected upon their living standards at different points of their lives. As Britain was yet a poor society with high levels of hunger and want, the subject of food was one to which autobiographers returned time and again, thus the autobiographical record contains a rich seam of relevant evidence that historians are yet to exploit.

The most salient feature of this autobiographical material is the sheer extent of hunger before 1850. In all, just over one-third of the writers – 115 in all – recalled experiencing hunger at some point in their life. In addition to providing a global figure for the extent of hunger, however, we can also use the autobiographies to analyse the distribution of hunger among different occupational groups. It is clear that the autobiographical evidence provides strong confirmation for the existence of different diets for these two groups as illustrated by the family budgets. Among the industrial and mining families, around 18 per cent of writers recollected having experienced hunger at some point in their life. In the agricultural families this figure was more than double – 42 per cent.

The agreement between the two sets of sources – the investigations and the autobiographies – is comforting. With higher family incomes and a lower proportion of income spent on food we would expect the industrial families to be at a lower risk of going hungry than their agricultural counterparts. But if we look more closely at the context and causes of hunger among the autobiographies a number of conflicts with the budget data start to emerge. Firstly, the profile of hunger looks different. In the rural context, though hunger is concentrated in childhood, it nonetheless continued across the life cycle. Children were at the greatest risk of going hungry, but adolescents and even adult males were also at risk of being unable to procure sufficient food. This finding is not surprising given the evidence contained in the family budgets. As we saw there, the margins were so tight in the agricultural families that one bad harvest leading to a spike in food prices could have a catastrophic effect on the whole family. In the industrial districts, however, just one writer recalled going hungry as an adult and none during adolescence.[40] In other words, hunger has been almost entirely relocated as a childhood experience. Furthermore, whereas most of the hunger in agricultural families can be explained by low incomes, this does not appear to be a significant factor in families deriving their income from industry. Two of the fifteen cases of were owing to low family incomes – both were prompted when a male breadwinner was temporarily unemployed,[41] and one was owing to the death of a wage-earning father.[42] In none of the remaining twelve cases where an autobiographer living in an industrial region recalled going hungry were low incomes given as a cause. The question thus arises, why had industrialisation not managed to eliminate hunger altogether? Why did the higher wages that were available not prevent children from still going hungry?

The answer is that a new cause had emerged, and its origin lay not in the level of family income, but in its use and distribution. These individuals went hungry not because their fathers were unable to earn sufficient income, but rather owing to a breakdown in the process of transforming male earnings into food for the family. A fairly colourful array of problems arises: two fathers had deserted their family, another was in prison for assault, another was unable to provide owing to a series of poor business decisions, and in the fifth case a mother and stepfather were simply failing to make proper provision for the children.[43] The single greatest cause of hunger, however, was drunkenness: seven autobiographers reported that their own or their father's drinking had at some point in their lives left them (or their families) without food to eat.[44] Had these twelve adults made regular provision for their families, the overall proportion of writers from the industrial districts experiencing hunger would drop to just 4 per cent – a vivid illustration both of the remarkable power of the new industrial sector to raise family incomes and lift families out of poverty, and of the hopeless inadequacy of the breadwinner model as a mechanism for achieving these aims.

And this takes us to the heart of what happened to working people with the onset of industrialisation and helps to explain why there has been so much confusion and disagreement over the effect of industrialisation on working people ever since the process begun at the end of the eighteenth century. Whether we turn to the family budgets or the autobiographical data, we find that rural Britain was largely unchanging and remarkably equal, but only because the great majority of workers had so miserably little in the first place. The incomes of industrial families were, on average, rising, but they were also much more divergent. Furthermore, when we look at the autobiographies we see another layer of complexity superimposed upon the industrial experience. As male incomes rose, their breadwinning become more unreliable, which created new forms of inequality lying at the heart of the family. The evidence is clear: industrialisation did help to raise incomes, but it also ushered in a far more complex, and unequal, society than that which it replaced. And the pattern that emerged in Britain during the late eighteenth and nineteenth centuries has since been repeated in many other parts of the globe. Industrialised economies are always both larger and more complex than the agricultural or hunter–gatherer economies that preceded. As a result, their capacity to generate greater wealth does not in itself guarantee welfare gains for every member of society.

There has been a strong tradition in historical writing to romanticise the simpler and more sustainable pace of life in pre-industrial societies. It is certainly correct that modern industrial societies are far less sustainable than anything the world has known previously. The difficulty, as we have seen here, is that that earlier economic models maintained their sustainability through a combination of higher mortality rates and lower living standards. They thus do not provide a template that can be adopted wholesale in the modern world where much lower mortality rates have become the norm. At the same time, however, we clearly have to reject the notion that industrialisation ushered in a simple process of linear progress. Industrialisation did raise wages but it was not simply an economic process. It was also a socio-cultural process and along with higher male wages came a series of changes that resulted in a society that was not only richer, but that was also more unequal. The twentieth century has shown us that political solutions – minimum wages, welfare states, and national health services, and so forth – have been necessary to ensure that economic growth guarantees welfare gains for all. The Industrial Revolution increased wealth, but it took political will to improve living standards for all as well.

Notes

1. E. A. Wrigley, "The Transition to an Advanced Organic Economy: Half a Millennium of English Agriculture", *Economic History Review* 59, no. 3 (2006) 435–80; Emma Griffin, *A Short History of the Industrial Revolution* (London: Palgrave, 2010).

2. IPCC, *Climate Change 2014: Synthesis Report*, Contribution of Working Groups I, II and III to the Fifth Assessment Report of the Intergovernmental Panel on Climate Change, R. K. Pachauri and L.A. Meyer (eds) (Geneva: IPCC, 2014).

3. Thomas Carlyle, *Past and Present*, 2nd edition (London, 1845).

4. See the discussion in Emma Griffin, *Liberty's Dawn: A People's History of the Industrial Revolution* (New Haven, CT: Yale University Press, 2013).

5. Carlyle, *Past and Present*, 4.

6. Friedrich Engels, *The Condition of the Working Class in England*, ed. with introduction Victor G. Kiernan (Harmondsworth: Penguin, 1987), 51.

7. Quoted in T. S. Ashton, "The Standard of Life of the Workers of England, 1790–1830", *Journal of Economic History*, Supplement IX (1949), 19–38, quote p. 20.

8. Arnold Toynbee, *Lectures on the Industrial Revolution of the Eighteenth Century in England* (London, 1884), 5.

9. Sidney Webb and Beatrice Webb, *Industrial Democracy*, 2 vols (London, 1898); J. L. Hammond and Barbara Hammond, *The Town Labourer, 1760–1832* (London, 1917); J. L. Hammond and Barbara Hammond, *The Skilled Labourer, 1760–1832* (London, 1919).

10. J. H. Clapham, *An Economic History of Modern Britain*, i (Cambridge: Cambridge University Press, 1926), 128, 561.

11. Phyllis Deane and W. A. Cole, *British Economic Growth, 1688–1959* (Cambridge, 1964), 27; R. M. Hartwell, "The rising standard of living in England, 1800–1850", *Economic History Review*, 1961, 397–416; Peter H. Lindert and Jeffrey G. Williamson, "English Workers' Living Standards During the Industrial Revolution: A New look", *Economic History Review*, 36, no. 1 (1983), 1–25.

12. E. J. Hobsbawm, "The British Standard of Living, 1790–1850", *Economic History Review*, 10 (1957), 46–68; E. P. Thompson, *The Making of the English Working Class* (London, 1963).

13 Robert Woods, *The Demography of Victorian England and Wales* (Cambridge, 2000), table 9.3, and p. 365.

14 P. Huck, "Infant Mortality and the Living Standards of English Workers During the Industrial Revolution", *Journal of Economic History*, 55, no. 3 (1995), 528–50.

15 Simon Szreter and Graham Mooney, "Urbanisation, Mortality, and the Standard of Living Debate", *Economic History Review*, 51, no. 1, (1998), table 6, p. 104.

16 Joel Mokyr, "Is There Still Life in the Pessimist Case? Consumption During the Industrial Revolution, 1790–1850", *Journal of Economic History*, 48, no. 1 (1988), 69–9, 90.

17 Roderick Floud et al., *The Changing Body: Health, Nutrition, and Human Development in the Western World since* (Cambridge University Press, 2011).

18 Nicholas F. R. Crafts, "Some Dimensions of the 'Quality of Life' During the British Industrial Revolution", *Economic History Review*, 50 (1997) 617–39.

19 Gregory Clark, "The Condition of the Working-Class in England, 1209–2004", *Journal of Political Economy*, 113, no. 6 (2005), 1307–1340; Idem, "Farm Wages and Living Standards in the Industrial Revolution: England, 1670–1869", *Economic History Review*, 53, no. 3 (2001), 477–505.

20 Loomis, William T., *Wages, Welfare Costs and Inflation in Classical Athens* (Ann Arbor, MI: University of Michigan Press, 1998); Peter H. Lindert and Jeffrey G. Williamson, "Measuring Ancient Inequality", Working Paper No. 13550, NBER, October 2007; Alfonso Moreno, *Feeding the Democracy: The Athenian Grain Supply in the Fifth and Fourth Centuries BC* (Oxford: Oxford University Press, 2007); Ian Morris, "Economic Growth in Ancient Greece", *Journal of Institutional and Theoretical Economics* 160 (2004), 709–42; Ian Morris, "Archaeology, Standards of Living, and Greek Economic History", in *The Ancient Economy: Evidence and Models*, ed. Joseph G. Manning and Ian Morris (Stanford, CA: Stanford University Press, 2005), 91–126; Robert C. Allen, "How Prosperous Were the Romans? Evidence from Diocletian's Price Edict (301 AD)", University of Oxford, Department of Economics, Discussion Paper Series No. 363, October 2007; Walter Scheidel, "Real Wages in Early Economies: Evidence for Living Standards from 1800 BCE to 1300 CE", *Journal of the Economic and Social History of the Orient*, 53, no. 3 (2010), 425–62.

21 Süleyman Özmucur and Sevket Pamuk, "Real Wages and Standards of Living in the Ottoman Empire, 1489–1914", *Journal of Economic History* 62 (2002), 293–321; Warren Treadgold, "The Struggle for Survival (641–780)", in *The Oxford History of Byzantium*, ed. Cyril Mango (Oxford: Oxford University Press, 2002), 129–50; Cecile Morrisson, and Jean-Claude Cheynet, "Prices and Wages in the Byzantine World", in *The Economic History of Byzantium: From the Seventh through the Fifteenth Century*, ed. Angeliki Laiou (Dumbarton Oaks, 2002), 807–70; Branko Milanovic, "An Estimate of Average Income and Inequality in Byzantium around Year 1000", *Review of Income and Wealth* 52 (2006), 449–70.

22 Jan Luiten Van Zanden, "Wages and the Standards of Living in Europe, 1500–1800", *European Review of Economic History* 3 (1999), 175–98; Robert C. Allen, "The Great Divergence: Wages and Prices from the Middle Ages to the First World War", *Explorations in Economic History*, 38 (2001), 411–47; Robert C. Allen, "Real Wages in Europe and Asia: A First Look at the Long-Term Patterns", in *Living Standards in the Past: New Perspectives on Well-Being in Asia and Europe*, ed. Robert C. Allen, Tommy Bengtsson, and Martin Dribe (Oxford: Oxford University Press, 2005), 111–30; Jean-Pascal et al., *Wages, Prices, and Living Standards in China, Japan, and Europe, 1738–1925*, Working Paper No. 1 (GPIH, October 2005).

23 Massimo Livi Bacci, et al., eds. *The Population of Europe: A History*, trans. Cynthia De Nardi Ipsen and Carl Ipsen (Oxford, 2000), table 6.4, p. 135.

24 E. A. Wrigley, R. S. Schofield, Roger Schofield, *The Population History of England 1541–1871* (Cambridge: Cambridge University Press, 1989).

25 Dionysios C. Stathakopoulos, *Famine and Pestilence in the Late Roman and Early Byzantine Empire: A Systematic Survey of Subsistence Crises and Epidemics* (Aldershot: Ashgate, 2004); J. C. Riley, *Rising Life Expectancy: A Global History* (Cambridge: Cambridge University Press, 2001); Robert William Fogel, *The Escape from Hunger and Premature Death, 1700–2100: Europe, America, and the Third World* (Cambridge, 2004); R. Schofield, D. Reher, S. Bideau, *The Decline of Mortality in Europe* (Oxford: Oxford University Press, 1991); Lester K. Little, ed., *Plague and the End of Antiquity: The Pandemic of 541–750* (Cambridge: Cambridge University Press, 2007).

26 This thesis was originally laid out in: Marshall Sahlins, *Stone Age Economics* (Chicago, IL: Aldine, 1972).

27 See Nancy J. Turner, *Ancient Pathways, Ancestral Knowledge: Ethnobotany and Ecological Wisdom of Indigenous Peoples of Northwestern North America* (McGill-Queen's University Press, 2014).

28   K. Hill, A.M. Hurtado, R.S. Walker, "High Adult Mortality among Hiwi Hunter–Gatherers: Implications for Human Evolution", *Journal of Human Evolution*, 52, no. 4 (2007) 443–54; Oskar Burger, Annette Baudisch, and James W. Vaupel, "Human Mortality Improvement in Evolutionary Context", *Proceedings of the National Academy of Sciences* 109 (2012), 44.

29   See Anthony Brewer, *The Making of the Classical Theory of Economic Growth* (New York: Routledge, 2010).

30   Graeme Donald Snooks, *Was the Industrial Revolution Necessary?* (New York: Routledge, 1994), figure 3.3, p. 65.

31   Clark, "Condition of the Working Class", table A2, pp. 1324–5.

32   Jack A. Goldstone, "Efflorescences and Economic Growth in World History: Rethinking the 'Rise of the West' and the Industrial Revolution", *Journal of World History*, 13, no. 2 (2002), 323–389.

33   E. A Wrigley, *People, Cities and Wealth* (New York, 1987), esp. 2–4, 21–2; Charles I. Jones, "Was an Industrial Revolution Inevitable? Economic Growth Over the Very Long Run", *Advances in Macroeconomics* 1, no. 2 (2001).

34   Clark, "Condition of the Working Class", table A2, pp. 1324–5.

35   C. H. Feinstein, "Capital Formation in Great Britain", in *The Cambridge Economic History of Europe*, ed. P. Mathias and M. M. Postan, i (Cambridge, 1978), 28–96; R. C. O. Matthews, C. H. Feinstein, J. C. Odling-Smee, *British Economic Growth, 1856–1973* (Stanford, 1982).

36   Sir F. M. Eden (A. G. L. Rogers, ed.), *The State of the Poor: A History of the Labouring Classes in England, with Parochial Reports* (London, 1928); David Davies, *The Case of the Labourers in Husbandry Stated and Considered* (London, 1795).

37   *Poor Law Amendment Act, Weekly Diet* (Parliamentary Papers [hereafter P.P.] 1837/8, XXXVIII); *Report from the Commissioners on the Poor Law* (P.P. 1834, XXVIII); *Report from the Select Committee of the House of Lords* (P.P. 1838); *Children's Employment Commission, Mines* (P.P. 1842, XVI); *Children's Employment Commission, Mines* (P.P. 1842, XVII); *Reports of Special Assistant of Poor Law Commissioners* (P.P. 1843, XII).

38   *Children's Employment Commission, Mines* (P.P. 1842, XVII); *Reports of Special Assistant of Poor Law Commissioners* (P.P. 1843, XII).

39   E. Howe, *The London Compositor 1785–1900* (Oxford, 1947); W. Neild, "Comparative Statement of the Income and Expenditure of Certain Families of the Working Classes in Manchester and Dukinfield, in the Years 1836 and 1841", *Journal of the Statistical Society of London*, 4 (1842), 320–34; Mrs Rundell, *A System of Practical Domestic Economy* (1824); H. Ashworth, "Depression of Trade at Bolton", *Journal of the Statistical Society of London*, 5 (Apr. 1842), 74–81; S. R. Bosanquet, *The Rights of the Poor and Christian Alms Giving Vindicated* (1841).

40   "Life of a Cotton Spinner, Written by Himself", *The Commonwealth* (27 December 1856).

41   Ibid., and Charles Shaw, *When I was a Child, by "an Old Potter"* (London, 1903; facs. repr. Wakefield).

42   George Marsh, "A Sketch of the Life of George Marsh, a Yorkshire Collier, 1834–1921", B920 MAR, Barnsley Archives and Local Studies.

43   John Hemmingway, "The Character or Worldly Experience of the Writer from 1791 to 1865", MC 766/1, 795X5, Norfolk Record Office; A Miner, "A Trade Union Solitary: Memoir of a Mid-Nineteenth-Century Miner", *History Workshop Journal* 25, no. 1 (1988), 148–65; Joseph Hodgson, *Memoir of Joseph Hodgson, Glazier, a Native of Whitehaven, Cumberland* (Whitehaven, 1850); Thomas Whittaker, *Life's Battles in Temperance Armour* (London, 1884; repr. 2009); Edward Allen Rymer, "The Martyrdom of the Mine, or 60 Years' Struggle for Life", ed. with an introduction by Robert G. Neville, *History Workshop Journal*, vols I and II (1976), 220–44.

44   George Allen, *The Machine Breaker; Or, the Heart-rending Confession of George Allen ... Written by Himself* (London, [1831?]); Charles Bent, *Autobiography of Charles Bent, a Reclaimed Drunkard* (Sheffield, 1866); James Dunn, *From Coal Mine Upwards, or Seventy Years of an Eventful Life* (London, [1910]); Anonymous, *An Exposition of the Nefarious System of Making and Passing Spurious Coin ... Being the Confessions of a Coiner* (Preston, undated); Edward G. Davis, *Some Passages from My Life* (Birmingham, 1898); George Mitchell, "Autobiography and Reminiscences of George Mitchell, 'One from the Plough'", in *The Skeleton at the Plough, or the Poor Farm Labourers of the West: with the Autobiography and Reminiscences of George Mitchell*, ed. Stephen Price (London, [1875?]).

References

Allen, George. *The Machine Breaker; Or, the Heart-Rending Confession of George Allen ... Written by Himself.* London, [1831?].

Allen, Robert C. "The Great Divergence: Wages and Prices from the Middle Ages to the First World War". *Explorations in Economic History* 38 (2001), 411–47.

Allen, Robert C. "How Prosperous Were the Romans? Evidence from Diocletian's Price Edict (301 AD)". University of Oxford, Department of Economics, Discussion Paper Series No. 363, October 2007.

Allen, Robert C. "Real Wages in Europe and Asia: A First Look at the Long-Term Patterns". In *Living Standards in the Past: New Perspectives on Well-Being in Asia and Europe*, ed. Robert C. Allen, Tommy Bengtsson, and Martin Dribe, 111–30. Oxford: Oxford University Press, 2005.

Anonymous, *An Exposition of the Nefarious System of Making and Passing Spurious Coin ... Being the Confessions of a Coiner.* Preston, undated.

Ashton, T. S. "The standard of life of the workers of England, 1790–1830". *Journal of Economic History*, Supplement IX (1949), 19–38.

Ashworth, H. "Depression of Trade at Bolton". *Journal of the Statistical Society of London*, 5 (April 1842), 74–81.

Bacci, Massimo L., et al., eds. *The Population of Europe: A History*. Trans. by Cynthia De Nardi Ipsen and Carl Ipsen. Oxford: Blackwell, 2000.

Bassino, Jean-Pascal, Debin Ma, Christine Moll-Murata, and Jan Luiten van Zanden. *Wages, Prices, and Living Standards in China, Japan, and Europe, 1738–1925*. Working Paper No. 1, GPIH, October 2005.

Bent, Charles. *Autobiography of Charles Bent, a Reclaimed Drunkard*. Sheffield, 1866.

Bosanquet, S. R. *The Rights of the Poor and Christian Alms Giving Vindicated*. 1841.

Brewer, Anthony. *The Making of the Classical Theory of Economic Growth*. London: Routledge, 2010.

Burger, Oskar, Annette Baudisch, and James W. Vaupel. "Human Mortality Improvement in Evolutionary Context". *Proceedings of the National Academy of Sciences* 109, no. 44 (2012), 18, 210–14.

Carlyle, Thomas. *Past and Present*. 2nd edition. London, 1845.

Children's Employment Commission, Mines (P.P. 1842, XVI).

Children's Employment Commission, Mines (P.P. 1842 XVII).

Clapham, J. H. *An Economic History of Modern Britain*. Cambridge: Cambridge University Press, 1926.

Clark, Gregory. "The Condition of the Working-Class in England, 1209–2004". *Journal of Political Economy*, 113, no. 6 (2005), 1307–40.

Clark, Gregory. "Farm Wages and Living Standards in the Industrial Revolution: England, 1670–1869". *Economic History Review*, 53, no. 3 (2001), 477–505.

Crafts, Nicholas F. R. "Some Dimensions of the 'Quality of Life' During the British Industrial Revolution". *Economic History Review*, 50 (1997), 617–39.

Davies, David .*The Case of the Labourers in Husbandry Stated and Considered*. London, 1795.

Deane, Phyllis and W. A. Cole. *British Economic Growth, 1688–1959*. Cambridge, 1964.

Davis, Edward G. *Some Passages from My Life*. Birmingham, 1898.

Dunn, James. *From Coal Mine Upwards, or Seventy Years of an Eventful Life*. London, 1910.

Eden, Sir F. M. *The State of the Poor: A History of the Labouring Classes in England, with Parochial Reports*, ed. A. G. L. Rogers. London, 1928.

Engels, Friedrich. *The Condition of the Working Class in England*, ed. and intro. Victor G. Kiernan. Harmondsworth: Penguin, 1987.

Feinstein, C. H. "Capital Formation in Great Britain". In *The Cambridge Economic History of Europe*, ed. I. P. Mathias and M. M. Postan, 28–96. Cambridge: Cambridge University Press, 1978.

Floud, Roderick, Robert W. Fogel, Bernard Harris and Sok Chul Hong. *The Changing Body: Health, Nutrition, and Human Development in the Western World Since*. Cambridge: Cambridge University Press, 2011.

Fogel, Robert W. *The Escape from Hunger and Premature Death, 1700–2100: Europe, America, and the Third World*. Cambridge: Cambridge University Press, 2004.

Goldstone, Jack A. "Efflorescences and Economic Growth in World History: Rethinking the 'Rise of the West' and the Industrial Revolution". *Journal of World History*, 13, no. 2 (2002), 323–89.

Griffin, Emma. *A Short History of the Industrial Revolution*. Basingstoke: Palgrave, 2010.

Griffin, Emma. *Liberty's Dawn: A People's History of the Industrial Revolution*. New Haven, CT: Yale University Press, 2013.

Hammond, J. L. and Barbara Hammond, *The Town Labourer, 1760–1832*. London, 1917.

Hammond, J. L. and Barbara Hammond, *The Skilled Labourer, 1760–1832*. London, 1919.

Hartwell, R. M. "The Rising Standard of Living in England, 1800–1850". *Economic History Review*, (1961), 397–416.

Hemmingway, John. "The Character or Worldly Experience of the Writer from 1791 to 1865". MC 766/1, 795X5, Norfolk Record Office.

Hill, K., A. M. Hurtado, and R. S. Walker. "High Adult Mortality Among Hiwi Hunter–Gatherers:

Implications for Human Evolution". *Journal of Human Evolution* 52, no. 4 (2007), 443–54.

Hobsbawm, E. J. "The British Standard of Living, 1790–1850". *Economic History Review*, 10 (1957), 46–68.

Hodgson, Joseph. *Memoir of Joseph Hodgson, Glazier, a Native of Whitehaven, Cumberland*. Whitehaven, 1850.

Howe, E. *The London Compositor 1785–1900*. Oxford, 1947.

Huck, P. "Infant Mortality and the Living Standards of English Workers During the Industrial Revolution". *Journal of Economic History*, 55, no. 3 (1995), 528–50.

IPCC. *Climate Change 2014: Synthesis Report*. Contribution of Working Groups I, II and III to the Fifth Assessment Report of the Intergovernmental Panel on Climate Change, ed. R. K. Pachauri and L.A. Meyer. Geneva: IPCC, 2014.

Jones, Charles I. "Was an Industrial Revolution Inevitable? Economic Growth Over the Very Long Run". *Advances in Macroeconomics* 1, no. 2 (2001), 1–44.

"Life of a Cotton Spinner, Written by Himself". *The Commonwealth* (27 December 1856).

Lindert, P. H. and Jeffrey G. Williamson. "English Workers' Living Standards During the Industrial Revolution: A New Look". *Economic History Review*, 36, no. 1 (1983), 1–25.

Lindert, Peter H. and Jeffrey G. Williamson. "Measuring Ancient Inequality". Working Paper No. 13550, NBER, October 2007.

Little, Lester K., ed. *Plague and the End of Antiquity: The Pandemic of 541–750*. Cambridge: Cambridge University Press, 2007.

Loomis, William T. *Wages, Welfare Costs and Inflation in Classical Athens*. Ann Arbor, MI: University of Michigan Press, 1998.

Marsh, George. "A Sketch of the Life of George Marsh, a Yorkshire Collier, 1834–1921". B920 MAR, Barnsley Archives and Local Studies.

Matthews, R. C. O., C. H. Feinstein, and J. C. Odling-Smee. *British Economic Growth, 1856–1973*. Stanford, 1982.

Milanovic, Branko. "An Estimate of Average Income and Inequality in Byzantium around Year 1000". *Review of Income and Wealth*, 52 (2006), 449–70.

Miner, A. [A Miner]. "A Trade Union Solitary: Memoir of a Mid-Nineteenth-Century Miner". *History Workshop Journal* 25, no. 1 (1988), 148–65.

Mitchell, George. "Autobiography and Reminiscences of George Mitchell, 'One from the Plough'". In *The Skeleton at the Plough, or the Poor Farm Labourers of the West: with the Autobiography and Reminiscences of George Mitchell*, ed. Stephen Price. London, [1875?].

Mokyr, Joel. "Is There Still Life in the Pessimist Case? Consumption During the Industrial Revolution, 1790–1850". *Journal of Economic History*, 48, no. 1 (1988), 69–92.

Moreno, Alfonso. *Feeding the Democracy: The Athenian Grain Supply in the Fifth and Fourth Centuries BC*. Oxford: Oxford University Press, 2007.

Morris, Ian. "Economic Growth in Ancient Greece". *Journal of Institutional and Theoretical Economics* 160 (2004), 709–42.

Morris, Ian. "Archaeology, Standards of Living, and Greek Economic History". In *The Ancient Economy: Evidence and Models*, ed. Joseph G. Manning and Ian Morris, 91–126. Stanford, CA: Stanford University Press, 2005.

Morrisson, Cecile and Jean-Claude Cheynet. "Prices and Wages in the Byzantine World". In *The Economic History of Byzantium: From the Seventh through the Fifteenth Century*, ed. Angeliki Laiou, 807–70. Washington, DC: Dumbarton Oaks, 2002.

Neild, W. "Comparative Statement of the Income and Expenditure of Certain Families of the Working Classes in Manchester and Dukinfield, in the Years 1836 and 1841". *Journal of the Statistical Society of London*, 4 (1842), 320–34.

Özmucur, Süleyman and Sevket Pamuk. "Real Wages and Standards of Living in the Ottoman Empire, 1489–1914". *Journal of Economic History* 62, no. 2 (2002), 293–321.

Poor Law Amendment Act, weekly diet (P.P. 1837/8, XXXVIII).

Report from the Commissioners on the Poor Law (P.P. 1834, XXVIII).

Report from the Select Committee of the House of Lords (P.P. 1838).

Reports of Special Assistant of Poor Law Commissioners (P.P. 1843, XII).

Riley, J. C. *Rising Life Expectancy: A Global History*. Cambridge: Cambridge University Press, 2001.

Rundell, Mrs. *A System of Practical Domestic Economy*. 1824.

Rymer, Edward Allen. "The Martyrdom of the Mine, or 60 Years' Struggle for Life". Robert G. Neville, ed. and intro. *History Workshop Journal*, 1–2 (1976), 220–44.

Sahlins, Marshall. *Stone Age Economics*. Chicago, IL: Aldine, 1972.

Scheidel, Walter. "Real Wages in Early Economies: Evidence for Living Standards from 1800 BCE to 1300 CE". *Journal of the Economic and Social History of the Orient*, 53, no. 3 (2010), 425–62.

Schofield, R., D. Reher, and S. Bideau. *The Decline of Mortality in Europe*. Oxford: Oxford University Press, 1991.

Shaw, Charles. *When I was a Child, by "an Old Potter"*. London, 1903; facs. repr. Wakefield.

Snooks, Graeme D. *Was the Industrial Revolution Necessary?* London: Routledge, 1994.

Stathakopoulos, Dionysios C. *Famine and Pestilence in the Late Roman and Early Byzantine Empire: A Systematic Survey of Subsistence Crises and Epidemics*. Aldershot: Ashgate, 2004.

Szreter, Simon and Graham Mooney. "Urbanisation, Mortality, and the Standard of Living Debate". *Economic History Review*, 51, no. 1 (1998), 84–112.

Thompson, E. P. *The Making of the English Working Class*. London, 1963.

Toynbee, Arnold. *Lectures on the Industrial Revolution of the Eighteenth Century in England*. London, 1884.

Treadgold, Warren. "The Struggle for Survival (641–780)". In *The Oxford History of Byzantium*, ed. Cyril Mango, 129–50. Oxford: Oxford University Press, 2002.

Turner, Nancy J. *Ancient Pathways, Ancestral Knowledge: Ethnobotany and Ecological Wisdom of Indigenous Peoples of Northwestern North America*. McGill-Queen's University Press, 2014.

Van Zanden, Jan Luiten. "Wages and the Standards of Living in Europe, 1500–1800". *European Review of Economic History* 3(1999), 175–98.

Webb, Sidney and Beatrice Webb, *Industrial Democracy*. 2 vols. London, 1898.

Whittaker, Thomas. *Life's Battles in Temperance Armour*. London, 1884; repr. 2009.

Woods, Robert. *The Demography of Victorian England and Wales*. Cambridge, 2000.

Wrigley, E. A., *People, Cities and Wealth*. New York, 1987.

Wrigley, E. A. "The Transition to an Advanced Organic Economy: Half a Millennium of English Agriculture". *Economic History Review*, 59, no. 3 (2006), 435–80.

Wrigley, E. A., R. S. Schofield, Roger Schofield. *The Population History of England 1541–1871*. Cambridge: Cambridge University Press, 1989.

9

ISLAM AND SUSTAINABILITY

The norms and the hindrances

Tarik Masud Quadir

Reverence for the environment has been integral to the Islamic tradition beginning with the Qur'an and the Hadith (sayings and deeds of the Prophet Muhammad). The development of an active "environmentalist" tradition within Islam was necessitated further by the fact that water, vegetation, and fertile lands have always been scarce resources in Arabia. Hence, the Islamic tradition offers much that is relevant to today's world concerned with maintaining a sustainable global environment.

Islamic environmentalism consists of Islam-inspired thoughts and actions that would ensure environmental sustainability. The Islamic environmentalists share the outward goals of mainstream secular environmentalists of the West and of their counterparts in the Muslim world in reducing carbon emissions, wastage of food or any other natural resource, and on the need for an economy that is not exploitative. But Islamic environmentalists are not motivated *primarily* by worldly concerns, but at least equally, by the Islamic vision of our intimate relationship with God and all His creations and the existing ethical directives in this regard. For them, environmentalism as such is an essential aspect of being a good Muslim.

The three categories of environment-sustaining norms

As veteran environmental activist Fazlun Khalid has observed, Islamic environmentalism is "not of the kind [of modern secular environmentalism] we are familiar with today which is a reaction to our excess, rather the laying down of patterns of behaviour which are norms of [an Islamic] society."[1] With that in mind, we may categorize the environment-sustaining norms of the Islamic tradition as follows:

1   Islamic worldview on the nature of reality.
2   Environmental ethics, *Shariah* laws, legal principles, and institutions.
3   The living tradition of Islamic inner path or Sufism (*tasawwuf*), which aids one to directly experience the metaphysical verities of the worldview Islam proclaims.

Islamic worldview on the nature of reality

Our conscious interactions with the natural world around us are determined primarily by how we perceive its reality—it cannot be otherwise—which is the concern of any worldview. Hence, the Islamic worldview is pertinent to how faithful Muslims may be motivated to interact with the natural world and choose environmentally sustainable practices.

The foundation of the whole Islamic tradition and its worldview is the overarching doctrine that God is One, Omnipresent, and ever engaged in every detail of nature (Qur'an 6:59). Since Allah refers to Himself in the Qur'an as the Creator (*al-Khaliq*), the Mighty (*al-Aziz*), Master of Sovereignty (*Malik al-Mulk*), the Merciful (*al-Rahim*), the Sustainer (*al-Muqit*), the Wise (*al-Hakim*), the Reckoner (*al-Hasib*), the Seeing (*al-Basir*), the Hearing (*al-Sami*), the Near (*al-Qarib*), the Loving (*al-Wadud*) as well as the Real (*al-Haqq*), the Light (*al-Nur*), the Subtle (*al-Latif*), the All-Encompassing (*al-Muhit*), the Inner (*al-Batin*), the Outer (*al-Zahir*), and so forth, God in Islam can be conceived in diverse ways: "Allah" can mean the transcendent, the personal God, as well as the immanent and the impersonal one Reality beyond any form or gender, depending on the context. Because all entities have been created by the one and the same Allah, they are all interrelated. All entities being part of the one reality, they are inseparable from the One. However, entities of nature are not to be equated with Allah; rather they are His creations, and from Islamic metaphysical perspective, they are Allah's Self-Disclosures at various levels.[2]

Indeed, the Qur'an repeatedly refers to the natural world or creations as the "signs of God" (*ayat*) which demonstrate His infinite wisdom, generosity, justice, power, and other divine qualities. Thus, every natural entity has a reality beyond its purely physical characteristics and that helps us to know God by revealing His qualities. As such, every entity in nature has a meaning and purpose; nature is a boon (Qur'an 55:13-77) and a support not only for our physical sustenance but also for our quest for God. Moreover, the Qur'an refers to the verses of the Qur'an by the same term, the signs of God (*ayat*). Indeed, frequently the Qur'an explains itself by urging Muslims to observe something in nature (Qur'an 24:39-40; 14:6; 2:26), and almost a quarter of all the chapters of the Qur'an are named after one or another natural phenomenon or entity. In other words, Muslims need both nature and the Qur'an to understand God and His purpose. We might say that from the Islamic point of view, nature is the grand revelation containing verses in the forms of all entities in nature, including the oceans, mountains, forests, clouds, and so forth. It is this "revelation" that has been there from time immemorial to remind human beings of the underlying eternal reality of God before He sent verbal revelations to mankind.

Every entity has a purpose (Qur'an 44:38; 21:16; 23:115) in God's plan and all entities are created and interrelated in perfect balance (*al-mizan*)[3] that we must not transgress.[4] All non-human entities in the universe are conscious of and praise God unceasingly whether we can perceive them as such or not: "The seven heavens and the earth and all things therein declare His glory. There is not a thing but celebrates His adoration" (Qur'an 17:44 and Q 22:18; 24:41). Furthermore, every non-human species in existence is a community like us (Qur'an 6:38). Unlike many human beings who deny or disobey God, every non-human entity in the universe is fully submitted to God always, and in that sense "the creation of the heavens and the earth is greater than the creation of mankind" (Qur'an 40:57).[5]

What is the reality and role of the human being? God created the human being in His own "image,"[6] taught him all the "names" (Qur'an 2:31), and charged him with the responsibility to be His vicegerent or representative (*khalifah*) on earth (Qur'an 35:39). Naturally, the human being can function as God's true representative only as His servant; even the prophets can fulfil their roles only as His perfect servants (Qur'an 21:26). In Islamic metaphysics, "names" in verse 2:31 refer to the names or qualities of God, which are reflected in nature, but *can* be manifest altogether and in perfect harmony only in a human being (Quadir, 2013, p. 77). By the same token, the human being is the microcosm of the universe and, as such, ontologically related to every entity (Qur'an 41:53; 51:20-21).[7]

Islamic environmental ethics, laws, and institutions

The Islamic worldview briefly summarized above is the foundation of Islamic values for universal good and what we might call the environmental ethics and laws of Islam. First of all, this worldview calls for reverence of nature, which consists of conscious God-worshipping entities who form communities like us, who are intimately related *to each other and to us*, and who, as the signs of God reflecting His qualities, can help us know God. Second, as representatives (*khalifah*) of God, the Most Merciful (*Al-Rahim*), the Most Compassionate (*al-Rahman*), and the Wise (*al-Hakim*), we human beings are charged with the responsibility to treat nature in a manner that is harmonious with His qualities and not as we wish.

The Qur'an and the Prophet Muhammad's sayings forbid wastage numerous times (Qur'an 6:141; 7:31; 17:26-27). In fact, the very interdependence of all entities observed in nature suggests a lack of wastage therein, and supports the Qur'an's claim of perfect balance in creation as well. The Prophet Muhammad made it clear that the prohibition against wastage of any resource applies even when it is available in plenty and is being used for an especially sacred cause.[8] In the same vein, sharing or charity given freely to those in need is strongly encouraged. The Prophet Muhammad famously said, "There is no believer who plants a tree or sows a field, and a human, bird, or an animal eats from it, but it shall be reckoned as charity from him (*Sahih Bukhari*, vol. 3, book 3, no. 513)," and that "people are partners in water, plants and fire (*Sunan Abu Dawud*, no. 3477)," three of the most common natural elements necessary to sustain human life.[9] However, the *shariah* allows an owner's exclusive rights to all the benefits from any privately owned limited resource—for instance, a well, fruit-bearing trees, or fields of crops and vegetables—except in times of crisis or deprivation when sharing one's wealth with the needy becomes an Islamic duty.[10] Indeed, not to share the excess with the needy is to waste the bounty of God. To be clear, Islam forbids extracting, consuming, or using the earth's resources in a manner that amounts to wastage or misuse and thus upsets the *balance* in nature where wastage is absent.

The Qur'an makes it abundantly clear that the bounties of the earth must be shared with all other living creatures (Qur'an 80:24-32; 25:48-49; 79:31-33). It warns of serious consequences for mistreating the signs of God by way of the example of an ancient community that was severely punished for refusing to give a she-camel her share of water and fodder, and then killing her (Qur'an 7:73). The Prophet Muhammad spoke of feeling the love of trees, mountains and other elements of nature, warned that God would have been more severe with human communities for their numerous sins if it were not for the presence of the pious elderly, babies, animals, and plants in our midst, and pleaded to God for rain at least for the sake of plants and animals.[11]

There are numerous other examples of the Prophet's kindness towards non-human entities. His general advice on treating others may be summarized in his own famous words: "Show mercy to those on earth, and He in Heaven will show mercy to you" (*Sunan Abu Dawud*, no. 4941). The Prophet emphasized the need to be kind to animals by citing the stories of someone who would be sent to hell for starving a cat to death (*Sahih Bukhari*, book 12, no. 712), of a prostitute who was forgiven of all her sins for quenching the thirst of a dog (*Sahih Bukhari*, book 54, no. 538), and of an ancient prophet who was chastised for burning of "a whole community that glorified God" when stung by a single ant (*Sahih Muslim*, book 26, no. 5567). Based on the Prophet's explicit sayings and examples, the *shariah* forbids overburdening animals, not providing them with adequate nourishment even when they are old and no longer useful to their owner, mutilating or branding bodies of animals in any way,[12] using them for target practice, inciting them to fight against each other, killing them except in genuine need of sustenance, killing animals by any but the least painful means or near their parents, using the skins of wild animals, castrating them, and keeping males and females separate during mating seasons.[13] Even at times of war, the *shariah* forbids destroying crops and killing of domestic animals belonging to the enemy.

Guided by the Prophet's example and advice, there are *shariah* laws against any pollution of land, water, and air[14] and to ensure just sharing of scarce natural elements and resources.[15] According to Islamic jurisprudence, no natural resource in its natural location can be anyone's exclusive property (even if on privately owned property); individuals or groups gain rights to use only according to their determined need. However, those who make efforts to ensure sustainability of a resource, such as a well, a stream, or a land, can earn greater rights to the use of that resource. *Shariah* laws on water resources are particularly well developed to ensure fair distribution and sustainable use of the sources.[16] Further, these laws are still operative, if not as strictly and widely as in the past, in many rural areas of the Arab world where water is a scarce resource.

In addition, based on the Prophet Muhammad's directly relevant example and statements, his immediate successors established two remarkable institutions for the conservation of natural resources:

1 Public reserves (*hima*) for the conservation or sustainable growth of vegetation, trees, and wildlife.
2 Inviolable zones (*harim*), where any development is forbidden in order to prevent impairment of water resources and public utilities, such as roads and shades of trees.

Until the last quarter of the twentieth century these two institutions played a major role in the preservation of natural resources, especially in Arabia. Only since the 1960s, the number and size of *himas* have declined sharply due to changing socioeconomic conditions, mechanization of agriculture, and other "development" projects on the same lands.[17] Over the same period, *harim* restrictions also have been gradually relaxed, and in the case of water resources, often ignored altogether.

One of the greatest tragedies with regard to Islam's environment-sustaining tradition has been the way the Prophet's established inviolable sanctuaries for wildlife, trees, and herbage around Mecca and Medina have been allowed to be "developed" since the twentieth century. The Prophet Muhammad declared that Mecca's precincts—estimated to be at least 600 square kilometers in area—were made as "a sanctuary by Allah (until the Day of Resurrection) … It is not allowed to uproot its shrubs or to cut its trees, or to chase its game (*Sahih Bukhari*, vol. 3, book 29, No. 59). Likewise, with the same stipulations the Prophet announced, "I have made Medina a sanctuary between its (Harrat) mountains" (*Sahih Bukhari*, vol. 3, book 30, no. 91). Othman Llewellyn reports that indeed "Hunting does not take place within the two inviolable sanctuaries, but otherwise the *sharia*'s strict environmental rulings pertaining to them are largely suspended in practice" in the name of "development" and by the logic of "dire necessity."[18]

In the Qur'an and the Hadith, unnecessary killing, denying nourishments to animals, and wasting, exploiting, dirtying, or polluting any natural resource are forbidden as a form of disobedience of God; likewise, committing injustice, disrespecting the signs of God, upsetting the balance, or corrupting (*fasad*) of nature are acts condemned by God in the strongest terms. As a famous verse warns, "Corruption (fasad) has appeared on the land and sea because of that which men's hands have earned, that He may let them taste some of that which they have done, that haply they might return [to the path of God]" (Qur'an 30:41; also 2:205-6). In light of the above, the Qur'an's statement that all entities were created subservient to the human being (45:13) cannot be taken as a license to treat nature as we wish in defiance of God and the Prophet Muhammad; God entrusted all entities to us with the charge to treat them according to His will.

The *shariah*'s scope and means for responding to changing conditions

A brief look at the objectives (*maqasid*) of the *shariah* and various methods of Islamic jurisprudence (*fiqh*) further enables us to see the comprehensive scope of the principles and laws of Islam for dealing with issues pertaining to sustainability in our times.

To begin with, Islamic jurisprudence (*fiqh*) seeks to maximize benefits and minimize detriments to society. Furthermore, according to traditional jurists, including the renowned Ibrahim al-Shatibi (d.1388), the ultimate stated objective of *shariah* is to maximize the welfare of *all* of creation, not just of human beings.[19] Most influential contemporary Islamic jurists both in the Sunni and Shia communities, including Yusuf al-Qaradawi and Mohammad Hashim Kamali, agree that the conservation of the natural world is one of the highest objectives (*maqasid*) of the *shariah*.[20]

Based on precedents in the Qur'an and Hadith (the two primary sources of law), Islamic jurisprudence uses several legal principles, among other tools, to determine right actions. Two of the most central of these principles are directly relevant to the conservation of nature:

1   interests of society as a whole take priority over interests of any individual or group;
2   avoiding harm takes precedence over the acquisition of benefits.

Moreover, there are several other legal methods to determine human actions in situations not directly addressed in the Qur'an and the Hadith that require *ijtihad* (informed rational analysis) by jurists. As we will see, these additional established methods of Islamic jurisprudence, along with existing rulings clearly based on the primary sources, make the *shariah* for the environment immensely powerful and able to respond to novel challenges in any age. The first of these methods, known as *qiyas* (juristic analogy), is to judge a contemporary situation by looking back at an analogous situation, if available, covered in the two primary sacred sources. Given the precedence of the Prophet Muhammad's prohibition against waste and impairment of water, the juristic tool of *qiyas* enables us to regard all kinds of pollution of water, air, and land as prohibited practices in Islam. Likewise, by applying juristic analogy (*qiyas*), *hima* and *harim* institutions can also be strengthened and extended to protect various natural commons from corporate or commercial incursions. In light of the Qur'an's warning against changing God's creation (4:119) and the Prophet Muhammad's prohibition against cruel treatments of animals, application of *qiyas* would render much of the contemporary practices of genetic engineering, animal vivisection, and industrial system of meat production and its consumption Islamically illegitimate.[21] Further, in light of the Qur'an's warning against transgressing the balance in creation, many of our contemporary practices are Islamically unacceptable. As Kamali has observed:

> When the cattle and grass-eating animals are fed with animal-sourced protein until it is manifested in such problems as "mad-cow disease," and when genetically modified fruits overtake the natural variety for commercial gain, the God-ordained balance in them is no longer immune against distortion.[22]

In these and in numerous other respects, Islamic goals for environmental sustainability are quite in sync with contemporary organic movements.

The second method, known as *ijma* (consensus), is to judge an action based on its unanimous acceptance by Muslim legal experts. The third, known as *al-masalih al-mursalah* (public welfare), is to determine if a particular action would benefit the society as a whole. Unfortunately, too often in the Muslim world today, environmentally degrading "development" projects are carried out in the name of *al-masalih al-mursalah* (public welfare) in the more religiously conservative societies, or in the name of "progress" in the relatively liberal ones, or both. The Aswan High Dam of Egypt built in 1964 by Soviet engineers on the Nile River, may be only the most glaring example of such damaging projects. This dam "disrupted the most stable agricultural environment on Earth" and created the "need" for chemical fertilizers where none had been needed.[23] One of the latest such ill-conceived projects is the coal-fired Rampal Power Plant that the Bangladesh government plans to build in a joint venture with India, despite widespread public protest. This plant is certain to have a devastating impact on the ecology of the nearby Sundarbans, the world's largest mangrove forest and home to the Royal Bengal Tigers and many other endangered species. UNESCO has urged the Bangladesh government to abandon the project but so far the government remains determined to go ahead with its own plan.[24] Obviously, such misadventures by the powers that be must be protested and prevented by promoting greater awareness of both the environmental consequences of such projects and of the *shariah*'s objectives and requirements to protect the environment. The fourth, known as *al-'urf al-salih* ("customary wholesome practices") is to include in the *shariah* customary practices of any given society which are in harmony with the *shariah* objectives. Fifth, known as *saad al-dharai* ("closing the gate to evil"), is to prevent a "legitimate" action if it is used as a cover for unwholesome ends. I believe that many "development" projects in the Muslim world that are degrading our natural resources should be halted on the grounds of "closing the gate to evil," and the spurious logic of "dire necessity" to justify developments in the areas of Mecca and Medina which the Prophet had designated as sanctuaries, should be utterly condemned and rejected.

The role of Sufism (tasawwuf)

All the normative practices of Islam are supposed to be performed with love and devotion to God. But the world can and often does distract us from maintaining that inner attitude even when we go through the religious routines of daily prayers, the Ramadan fasting, etc. Sufism seeks to correct this situation by additional practices that help to keep our mind focused on the reality of God's presence, power, mercy, generosity, beauty, wisdom, justice, and so forth, at any time and place. The Sufi path is environment-sustaining in at least three very significant ways:

1   Nature features prominently on this path as a reminder of God's presence and qualities.
2   This path seeks to cultivate the qualities of contentment, simplicity, and the spirit of charity—qualities that go against consumerism and wastage.
3   It enables many to have direct spiritual experiences that confirm at the heart-level at least some aspects of the Islamic worldview, and thus provides a transcendental justification for the practice of environmental ethics.[25]

Understanding the four hindrances and how they may be overcome

A careful look at the above outline of environment-protective teachings contained in the Qur'an, Hadith, and the established legal principles and institutions should leave no doubt about the potential of Islam in motivating Muslims—a religion that still has a much stronger hold on a typical Muslim's mind, especially in Muslim majority countries, than other religions have on most Westerners—to engage in the cause of a sustainable environment. It is not necessary to reinterpret the Islamic tradition in order to make it environmentally friendly.[26] Instead, we need to take three essential steps:

1 Draw attention to Islam's clear existing directives to treat nature with respect and kindness, contemplate on its symbolic meanings, and to learn from it.
2 Transform our everyday habits accordingly.
3 Learn from practical experiences of the outstanding leaders of the Islamic eco-*pesantrens* of Indonesia[27] and from veteran Islamic environmental activists, such as Fazlun Khalid of the International Islamic Foundation of Ecology and Environmental Sciences (IFEES), Othman Abd ar-Rahman Llewellyn at the Saudi Wildlife Authority[28] and Dr. Massoumeh Ebtekar, the former vice president of Iran and the leading Islamic environmentalist in that country.[29]

Today, the Islamic environmentalists seek to inspire faithful Muslims across the world to actively engage in environmentalism on the basis of the Islamic norms that I have briefly outlined above, though with varying emphasis.[30] This approach to environmentalism has been widely used particularly in Indonesia,[31] the largest Muslim country in the world, and in Iran.[32] These efforts have had clear benefits, but the potential for the Muslim world's environmental activism and contribution is immensely greater. There are four *interrelated* major factors that have hindered environmental activism in the Muslim world:

1 lack of sufficient awareness of the environmental crisis and about Islamic environmental teachings;
2 not seeing the crisis as Muslims' responsibility to solve;
3 modern capitalism; and
4 scientism, the assumption that modern science is the most reliable, if not the only means of true knowledge.

In the next few pages, I will examine these hindrances in the context of Muslim communities, and demonstrate how they can be overcome to make the environment-sustaining tradition within Islam an *effective* force for restoring the global environment.

First, the vast majority of the population in Muslim majority countries are not sufficiently aware either of the disastrous and global nature of today's environmental crisis, or of the Islamic teachings about how one must view and treat nature. Such ignorance is not desirable but not surprising in the present context. While the West was undergoing large-scale industrialization, much of the Muslim world was enduring colonial rule because of that, and Muslims were preoccupied with efforts—usually with the loss of countless lives—to regain political and economic independence. By the 1960s, after many decades, and in several cases more than a hundred years of colonial rule, the last of the Muslim nations, Algeria, was finally free of direct Western rule. The 1960s were also the time that the modern secular environmental activism began in earnest in America before it spread to other countries. However, the Muslim world, though technically in a post-colonial state, was not free yet from political, economic, military, and cultural interferences by the former colonial powers and the Soviet Union. A combination of these interferences and the struggles by Muslim nations to chart their own course forward in a global system not of their making, have resulted in these nations being gripped by one major crisis after another. Not surprisingly, in these times of crises, while the vast majority of

Muslims have held on to the basic articles and practices of their faith, their grasp of the Islamic intellectual and spiritual heritage has been much weakened by numerous Western ideological influences including scientism and scientific progressivism—the conviction that modern science is the most important means of human progress—which have facilitated the modern ways and means of perpetrating major environmental degradations.[33] The upshot, as far as what concerns us here, is that the vast majority of Muslims have not only remained unaware of the extent of the environmental crisis and climate change, but also of the immensely rich treasure of environment-sustaining teachings and institutions inherent in their own tradition.

Without further delay, Muslim educators and policymakers must educate themselves on the catastrophic nature of the current global environmental crisis and the Islamic environmental ethics and institutions that could help counter its devastating march. Then, they must take measures to ensure propagation of that knowledge to all sectors of Muslim societies. Mosques, schools, and the media must join hands in the dissemination of accurate and up-to-date knowledge on the subject. This is the most urgent step Muslims need to take in the next decade before it is too late.

Second, even when they become aware of the global environmental crisis, many Muslims don't see the crisis as their responsibility to solve. It is not surprising that it was in America and in other highly industrialized countries that environmental pollution and degradation was first observed and protested; large-scale industrial pollution began in these countries. The Muslim countries, though guilty of overlooking their own environmental degradations over the last several decades of industrialization, have thus far had much less negative impact on global environmental conditions. However, given the fact that we are in a race against time to save the environment of the *whole* world, this feet-dragging on the part of many Muslims is *indefensible and unforgiveable*, and undeniably so, in light of the Islamic teachings on our duty to protect nature. It is incumbent upon us, as human beings sharing this planet with others and as Muslims, to do *our part*, at the very least, in this time of dire threat to our global environment, in every way we can.

Third, the norms and the materialistic values of the prevailing modern, capitalist, economic system have mostly opposed the environmentalist spirit in the Muslim world, as elsewhere. We should not forget that modern capitalism and industrialization were initially imposed and administered in almost every Muslim country by the colonial powers until, by the first half of the twentieth century, these moves altered the indigenous economies and modes of work for the masses in much of the Muslim world, as elsewhere.[34] Along with these changes, the colonized societies had to conform increasingly to the norms and the materialistic values of the prevailing capitalist economic system. This economic system was founded on the premise that human welfare can be achieved mainly by greater accumulation of material wealth through the exploitation of natural resources with the help of ever more powerful technologies. Moreover, this extractivist economic system could thrive—by the need to maximize profit—*only by promoting insatiable consumerism and selfish individualism*, values that have been strongly discouraged by all of the world religions. Indeed, ever since the Industrial Revolution, progress or development of any nation has been measured primarily by the extent of the establishment of institutions that could further its accumulation of material wealth. The fate of nature has not been of much concern to the capitalist elite or the powers that control them, until very recently. Hence, the inherent environment-sustaining spirit of the Islamic tradition has had to struggle against the modern capitalist ethos in Muslim countries, as elsewhere. The question is not whether we have reaped numerous benefits from modern capitalism and industrialization, but whether such benefits are really worth the catastrophic climate change and/or a complete corruption of our environment.

Most Islamic environmentalists have sought to persuade Muslims to observe Islamic environmental ethics within the existing economic order. However, some have gone further by suggesting the need to overhaul the prevailing economic system by eliminating the practice of charging interest (*riba*) on money. Charging interest is a cornerstone of modern capitalism, but strongly condemned in the Qur'an (2: 275-279; 3:130; 4:161; 30:39) as well as in the Old and the New Testament. A banking system created with *riba* is focused not on serving public interest, but on maximizing profits for a few.[35] Indeed, the *riba*-based banking system propelled the Industrial Revolution and still enables financing countless so-called "development" projects across the world which wreak havoc on the environment and, at the same time, burden communities and nations with huge debts.[36] The revival of an Islamic banking system four decades ago and a renewed interest in charitable foundations (*awqaf*) are Islamic attempts to address the problem.[37] However, Islamic banks could do much more by being more proactive in investing directly in environment-nurturing projects.[38]

Muslims must also learn from other market options being discussed by mainstream environmentalists. Polluters must compensate, as the *shariah* stipulates, for the damages to the environment,[39] or, we must not allow businesses to externalize environmental costs. Muslim governments should invest more heavily in building infrastructures for harnessing alternative sources of energy such as solar and wind power. The big oil-producing countries have the resources to make such changes and simultaneously scale back oil production. What is required of the oil producers is to give up their addiction to cheap oil-money for the sake of the whole world. Other Muslim countries (Indonesia, Malaysia, Bangladesh, Pakistan, Afghanistan, Turkey, Egypt, etc.) should also minimize their dependence on fossil fuels by vigorously pursuing the means for harnessing alternative energy. In Muslim countries, these proposed changes cannot be accomplished without *tremendous* social pressure on the governments and the wealthy elites that control them. For this reason, we need far greater awareness of the devastating effects of following the business as usual approach, and be *unrelenting* in re-educating the faithful Muslims to recognize the demands of their faith and *cease* tolerating the blatant and unforgiveable destruction of the environment that God entrusted to our care.

The fourth hindrance is the prevalence of scientism. I believe that scientism is the greatest obstacle to a religion-inspired environmentalism. Since the advent of the Age of Enlightenment following the Scientific Revolution, scientism has been undermining the religion-based views of nature and the universe. The European fatigue from the wars between Catholics and Protestants (1524–1648) was followed by modern science's success at the end of the seventeenth-century in presenting a strictly mathematical, mechanical, and thus, verifiable worldview (based exclusively on the empirical evidence from the "seen," but not the realms of the "unseen") that *contradicted* the traditional Christian worldview. In contrast to that of modern science, the traditional Christian worldview was deemed dogmatic and untrue; and the symbolic aspect of the Christian worldview was not adequately defended.[40] This perceived failure of Christianity unleashed an enthusiasm for modern science that metamorphosed, unfortunately, into scientism.[41] Since the nineteenth-century, scientism has been helped further by the inventions of technologies directly based on scientific research and which greatly eased numerous human inconveniences. After the invention of each such labor-saving technology, one of the effects was even greater scientism, which in turn produced further enthusiasm for modern science.

Modern science is a purely *materialistic* science and can be true at the physical level of reality. However, scientism is not modern science; the former ignores the blatant truth that the later can never provide us any more than the physical data of our surroundings. Being limited to the physical domain, modern science cannot prove or disprove any non-material truth-claims by religion, spirituality, or culture.[42] By the same token, modern science cannot tell us anything about meanings, values, and purposes of any entity except in purely material, and therefore, radically limited terms. Thus modern science cannot help us to understand ultimate meanings and purposes of either natural entities or of human beings. By ignoring these limitations, scientism vastly overextends the scope of modern science's legitimate parameters. So, when modern scientific data is used to define the *whole* of the reality of natural entities, it is an instance of scientism that makes us miss out on their spiritual and symbolic meanings and view the natural world as pure *matter*, and thus reconceptualize nature as something inert, mechanical, and easily exploitable.[43] Many Islamic and other thinkers see this materialistic transformation of the European worldview, which began with the Age of Enlightenment, as the root cause behind industrialization and modern capitalism's ruthless exploitation of natural resources.[44]

Scientism is still the prevalent worldview of academia, and embedded as a pre-analytic *a priori* assumption with an iron grip on the opinions of most policymakers, and therefore, in the setting of much of the contemporary goals and priorities of the modern world. Any criticism of scientism is too often mistaken as a rejection of modern science. The prevalence of scientism in almost every institution of the modern world renders any religious or sacred view of nature discussed earlier inconsequential, except perhaps as fodder for poetic musings. In such a skewed intellectual climate, religions may claim a role for saving our own human soul—an intangible reality whose existence Muslims and other believers cannot easily deny– but religions are not allowed to influence any "serious" discussion of the nature of reality of our surroundings which are deemed tangible, and thus, within the exclusive purview of modern science. Thus, as scientism is popularized, particularly with respect to the natural world, the Islamic worldview, or any religious worldview that believes in a higher reality beyond the material plane, suffers proportionately. Modern science's ability to help us understand the extent of the physical damage already inflicted on the environment and the potential for catastrophic damage in the future should not delude us into thinking that it is also the *main* solution to this crisis. The Islamic tradition clearly has a role to play in the emerging sustainability movement, and one should emphasize the many commonalities between Islamic and non-Islamic approaches to conservation, waste, sustainable use of resources, and so on.

From the modern scientific perspective, unless we are aware of its limitations, it makes no sense to speak of nature as the signs of God that are conscious God-adoring entities, and ontologically connected to each other and to human beings. Accordingly, by contradicting religious or spiritual claims, the *prevalence* of modern science gives rise to scientism which rejects the authenticity of the Islamic or any religious worldview, and hence, vastly weakens any logical basis in religion for observing the earlier mentioned Islamic environmental ethics, *shariah* laws, and the legal instruments like the *hima* and the *harim*. However, scientism, to the extent accepted by many Muslims today, did not initiate the undermining of the *shariah* and the other traditional Islamic institutions in the Muslim world; that process was initiated by administrative decisions made by Western colonial powers in much of the Muslim world.[45] For instance, to facilitate governance, the colonial administrators largely replaced *shariah* laws in public with modified versions of civil codes used back in their homelands in Europe.

The problem with scientism, though acknowledged by most of the prominent Islamic environmentalists, is ignored by the vast majority of Muslim and non-Muslim environmentalists as something too fundamental to the prevailing modern civilization to be questioned. Yet, there can be no denying that scientism induces materialism, undermines all religious or spiritual worldviews, and thus, *severely* undermines Islam's or any other religion's rationale for environmental ethics. Put differently, scientism, which in many ways is at the root of the current environmental crisis, cannot simultaneously serve as the intellectual foundation for *solving* the self-same crisis. We cannot tap the immense potential of religiously inspired environmentalism without exposing the glaring falsity of scientism. The fallacy of scientism can be exposed in two powerful ways within the Islamic world: by Islamic philosophical arguments and by the promotion of Sufism. Islamic philosophical arguments can demonstrate that scientism violates three cardinal doctrines of Islam:

1   Reality is one.
2   Reality is hierarchical, from the reality of God's Essence at the top, down to the angelic plane, psychic plane, and finally to the material plane, the only plane recognized by modern science.
3   Meaningfulness and purposefulness of all entities in the natural world.[46]

In addition, the living tradition of Sufism can be the most powerful weapon against scientism, because this path enables many Muslims to have inner *experience* of at least some of the professed spiritual realities that are *beyond the material plane.* These inner experiences too must be regarded as "empirical" evidence against modern scientistic claims.

Muslims cannot save the environment without active cooperation with other faith communities that span the world. Hence, a truly Islamic environmentalism should also revive the traditional Islamic spirit of respect for non-Islamic religions in acknowledging their validity— as long as they promote kindness, justice, righteous conduct, and the ultimate oneness of God or Reality—even if imperfect, ways to God, as the Qur'an (49:13; 29:46; 2:62; 2:111-12; 2:256; 3:84; 5:48; 22:67; 5:68-69; 10:47; 14:4; 4:164; 16:125; 22:40;) and the example of Prophet Muhammad suggest.[47] Now, more than ever, the faith communities need to stand together to face the common danger that confronts us all. Certainly, the essential truths of the great world-religions have much, much more in common than the outward differences that mostly divide them (Qur'an 22:67; 42:13). Further, as alluded to at the outset, there are tremendous commonalities between religious and secular approaches to environmental sustainability.

In summary, Islam has had a great "environmentalist" tradition that was embedded in it through the Qur'an and the Hadith. Now, Muslims must revive the tradition of environmental protection and engage the attention of all faithful Muslims of their religious duty in this regard. The flame of greed—read consumerism—is deeply ingrained in human souls. From a religious perspective, only that which can reach our souls at a deeper level—such as divine directives or inspirations—and not *just* better economic models, policy changes, and more efficient technology, however well intended, can keep the flame of greed sufficiently restrained to let nature survive and even thrive. Muslim educators and policymakers must recognize and address all the hindrances against environmental activism and follow our recommendations (or develop better ones) for overcoming them. The hindrances are *interrelated* and therefore must be addressed *simultaneously* by coordinated efforts by educators, religious leaders, the media, and government policymakers in order to overcome them all. As Muslims trusting in God's power, mercy, and purposefulness, we cannot either loose hope or give up our duty till the end of the world. As the Prophet advised, "If the Final Hour comes while you have a palm-cutting in your hands and it is possible to plant it, you should plant it" (*Al-Adab al-Mufrad*, book 27, no. 479).

Notes

1   F. Khalid, "Islam and the Environment: Ethics and Practice," paper presented at the 15th General Conference, Royal Aal al-Bayt Institute for Islamic Thought, Amman, Jordan, September 27–29, 2010, retrieved on November 15, 2016 from www.aalalbayt.org/EnvConference/018.pdf.

2   T. M. Quadir, *Traditional Islamic Environmentalism: The Vision of Seyyed Hossein Nasr.* (Lanham, MD: University Press of America, 2013), 74–6.

3   "We have created therein [the Earth] all manner of things in due balance" (Qur'an 15:19; 54:49; 13:8).

4   "Heaven He has raised and the balance He has set, that you transgress not in the balance" (Qur'an 55:7-8). And then again, "Call upon your Lord humbly and in secret. Truly He loves not the transgressors. And work not corruption after it has been set aright" (7:55-56). Please keep in mind that the serial numbers of Qur'anic chapters (*surah*) are not in the order they were revealed to the Prophet Muhammad.

5   This verse should be understood in the context of the previous verse, which condemns human pride that makes many disregard the signs of God.

6   According to a well-known saying of the Prophet Muhammad, "Allah created Adam in His own image."

7   On human status as God's representative and his ontological link with all of creation, see M. Murad, "Vicegerency and Nature," *Critical Muslim* 19 (July–September 2016), 65–75.

8   M. I. Dien, *Environmental Dimension of Islam* (Cambridge: Lutterworth Press, 2000), 32–3.

9   *Sahih Bukhari, Sahih Muslim, Sunan Abu Dawud* and *Al-Adab al-Mufrad* are among the most rigorously authenticated collections of the reports of the Prophet Muhammad's sayings and deeds (Hadith) and sourced for this essay. Citations to these documents are given in the text.

10   Dien, *Environmental*, 32–4.

11   Ibid., 104–5.

12   According to the Qur'an, it is Satanic to "alter God's creation" (4:119).

13   A. B. A. Masri, *Animal Welfare in Islam* (Markfield: The Islamic Foundation, 2007), 33–50; U. F. Moghul, and H. K. Safar-Aly, "Green Sukuk: The Introduction of Islam's Environmental Ethics to Contemporary Islamic Finance," *Georgetown International Environmental Law Review* 27, 1 (2014–15): 20–23.

14   Moghul and Safar-Aly, "Green Sukuk," 20.

15   O. A. Llewellyn, "The Basis for a Discipline of Islamic Environmental Law," in *Islam and Ecology: A Bestowed Trust*, R. C. Foltz, F. M. Denny, A. and Baharuddin, eds. (Cambridge, MA: Harvard University Press, 2003), 198–207; Dien, *Environmental*, 30–42.

16   Llewellyn, "The Basis," 203–5.

17   Ibid., 213–15.

18   Ibid., 209.

19   O. A. Llewellyn, "Islamic Jurisprudence and Environmental Planning," *Journal of Research in Islamic Economics* 1, 2 (1984): 24.

20   M. H. Kamali, *The Middle Path of Moderation in Islam* (New York: Oxford University Press, 2015), chapter 15.

21   See Masri, *Animal Welfare*, 39–51.

22   Kamali, *Middle Path*, 144.

23   See D. R. Montgomery, *Dirt: The Erosion of Civilizations* (Berkeley, CA: University of California Press, 2012), 43.

24   See I. Mahmoud, "UNESCO Calls for Shelving the Rampal Project," retrieved on January 24, 2017 from en.prothom-alo.com/environment/news/122299/Unesco-calls-for-shelving-Rampal-project.

25   Quadir, *Traditional*, 87–97.

26   See M. S. Islam, "Old Philosophy, New Movement: The Rise of Islamic Ecological Paradigm in the Discourse of Environmentalism," *Nature and Culture* 7, 1 (Spring 2012): 72–94.

27   See A. M. Gade, "Tradition and Sentiment in Indonesian Environmental Islam," *Worldviews* 16 (2012): 263–85.

28   See F. Khalid, "Applying Islamic Environmental Ethics," in *Environmentalism in the Muslim World*, R. C. Foltz, ed. (New York: Nova Science Publishers, 2005), 87–111.

29   A. M. Schwencke, *Globalized Eco-Islam: A Survey of Global Islamic Environmentalism* (Leiden Institute of Religious Studies: Leiden University, 2012), 35–6.

30   M. Arnez, "Shifting Notions of Nature and Environmentalism in Indonesian Islam," in *Environmental and Climate Change in South and Southeast Asia*, B. Schuler, ed. (Leiden: Brill Academic Publishers, 2014), 75–101; see also, N. Mohammad, "Islamic Education, Eco-ethics and Community," *Studies in Philosophy Education* 33 (2014): 315–28; Islam, "Old Philosophy"; F. M. Mangunjaya and J. E. McKay, "Reviving an Islamic Approach for Environmental Conservation in Indonesia," *Worldviews* 16 (2012): 286–305.

31 See Arnez, "Shifting."

32 See R. C. Foltz, "Iran," in *Environmentalism in the Muslim World*, R. C. Foltz, ed. (New York: Nova Science Publishers, 2005), 3–16.

33 Quadir, *Traditional*, chapter 4.

34 See, for instance, the effect of colonization and industrialization in the case of undivided South Asia, home to the largest ethnic group of Muslim population, and in Egypt, the largest and most influential of Arab Muslim countries in P. Bent, "Historical Perspective on Precarious Work: The Cases of Egypt and India under British Imperialism," *Global Labour Journal* 8, 1 (2017), retrieved on June 6, 2017 from https://mulpress.mcmaster.ca/globallabour/article/view/2716. Bent concludes, "The historical record shows that the industrialization that did occur in colonial Egypt and India was highly disruptive to existing social and economic systems. These changes resulted in the creation of working arrangements that were unstable, insecure, and contingent—in a word, precarious." Also see Quadir, *Traditional*, 109; D. Commins, "Hasan al-Banna," in *Pioneers of Islamic Revival*, ed. A. Rahenma (London: Zed Books, 1994), 125–8.

35 J. G. Speth, *America the Possible: Manifesto for a New Economy* (New Haven, CT: Yale University Press, 2012), 120–21.

36 See Y. Dutton, "The Environmental Crisis of Our Time: A Muslim Response," in *Islam and Ecology: A Bestowed Trust*, R. C. Foltz, F. M. Denny, and A. Baharuddin, eds. (Cambridge, MA: Harvard University Press, 2003), 323–40.

37 See M. Cizakca, "*Awqaf* in History and its Implications for Modern Islamic Economies," *Islamic Economic Studies* 6, 1 (1998): 43–70.

38 See B. S. Sairally, "Integrating Environmental, Social and Governance (ESG) Factors in Islamic Finance: Towards the Realization of *Maqasid al-Shariah*." *ISRA International Journal of Islamic Finance* 7, 2 (2015): 145–54.

39 Moghul and Safar-Aly, "Green Sukuk," 23–4.

40 W. Smith, *Cosmos and Transcendence: Breaking through the Barrier of Scientistic Belief* (Peru, IL: Sherwood Sugden & Company Publishers, 1984), 43–58; also, Keith Thomas, *Man and the Natural World: Changing Attitudes in England, 1500–1800* (London: Allen Lane, 1983).

41 On the many forms of scientism, see Mikael Stenmark, "What is Scientism?," *Religious Studies* 33 (1997): 15–32.

42 See I. Hutchinson, *Monopolizing Knowledge: A Scientist Refutes Religion-Denying, Reason-Destroying Scientism* (Belmont, MA: Fias Publishing, 2011).

43 Quadir, *Traditional*, 57–8; on scientism and the Enlightenment, see J. L. Caradonna, *Sustainability: A History* (Oxford: Oxford University Press, 2014).

44 S. H. Nasr, The *Encounter of Man and Nature: The Spiritual Crisis of Modern Man* (London: George Allen and Unwin, 1968); O. Bakar, *Environmental Wisdom for Planet Earth: The Islamic Heritage* (Kuala Lumpur: Center for Civilizational Dialogue, 2007); Quadir, *Traditional*, 56–9.

45 Quadir, *Traditional*, 109.

46 Ibid., 158–67.

47 See R. Shah-Kazemi, *The Other in the Light of the One: The Universality of the Qur'an and Interfaith Dialogue* (Cambridge: Islamic Text Society, 2006).

References

Arnez, M. "Shifting Notions of Nature and Environmentalism in Indonesian Islam." In *Environmental and Climate Change in South and Southeast Asia*, B. Schuler, ed., 75–101. Leiden: Brill Academic Publishers, 2014.

Bakar, O. *Environmental Wisdom for Planet Earth: The Islamic Heritage*. Kuala Lumpur: Centre for Civilizational Dialogue, 2007.

Bent, P. "Historical Perspective on Precarious Work: The Cases of Egypt and India under British Imperialism." *Global Labour Journal* 8, 1 (2017). Retrieved on June 6, 2017 from https://mulpress.mcmaster.ca/globallabour/article/view/2716.

Caradonna, J. L. *Sustainability: A History*. Oxford: Oxford University Press, 2014.

Commins, D. "Hasan al-Banna," in *Pioneers of Islamic Revival*, A. Rahenma, ed. London: Zed Books, 1994.

Cizakca, M. "*Awqaf* in History and its Implications for Modern Islamic Economies." *Islamic Economic Studies* 6, 1 (1998): 43–70.

Dien, M. I. *Environmental Dimension of Islam*. Cambridge: Lutterworth Press, 2000.

Dutton, Y. "The Environmental Crisis of Our Time: A Muslim Response." In *Islam and Ecology: A Bestowed Trust*, R. C. Foltz, F. M. Denny, and A. Baharuddin, eds., 323–40. Cambridge, MA: Harvard University Press, 2003.

Foltz, R. C. "Iran." In *Environmentalism in the Muslim World*, R. C. Foltz, ed., 3–16. New York: Nova Science Publishers, 2005.

Gade, A. M. "Tradition and Sentiment in Indonesian Environmental Islam." *Worldviews* 16 (2012): 263–85.

Hutchinson, I. *Monopolizing Knowledge: A Scientist Refutes Religion-Denying, Reason-Destroying Scientism.* Belmont, MA: Fias Publishing, 2011.

Islam, M. S. "Old Philosophy, New Movement: The Rise of Islamic Ecological Paradigm in the Discourse of Environmentalism." *Nature and Culture* 7, 1 (Spring 2012): 72–94.

Kamali, M. H. *The Middle Path of Moderation in Islam.* New York: Oxford University Press, 2015.

Khalid, F. "Applying Islamic Environmental Ethics." In *Environmentalism in the Muslim World*, R. C. Foltz, ed., 87–111. New York: Nova Science Publishers, 2005.

Khalid, F. "Islam and the Environment: Ethics and Practice." Paper presented at the 15th General Conference, Royal Aal al-Bayt Institute for Islamic Thought, Amman, Jordan, September 27–29, 2010. Retrieved on November 15, 20116 from www.aalalbayt.org/EnvConference/018.pdf.

Llewellyn, O. A. "The Basis for a Discipline of Islamic Environmental Law." In *Islam and Ecology: A Bestowed Trust*, R. C. Foltz, F. M. Denny, and A. Baharuddin, eds., 185–247. Cambridge, MA: Harvard University Press, 2003.

Llewellyn, O. A. "Islamic Jurisprudence and Environmental Planning." *Journal of Research in Islamic Economics* 1, 2 (1984): 27–46.

Mahmoud, I. "UNESCO Calls for Shelving the Rampal Project." Retrieved on January 24, 2017 from en.prothom-alo.com/environment/news/122299/Unesco-calls-for-shelving-Rampal-project.

Mangunjaya, F. M. and J. E. McKay. "Reviving an Islamic Approach for Environmental Conservation in Indonesia." *Worldviews* 16 (2012): 286–305.

Masri, A. B. A. *Animal Welfare in Islam.* Markfield: The Islamic Foundation, 2007.

Moghul, U. F. and H. K. Safar-Aly. "Green Sukuk: The Introduction of Islam's Environmental Ethics to Contemporary Islamic Finance." *Georgetown International Environmental Law Review* 27, 1 (2014–15): 1–60.

Mohammad, N. "Islamic Education, Eco-ethics and Community." *Studies in Philosophy Education* 33 (2014): 315–28.

Montgomery, D. R. *Dirt: The Erosion of Civilizations.* Berkeley, CA: University of California Press, 2012.

Murad, M. "Vicegerency and Nature." *Critical Muslim* 19 (July–September, 2016): 65–75.

Nasr, S. H. *The Encounter of Man and Nature: The Spiritual Crisis of Modern Man.* London: George Allen and Unwin, 1968.

Quadir, T. M. *Traditional Islamic Environmentalism: The Vision of Seyyed Hossein Nasr.* Lanham, MD: University Press of America, 2013.

Schwencke, A. M. *Globalized Eco-Islam: A Survey of Global Islamic Environmentalism.* Leiden: Institute of Religious Studies, Leiden University, 2012.

Sairally, B. S. "Integrating Environmental, Social and Governance (ESG) Factors in Islamic Finance: Towards the Realization of *Maqasid al-Shariah*." *ISRA International Journal of Islamic Finance* 7, 2 (2015): 145–54.

Shah-Kazemi, R. *The Other in the Light of the One: The Universality of the Qur'an and Interfaith Dialogue.* Cambridge: Islamic Text Society, 2006.

Smith, W. *Cosmos and Transcendence: Breaking through the Barrier of Scientistic Belief.* Peru, IL: Sherwood Sugden & Company Publishers, 1984.

Speth, J. G. *America the Possible: Manifesto for a New Economy.* New Haven, CT: Yale University Press, 2012.

Stenmark, M. "What is Scientism?" *Religious Studies* 33 (1997): 15–32.

Thomas, K. *Man and the Natural World: Changing Attitudes in England, 1500–1800.* London: Allen Lane, 1983.

PART V

The recent history of sustainability

10

THE US ENVIRONMENTAL MOVEMENT OF THE 1960S AND 1970S

Building frameworks of sustainability

Erik W. Johnson and Pierce Greenberg

Introduction

The US environmental movement has been a vibrant agent of societal change for the past 50 years. Since the first Earth Day in 1970, the movement has grown dramatically in both size and the diversity of tactics, membership, and the issues it represents.[1] Over this period, the movement has achieved significant legislative gains,[2] environmental education has blossomed, the environment has gone "from Heresy to Dogma" among business organizations,[3] and public support for the environment has generally remained strong and broad based.[4] One key to sustained vibrancy and effectiveness within American environmentalism was the turn towards sustainability issues as an arena of focus and conceptual innovation. Within American environmentalism (and across much of the world), sustainability is an increasingly central organizing framework that operates across a wide variety of institutional domains. The elaboration and extension of various perspectives on sustainability represented in this volume, and the movement of these concepts across various social institutions and sub-cultures, has its intellectual roots in the "new" environmental movement of the 1960s and 1970s. To better understand today's sustainability movement, this chapter focuses on the groundwork laid during this emergent phase of a new national environmental movement. The late 1960s and 1970s environmental movement changed cultural understandings about the interplay between human and biophysical systems. During this period, Americans began to see humans and human systems as fitting within natural systems, rather than seeing human society and economy as separate and nature an object to dominate. This shift was facilitated by a new environmental movement, which laid the intellectual groundwork for the sustainability focus that has grown in significance since the early 1990s.

We also argue that sustainability has been such a powerful conceptual force in part because of how it fits within the broader US environmental movement. That is, sustainability is somewhat unique in being both strongly resonant and adaptable enough to appeal to both the "new" wing that emerged in the 1960s/1970s and the much larger, more resourced and more venerable wildlife and wildlands conservation wing of the movement. One consequence is that concepts related to sustainability help to facilitate more diverse and effective coalitions in the environmental movement today. In this and other ways there are parallels between the current focus on sustainability and the more limited issues of pesticides, which fostered similar cross-movement coalitions in the 1960s and 1970s, while simultaneously introducing Americans to the science of ecology and greater sensitivity to the ways in which human socio-economic and biophysical systems interact. As we see it, the environmental movement and the sustainability movement are tightly intertwined, but the latter would not have come into existence without the former.

Ecological ways of thinking had been developing for many decades in the sciences, but it was the environmental movement that brought ecological education to the wider public and made ecological issues prominent on governmental, business, and other institutional agendas. A central achievement was re-theorizing the man–nature relationship by making commonplace the notion that both economic systems of production and human well-being are inextricably linked with the health of ecological systems. By linking human well-being and economic development to ecological systems, the emergent environmental movement set the stage for the various theoretical approaches to sustainability that are the focus of this volume.

This chapter begins by briefly sketching the development of, and connections between, the conservation and "new ecology" wings of the US environmental movement. We highlight in particular some of the key figures of the US environmental movement in the 1960s and 1970s who were influential in pushing ecological systems thinking to the forefront of the environmental agenda. We then discuss the issue of pesticides and how in the 1960s and 1970s it (1) introduced ecological systems theory to the American public at large and (2) provided the connective tissue for wilderness conservation groups to become increasingly concerned about human well-being and the interconnectedness of nature and humans. There are strong parallels between the issues highlighted by pesticides and the workings of sustainable development today. We conclude by making broad and explicit linkages between the developments in the 1960s/1970s environmental movement and conceptions of sustainability that are prevalent today.

From a wilderness conservation to a "new" ecology movement

Standard histories of the US environmental movement begin with progressive era (1880–1920) resource and wilderness conservation and preservation efforts, and conflicts between the two camps. Conservationists such as Gifford Pinchot sought to maximize renewable—what we might today call "sustainable"—human use of wilderness areas while preservationists, led by John Muir, sought to maintain wilderness in a pristine setting by setting aside relatively untouched parcels of land. Despite the often-bitter conflicts between these perspectives (and towering figures), there were many similarities. Both types of groups were composed primarily of economic and political elites of the era, who built the organizational infrastructures that remain central to the environmental movement today. For example, a leading conservationist group of the era, and today still, is the Boone and Crockett Club. When the organization was established at a dinner party held by Theodore Roosevelt in 1887 it included roughly 20 social elites as founding members, and included Roosevelt, George Bird Grinnell (editor of *Forest and Stream* magazine), Henry Cabot Lodge and J. Pierpont Morgan.[5] Today its membership continues to be drawn from the milieu of well-heeled sportsmen and the group is the organization of record for trophy hunting, in addition to focusing on the conservation of habitat and wildlife populations. Indeed, Teddy Roosevelt is emblematic of the top-down, keep-nature-pristine-for-the-recreational-use-of-social-elites ethos that drove first-wave environmentalism.

The leading preservationist organization of the period, the Sierra Club, focused narrowly on preserving large tracks of "untouched" landscape (the legacy of native peoples was often ignored) in the Sierra Nevada mountains east of San Francisco. Founded in 1892 by John Muir, who now graces the California state quarter, and academics from University of California–Berkeley and Stanford University, the early Sierra Club was an elite membership organization dominated by prominent academics, politicians, and business leaders and required the sponsorship of two existing members to join.[6] The focus was on fostering both protection and the recreational (e.g. skiing and mountaineering) use of the Sierra wilderness, and the Club was largely influential in establishing Yosemite and Sequoia National Parks in 1890.

Americans during this pre-war period were, in the words of pioneering environmental sociologists William Catton and Riley Dunlap, adhered to a human exemptionalist paradigm that viewed humans as existing in a different realm than their environment and acting upon it.[7] Given the technological advancement associated with industrialization, the prominent "Western" worldview, with deep roots in European intellectual and scientific culture, involved four key tenets:

1   people were separate from nature, over which they had dominion;
2   people choose their own destiny, of which there are no limits;
3   the vast world provides unlimited opportunities for human growth and pursuit; and
4   for every future problem, there will be a solution.[8]

This perspective is reflected, to varying degrees, in the nature conservation and preservation movements of the 1800s and early 1900s, which focused on preserving or conserving isolated chunks of nature (i.e. nature separate from people). In the US especially, these chunks of nature were often set aside in a unique national park and wildlands system.[9] Although the battle between Pinchot and Muir over the plan to build the Hetch Hetchy dam in Yosemite had an obvious connection to energy concerns—the dam, which was eventually built, provided electricity to the city of San Francisco—the debates over land conservation around 1900 often took place in isolation of wider social, economic, and environmental problems.

After a period of relative dormancy during and between the World Wars, a "new" post-war environmental movement developed in response to accelerated ecological disruptions and a qualitative shift in the nature of environmental hazards (see Figure 10.1). A post-war petro-chemical-based economy led to a whole "new species" of troubles centered on risks from nuclear energy and war, air and water pollutants, and hazardous wastes.[10] The environmental movement that mobilized during the 1960s and 1970s focused on these new classes of pollution and related human health issues[11] and a "new breed" of national environmental organizations focused not on wilderness or wildlife protection, but ameliorating threats from environmental pollutants.[12]

Intellectual origins of "new environmentalism"

The incalculably influential writings of Rachel Carson were pivotal in bringing the new ecological consciousness to the American public. Carson (1907–64) earned an MA in zoology from Johns Hopkins University and worked for many years as a scientist in the US Fish and Wildlife Service. She also wrote extensively for popular consumption, translating science and natural history into high-quality and widely accessible prose. Her nature writing featured three monographs about natural biophysical processes in the ocean, including the highly acclaimed international bestseller *The Sea Around Us*.[13] In 1962, Carson published *Silent Spring*—a

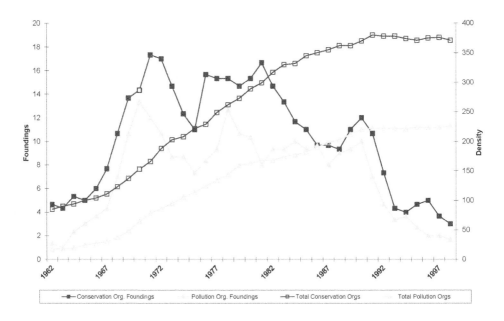

Figure 10.1 Foundings and population densities of US conservation and new ecology organizations, 1962–98

momentous book that many credit with birthing the modern environmental movement, and which has been called the fifth most important non-fiction book of the twentieth century.[14] In it, Carson moves her ecological lens from the ocean to the world of birds, and focuses not just on explicating natural systems but on how human health and well-being is inter-dependent with ecological systems and how human actions had caused massive disruptions to both humans and the natural world. In particular, Carson focused on DDT and other persistent pesticides that, at the time, were being prescribed for widespread use. Carson showed how human disruption of these ecological systems was likely to reverberate back on human systems through heightened rates of cancer and birth defects. She questioned, in short, the costs associated with the post-war chemical revolution in American manufacturing by arguing that nature could not be controlled as some system separate from humans, but that we are inextricably connected and changes in one affect the other. Perhaps more than any other author in history, it was Rachel Carson who popularized the science of ecology by calling attention to the way pesticides disrupted the "web of life" and implications of this disruption for human and wildlife health. The issue of pesticides proved pivotal as an organizing force that combined traditional concerns with wildlife protection and new concerns about the nature of industrial production in a chemical–industrial economy. Although she died during the controversy that ensued, her effort led eventually to the banning of DDT and increased regulatory oversight of pesticides.

Another source of intellectual leadership in the early environmental movement was Barry Commoner. Commoner was as a professor of plant physiology at Washington University in St. Louis for most of his career. He became famous in the 1950s for his opposition to nuclear weapons testing. His research helped to introduce a type of citizen-science, as well as the specific technique of biomonitoring, which has become a common form of environmental activism today.[15] Commoner's most famous research study was conducted under the auspices of a citizens' group he helped found called the Committee for Nuclear Information. In December of 1958 the Committee put out a public call to collect baby teeth for a study of nuclear fall-out, receiving tens of thousands of teeth over the next few years and providing participants who mailed in their teeth a certificate confirming "I gave my tooth to science." Results published in *Science* magazine in November 1961 demonstrated unequivocally that nuclear fall-out from testing, even though it was conducted in remote locations, reverberated back on human systems through the absorption of Strontium 90 released in the atmosphere.

As an ecologist, this was part of a more general maxim Commoner preached, the first law of ecology, that "everything is connected to everything else." This is a common notion in today's world, where the butterfly effect is invoked in popular media and talk. But in 1970, the first law of ecology remained somewhat novel. The second of Commoner's laws of ecology is that wastes from economic production do not simply go "away." Instead, there is no such thing as waste. When nature is converted for human production processes, everything that is generated will remain part of biophysical reality and subject to reabsorption in those systems. For example, CO_2 from fossil fuel emissions do not simply "dissolve" into thin air—instead, they trap heat in the atmosphere and contribute to global climate change, unless sequestered in some form. Commoner's third law was the notion that "nature knows best"—another fundamental tenet of sustainability theory. That is, production and development processes should be aligned with ecological systems, and technologies that seek to "improve upon" nature are largely misguided and may cause lasting damage to the environment.

By 1970, Commoner's focus had expanded from the nuclear threat to include an expanding class of environmental threats tied to modern petro-chemical industrial production. He argued for fundamental restructuring of the economy so that production processes better align with the laws of ecology. Commoner's writing was highly critical of reductionist thinking and focused on the need to see interrelationships with systems as well as to highlight what he referred to as "external diseconomies," or what economists today call externalities of produc-tion processes—the social and environmental consequences that are nudged out of economic models. Commoner focused considerable attention on how polluting industries were subsi-dized by societies that absorb the costs of "external diseconomies," such as increased cancer and asthma rates resulting from the release of pollutants during production processes. As such, his writings remain foundational to modern approaches to sustainability.

Activists promoting ecological consciousness, the connectedness of human and biophysical systems, and the real limits that are imposed by this interconnection, built also on vivid imagery from a famous picture "Earth Rise" taken on Christmas Eve 1968 by astronaut William Anders from the moon. The photo, depicting our planet as a small and seemingly vulnerable ball float-ing in a vast universe, had a great impact on the American public and indeed on the entire world.[16] A prominent imagery invoked during this period was of a spaceship earth metaphor, emphasizing the ways in which human societies exist in interaction with ecological systems with very real physical limits.[17]

A particularly influential attempt to explore ecological limits, and the ways in which social systems interact with these limits, was the Club of Rome's enduringly important *Limits to Growth* (1972).[18] The Limits to Growth Study was an early application of computer models to simulate societal growth scenarios (*not* predictions), and the likelihood of overshooting the limits of a finite system of ecological resources and sinks. In two of the three growth scenarios presented, the result was the collapse of ecological and thus global socio-economic systems. In the third, the world collaborates on technological fixes and progressive social policies in order to keep development sustainable. This high visibility study became a foundational text in environmental and sustainability studies curriculum and helped to propagate a fundamental operating assumption in sustainability, which is that there are real biophysical limits to economic growth. The book, which caused a huge controversy in a post-War society addicted to growth, has been updated and republished several times, and recent research has tended to validate its findings.

Around the same time, British economist E. F. Schumacher published his book that popularized the phrase *Small is Beautiful*.[19] In this book, and other writings, Schumacher argued that the modern economy was unsustainable because it did not recognize the finite nature of ecological resources and pollution sinks. In this way, his work built on both the limits research and the spaceship earth imagery. Schumacher, a Chief Economic Advisor in the UK, focused on the need for a philosophical change in our orientation to the market and growth, and adopting technologies of appropriate scale for harmony with ecological and social systems, rather than geared towards mass production. An early skeptic of the validity of GNP (gross national product) as a measure of human well-being, Schumacher argued that the focus should be not on the economy itself, but what the economy is theoretically supposed to enhance: human well-being. Schumacher was an early advocate for the need for new metrics that connected environmental realities to social and economic well-being. The critique of the GNP, later echoed by Herman Daly, was that the GNP counted equally all economic activity, whether it had been beneficial or not to human populations and the environment. Progress on these measurement issues, and theorizing a sustainable economy, was made by Daly, and John Cobb, Jr.,[20] and others, but it was not until the 1990s and beyond that comprehensive sustainability metrics and indicators really began to evolve, including the Genuine Progress Indicator, The Happy Planet Index, and Ecological Footprint Analysis. The latter metric was developed by Bill Rees and Mathis Wackernagel and has become a standard in scientific analyses of the impacts of human populations.[21]

The works reviewed above were crucial to establishing the linkages between human and ecological systems. The early work of figures such as Carson and Commoner communicated the basic tenets of sustainability at a time when the American public was increasingly exposed to environmental problems. Visible environmental hazards—such as smog, oil spills, nuclear accidents, and water pollution—proliferated throughout the post-War period. The work of public intellectuals helped contextualize those hazards within a broader framework that considered the deep interconnections between human action and environmental outcomes. The emergence of a modern environmental movement helped to spread concepts of ecology broadly.

From wilderness conservation to human health

A new "ecology" wing of the environmental movement that emerged during the 1960s and 1970s propagated the intellectual seeds of modern sustainability theory. The focus of this new wing was not conservation but issues of environmental quality and human health (e.g. smog, water pollution, pesticides). The negative human health impacts of pollution were not newly discovered during the 1960s,[22] but the qualitative shift in the types of persistent pollutants generated by post-War petro-chemical manufacturing processes introduced new types of threats to human health. The shifting focus from saving trees and wildlife "out there," which was often associated with the influential head of the Department of the Interior in the 1960s, Stewart Udall, and towards limiting environmental pollutants, served to broaden the scope of the environmental movement well beyond the narrow swath of elites who dominated earlier conservation and preservation movements.[23] It also highlighted the interconnection of human and biophysical systems and in so doing contributed toward a broader public shift from the human exemptionalist paradigm and towards a new ecological paradigm.[24] This new paradigm tore down past conceptions of the natural environment as separate from the human or built environment. The growing recognition of the interconnectedness of ecology—and the harmful impacts of technological and economic activities—gained a substantial foothold as a shared perspective across the American public and among environmentalists and environmental organizations.

Within just a few short years a venerable movement for the conservation of wildlife and wild places had been subsumed within a new ecological perspective and movement.[25] The speed with which the new perspective proliferated was dramatic, as evinced in contemporary media coverage. Figure 10.2 shows counts of stories in the Readers' Guide to Periodical Literature index that are listed under different keywords associated with conservation or pollution/human health issues, and on the right-hand axis the ratio of conservation to pollution articles.[26] There is both a large increase in the numbers of articles published in 1970, and a rapid change in the proportional breakdown. Before 1970, there were always at least two news articles published on wildlife and wildlands conservation for every one article on pollution issues, often quite more. After 1970, other than a few blips during the 1980s, the ratio is generally below 1.5:1 and sometimes the topics receive equal coverage. There is no doubt that an ecological understanding and framing of both conservation and pollution issues came to prominence in the environmental movement,[27] among the US public,[28] and even world society.[29] This shift in orientation towards seeing human and ecological systems as intertwined and environmental issues as inextricably linked to human health laid seeds for both sustainability and the strong justice orientation that has come to define much work on the environment today.

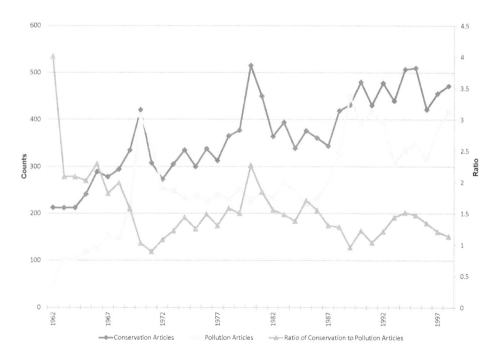

Figure 10.2 Counts of conservation and pollution articles from the Readers' Guide to Periodical Literature index

1970 is a clear turning point in the Readers' Guide data because it marks a momentous event in the history of the US environmental movement that had important influences on the orientation of the American political, media, and civil society to environmental systems and how humans interact with them. Earth Day 1970, billed at the time as the largest protest event ever, was both a high point of the early environmental movement in the United States and an important launching pad that helped to drive civic participation in environmental issues around the country and created an entire cadre of influential environmental activists.[30] The ecological way of thinking promoted by Carson and Commoner was a distinguishing feature of the first Earth Day events. In place of traditional wildlife and resource conservation concerns, the focus of Earth Day was on cultural orientations to nature, and how human economic and ecological systems interact. Indeed, the interaction between political and economic systems took center stage, with politicians, representatives of state regulatory agencies, and business interests commanding attention at the first events.[31] Conceived of originally as a teach-in modeled off of the civil rights movement, Earth Day brought the science of ecology to large swaths of the American public. A generation of baby boomers came to increasingly conceptualize humans as part of the environment and socio-political and economic systems of production as both impinging and relying upon biophysical systems.

Pesticides and convergence across the environmental movement

Perhaps no issue played a more important role in motivating activism within this new ecology wing of the environmental movement than pesticides. Pesticides were emblematic of the post-War industrial economy, including the rise of industrial agriculture that now dominates our food system. Pesticides were a key mobilizing issue that clearly demonstrated how human and biophysical systems reverberate upon one another. It was pesticides, and especially the use of DDT, that was the focus of Carson's most influential writing, and concern about pesticides was also the original focus of many leading new-ecology groups. Most prominently, the Environmental Defense Fund emerged out of a local attempt by bird lovers on Long Island, inspired by the writing of Carson, to halt DDT spraying for mosquito control in Suffolk County, NY. Working with lawyer Victor Yannacone, the group initiated a successful lawsuit that helped to initiate waves of litigation, which served as one defining tactic of the modern environmental movement. Building on interest in that case the Environmental Defense Fund (EDF), with financial support from the Audubon Society, the Conservation Foundation, and later the Ford Foundation, incorporated in 1967. Initially focused on pesticides, the group quickly expanded into a number of both wilderness and pollution issue domains. The EDF embraced the scientific-law organizing model and quickly emerged as a leading "full-service"[32] environmental advocacy organization attending to a broad range of pollution and wildlife-related issues. A member of the select "group of ten" that dominated the mainstream environmental movement in the 1980s and 1990s, the EDF remains broadly focused topically today and continues to span across wildlife and pollution issues. The group is today perhaps best known for working closely with corporate interests to identify environmental strategies with demonstrable economic benefits for producers.

Pesticides proved such a powerful mobilizing issue due to its unique ability to bridge across diverse segments of the environmental movement and bring a wide variety of actors and coalitions to the table. That is, pesticides were unique because both the conservation and "new ecology" wings of the environmental movement could readily organize and advocate around it. Pesticides, as Carson demonstrated, were a grave threat to the wildlife that American naturalists already cared about and to human health.

One consequence was, at the same time as the emergence of the "new" ecology wing of the environmental movement, some venerable wilderness and wildlife protection groups began to integrate pollution issues into core organizational agendas. It was often pesticides that opened the door. Most prominently, the Sierra Club, the leading national voice of the preservation movement, was by the mid-1970s consciously attempting to balance the organization's national issue agenda 50/50 between traditional conservation issues and newer pollution and human health concerns.[33] It was a resolution on the use of pesticides in national forests and parks that was "the first clear and thoroughly articulated environmental policy stand made by the Club,"[34] and which opened the door for the Sierra Club's movement from strict wilderness protection organizations to a modern environmental organization. The Sierra Club story illustrates larger trends in national environmental organizations, shifting from a nearly exclusive conservation focus to incorporate issues of pollution. In a sense, the sustainability movement built off of this logic and extended "environmental concerns" to the more complex business of creating sustainable social, economic, and environmental systems.

Pesticides became a central interest for both new ecology and conservation groups. Because of this, it united the dominant wings of the environmental movement and fostered productive activist coalitions. The importance of these cross-movement coalitions should not be underestimated because, while the new ecology wing of the environmental movement generally enjoys greater media and scholarly attention, the wildlife and wilderness conservation branch of the movement remains dominant in terms of both numbers (as reflected in Figure 10.2) and the amount of resources available. The new ecology wing has always been and remains small in comparison to the much larger and more venerable conservation groups in the environmental movement.[35] That is, wilderness and wildlife conservation groups are both more numerous and control considerably more resources on average (in terms of membership numbers, budgets, and staff employed), even when the financial outlier that is the massive Nature Conservancy (with contributions and grants of more than $958 million in 2014) is removed from analysis. In part, the lesson of pesticides demonstrates why sustainability as a concept is so important: it links the smaller but more prominent "new" environmental movement with the considerably larger and more resourceful wilderness and wildlife conservation wing by showing how change in human systems is linked to but transcends the preservation of wildlife and wild places.

Seeds of sustainability

The modern sustainability movement builds on intellectual foundations created by the 1960s and 1970s environmental movement. The merger of environment and economic development at the international level, however, fostered the move towards sustainability over the course of the following decade and culminating in the landmark 1992 Earth Summit.

The domestic political context was challenging for environmentalists during the 1980s. Command and control pollution regulations established during the previous decade were under assault from dominant free-market economic logics ushered in during the 1980s and the elections of Ronald Reagan in the US and Margaret Thatcher in England. The political context of the 1980s posed a major challenge to ideas about the limits of growth, or to mandatory limits on corporate behavior. However, the concept of sustainability—and its focus on self-interested movement towards production processes with both ecological and economic benefits—managed to gain a foothold on the international stage through the merger of environment and development concerns. The United Nations Environmental Programme (UNEP)-commissioned *World Conservation Strategy* report in 1980 was one of the first to use the phrase "sustainable development," and signaled a "shift in environmental consciousness from strict conservationism … to a more constructive philosophy of social transformation and a more dynamic appreciation of the interplay between the environment, the economy, and human well-being."[36] Then in 1987, the UN World Commission on Environment and Development (WCED) released the influential *Our Common Future* report, which helped set the stage for the landmark Earth Summit in 1992, and drew stark connections between the limits of ecological systems and the massive global disparities in resource use between developed and less-developed countries. There was a global desire to protect important ecological resources located in less developed countries, such as the Amazon rainforest, but that protection was inextricably linked with expectations of reviving economic growth—a troublesome association that has hindered and divided the sustainability movement for decades.

The UN Earth Summit (1992) in Rio de Janeiro, Brazil marked a watershed moment in linking environmental protection with international development. The meeting, formally dubbed the United Nations Conference on Environment and Development (UNCED), was the third in a series of international UN environmental summits that began in 1972. Rio stood out, however, for being "the largest and most public UN conference ever held. Attended by record numbers of governments, international agencies, and journalists, UNCED … was also the center of a massive international NGO lobby."[37] The conference drew representatives from nearly all of the world's countries, including 120 heads of state, to discuss how to navigate the relationship between the planet's conjoined biophysical and socio-economic systems. The focus was on both conservation and pollution issues that had defined the environmental movement since the first Earth Day, and the need to protect the Earth within the context of economic development for the world's poorest peoples. The conference was unique in that it directly addressed the ties between social, economic, and environmental justice on a global scale, and connected transportation, energy, food, and much more to the cause of sustainable development.

The Earth Summit was also uniquely important because of the extent to which it served as a major mobilizing force for environmental movements around the globe. Parallel to the official UN Summit was a Global Forum that drew representatives from an estimated 9,000 non-governmental organizations.[38] A signature achievement of the conference was the inclusion of NGOs in the preparation and implementation of conference outcomes. The Johnson data on US national EMOs shows a temporary decline in rate of creation for national EMOs, and a bump in organizational growth in the lead-up and immediate aftermath of the 1992 Earth Summit (see Figure 10.2). This is also the point at which a focus on sustainable development spread rapidly among environmental organizations in the US. This can be seen in our organizational data where "sustainable development" hardly registers as a relevant issue during the 1980s, whereas nearly 10 percent of the population of EMOs were focusing on sustainable development issues by 2000. As with pesticides, sustainable development has been of interest to both conservation and new ecology groups.[39] Moreover, local and national groups around the world were brought towards sustainability by participating in one of the major outcomes of the conference, called Agenda 21, which set out non-binding, voluntary guidelines for businesses and governments throughout the world. Agenda 21 provided the conceptual hook for local civil society actors and governments around the world to implement provisions of sustainable development.[40]

Conclusion

There are clear connections between the environmental movement in the 1960s and 1970s and the modern sustainability movement. The primary connection centers on the environmental movement's success in advancing the principles of ecology—a fundamental shift in how society–nature interactions are conceived in the US. The issue of pesticides—brought to light by Carson's *Silent Spring*—played a particularly important role in introducing issues of human health and pollution to well-resourced wildlife conservation groups. The groundswell of interest in the connection between "limits to growth," wildlife concerns, and human well-being was due in part to the environmental movement—and was carried over into an international conversation about sustainable development. In sum, the sustainability movement is best viewed as a dynamic elaboration of the environmental movement; an attempt to learn from environmentalists and environmental science and incorporate environmental concerns into a far-reaching vision for a sustainable society.

Notes

1   See Christopher J. Bosso, *Environment, Inc.: From Grassroots to Beltway* (Lawrence, KS: University Press of Kansas, 2005); Erik W. Johnson and Scott Frickel, "Ecological Threat and the Founding of US National Environmental Movement Organizations, 1962–1998," *Social Problems* 58 (2011): 305–29; R. E. Dunlap and A. G. Mertig, "The Evolution of the US Environmental Movement from 1970 to 1990: An Overview," *Society and Natural Resources* 4 (1991): 209–18.

2   Richard N. L. Andrews, *Managing the Environment, Managing Ourselves: A History of American Environmental Policy* (New Haven, CT: Yale University Press, 1999).

3   A. J. Hoffman, *From Heresy to Dogma: An Institutional History of Corporate Environmentalism* (Palo Alto, CA: Stanford University Press, 2001).

4   Fred C. Pampel and Lori M. Hunter, "Cohort Change, Diffusion, and Support for Environmental Spending in the United States," *American Journal of Sociology* 118 (2012): 420–48.

5   Robert Gottlieb, *Forcing the Spring: The Transformation of the American Environmental Movement* (Washington, DC: Island Press, 1993), 327.

6   Michael P. Cohen, *The History of the Sierra Club, 1892–1970* (San Francisco, CA: Sierra Club Books, 1988).

7   William R. Catton and Riley E. Dunlap, "A New Ecological Paradigm for Post-Exuberant Sociology," *American Behavioral Scientist* 24 (1980): 15–47.

8   Ibid., 17–18.

9   R. Nash, "The American Invention of National Parks," *American Quarterly* 22 (1970): 726–35.

10 Kai Erikson, *A New Species of Trouble: Explorations in Disaster, Trauma, and Community* (New York: W. W. Norton & Co., 1994).

11 See Bosso, *Environment, Inc.*; Gottlieb, *Forcing the Spring*; Dunlap and Mertig, "The Evolution of the US Environmental Movement"; Mark Dowie, *Losing Ground: American Environmentalism at the Close of the Twentieth Century* (Cambridge, MA: MIT Press, 1995); Samuel P. Hays, *Beauty, Health, and Permanence: Environmental Politics in the US, 1955–1985* (Cambridge: Cambridge University Press, 1987); Paul McLaughlin and Marwan Khawaja, "The Organizational Dynamics of the US Environmental Movement: Legitimation, Resource Mobilization, and Political Opportunity," *Rural Sociology* 65 (2000), 422–39; R. C. Mitchell, A. G. Mertig, and R. E. Dunlap, "Twenty Years of Environmental Mobilization: Trends Among National Environmental Organizations," in *The US Environmental Movement, 1970–1990*, R. E. Dunlap and A.G. Mertig, eds. (Philadelphia, PA: Taylor and Francis, 1992); Ronald G. Shaiko, *Voices and Echoes for the Environment* (New York: Columbia University Press, 1999).

12 Mitchell et al., "Twenty Years"; Johnson and Frickel, "Ecological Threat."

13 Rachel Carson, *The Sea Around Us* (Oxford: Oxford University Press, 1951).

14 Rachel Carson, *Silent Spring* (Boston, MA: Houghton Mifflin, 1962).

15 Scott Frickel, "Just Science? Organizing Scientist Activism in the US Environmental Justice Movement," *Science as Culture* 13 (2004): 449–69; Sabrina McCormick, *Mobilizing Science: Movements, Participation, and the Remaking of Knowledge* (Philadelphia, PA: Temple University Press, 2009).

16 Robert Poole, *Earthrise: How Man First Saw the Earth* (New Haven, CT: Yale University Press, 2008).

17 K. E. Boulding, "The Economics of the Coming Spaceship Earth: Environmental Quality in a Growing Economy," in *Essays From The Sixth RFF Forum*, Henry Jarrett, ed. (Baltimore, MD: Johns Hopkins University Press, 1966); Richard Buckminster Fuller, *Operating Manual for Spaceship Earth* (Carbondale, IL: Southern Illinois University, 1968).

18   Donella H. Meadows, Dennis L. Meadows, Jørgen Randers, and William W. Behrens III, *The Limits to Growth: A Report for the Club of Rome's Project on the Predicament of Mankind*, commissioned by Club of Rome (New York: Universe Books, 1972).

19   Ernst F. Schumacher, *Small is Beautiful: Economics as if People Mattered* (New York: Harper & Row, 1975).

20   Herman E. Daly and John B. Cobb, *For The Common Good: Redirecting the Economy Toward Community, the Environment, and a Sustainable Future* (Boston, MA: Beacon Press, 1989).

21   Mathis Wackernagel and William E. Rees, *Our Ecological Footprint: Reducing Human Impact on the Earth,* (Philadelphia, PA: New Society Publishers, 1996).

22   See Gottlieb, *Forcing the Spring*; Martin V. Melosi, *Garbage in the Cities: Refuse, Reform, and the Environment* (Pittsburgh, PA: University of Pittsburgh Press, 1981); Joel A. Tarr, ed., *Devastation and Renewal: An Environmental History of Pittsburgh and Its Region* (Pittsburgh, PA: University of Pittsburgh Press, 2004).

23   Pampel and Hunter, "Cohort Change."

24   Catton and Dunlap, "A New Ecological Paradigm for Post-Exuberant Sociology."

25   See Robert J. Brulle, *Agency, Democracy, and Nature: The US Environmental Movement From a Critical Theory Perspective* (Cambridge, MA: MIT Press, 2000); Russell J. Dalton, *The Green Rainbow: Environmental Groups in Western Europe* (New Haven, CT: Yale University Press, 1994); Gottlieb, *Forcing the Spring*; Hays, *Beauty, Health, and Permanence*; Angela G. Mertig, Riley E. Dunlap, and Denton E. Morrison, "The Environmental Movement in the United States," *Handbook of Environmental Sociology* (2002): 448–81; David Schlosberg, *Environmental Justice and the New Pluralism: The Challenge of Difference for Environmentalism* (Oxford: Oxford University Press, 2002).

26   For more details on the coding see Erik W. Johnson, "Social Movement Size, Organizational Diversity and the Making of Federal Law," *Social Forces* 86 (2008): 967–93.

27   See Gottlieb, *Forcing the Spring*; Johnson and Frickel, "Ecological Threat."

28   Hays, *Beauty, Health, and Permanence*; Thomas R. Dunlap, *Nature and the English Diaspora: Environment and History in the United States, Canada, Australia, and New Zealand.* (Cambridge: Cambridge University Press, 1999).

29   John W. Meyer, John Boli, George M. Thomas, and Francisco O. Ramirez, "World Society and the Nation-State," *American Journal of Sociology* 103 (1997): 144–81.

30   Adam Rome, *The Genius of Earth Day: How a 1970 Teach-In Unexpectedly Made the First Generation* (New York: Hill & Wang, 2013).

31   Ibid.

32   Shaiko, *Voices and Echoes for the Environment*.

33   Michael McCloskey, *Sierra Club Executive Director: The Evolving Club and the Environmental Movement, 1961–1981* (Berkeley, CA: Regional Oral History Office, Bancroft Library, University of California, 1983), 129.

34   Cohen, *History of the Sierra Club*, 338.

35   Johnson, "Social Movement Size".

36   Jeremy L. Caradonna, *Sustainability: A History* (New York: Oxford University Press, 2014), 141.

37   Alison Van Rooy, "The Frontiers of Influence: NGO Lobbying at the 1974 World Food Conference, the 1992 Earth Summit and Beyond," *World Development* 25 (1997), 98.

38   Ibid., 101.

39   The methodology for coding this data vastly underestimates the invocation of sustainability language among these groups. It codes only a narrow subset that is focused on the integration of economic and technological reform and development.

40   William M. Lafferty and Katarina Eckerberg, *From the Earth Summit to Local Agenda 21: Working Towards Sustainable Development* (Hoboken, NJ: Taylor and Francis, 2013); Sean Southey, "Accelerating Sustainability: From Agenda to Action," *Local Environment* 6 (2001): 483–89.

References

Andrews, Richard N. L. *Managing the Environment, Managing Ourselves: A History of American Environmental Policy.* New Haven, CT: Yale University Press, 1999.

Bosso, Christopher J. *Environment, Inc.: From Grassroots to Beltway.* Studies in Government and Public Policy. Lawrence, KS: University Press of Kansas, 2005.

Boulding, K. E. "The Economics of the Coming Spaceship Earth: Environmental Quality in a Growing Economy." In *Essays From The Sixth RFF Forum*, edited by Henry Jarrett. Baltimore, MD: Johns Hopkins UniversityPress, 1966.

Brulle, Robert J. *Agency, Democracy, and Nature: The US Environmental Movement from a Critical Theory Perspective*. Cambridge, MA: MIT Press, 2000.

Caradonna, Jeremy L. *Sustainability: A History*. New York: Oxford University Press, 2014.

Carson, Rachel. *The Sea Around Us*. Oxford: Oxford University Press, 1951.

Carson, Rachel. *Silent Spring*. Boston, MA: Houghton Mifflin, 1962.

Catton, William R., and Riley E. Dunlap. "A New Ecological Paradigm for Post-Exuberant Sociology." *American Behavioral Scientist* 24, no. 1 (1980): 15–47.

Cohen, Michael P. *The History of the Sierra Club, 1892–1970*. San Francisco, CA: Sierra Club Books, 1988.

Dalton, Russell J. *The Green Rainbow: Environmental Groups in Western Europe*. New Haven, CT: Yale University Press, 1994.

Daly, Herman E., and John B. Cobb. *For the Common Good: Redirecting the Economy toward Community, the Environment, and a Sustainable Future*. 2nd edition, updated and expanded. Boston, MA: Beacon Press, 1989.

Dowie, Mark. *Losing Ground: American Environmentalism at the Close of the Twentieth Century*. Cambridge, MA: MIT Press, 1995.

Dunlap, Riley E., and Angela G. Mertig. "The Evolution of the US Environmental Movement from 1970 to 1990: An Overview." *Society and Natural Resources* 4, no. 3 (July 1, 1991): 209–18.

Dunlap, Thomas R. *Nature and the English Diaspora: Environment and History in the United States, Canada, Australia, and New Zealand*. Studies in Environment and History. Cambridge: Cambridge University Press, 1999.

Erikson, Kai. *A New Species of Trouble: Explorations in Disaster, Trauma, and Community*. 1st edition. New York: W. W. Norton & Co., 1994.

Fuller, Richard Buckminster. *Operating Manual for Spaceship Earth*. Carbondale, IL: Southern Illinois University, 1968.

Frickel, Scott. "Just Science? Organizing Scientist Activism in the US Environmental Justice Movement." *Science as Culture* 13, no. 4 (December 1, 2004): 449–69.

Gottlieb, Robert. *Forcing the Spring: The Transformation of the American Environmental Movement*. Washington, DC: Island Press, 1993.

Hays, Samuel P. *Beauty, Health, and Permanence: Environmental Politics in the United States, 1955–1985*. Studies in Environment and History. Cambridge: Cambridge University Press, 1987.

Hoffman, Andrew J. *From Heresy to Dogma: An Institutional History of Corporate Environmentalism*. Expanded edition. Stanford, CA: Stanford Business Books, 2001.

Johnson, Erik W. "Social Movement Size, Organizational Diversity and the Making of Federal Law." *Social Forces* 86, no. 3 (2008): 967–93.

Johnson, Erik W., and Scott Frickel. "Ecological Threat and the Founding of US National Environmental Movement Organizations, 1962–1998." *Social Problems* 58, no. 3 (2011): 305–29.

Lafferty, William M, and Katarina Eckerberg. *From the Earth Summit to Local Agenda 21: Working towards Sustainable Development*. Hoboken, NJ: Taylor & Francis, 2013.

McCloskey, Michael. *Sierra Club Executive Director: The Evolving Club and the Environmental Movement, 1961–1981*. Berkeley, CA, 1983.

McCormick, Sabrina. *Mobilizing Science: Movements, Participation, and the Remaking of Knowledge*. Philadelphia, PA: Temple University Press, 2009.

McLaughlin, Paul, and Marwan Khawaja. "The Organizational Dynamics of the US Environmental Movement: Legitimation, Resource Mobilization, and Political Opportunity." *Rural Sociology* 65, no. 3 (September 1, 2000): 422–39.

Meadows, Donella H., Dennis L. Meadows, Jørgen Randers, and William W. Behrens III. *The Limits to Growth: a Report for the Club of Rome's Project on the Predicament of Mankind*. New York: Universe Books, 1972.

Melosi, Martin V. *Garbage in the Cities: Refuse, Reform, and the Environment*. History of the Urban Environment. Pittsburgh, PA: University of Pittsburgh Press, 1981.

Mertig, Angela G., Riley E. Dunlap, and Denton E. Morrison. "The Environmental Movement in the United States." *Handbook of Environmental Sociology*, 2002, 448–81.

Meyer, John W., John Boli, George M. Thomas, and Francisco O. Ramirez. "World Society and the Nation State." *American Journal of Sociology* 103, no. 1 (1997): 144–81.

Mitchell, R. C., A. G. Mertig, and R. E. Dunlap, "Twenty Years of Environmental Mobilization: Trends Among National Environmental Organizations." In *The US Environmental Movement, 1970–1990*, R. E. Dunlap and A.G. Mertig, eds. (Philadelphia, PA: Taylor and Francis, 1992); Ronald G. Shaiko, *Voices and Echoes for the Environment*. New York: Columbia University Press, 1999.

Nash, Roderick. "The American Invention of National Parks." *American Quarterly* 22, no. 3 (1970): 726–35.

Pampel, Fred C., and Lori M. Hunter. "Cohort Change, Diffusion, and Support for Environmental Spending in the United States." *AJS; American Journal of Sociology* 118, no. 2 (September 1, 2012): 420–48.

Poole, Robert. *Earthrise: How Man First Saw the Earth*. New Haven, CT: Yale University Press, 2008.

Rome, Adam. *The Genius of Earth Day: How a 1970 Teach-in Unexpectedly Made the First Green Generation*. First edition. New York: Hill & Wang, 2013.

Schlosberg, David. *Environmental Justice and the New Pluralism: The Challenge of Difference for Environmentalism*. Oxford: Oxford University Press, 2002.

Schumacher, Ernst F. *Small Is Beautiful: Economics as If People Mattered*. New York: Harper & Row, 1975.

Shaiko, Ronald G. *Voices and Echoes for the Environment: Public Interest Representation in the 1990s and Beyond*. New York: Columbia University Press, 1999.

Southey, Sean. "Accelerating Sustainability: From Agenda to Action." *Local Environment* 6, no. 4 (November 1, 2001): 483–89.

Tarr, Joel A., ed. *Devastation and Renewal: An Environmental History of Pittsburgh and Its Region*. Pittsburgh, PA: University of Pittsburgh Press, 2004.

Van Rooy, Alison. "The Frontiers of Influence: NGO Lobbying at the 1974 World Food Conference, the 1992 Earth Summit and beyond." *World Development* 25, no. 1 (January 1, 1997): 93–114.

Wackernagel, Mathis, and William E. Rees. *Our Ecological Footprint: Reducing Human Impact on the Earth*. New Catalyst Bioregional Series. Philadelphia, PA: New Society Publishers, 1996.

11

SUSTAINABLE DEVELOPMENT AND THE UNITED NATIONS

Iris Borowy

Introduction

It would be difficult to overestimate the importance of the UN for the rise of sustainable development (SD) as a policy goal. It was the cradle of the term, if not actually of the concept in its modern form, and through the Sustainable Development Goals it continues to be a driving force in global efforts to translate this idea into policy.

This dominant role is hardly coincidental. Both in terms of problem recognition and efforts at problem solving, the UN offers major strengths. In the age of climate change, biodiversity loss, a shrinking ozone layer, but also of staggering economic inequality with a Gini coefficient of 0.7,[1] the issues at hand frequently only become visible when viewed on a global scale. More importantly, in a globalized world, as economic decisions taken in one part of the world affect social and environmental developments in other parts, these connections risk remaining undetected unless concerned groups draw attention to them and find a platform from which their voices are heard. Such arguments, in turn, risk remaining without consequences unless they have access to a forum designed to accommodate global debate, negotiations and, ideally, practical cooperation. The evolution of SD, both as a concept and as a budding policy has required that people from different continents, different world views, and different expertise, but with a shared sense of the political sensitivities involved, find an institutionalized framework in which arguments lead to debate, debate to negotiations, negotiations lead to policy decisions, and such decisions eventually lead to action. This is not to say that this process has been smooth or outstandingly successful, particularly when it comes to action. Nor does it argue that the UN was essential to challenge mainstream ideas of economic development. Given the obvious shortcomings of the existing form of global development, alternative economic models could and did emerge in various places. But this paper does argue that without an infrastructure of institutionalized connection between scientific study, diplomatic negotiations, and political agency, the concept of development would look very different today. It took the combination of international environmental negotiations, based on the Stockholm Conference in 1972, the debates on the "New International Economic Order" during the 1970s, and a UN tradition of forming independent Commissions in charge of studying world problems to produce the idea of SD as we know it today.

Any evaluation of what the UN did or did not "achieve" in this process is bound to be controversial, with findings depending on assumptions on what could or should have been achieved. At its core, SD is about addressing competing interests: those of people living today versus those living in the future, the rich versus the poor, and the North versus the South—and between all those shades in between. Consequently, the history of SD at the UN is largely the history of attempts to mediate between these interests. It is a story of debates on who bears more responsibility for problems and, consequently, for problem solving, but also of power asymmetries, political calculations and psychological inertia, all of which have influenced the outcome of negotiations as much as or more than factual arguments. Depending on perspective, the result reflects a history of missed opportunities or of a dogged continuation of efforts.

This paper traces the tension between creating, seizing, missing, and creating new opportunities.

The 1970s

The tension between environmental and developmental concerns surfaced powerfully during preparations for the UN Conference on the Human Environment, held 1972 in Stockholm, the first large-scale conference dedicated to the environment. While many people in the global North welcomed the meeting as an event that finally put the environment onto the global agenda, policymakers in the South initially rejected the initiative, suspecting a neo-colonial strategy to obstruct their industrialization and economic growth under the guise of environmental concerns. After all, Europe, North America, and Japan had all "sacrificed their air, forests, and water to rapid industrial development" and it seemed unreasonable that they should have the right to lecture the rest of the world about reasons why it should not be allowed to do the same.[2] It took a preparatory meeting of representatives of low-income countries and prominent developmental experts at the Swiss resort of Founex to even make the Stockholm Conference possible. The Founex meeting produced a report which addressed key elements of Southern concerns about environmental debates, including the primary responsibility of industrialized countries for most environmental degradation and the need to address the environment in ways that neither infringed on national sovereignty nor impeded economic development for low-income countries.[3]

This report fed into the Stockholm Conference, and its lengthy "Declaration" and "Action Plan," which made clear the basic dilemma of environmental protection in the face of widespread poverty and need for economic development in large parts of the world. Thus, there were basic requirements of human existence such as food, shelter, work, education, and healthcare which developing countries could not afford to ignore in the interest of "uncertain future needs." Consequently, the "problem was how to reconcile those legitimate immediate requirements with the interests of generations yet unborn."[4] It was probably the clearest enunciation of the crucial challenge of finding a way to diffuse the tension between social justice and intergenerational justice, i.e. between the needs of poor people today and of all people in the future. As development-related policies, the Action Plan proposed the collection and distribution of information in order to identify environmental problems and spread environmentally friendly technology. It also declared that environmental consideration must not be used to inhibit development through discriminatory regulations.[5]

This uneasy combination of concerns about environmental and developmental challenges shaped international debates during the following years, and different groups chose different priorities. One strand evolved along negotiations at the UN about a New International

Economic Order (NIEO). This demand reflected increasing frustration among low-income countries in the global South at what they experienced as systemic obstructions which kept them from developing their economic potential. Perspectives on what exactly NIEO encompassed differed, ranging from changes in financial and trade regulations to a more far-reaching restructuring of the economic, political, and cultural bases of power asymmetries. Specific demands included more stable prices for commodities or raw materials exported by low-income countries; the transfer of resources from richer to poorer states through increased development aid and/or debt relief; regulations for transnational corporations (TNCs); free or much expanded access to the markets of industrial countries for products of developing countries; reforms of the international monetary system, and generally more influence in international platforms where political and economic questions were negotiated.[6] In May 1974, the UN General Assembly adopted a Declaration on the Establishment of a New International Economic Order, which called for a long list of respective measures.[7] For some years, there were serious negotiations between representatives of high- and low-income countries along those lines, but by the early 1980s, they had all but faded in the face of rising neoliberalism in Britain and the USA and the eruption of the debt crisis in Latin America and Africa.[8] However, the demands and the frustrations related to the failure of this initiative lingered and would weigh heavily on subsequent considerations regarding SD.

In a separate strand, considerations about "how to continue reaping the benefits of economic growth, while avoiding undesirable or unacceptable damage to the environment" were taking place at the Organisation for Economic Co-operation and Development (OECD).[9] In the 1970s, many countries introduced environmental regulations, typically forcing industrial producers to take end-of-pipe measures to mitigate the pollution and other environmental damage of their activities. In preparation of a 1979 conference of ministers of the environment, a background paper discussed the economic effect of these measures and concluded that "environmental aims and economic growth objectives are or should be made complementary" and that measures designed to decrease the burden on the environment had "made significant contributions to improvements in welfare in OECD communities."[10] In what came to be the guiding principle of SD, the report argued that environmental and economic goals could be further reconciled if the environment was no longer treated as an extraneous component of economic development but as one of its constitutive elements, The new value was that policies regarding energy, raw materials, employment, and urban planning would be coordinated with environmental measures.[11]

These developments formed the background for a meeting commemorating the ten-year anniversary of the Stockholm Conference in 1982. Realistically, the record of global environmental degradation was sobering. As several delegates pointed out, the Action Plan had been partially implemented at best, and though environmental issues enjoyed a higher profile than in 1972, grave environmental destruction persisted.[12] It was the type of complex global challenge for which the UN had already created international independent commissions in the past.[13] This precedent encouraged a UN resolution calling for the creation of a design "to propose long-term environmental strategies for achieving sustainable development by the year 2000 and beyond" and "to recommend ways in which concern for the environment may be translated into greater co-operation among developing countries and between countries at different stages of economic and social development and lead to the achievement of common and mutually supportive objectives which take account of the interrelationships between people, resources, environment and development."[14] It was an impossible task, but also an immense opportunity.

The Brundtland Commission

In May 1984, the World Commission on Environment and Development (WCED, better known as Brundtland Commission) came together in Geneva, the first of eight meetings over a three-year period. Commissioners came from 21 countries in different parts of the world, a majority being from Southern countries, all with a background of working in or with international organizations. It was headed by Gro Harlem Brundtland, former Environmental and Prime Minister of Norway. Its Secretary-General, the Canadian Jim MacNeill, had until then headed the Environmental Directorate at the OECD, and he brought with him the principles on which Commission discussions would build: the environment and the economy should and could be rendered compatible, and policies required a comprehensive, upstream approach, looking at underlying causes rather than symptoms of problems. The Secretariat solicited background papers and reports on the topics, and in addition numerous NGOs, institutions, or individual people contributed information, either via mail or during public meetings, and collectively this body of papers formed the basis for Commission deliberations.

Commissioners were not so much expected to provide factual expertise but to engage in the difficult process of finding agreement and compromise for a consensus development concept. This task was far from easy. Given their diverse national, cultural, and professional backgrounds, Commissioners disagreed on a list of issues. A recurring object of controversy was nuclear energy, an emotional issue that gained additional relevance after the Chernobyl disaster in 1986. But most disagreements were tied to a North–South divide in worldviews: generally speaking, Northern Commissioners tended to focus on local economic weaknesses and environmental burdens for which the respective societies deserved support. By contrast, Southern Commissioners were more likely to see local problems as tied to a global economic system that privileged the affluent of the North and forced Southern societies to engage in environmentally destructive procedures. By contrast, all Commissioners agreed that poverty was key to any solution. But this unanimity facilitated discussions only up to a point. Inevitably, Commissioners faced the vicious circle which, in one way or another, had haunted talks since before Founex: Poverty was unsustainable, and poverty reduction required economic growth in poor countries. However, economic growth in poor countries was difficult to implement without economic growth also in industrialized countries, which, in turn, would presumably push the world beyond global environmental limits.

It was an intractable confrontation of physical against political impossibilities: If poor countries needed to grow in order to reduce poverty, but the growth of the entire global economy was constrained by physical limits on a finite planet, then the logical consequence was that wealth needed to be redistributed from rich to poor (i.e. from the North to the South), or at least further economic growth in the North with its mass production, consumption, and waste needed tight restrictions. Inevitably, these proposals were patently unacceptable in industrialized societies where the expectations, jobs, and futures of many people depended heavily on rising incomes through ongoing economic growth. However, in the absence of substantial changes in industrialized countries, there was little reason for people in low-income countries to accept restricting regulations for their development.[15] At some point, environmental advisor Nitin Desai commented in exasperation:

> It sounds often as if I'm on a bicycle and somebody would just pass me on a motorcar and tells me: look how terrible motorcars are, don't try and get them at all. I don't think any description of sustainable assessment will ever be complete unless the essential issue of sustainable development on a global scale is addressed. And that

central issue is that of the impact of activities in the industrial countries on what is happening at a global level.[16]

To Desai, it made no sense to see "development" as something that had to happen in Southern countries although it was mainly decisions taken in industrialized countries that affected global development. To Commissioners from Northern countries, by contrast, the developmental deficiencies in the global South seemed obvious enough.

This dilemma tied into the question about what character the final report should take. Should it be a profound and alarming analysis of the status quo with recommendations for the radical changes that the situation called for? Or would it be better to present milder recommendations that did not require revolutionary changes and stood a better chance of acceptance and implementation? Would the report lose readership if it was too scientific, and credibility if it was insufficiently scientific?[17]

There were no easy answers, and every choice came with a price. Brundtland insisted on unanimity, arguing that anything else would reduce the report from a powerful global position to a meaningless collection of individual opinion.[18] Her position made eminent sense, but it required a degree of compromise not everyone was ready to make. Midway through the process, the Mexican member, who came from a dependency theory[19] background, left the Commission.[20] The remaining Commissioners decided that the issue at hand was sufficiently important to try to arrive at painful compromise, where possible, and to remain vague where it was not. This strategy gave rise to the often-cited definition of sustainable development as "development that meets the needs of the present without compromising the ability of future generations to meet their own needs."[21] Crucially, the satisfaction of these needs should give priority to the needs of the poor and recognize overall environmental limits, "imposed by the state of technology and social organization."[22] In addition, Commissioners emphasized those aspects on which they could agree, notably the need for poverty reduction and for the spread of more environmentally friendly technology, but also legal and political reforms that would introduce new coordinating and regulatory features into international relations which would indirectly affect international power relations. By contrast, they downplayed the environmental pressures that resulted from wealth and from high-consumption lifestyles. Though this bias led to a somewhat slanted perspective, the final report still ended with a remarkably multifaceted understanding of sustainable development.

For the Commissioners, the centerpiece of the report was the need for a transition to a sustainable form of development, which meant integrating considerations in decision making. This would "require a change in attitudes and objectives and in institutional arrangements at every level."[23] It was the logical continuation of the concepts of the OECD and of Stockholm: rather than introduce environmental measures as an add-on to reduce the damage done by economic production designed to maximize wealth, environmental consideration should be part of production planning from the beginning, thereby changing the basic rationale of economic activities. Sustainable development, in other words, would require a fundamental change in public and private decision-making. These changes were to be supported by various new institutions. Internationally, the Commission recommended the establishment of a UN board for sustainable development, a global risk assessment program, a special international banking facility for development and for the protection of critical habitats and ecosystems, and increased responsibility of the International Court of Justice for environmental and resource management problems. Nationally, the report urged for reformed legal structures which would make it a state responsibility to "observe the principle of optimum sustainable yield in the exploitation of living natural resources and ecosystems," to make available all relevant information to the public and to assess all major new policies regarding their effect on sustainable development.[24] Arguably, the most important proposal addressed automatic financing on an international level: thus, revenue could be raised from the use of international commons like ocean fishing, transportation, sea-bed mining, Antarctic resources or from the use of the orbit for satellite positioning; taxes in international trade, both general and specific; or fees on international financial transactions such as IMF drawing rights.[25] These payments would affect mostly high-income countries and the revenues should be used to ease the transition to SD in low-income countries.

If implemented, this program would have resulted in a fundamentally changed world in which private actors, national authorities, and international organizations all geared their policy decisions to considerations of sustainable development, and were legally obliged to do so. It was the closest any agency had come so far to a comprehensive theory that incorporated elements of wealth, limits, and distribution. By developing the concept of SD and making it widely known, the Brundtland Commission seized its opportunity.

The initial response to the report was enormous. Virtually all UN organizations, the World Bank, NATO, and various regional organizations discussed the Brundtland report in special meetings and/or prepared reports on how they intended to integrate elements of sustainable development into their work. Similarly, in numerous countries, parliaments, special task forces, or governmental conferences discussed this question. NGOs and academic institutions around the world dedicated events, conferences, classes, or special programs to the presentation and discussion of the Brundtland report and sustainable development. The publisher Earthscan issued a Reader's Guide to the report in English, Spanish, French, Hindi, Chinese, and Urdu.[26] If, in 1980, "sustainable development" had been an obscure expression familiar only to a small fringe group of scholar-activists, ten years later it had become difficult not to be familiar with the term.

However, while the Brundtland Commission was very successful at spreading the name of "sustainable development," it did not succeed in endowing it with an unequivocal meaning. Increasingly, the broad range of topics and recommendations in the report turned out to be its weakness as much as its strength. While it did justice to the complexity of real life and to the comprehensive nature of any realistic strategy, it included such a variety of elements that it provided ample ground for selective reading and endorsement. Rather than a comprehensive concept that derived its value from bringing together the disparate elements of development, SD was increasingly perceived as a collection of suggestions, offering "something for everyone."[27] Most governments chose to understand SD in narrowly environmental terms, effectively overlooking a large portion of what the Brundtland Commission had recommended, and some did little more than rename existing environmental programs.[28] The need for increased wealth through economic growth was most readily endorsed by many mainstream institutions, but it also attracted the most vehement protests from environmental activists, who felt that it was tantamount to greenwashing capitalism and environmental destruction. In the process, both sides overlooked the profound changes that the Brundtland report saw as necessary to make economic growth compatible with environmental protection. Others did note the comprehensive inclusion of potentially contradictory aspects but discarded it as intellectually dishonest inconsistencies or vagueness.[29]

Overall, the widespread endorsement of SD presented a chance for more far-reaching discussions on what it would take to make development sustainable. But the collective reaction by policymakers, organizations, and much of the public was that, by not engaging in a serious discussion of the pros and cons of SD in the existing world, SD was reduced from a complex concept to a soundbite, and constituted a missed opportunity to place modern life on a more sustainable path.

The Rio Earth Summit

Most of the recommendations of the Brundtland report were never implemented, but the proposal for a follow-up meeting was. In 1992, the UN Conference on Environment and Development (UNCED), soon dubbed "Earth Summit," took place at Rio de Janeiro. The best hope was to strike a "global bargain" in which the industrialized countries of the North would contribute commitments on finance, technology transfer, and a moderation of their high-consumption life-style in exchange for Southern commitments to policies regarding greenhouse gas restrictions, rain forest conservation, and "sustainable development."[30] In other words: Northern acceptance of redistribution and some restriction of affluence would buy Southern acceptance of a certain restriction of their potential wealth, in that they would preserve part of their forests as sinks for Northern (and Southern) waste instead of turning them into man-made capital, thus stretching limits for Northern countries. In essence it was the idea of mutual subsidies: Northern money would subsidize Southern gains towards more wealth and Southern sinks would subsidize the continuation of Northern wasteful affluence. It can also be seen as a bargain of a redistribution of man-made capital against a redistribution of natural capital. On balance, the hope was that the deal would enable the worldwide economy to remain within (or not as far beyond) global limits.

If it had worked, the bargain would have profoundly changed the way the North and the South, the rich and the poor, related to one another. At its heart, it would have meant that all countries, collectively and individually, contribute efforts towards meeting the challenge of creating equitable affluence without taxing the production and waste absorption capacity of the Earth beyond its limits. It would have meant that a critical mass of governments accept the logic that, in the long run, it was worthwhile to accept the drawbacks of such a bargain in order to reap its benefits.

Some sincere efforts were made, and in long hours of negotiations the delegates of the diverse governments of the world may have come closer to such a deal than at any time before or since, though eventually, too many governments aimed too much at maximizing gains while minimizing losses. Nevertheless, several important agreements were hammered out. Arguably, the main outcome at Rio was Agenda 21, which could be perceived as the detailed action plan to implement the Brundtland Commission ideas of SD. Agenda 21 combined analysis and detailed action plans on 115 program areas. Its estimated cost added up to a staggering $600 billion per year, on average, until the year 2000, whereby $125 billion should be provided by high-income countries as grants, the rest would come from low-income countries, aided by foreign credits.[31] After prolonged acrimonious negotiations, several OECD countries pledged significant funds, though nowhere near $125 billion.[32] In return, Southern countries remained intransigent about their part of any large environmental bargain, effectively stonewalling any commitments on forest conservation. Their unwillingness to renounce material economic benefits was matched by a similar unwillingness among OECD countries, above all the USA, to rein in their consumption patterns—or even acknowledge them as a problem.[33] In the end, Agenda 21 did contain very frank language about an "unsustainable pattern of consumption and production, particularly in industrialized countries" as "the major cause of the continued deterioration of the global environment" and "a matter of grave concern, aggravating poverty and imbalances."[34] But this language was no substitute for tangible commitments, let alone tangible action.

Agenda 21 confirmed "[e]conomic growth, social development and poverty eradication" as "the first and overriding priorities in developing countries and … essential to meeting national and global sustainability objectives."[35] This commitment to wealth as the paramount element of development was mitigated by references to redistribution through the transfer of additional financial resources from high- to low-income countries. A long list of possible measures illus-trated where these new and additional financial resources could come from, ranging from private donations, Official Development Assistance (ODA), and grants from development banks and agencies to debt relief, investment incentives, and support of R&D.[36] Of this combined endorsement of affluence and redistribution, the latter entailed no commitment and was destined to verge on dead letter. While the idea of limits remained the underlying justification for the conference, the casual reader could be excused for overlooking it. The word appeared only twice in the entire 500-page text.[37]

The outcome of the Earth Summit was received critically at the time, especially by envi-ronmentalists, as falling far short of expectations.[38] However, without doubt the event spread the word about SD. Within a year, at least 120 international meetings on relevant topics had taken place and a scientific journal, named *Sustainable Development* had been born.[39] The use of the expressions virtually exploded in published texts. This proliferation of the term reflected the extent to which it satisfied the need for a concept that would reconcile material well-being with a healthful and supportive natural environment. But the lack of effective implementation reflected a widespread unwillingness to accept the consequences of this concept. In large part, the world community acted like people who try to lose weight by buying a book about a healthy, long-term diet and placing it on their shelves.

Sustainable development: Recent strands

After 1992, the concept of SD was largely absorbed into a conventional industrialization discourse of development with a focus on poverty. Poverty reduction was what everybody could agree on. The focus also encouraged a shift of perspective from SD as a global concern to SD as a component of a strategy to make Southern low-income countries more like North-ern high-income countries. In the process, SD shrunk from a development theory in its own right to an adjunct of a mainstream model, whereby increasing wealth was recognized as a key goal with distribution and limits playing a secondary role at best.

The process could be observed in the Millennium Development Goals (MDGs). In 2000, the UN General Assembly adopted a long catalog of principles and commitments for a better world. Building on the combined activities of the UN and the OECD, it was a list of highly diverse measures such as peace-building, strengthening the UN and the International Court of Justice, and reducing poverty. These plans came to be known as the Millennium Development Goals. By 2005, they had become categorized into eight goals, specified by 18 targets and 48 quantifiable indicators (21 targets and 60 indicators in the UNDP format).[40] Numerous aspects of the Millennium Declaration were not adopted into the MDGs, including disarmament and the implementation of Agenda 21, and these omissions changed the overall conceptualization of what constituted development from changing the state of the world to changing the state of Southern low-income countries.[41]

Goal seven referred to ensuring "environmental sustainability," whereby this choice of title already betrayed that this had little to do with SD in a Brundtland sense. The qualifier of "envi-ronmental" already indicated that this goal was not about SD, which was, by definition a comprehensive concept, and the targets confirmed that the use of the term "sustainability" entailed an element of false advertising. Two targets referred to access to sanitation and clean drinking water and to unidentified improvements for slum dwellers, demands which seemed more relevant to the health-related goals four, five, and six (to reduce child mortality, improve maternal health, and to combat HIV/Aids, malaria, and other diseases). Target 7a, to "integrate the principles of sustainable development into country policies and programmes" took up the central demand in the Brundtland report, but it was difficult to see how such an integration could be measured by the amount of land covered by forest, per capita CO_2 emissions, or by species extinction, the indicators listed. Ironically, targets that captured the controversial nature of SD, as discussed by the Brundtland Commission, were found in goal eight, to "develop a global partnership for development." Its targets and indicators took up old NIEO demands of Southern countries for an "open, rule-based, predictable, non-discriminatory trading and financial system," measured by the access low-income countries had to markets and in tariff reduction, in overall ODA and in ODA dedicated to social services. It also included debt management, universal access to essential drugs, and modern communication technology.

During the following years, progress on the MDGs turned out to be substantial but uneven, with most success claimed for reductions in poverty, child mortality, incidence of malaria and tuberculosis, and for access to sanitation. These improvements were read and largely related to (or at least compatible with) increases in wealth. However, the record also showed shortcomings of global development such as the increase of global emissions of carbon dioxide by over 50 percent. Potentially even more revealing are those aspects on which the final report remains silent. Five out of 60 indicators are not mentioned, including 1.2 (poverty ratio gap), 1.3 (share of poorest quintile in national consumption) and 1.4 (growth rate of GDP per person employed), all related to distributional (in-)equality. Clearly, among competing goals in SD, improvements in wealth and living standards won while environmental limits and distributional justice lost out.[42]

This uneven outcome was clear years before the end of the MDGs. In 2012, at the twentieth anniversary of UNCED, participants at a Rio+20 conference, again in Rio de Janeiro, reviewed discussions since the Stockholm Conference in 1972 and decided to replace the MDGs with Sustainable Development Goals (SDGs) in 2015, thereby returning sustainability considerations to the center of development thinking.[43] UN Secretary-General Ban Ki-moon launched the UN Sustainable Development Solutions Network (SDSN), designed to mobilize a broad range of global scientific and technological expertise and to give the impression of maximum openness, transparency and outreach, with the internet as main medium.[44] In January 2016, the SDGs officially came into force. Compared to the MDGs, the SDGs are not only broader but far more committed to connecting socio-economic improvements with their environmental contexts and with socio-economic development elsewhere. Additional factors include inequalities, economic growth, decent jobs, cities and human settlements, industrialization, oceans, ecosystems, energy, climate change, sustainable consumption and production, peace and justice.[45] Many addressed aspects whose ongoing trends are uncertain or downright negative, such as climate change, global arms expenditures, deforestation, desertification, waste production, or road traffic deaths. Most—or, indeed, all—of the demands formulated in the 169 targets had been voiced before, notably in the Millennium Declaration in 2000, but also earlier at the UN conferences of the 1990s, including the 1992 Rio conference, and many already by the Brundtland Commission and before.

Thus, the SDGs could be seen as old wine in new bottles, but in this case the bottles make a lot of difference. The format of firmly established and quantifiable development goals endowed this program with a new commitment and perceived achievability. It also turned the SDGs into a fascinating hybrid, combining the contents of broad, global concept of development of 1990s UN conferences with the shape of MDGs, whose primary effect had been to transform this narrow approach to a regionally focused concept, demanding and promising change in the South. Though the MDGs seemed to obstruct a truly global view on development in 2000, ironically they provided the impetus, the infrastructure, and the vocabulary for this developmental concept to return with a vengeance in 2015.

Conclusions

Since its emergence in the early 1980s, SD has had a contradictory record of, on the one hand, extreme success as an idea, which became universally known within relatively few years, and of little apparent success in mitigating global inequity and threats of overshooting global limits, on the other. This discrepancy had led to widespread disenchantment with the concept and accusations of its ineffectiveness. However, blaming the concept of "sustainable development" for the absence of sustainability in the real world seems disingenuous, similar to blaming the concept of "love" for the continuation of warfare. Instead, there is a clear mismatch between the attraction of a concept, which promises to solve the major problem of the present and the future, combined with the unwillingness of a critical mass of institutions and people to accept the profound changes that this concept demands. So far, few people have been willing to give up the existing development model and its promise of ever-increasing wealth for a strange and frightening new model. The result has been a constant struggle for the construction of SD, whereby actors routinely try to establish an interpretation that combines a maximum of problem–solution power with a minimum of painful change.

To what extent the outcome of the last 45 years of SD discussions at the UN are viewed as a succession of missed opportunities or a series of activities in which one initiative built on the achievements of the last, however limited, is open to interpretation. In many ways, the outcome of SD at the UN is impressive. Over the decades of activities in the field, SD has become a universally accepted and near-ubiquitous goal. Its use has shown a tendency from passive descriptions of problems and policy concepts to active programs, and from high-profile UN events with limited and separate civil society activities to full horizontal and multi-sector cooperation. At the time of writing, in October 2016, the UN operated an SD knowledge platform, which listed well over 2,100 registered voluntary commitments and partnerships from institutions ranging from Photographers without Borders to East Africa Breweries Limited, Portland Community College, and various World Bank groups.[46] On the other hand, the effect on the ground is hardly visible: the world population still uses the resources and the waste absorption capacities per year which the Earth needs 18 months to reproduce or regenerate.[47]

To what extent future generations will regard the UN as an organizer of missed opportunities or as the driving force behind a global transformation towards a sustainable world will not be clear for decades. Perhaps it will never be clear. In any event, the responsibility for performance will lie with people in many positions and many places, but any effective implementation of SD in the real world will require policy coordination on a global scale. In the absence of any realistic alternative, it is safe to assume that the UN will continue to play a central role in SD for many years to come.

Notes

1   The Gini coefficient, introduced by Italian statistician Corrado Gini in 1912, is a widely used measurement of income (in-)equality. Its number lies between 0, whereby everyone would earn the same, and 1, where one person would receive all income of a given society.
2   Craig Murphy, *The United Nations Development Programme: A Better Way?* (Cambridge: Cambridge University Press, 2006), 260.
3   UN, *The Founex Report on Development and Environment*, 1971, retrieved on November 10, 2016, from www.stakeholderforum.org/fileadmin/files/Earth%20Summit%202012new/Publications%20and%20Reports/founex%20report%201972.pdf.
4   UN, *Report of the United Nations Conference on the Human Environment*, Brief Summary of the General Debate (Stockholm: UN, 1972), article 36.

5   UN, Action Plan for the Human Environment, B5, Recommendations 102–8, retrieved on June 7, 2017 from www.un-documents.net/aphe-b5.htm.

6   Paul Streeten, "The New International Economic Order," *International Review of Education* 28, no. 4 (1982): 409–13; Nils Gilman, "The New International Economic Order: A Reintroduction," *Humanity* 6, no. 1 (2015): 1–16.

7   UN, Declaration on the Establishment of a New International Economic Order, A/RES/S-6/3201, May 1, 1974, retrieved on September 14, 2016, from www.un-documents.net/s6r3201.htm.

8   Gilman, "New International Economic Order," 1–16.

9   OECD, *The State of the Environment in OECD Countries* (Paris: OECD, 1979), 17.

10 OECD Environment Committee, *Environment and Current Economic Issues*, ENV(79)4 (OECD Archive, February 20, 1979), 39.

11 Ibid., 41.

12 UN, "Expectations" of the Governing Council for Consideration by the World Commission on Environment and Development, Annex II to UNEP/GC.13/3/Add.2, 1982, UN Library at Geneva (UNOGL).

13 Ramesh Thakur, Andrew F. Cooper and John English, eds., *International Commissions and the Power of Ideas* (Tokyo: United Nations University Press, 2000).

14 UNEP, Draft Resolution recommended by the Governing Council of UNEP at its 11th session, May 23, 1982, S-1051-0014-05, UN Archive.

15 Iris Borowy, *Defining Sustainable Development for Our Common Future: A History of the World Commission on Environment and Development (Brundtland Commission)* (Abingdon: Routledge, 2014), 55–116.

16 Ottawa Public Hearing, Evening Session, May 28, 1986, vol. 36, doc. 27, IDRC, 35.

17 Borowy, *Defining Sustainable Development*, 105 and passim.

18 Gro Harlem Brundtland, *Madam Prime Minister: A Life in Power and Politics* (New York: Farrar, Straus and Giroux, 2005), 215.

19 Dependency theory emerged in the 1950s based on studies by economists Raúl Prebisch and Hans Singer. Rejecting modernization theory, which assumed that low-income countries were merely late-comers on a path towards a more modern and prosperous economy, common to all countries, dependency theory argued that it was the weak position of low-income countries within a world system of economies that kept them in poverty. Thus, rather than emulate the development taken by industrialized countries, low-income countries would have to break out of their dependence on a global economic system whose rules were determined by high-income countries. See e.g. Jorge Larrain, *Theories of Development: Capitalism, Colonialism and Dependency* (Cambridge: Polity Press, 1989), 111–210.

20 Borowy, *Defining Sustainable Development*, 124.

21 World Commission on Environment and Development (WCED), *Our Common Future* (Oxford: Oxford University Press, 1987, reprinted 2009), 40.

22 Ibid., 44–5.

23 Ibid., 62.

24 Ibid., 308–42.

25 Ibid., 341–2.

26 Borowy, *Defining Sustainable Development*, 165–72.

27 Heather Smith, "The World Commission on Environment and Development: Ideas and Institutions Intersect," in *International Commissions and the Power of Ideas*, edited by Ramesh Thakur, Andrew F. Cooper and John English (Tokyo: United Nations University Press, 2000), 76–98, here p. 80.

28 OECD, *International Response to the Report of the WCED*, SG/WCED (88)2 (OECD Archive, October 25, 1988); UN, Implementation of General Assembly Resolutions 42/186 and 42/187, A/44/350, 27 July 1989, S-1051-0033-0006, UN Archive.

29 Hilkka Pietilä, "Environment and Sustainable Development," *IFDA Dossier* 77 (1990): 61–70; Wolfgang Sachs, "Global Ecology and the Shadow of 'Development'," in *Global Ecology: A New Arena of Political Conflict*, ed. Wolfgang Sachs (London: Zed Books, 1993), 2–21; Maria Mies and Vandana Shiva, eds., *Ecofeminism* (London: Zed Books, 1993); Subhabrata Banerjee, "Who Sustains Whose Development? Sustainable Development and the Reinvention of Nature," *Organization Studies* 24, no. 1 (2003): 143–80; John Robinson, "Squaring the Circle? Some Thoughts on the Idea of Sustainable Development," *Ecological Economics* 48 (2004): 369–84; Ted Trainer, "A Rejection of the Brundtland Report," *IFDA Dossier* 77 (1990), 84.

30   Stanley Johnson, "Introduction: Did We Really Save the Earth at Rio?," in *The Earth Summit, The United Nations Conference on Environment and Development*, ed. UNCED (London: Graham & Trotman/Martinus Nijhoff, 1993), 5–6.

31   Olav Stokke, *The UN and Development: From Aid to Cooperation* (Bloomington, IN: Indiana University Press, 2009), 358; Paul Little, "Ritual, Power and Ethnography at the Rio Earth Summit," *Critique of Anthropology* 15, no. 3 (1995), 271.

32   Geoffrey Palmer, "The Earth Summit: What Went Wrong at Rio?," *Washington University Law Review* 70, no. 4 (1992), 1020.

33   Johnson, "Introduction," 7.

34   UN, Agenda 21, 1992, retrieved on October 12, 2016 from www.unep.org/Documents.Multi lingual/Default.asp?DocumentID=52, §4.3.

35   Ibid., §33.3.

36   Ibid., §33.10.

37   Ibid.

38   Sachs, "Global Ecology and the Shadow of 'Development.'"

39   Andrew Jordan, "The International Organisational Machinery for Sustainable Development: Rio and the Road Beyond," *Environmentalist* 14, no. 1 (1994): 23–33.

40   The sites of the MDGs and of UNDP differ regarding goals 5, 6 and 7; see www.unmillenniumproject.org/goals and http://mdgs.un.org/unsd/mdg/Host.aspx?Content=Indicators/OfficialList.htm (retrieved October 12, 2016).

41   Najam Adil, "Unraveling of the Rio Bargain," *Politics and the Life Sciences* 21, no. 2 (2002): 48.

42   UN, *Millennium Development Goals Report 2015* (New York: UN, 2015).

43   UN, *Future We Want*, Outcome Document (New York: UN, 2012), retrieved on October 12, 2016 from https://sustainabledevelopment.un.org/rio20/futurewewant.

44   UN, Sustainable Development Knowledge Platform, retrieved on October 14, 2016 from https://sustainabledevelopment.un.org.

45   Ibid.

46   UN, "Partnerships for SDGs," retrieved on October 14, 2016 from https://sustainabledevelopment. un.org/partnerships.

47   Global Footprint Network, retrieved on October 14, 2016 from www.footprintnetwork.org/en/index.php/GFN/page/world_footprint.

References

Adil, Najam. "Unraveling of the Rio Bargain." *Politics and the Life Sciences* 21, no. 2 (2002): 48.

Banerjee, Subhabrata. "Who Sustains Whose Development? Sustainable Development and the Reinvention of Nature." *Organization Studies* 24, no. 1 (2003): 143–80.

Borowy, Iris. *Defining Sustainable Development for Our Common Future: A History of the World Commission on Environment and Development (Brundtland Commission)*. Abingdon: Routledge, 2014.

Brundtland, Gro Harlem. *Madam Prime Minister: A Life in Power and Politics*. New York: Farrar, Straus and Giroux, 2005.

Gilman, Nils. "The New International Economic Order: A Reintroduction." *Humanity* 6, no. 1 (2015): 1–16.

Global Footprint Network. Retrieved on October 14, 2016 from www.footprintnetwork.org/en/index.php/GFN/page/world_footprint.

Johnson, Stanley. "Introduction: Did We Really Save the Earth at Rio?." In *The Earth Summit, The United Nations Conference on Environment and Development*, ed. UNCED. London: Graham & Trotman/Martinus Nijhoff, 1993), 5–6.

Jordan, Andrew. "The International Organisational Machinery for Sustainable Development: Rio and the Road Beyond." *Environmentalist* 14, no. 1 (1994): 23–33.

Larrain, Jorge. *Theories of Development: Capitalism, Colonialism and Dependency*. Cambridge: Polity Press, 1989.

Little, Paul. "Ritual, Power and Ethnography at the Rio Earth Summit." *Critique of Anthropology* 15, no. 3 (1995): 271.

Mies, Maria, and Vandana Shiva, eds. *Ecofeminism*. London: Zed Books, 1993.

Murphy, Craig. *The United Nations Development Programme: A Better Way?* Cambridge: Cambridge University Press, 2006.

OECD Environment Committee. *Environment and Current Economic Issues*, ENV(79)4. OECD Archive, February 20, 1979.

OECD. *International Response to the Report of the WCED*, SG/WCED (88)2. OECD Archive, October 25, 1988.

OECD. *The State of the Environment in OECD Countries*. Paris: OECD, 1979.

Ottawa Public Hearing, Evening Session, May 28, 1986, vol. 36, doc. 27, IDRC, 35.

Palmer, Geoffrey. "The Earth Summit: What Went Wrong at Rio?" *Washington University Law Review* 70, no. 4 (1992), 1020.

Pietilä, Hilkka. "Environment and Sustainable Development." *IFDA Dossier* 77 (1990): 61–70.

Robinson, John. "Squaring the Circle? Some Thoughts on the Idea of Sustainable Development." *Ecological Economics* 48 (2004): 369–84.

Sachs, Wolfgang. "Global Ecology and the Shadow of 'Development'." In *Global Ecology: A New Arena of Political Conflict*, ed. Wolfgang Sachs. London: Zed Books, 1993, 2–21.

Smith, Heather. "The World Commission on Environment and Development: Ideas and Institutions Intersect." In *International Commissions and the Power of Ideas*, eds. Ramesh Thakur, Andrew F. Cooper and John English. Tokyo: United Nations University Press, 2000, 76–98.

Stokke, Olav. *The UN and Development: From Aid to Cooperation*. Bloomington, IN: Indiana University Press, 2009.

Streeten, Paul. "The New International Economic Order." *International Review of Education* 28, no. 4 (1982): 409–13.

Thakur, Ramesh, Andrew F. Cooper and John English, eds. *International Commissions and the Power of Ideas*. Tokyo: United Nations University Press, 2000.

Trainer, Ted. "A Rejection of the Brundtland Report." *IFDA Dossier* 77 (1990), 84.

UN. "Expectations" of the Governing Council for Consideration by the World Commission on Environment and Development, Annex II to UNEP/GC.13/3/Add.2, 1982, UN Library at Geneva (UNOGL).

UN. "Partnerships for SDGs." retrieved on October 14, 2016 from https://sustainabledevelopment.un.org/partnerships.

UN. Action Plan for the Human Environment, B5, Recommendations 102–8. Retrieved on June 7, 2017 from www.un-documents.net/aphe-b5.htm.

UN. Agenda 21. 1992. Retrieved on October 12, 2016 from www.unep.org/Documents.Multilingual/Default.asp?DocumentID=52, §4.3.

UN. Declaration on the Establishment of a New International Economic Order, A/RES/S-6/3201, May 1, 1974. Retrieved on September 14, 2016, from www.un-documents.net/s6r3201.htm.

UN. *Future We Want*, Outcome Document. New York: UN, 2012. Retrieved on October 12, 2016 from https://sustainabledevelopment.un.org/rio20/futurewewant.

UN. Implementation of General Assembly Resolutions 42/186 and 42/187, A/44/350, 27 July 1989, S-1051-0033-0006, UN Archive.

UN. *Millennium Development Goals Report 2015*. New York: UN, 2015.

UN. *Report of the United Nations Conference on the Human Environment, Brief Summary of the General Debate*. Stockholm: UN, 1972.

UN. Sustainable Development Knowledge Platform. Retrieved on October 14, 2016 from https://sustainabledevelopment.un.org.

UN. *The Founex Report on Development and Environment*. 1971. Retrieved on November 10, 2016, from www.stakeholderforum.org/fileadmin/files/Earth%20Summit%202012new/Publications%20and%20Reports/founex%20report%201972.pdf.

UNEP. Draft Resolution recommended by the Governing Council of UNEP at its 11th session, May 23, 1982, S-1051-0014-05, UN Archive.

World Commission on Environment and Development (WCED). *Our Common Future*. Oxford: Oxford University Press, 1987, reprinted 2009.

12

THE GROWTH PARADIGM

History, hegemony, and the contested making of economic growthmanship

Matthias Schmelzer

Introduction

The Organisation for Economic Co-operation and Development (OECD) has proclaimed that "[f]or a good portion of the 20th century there was an implicit assumption that economic growth was synonymous with progress: an assumption that a growing gross domestic product (GDP) meant life must be getting better."[1] Indeed, the dominance of the growth imperative is hard to ignore: Growth statistics regularly appear on the front pages of newspapers, play a key role in economic analyses, and pervade political debates, not only across the political spectrum but also in all countries. Since these numbers have come to form our very language, it seems almost impossible to think about economic issues without referring to GDP and its proxies. The recent global economic crisis has conspicuously demonstrated how dependent capitalist economies are on growth and how even minor reductions in growth rates were received with almost religious disappointment.[2]

Environmental historian John R. McNeill has argued that the "overarching priority of economic growth was easily the most important idea of the twentieth century."[3] Although this statement might at first seem exaggerated, there are good reasons that justify this view. Not only was the idea of economic growth at the core of the ideologies of the socio-economic and political systems whose competition marked the twentieth century, including both capitalism and communism in their different varieties, but the social and economic policies that were the result of the overarching priority of economic growth, or were justified by it, have fundamentally and irreversibly reshaped societies and the planet itself. Over the twentieth century, millions of people have come to take part in the production and consumption of ever increasing quantities of goods and services, even though these processes have been extremely uneven over time and space. At the same time, economic growth has caused environmental changes of unprecedented proportions that are threatening the livelihood of millions of people today, and even more so that of future generations. Ecologists, geologists, and historians have used the concept of the "Anthropocene" to mark the fundamental transformations related to the fact that through the global spread of capitalist modes of production and living humanity itself has become the dominant geological force on planet earth.[4]

In light of the sweeping acceptance of the pursuit of growth as a key policy goal around the world it is easy to forget that not only the reality of economic expansion, but even more so growth as a key category of economic and public discourse is a surprisingly recent phenomenon. Although a highly ambivalent and elusive term, the semantic core of economic growth is statistically fixed. It is generally defined as the annual increase in the monetary value of all the goods and services produced within a country, including the costs of producing all the services provided by the government, measured either as gross national product (GNP) or GDP.[5] Before the 1820s, when economic growth accelerated in the context of the Industrial Revolution, economic activity around the world had been characterized by periodic ups and downs, only expanding by an average of 0.05 percent annually—as far as this can be measured retrospectively—and this was largely due to the slow increase of populations. Even more recently, the term "economic growth" was not widely used before the middle of the twentieth century, but during the 1950s it advanced to become a key notion, not only within economics and other social sciences, but also in political discourses and everyday speech (see Figure 12.1). How can this be explained?

Scholars from a variety of fields, including renowned historians, have described growth as a "fetish" (John R. McNeill) or "obsession" (Barry Eichengreen, Hermann van der Wee), an "ideology" (Charles Maier, Alan Milward), "social imaginary" (Cornelius Castoriadis, Serge Latouche), or an "axiomatic necessity" (Nicholas Georgescu-Roegen). With diverging emphasis, these scholars have highlighted the quasi-religious adoration of growth by economists and

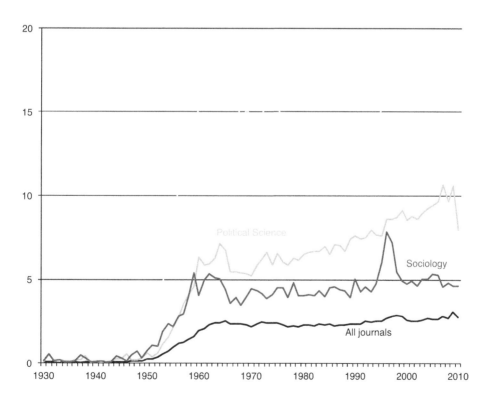

Figure 12.1   Percentage of articles published in all academic journals in the JSTOR database that contain the term "economic growth," by discipline, 1930–2010

Source: calculations based on data from Data for Research (http://dfr.jstor.org).

policy-makers, the underlying interests that are served and at the same time concealed by the dominance of the growth discourse, in particular in the context of postwar class conflicts and Cold War confrontations, or the general acceptance of growth as an incontestable dictum.

However, while studies on economic growth—explaining, assessing, and modeling its causes, effects, and various growth policies—constitute the core of both economics and economic history, there are strikingly few accounts on how economic growth became self-evidently regarded as the key goal of economic policy-making by social scientists, politicians, and the general public.[6] Building on studies of the so-called Post-Development school, I argue that the exceptional position of economic growth as a core policy goal is based on the hegemony of the "economic growth paradigm" and cannot be adequately understood without taking into account the complex structure and long-term historical evolution of this paradigm and its underlying power relations. Economic growth is, of course, one of the key features of capitalist societies, which are predicated upon the continuous accumulation of capital, and to some degree all states in a competitive state system pursue the national interest of increasing their wealth and thus their power. In fact, capitalist societies are dynamically stabilizing and reproducing themselves in a steady process of expansion and intensification with regard to space, time, and energy.[7] One could thus argue that growth is merely the name given to the automatic expansion produced by the economic and technological system.[8] Yet, similar to "development," growth was not merely a socio-economic or technological process or the result of power-relations, but also "a particular cast of mind … a perception which models reality, a myth which comforts societies, and a fantasy which unleashes passions."[9]

After setting this analysis in the context of the current debate about the relationships between GDP, welfare, and environmental sustainability, the chapter introduces a definition of the growth paradigm. Based on original archival research, the main sections sketch the historical making and remaking of the growth paradigm in postwar history by focusing on the debates within one of the least understood international organizations, the OECD. Founded in 1948 as the Organisation for European Economic Co-operation (OEEC) to administer Western European Marshall Plan aid, the OECD was in 1961 refounded as the key forum and think-tank for core capitalist countries, in which civil servants developed, harmonized, and collectively legitimated economic expertise, standards, norms, and policies.[10] In particular, the chapter highlights four discourses that collectively legitimated, universalized, and naturalized the growth paradigm and its underlying social and power relations: GDP as a measure; growth as panacea; growth as the universal yardstick; and growth without limits.

GDP, welfare, and limits: the current debate

Over the last 200 years, economic growth has been bound up with the most dramatic rise in living standards for millions of people, even though socially (class, race, gender) and geographically the effects have been very uneven, and it is still the rallying cry for millions of others who hope for a better life. As French economist Thomas Piketty (2014) has demonstrated in *Capital in the Twenty-First Century*, economic growth seems to be essential to counter capitalism's tendencies to increase inequality, since in times of slow growth inequalities in income and wealth increase as wages tend to grow much slower than returns on capital.

However, there are good reasons to question the desirability or possibility of further quantitative growth in industrialized countries. First, the universal merits of maximizing growth have become rather dubious. Studies in welfare and feminist economics, social history, and ecological economics have shown that the focus on GDP is "mismeasuring our lives."[11] In particular in recent years evidence has been mounting that as national income increases beyond a certain threshold (which most OECD countries reached in the 1970s), it ceases to translate into improvements in human wellbeing. These studies demonstrate that recent growth in rich nations did not alleviate poverty, nor has it been indispensable for human flourishing, for which other factors, most importantly the degree of equality, are much more important.[12]

Second, the ecological and social costs of economic growth are not negligible, especially in the context of achieving global social justice to overcome the North–South divide and repay the accumulated ecological debt of the rich countries. The fundamental promise of growth— to raise the living standard and consumption of soon to be nine billion people to Western levels through a continuous expansion of world GDP—has been irrevocably shattered by the ecological predicament, most prominently climate change. Economic analyses show that achieving equitable modes of living in the global South while staying within planetary boundaries will only be possible if the countries in the North drastically reduce their ecological impact, which most likely implies reductions of economic output to be effective.[13]

Third, the future possibilities of actually achieving further quantitative growth have become more and more precarious due to resource and energy scarcities (such as peak oil) and internal structural problems (such as over-accumulation and financialization), which have led to declining or stagnating growth rates.[14] In a growth society, in which all kinds of policies are predicated upon ever increasing production and consumption, slower growth reinforces all the social and economic problems associated with economic crises, such as rising inequality, unemployment, public debt, social tensions, and even an undermining of democracy. A growing array of economists and theorists thus demand to look "beyond growth," arguing that growth may currently be causing the same problems it was originally hoped to solve and that political responses need to adapt to the changing social, economic, and environmental circumstances.[15] While there is an ongoing debate about the question whether only degrowth—a planned contraction of economic activity aimed at increasing wellbeing and equality—can sufficiently reduce environmental impact in the global North, or whether we should simply start to ignore GDP numbers, the position that the political focus on growth is misleading and problematic is spreading rapidly.[16]

In light of these perspectives the pervasiveness of GDP as a measure of social wellbeing and of growth as a policy goal seem rather peculiar—a "puzzle" or "paradox" in need of explanation, that will be addressed historically in this article.[17]

The growth paradigm: towards a definition

The term "growth paradigm" was first introduced by ecological economist Herman Daly to characterize the preanalytic vision of mainstream economists that justified their belief in unlimited growth.[18] The term has since been employed rather vaguely by ecologists, political scientists, and in public discourse to describe the worldview associated with growthmanship. Thomas S. Kuhn's paradigm theory is helpful in understanding how "the entire constellation of beliefs, values, techniques and so on shared by members of a given community" create normalcy and repress conflicting ideas.[19] Yet his account, which is geared towards understanding paradigm maintenance and change in the natural sciences, has to be considerably broadened to capture the predominance of growthmanship that pervades not only a specific scientific community, but multiple academic, political, and social communities of experts and the wider public. Thus, even though loosely drawing on Kuhn's conception, growth has to be analyzed as a "social paradigm," in the making and legitimation of which the academic community of economists played a key role, but which is much more general. Rather, the "growth paradigm" as used here resembles Charles Maier's characterization of "the idea of sustained economic growth" as "the economic 'ideology' that came to play the greatest role in the non-Communist world" in the postwar era, or the use by Stephen Purdey and Gareth Dale, who emphasize the function of growth as an idealized refiguration of capital accumulation.[20]

Building on these accounts and broadening Kuhn's original conception, I use the term "growth paradigm" to describe a specific ensemble of societal, political, and academic discourses, theories, and statistical standards that jointly assert and justify the view that economic growth as conventionally defined is desirable, imperative, and essentially limitless. Collectively, these assumed (1) that GDP, with all its inscribed reductions, assumptions, and exclusions, adequately measures economic activity; (2) that growth was a panacea for a multitude of (often changing) socio-economic challenges; (3) that growth was practically the same as or a necessary means to achieve some of the most essential societal goals such as progress, well-being, or national power; and (4) that growth was essentially unlimited, provided the correct governmental and intergovernmental policies were pursued.

By examining the emergence and complex interplay of these mutually reinforcing strands within the OECD, this analysis goes beyond existing research by highlighting the transnational dimension and through its source-based and organizationally grounded approach, it argues that a transnational analysis is best suited to understand the significance of the standardization of economic statistics, of the internationalization of economics and related transnational transfers, and of international comparisons and competition between countries. By not only relying on published materials by professional economists and official policy statements, this approach sees economic ideas not as detached from particular socio-economic, political, or organizational contexts but as grounded and ingrained in statistical standards, international policy frameworks, and widely accepted norms. It thus emphasizes the politically contested evolution of the growth discourse and its continuous remaking.[21]

The idea of economic progress, preliminary growth theories, and macroeconomic policies geared towards expansion already emerged in the nineteenth and early twentieth century. However, the modern growth paradigm differed in three important ways from those earlier concepts. First, it was only with the development, standardization, and internationalization of national income accounting techniques that a uniform conception of "the economy" could take hold, a new economic matrix that framed what exactly was growing, and the techniques of quantification to make this measurable. Second, not until the 1950s did economic growth become the primary policy goal, the responsibility of governments, and the most salient indicator for national success and societal welfare. And third, it was only in the mid-1950s that the notion that long-term unlimited economic growth was actually achievable first gained widespread acceptance, in connection with the birth of the first modern growth theories, and only since then has the narrative of progress become bound up with continuing growth of GDP.[22]

The emergence of the growth paradigm in postwar history

GDP as the measure

How did growth become a self-evident concept in the postwar period? The following sections sketch the entangled emergence of the four discursive strands in postwar history. First, the growth paradigm was based on the claim that economic activity is adequately measured as the level of the national product. While presented as an objective, universal, and technical device, the history of the contested making of this international standard reveals not only the inscribed reductions, assumptions, and exclusions, but also its power to naturalize a particular mode of seeing the world and its potential for being turned into a universal metric of worth.

Based on a long tradition of measuring the riches of kings and countries, official government statistics started to be developed in the context of the Great Depression of the 1930s and the related move to state planning in many countries.[23] In this context, modern national income accounting, which transferred the technique of double entry bookkeeping from the firm level to the entire economy, was originally designed by John Maynard Keynes, James Meade, and Richard Stone in Britain and was adopted by Milton Gilbert in the US. By assuming the economy as a pre-existing and stable thing waiting to be measured by some best set of tools, many scholars have tended to misinterpret national income accounting as a mere technical device, ignoring how in this process "the economy" in its present-day meaning was constructed in the first place.[24] It is thus no wonder that national income accounting instruments have been described as "the most important new tools of economic analysis and policy-making in the second half of the 20th century."[25] It had a constitutive power in providing the epistemology of how economists, politicians, but also the general public came to see and think about economic problems and solutions and can thus "be understood as a construction of reality in which growing numbers of community members have come to accustom themselves with the concept of the macro-economy via a *lingua franca* of accounting."[26]

GDP accounting was developed for very specific purposes, at a specific time, and for specific types of economies. The accounting standard was produced for the developed capitalist economies of Western Europe and North America, tailored to their historical situation in the mid-twentieth century, and fundamentally facilitated state management in the Fordist regime. It fundamentally "evolved as a war-planning tool"[27] and became a powerful instrument in the estimation of militarization costs and economic planning in the Allies' fight against fascism during the Second World War and was then used to solve specific Western European problems of reconstruction in the postwar and Cold War period within a Keynesian framework. After the Second World War this method was internationally standardized and thus globalized through international organizations, in particular the OEEC and the United Nations (UN), and within merely a decade 60 countries published national accounts.[28]

In light of the exceptional rise of national income accounting to a Western and, since the 1990s, a global standard, it is important to highlight that this standard has always been disputed. Remarkably, most of the controversies that are currently waged about how GDP is "mismeasuring our lives" can actually be traced back to the period of the making and international standardization of GNP statistics in the late 1940s and early 1950s. Not only were early income statisticians extremely skeptical regarding the universality and the ability of their frameworks to measure welfare and make meaningful comparisons over time and space, they were also engaged in quite fundamental controversies about how to define economic output, whether to include non-monetary housework, and about the explanatory power of these numbers.

These early controversies around national income accounting were particularly pronounced in three areas. First, the welfare critique of national income accounting is as old as the statistics themselves. Already in 1934, Simon Kuznets warned the US Congress that the "welfare of a nation can scarcely be inferred from a measurement of national income."[29] When modern GNP figures were standardized in the late 1940s, the protagonists repeatedly emphasized that they were "not trying to measure welfare, but the value of production from a business point of view."[30] However, this cautious contextualization and qualification of the statistical framework was continuously undercut by contradictory statements by national accounting experts, in particular by economists and public officials, and was soon forgotten. By the mid-1950s, there was no question anymore as to whether GNP represented the "welfare" of a country; it was simply taken for granted.[31]

Second, most economists at that time argued against comparisons of GDP levels between countries (in particular between industrialized countries and what had just been defined as "under-developed" countries) and against the use of these figures for development economics and policies. The exclusion of unpaid labor, housework, and the entire non-market sector, as well as the concept of the "household" and anthropological assumptions about humans as "economic man"—all developed with the US and Western Europe in mind—made the application of GNP accounting to colonial or decolonizing subsistence economies highly problematic or even, many thought, impossible.[32]

A final key debate, which has been neglected in most of the histories of these events, centered on the exclusion of unpaid work from the accounts. Much later, from the 1970s onwards, feminist economists began to criticize GDP accounting for not accounting for non-monetary labor, which is done predominantly by woman in the household, and thus devalues female work. While this research has been vital in highlighting the gendered nature of these statistics, it has rested on the unquestioned assumption that the work of woman was just not taken into account by the makers of national income accounting.[33] However, the record shows that up until the Second World War, many economists and statisticians regarded household labor as part of the productive activities of societies. Domestic unpaid services, which were (and still are) predominantly done by women, were not only included in many of the early national income estimates, but during and after the Second World War they also became part of the official national accounts in countries such as Norway and Hungary.[34] However, existing approaches and discussions were deliberately homogenized and streamlined by the international standardization of national income accounting, which defined non-monetary domestic income as outside the production boundary.[35] Thus, domestic work was not just forgotten, it was regarded as self-evidently unproductive or non-economic. That is, it was explicitly written out of the accounts.

As can be seen from these controversies, the definition of the economy and its measurements were anything but self-evident.[36] A high-level academic debate at the annual meeting of the Econometric Society in September 1947 in Washington is highly illuminating regarding the general mistrust of the political power of a "single figure." At this prestigious meeting, key protagonists of the debate discussed very controversially the fundamental contradictions and imperfections in the existing accounting practices described above. In the face of these ambiguities and problems, the production of one powerful but potentially misleading figure for public use was seen as problematic. Yet the dynamics of international organizations and postwar reconstruction were not considerate of these academic doubts. As the US economist Arthur Smithies remarked:

> These figures have been produced and people use them. If we were starting afresh, I would have a great deal of sympathy with what has been said about not using a single figure, and not even producing one. But the way the thing stands now is that in every governmental problem where a multiplicity of regions or countries is involved, national-income figures are used. … Therefore, I think the statistician cannot bury his head in the sand in this matter. He should know the practical politicians will use his results and probably will misuse them. And therefore I do believe that it is imperative to make the best single figure that is possible and to use a few very simple rules for its application.[37]

Rather than building on a scientific consensus and statistical knowledge, it was the political usefulness of market-oriented income data, especially in the context of international cooperation during the Second World War and in the postwar and early Cold War era, and the essential function of the modern state in the production of social statistics, that made a process of standardization and international harmonization seemingly inevitable. It was a small network of predominantly American and British statisticians around Richard Stone and the OEEC that gained considerable leverage through international organizations, and this network was thus able to establish an international standard. Governments and in particular international organizations acutely required comparative statistics to manage member country contributions and international aid flows and thus cut short these disputes among academics and deliberately homogenized and streamlined existing approaches by standardizing a particular version of GDP accounting.[38] In sharp contrast to these cautious academic debates, the public use of these numbers in the following years turned GDP into a universal yardstick that came to form the basis for both macro-economic growth theories and modern political growthmanship.

Growth as panacea

The growth paradigm asserted that economic growth is a universal remedy for some of the most pressing challenges of modern societies and imperative to avoid economic and social crises. While the specific challenges that according to growth discourses could only be met by GDP growth continuously changed in the postwar period, depending on the socio-economic circumstances, the reliance on growth as a panacea has remained stable.

The idea of economic growth—of continuously increasing levels of national output—was conspicuously absent from policy debates in the immediate postwar years. Not only was its statistical foundation still in the making, but other policy concerns such as high employment, stability, and the restoration of prewar levels of production were uppermost in the minds of policy-makers and economists. Rather than decades of continued economic growth, until the early 1950s most economists, businessmen, and politicians expected a postwar recession. There was "hardly any trace of interest in economic growth as a policy objective in the official or professional literature of western countries before 1950," as historian Heinz Arndt has showed.[39] The official agreement forming the OEEC can be taken as symptomatic in identifying as the objective of economic policies a "strong and prosperous European economy," which, in 1948 was not thought of as a continuously expanding market for more and more goods and services, but the objective was "to achieve as soon as possible and maintain a satisfactory level of economic activity."[40]

However, hopes of plenty and plans of economic expansion were all in the air and the idea of growth was on the rise. In the West, it was first publicly stated in the late 1940s by US government advisors, and by the mid-1950s economic expansion had become a major policy goal in most countries and soon all around the world.[41] When the Secretary-General of the OEEC Robert Marjolin described in his memoirs what in the early 1950s became the aim of "economic policy for the future" in Europe, he proclaimed: "Sustained and as rapid as possible 'growth' was the supreme objective, to which [other policy objectives] had to be subordinated."[42]

The political focus on growth as the key goal first emerged in the US in the late 1940s and in Western Europe in the early 1950s as a (post)war response to the economic and social problems associated with rearmament, European reconstruction, political instability, colonial decline, and the Cold War. Most importantly, growth promised to turn difficult political conflicts over distribution into technical, non-political management questions of how to collectively increase GDP. By thus transforming class and other social antagonisms into apparent win-win situations, it provided what could be called an "imaginary resolution of real contradictions" and played a key role in producing the stable postwar consensus around embedded liberalism:[43] It helped integrate labor and the political Left, rendered rearmament feasible without a decline in living standards; it helped stabilize the Bretton Woods system, and in the context of global inequalities it offered the (post)colonial countries in the global South a possible route out of poverty towards what came to be defined as "progress." Most importantly, it helped to overcome the political focus on equality and redistribution. As noted by the American economist, advisor of President Eisenhower, and governor of the Federal Reserve Bank Henry Wallich, "Growth is a substitute for equality of income. As long as there is growth there is hope, and that makes large income differentials tolerable."[44] In fact, growth became presented as the common good, thus justifying the particular interests of those who benefitted most from the expansion of market transactions as beneficial for all.

It was not in terms of equality, emancipation, or employment that nation states around the world came to compete against each other, but in terms of rising quantities of goods and services produced. Already in 1958 Khrushchev proclaimed: "Growth of industrial and agricultural production is the battering ram with which we shall smash the capitalist system."[45]

Symptomatic for the political focus on growth in that era was what contemporaries have aptly termed "competitive targetry."[46] Driven by planning euphoria, technocratic optimism, international competition, and Cold War rivalry, national governments and international organizations around the world formulated bold numerical policy goals, most importantly growth targets. The most well-known numerical growth targets were set within the Soviet Union. At the twenty-first Party Congress in January 1959, Khrushchev presented a seven-year plan up to 1965, in which economic growth was declared the main task of economic policies, and in which he claimed that until 1970 the USSR would have a higher standard of living than the US. And at the twenty-second Party Congress in 1961, the Soviet government committed itself to far-reaching growth plans that aimed at raising production by 150 percent within ten years and by 500 percent within twenty years.[47]

However, not just countries in the Soviet bloc proclaimed their political goals in terms of economic expansion. Already in 1951 the OEEC declared its first expansion target, to increase the combined GNP of its member countries by 25 percent within five years. The main thrust behind this highly publicized manifesto were efforts to ramp up military spending and investments and to postpone current consumption to finance the rearmament for the Korean War and the cope with the Western European dollar shortages.[48] The OECD, which developed into what one of its directors adequately described as a "temple of growth for industrialized countries" in which "growth for growth's sake" became the supreme and largely unquestioned objective, was even more ambitious:[49] At its first Ministerial meeting in November 1961 the OECD proclaimed the aim, which epitomized the prevalent vision of human progress at that time: to increase the combined GNP of the OECD economies by 50 percent during the 1960s. In 1970, at a time when quantitative growth was heavily criticized, even within the OECD, Ministers even proclaimed a 65 percent increase of the combined GNP within a decade.[50] Similar target growth rates were also issued in the 1960s and 1970s in most countries, among them Yugoslavia, Japan, India, Sweden, France, and the United Kingdom.[51]

While the growth paradigm emerged in the Fordist postwar regime as a Keynesian and interventionist expression of "high modernism," it was flexible enough to encompass neoclassical and liberal schools of economic thinking and to adapt to the shifts in economic reasoning. Rather than undermining the growth paradigm, what has been called the monetarist "counter-revolution," the "marketization" of economics or the rise of "neoliberalism," merely re-articulated growthmanship in a new guise. Arguments justifying the benefits of growth were adapted to highlight the so-called "trickle-down" argument according to which not state-sponsored redistribution but the unhampered workings of growing markets would benefit even the most disadvantaged. Most fundamentally, instead of seeing government as the guarantor in charge of boosting growth in multiple policy fields and ensuring through welfare state policies that growth benefitted the majorities, this "antistatist growthmanship" deemed government interventions that did not enable free-market activities as obstacles to growth. In the neoliberal growth regime "growth" still carried the promises of employment, equality, welfare, and rising living standards, which had been so central to its rise in the postwar era. But "sustained non-inflationary growth," the new reasoning went, now depended on open trade, functioning markets, all-out liberalization, higher investments, and—most generally—higher "profitability."[52] While the "golden age" still acted as a legitimating force, higher rates of unemployment, declining wages, rising inequality, and welfare cuts were justified as necessary prerequisites of faster growth, which, so the promise went, would create more jobs and rising wages and living standards in the future.

Wendy Brown has recently argued that in the 1980s the state was radically economized in three ways:

> The state secures, advances, and props the economy; the state's purpose is to facilitate the economy, and the state's legitimacy is linked to the growth of the economy—as an overt actor on behalf of the economy. State action, state purpose, and state legitimacy: each is economized by neoliberalism.[53]

A focus on the rise of the growth paradigm shows that already during the 1950s the state had taken on this encompassing responsibility, which became in the context of growth politics its *raison d'état*, as can be demonstrated by focusing on the debates around the OECD's growth targets. The dependency of the state from economic expansion and the focus on solving all kinds of problems—ranging from inequality, unemployment and public debt to environmental depletion—through economic growth have turned the modern state into a growth state.

Growth as universal yardstick

The growth paradigm was reinforced by the belief and often-implicit assumption that economic growth was practically the same as, or a necessary means of, achieving some of the most essential ambitions of modern societies, such as social well-being for all, progress, modernity, societal dynamism, national power, or prestige. Steady growth and the prospect of ever increasing consumption helped resurrect the belief in progress, which had been so seriously eroded by the Great Depression, two World Wars, and the Shoah.[54] The technical, scientific, and politically neutral aura of growthmanship, which was underwritten by an array of tools for measuring, counting, predicting, and managing growth, could easily be contrasted to what had come to be seen as the irrational management of states in the 1930s, to nationalistic and imperial rivalries, and to the ideology of fascism.[55] Yet despite this technocratic appeal, during the 1950s, and culminating in the 1960s, the idea of "economic growth" became charged with multifaceted meanings, suffused with arresting symbolisms, and imbued with ardent assumptions, all of which produced the connections between GDP-growth and core societal values mentioned above.

These associations became so close that key societal objectives were to a considerable degree reduced to or identified with economic growth itself. If one wanted to assess the status of a country with regard to one of these objectives, it became common to take GDP, recent growth rates, or growth prospects as the most basic point of reference. Not only in governmental and intergovernmental documents on the general state of a country or region, but also in newspaper articles, within economics or progressively in other social sciences such as sociology or history, GDP growth became a "barometer" to measure the overall state of an economy. It was this process that finally lifted the extent or expansion of the monetary value of the market economy, measured as GDP, to the status of a powerful benchmark employed to assess success or failure of government policies, but also to symbolically determine the relative status of countries and power blocs within the international arena. Driven by the revolution of rising expectations, economic growth became the globally accepted yardstick of progress, not only in the capitalist industrialized countries, but also in the Soviet Union, China, and in the countries of the global South.

The debates around the making of the OECD growth target in 1961 provide a striking illustration. In the discussion among Ministers, the OECD growth target was not only discussed both by proponents and by opponents as the primary proxy for progress and for the general success of the capitalist free market system,[56] but GNP growth was also blended with the concept of increases in human wellbeing. For example, US Under Secretary of State for Economic Affairs George Ball stated that the aim was to "prove" that capitalism was better for people than all other economic systems:

> Countries had an opportunity to prove … that their empirical mixture of public and private enterprise was far more dynamic and, simultaneously, more conducive to human well-being, than any other economic arrangement that the world had devised. This was an opportunity … to re-affirm—both among Member countries themselves and in their relations with less-developed nations—a central thesis of their ethics: that self-interest was entirely consistent with a sincere devotion to the interest of others.[57]

Further, in the context of the Cold War, colonial decline, and international competition, economic growth became the most essential symbol and key foundation of national power. Growth became associated with vitality, vigor, and strength, and was contrasted with lethargy. The prestige of a country came to hinge on its rate of growth as compared to that of other countries, causing "national inferiority complex[es]" in countries with seemingly slow growth rates.[58] And it became central in defining the purpose and identity of countries. As one British Treasury official claimed: "Economic growth may become, properly handled, a synonym for the real national interest."[59]

One could even argue, following Alan Milward, that growth had come to take the space in the national imaginaries that had hitherto been occupied by territorial expansion. Both the draft communiqué proclaiming the growth target (written by none other than Robert Solow, the godfather of modern growth modeling and, at that time, economic adviser to John F. Kennedy) and the American speech at the meeting emphasized the advantages of adding the economic power of another country the size and wealth of the US to the Atlantic community. The characteristic style and formulations of these statements depict the notion of economic growth as a process of opening up or conquering new territories. George Ball explained bombastically and in a militarized language what the growth target was all about:

> It was within countries' power to achieve for the free world an unparalleled conquest—a conquest without sacrifice on the part of the people and without damage to spiritual or cultural values, a conquest achieved merely by the effective utilization of the inherent capabilities.[60]

US chief economist Walter Heller put it in a speech in a similar way, employing the metaphor of "frontier":

It might be possible to create and maintain an American society in a stationary economy, but it would surely be difficult. Much of what is best in the American character
is a reflection of growth—first through the external frontier as the Nation pushed
West, and now through the internal frontier of expanding educational, occupational
and economic opportunity.[61]

Since growth became *the* yardstick for power, progress, and prosperity, it had to be exalted to
become a "responsibility" states had to assume, an "imperative" that could not be evaded, and
thus, in the words of one critic of this development, "the *one* preeminent requisite and priority" in policy-making.[62] As James Tobin has argued in the keynote at the 1964 AEA meeting:
"Growth has become a good word. And the better a word becomes, the more it is invoked to
bless a variety of causes and the more it loses specific meaning." It had become "a new synonym
for good things in general" and "a fashionable way to describe other economic objectives."[63]

Growth as endless

Finally, a fundamental and often-implicit supposition underlying the growth paradigm was that
economic growth could potentially continue at least for decades, if not forever, provided the
correct governmental and intergovernmental policies were pursued. Most fundamentally, it
rested on the newly emerging conceptualization of "the economy" as a self-contained totality
of monetary flows forming the relations between production, distribution, and consumption
within national boundaries. This notion, which is nowadays largely taken for granted but
emerged only in the 1930s and 1940s in connection with the rise of oil, superseded a view of
economic processes conceptualized in terms of physical flows of resources, matter, and energy,
which suggested limits to growth. In contrast, the new measures such as GDP, which focused
on "the speed and frequency with which paper money changed hands," could expand without
increasing in physical or territorial size.[64]

  After the classical and neoclassical stagnation theories had been dismissed in the early 1950s,
economists and politicians largely took the infinite possibility of pursuing economic expansion
for granted, and from the late 1960s onwards they started to justify the credo of unlimited
growth with reference to technological progress and the power of the price mechanism and
competition.[65] In this regard, growthmanship was mutually reinforcing with the increasing
importance of economic knowledge-production as a key justificatory basis for policy-making
within the modern state. While their ability to measure, model, and steer growth made economists increasingly indispensable for managing modern societies based on growth, the
proliferation of economic approaches also strengthened the growth paradigm. The increased
responsibility of governments to boost growth not only advanced the authority of economic
experts in traditionally non-economic, social, and cultural realms such as science and education
policies, but also fundamentally intensified what has been discussed as the scientization or
economization of the social.[66]

  The predominantly unquestioned hegemony of growth was rather short-lived. It lasted only
from the mid-1950s to the late 1960s, thus spanning the exceptional period of high and stable
growth rates of the golden decades, after which in the wake of the social unrest of "1968," and
the rise of environmental movements, growth was profoundly criticized. For several years, some
of the key propositions of the growth paradigm were questioned in a debate that revolved
around the statistical codification of growth, the conflation of GDP growth with progress and
increasing wellbeing, and the ecological prospects of economic expansion on a finite planet.
Symptomatically, after the 50-percent growth target was over-achieved in 1970, the OECD's

Secretary-General Emile van Lennep proposed to focus on a rather profound questioning of the growth paradigm: "To what uses should this growth be put? If increased growth does not create improved conditions of life, will not growth become an illusion? What is the point of more unless more means also better?"[67]

Remarkably, it was a group of international bureaucrats within the OECD who, concerned about the negative effects of modernization and economic expansion, actually proved instrumental in launching the Club of Rome.[68] However, while this critical perspective prevailed in social movements and broader societal debates well into the 1980s, with the onset of economic turmoil, soaring energy prices, and stagflation in the mid-1970s, these critiques were—within the political establishment—dismissed or incorporated into the norm of "qualitative growth."[69] This "dialectical" concept superficially and rhetorically incorporated the social and ecological critique of quantitative "growth for growth's sake," yet the core economic outlook of the growth paradigm did not change much. Growth was not only deemed imperative to fight widespread unemployment, create economic and political stability, and because it embodied key societal objectives, but also to generate the resources to counter the ecological and social side effects of GDP growth. In the following decades similar concepts have continuously emerged that supposedly address the shortcomings of growthmanship, key among them "sustainable growth" in the 1980s, and since then "inclusive," "pro-poor" and "green" growth.[70]

Most recently, the OECD has again taken a more cautious approach to growth: On the one hand, the organization qualifies its aims as "stronger, cleaner, fairer economic growth," as "inclusive" and "green growth," and repeatedly claims that growth could not be an aim for policy-makers due to its inherent flaws and insurmountable ecological constraints. Yet at the same time, the OECD still advocates, in the words of Secretary-General Gurria, "to make growth the number one priority."[71] It has accordingly been criticized as behaving like "Dr. Jekyll and Mr. Hyde."[72] But whereas in 1961, when the OECD started to promote the highest growth possible, long-term growth seemed to imply, for the OECD's first Secretary-General Kristensen, "unbelievable" levels of production, fifty years later ecological disasters seem easier to imagine for most government experts than an end of growth. Current OECD projections for 2050 forecast a quadrupling of global GDP.[73]

Conclusion

Despite these optimistic projections, a specter is haunting not just Europe, but the entire industrialized world—the specter of secular stagnation. In the wake of the recent economic crisis at the end of the first decade of the twenty-first century, more and more economists have stated that they believe we have entered a new stage in the history of economic development—the end of growth for the countries that industrialized early. Going back to stagnationist theories that were formulated in the late 1930s, most prominently by US economist Alvin Hansen, the authors of this "new secular stagnation hypothesis" predict the demise of relevant growth rates in the coming decades and put forth various reasons—both technological, demographic, historical, and economic—for their claims. The term secular stagnation gained particular prominence through a November 2013 speech by Lawrence Summers, former President of Obama's National Economic Council, at the IMF Forum, but proponents of this end of growth thesis range from Tyler Cowen, author of *The Great Stagnation,* to such famous economists as Robert Solow, Paul Krugman, and Thomas Piketty.[74] Even though not yet common sense among economists, their arguments have gained considerable traction due to a continuous slack in economic output, in particular in the EU, and due to continuously low real interest rates close to or below zero. In the words of Barry Eichengreen: "The idea that America and the other advanced economies might be suffering from more than the hangover from a financial crisis resonated with many observers."[75]

In the long term, economic growth might just not develop in the form of the "hockey stick" we are used to imagining—being stagnant for most of human history and then speeding up very rapidly into an almost vertical rise following a J-curve. Rather, those regions that kicked off capitalist industrialization first seem to have been transitioning into a development more adequately described as an S-curve, in which rapid acceleration slows down and eventually comes to a halt. In the long term, the fast economic growth of Western societies from 1760–1970 might prove to be the historical exception.

So why is the possible end of growth a threatening specter and not—as in the vision of some of the greatest economists of the nineteenth and twentieth centuries—a welcome future? After all, as John Stuart Mill or John Maynard Keynes both famously argued, the end of economic growth does not imply the end of human improvements.[76] To start to come to terms with this conundrum, this article has analyzed the historical making and remaking of the economic growth paradigm, specifically focusing on economic and policy-making expertise.

Even though asserting universal validity for all places and in perpetuity, the growth paradigm and its statistical surrogate were invented in very specific social, spatial, and historical contexts for limited purposes and are to this day inherently shaped by these context, interests, and power constellations. By focusing on the debates within the industrialized countries' collective think-tank—the OECD—this article has analyzed this process by highlighting the historical evolution of four entangled discourses. These claimed that GDP, with all its inscribed reductions and exclusions correctly measures economic activity, that its growth serves as a magical ward to solve all kinds of often changing key societal challenges, that growth was

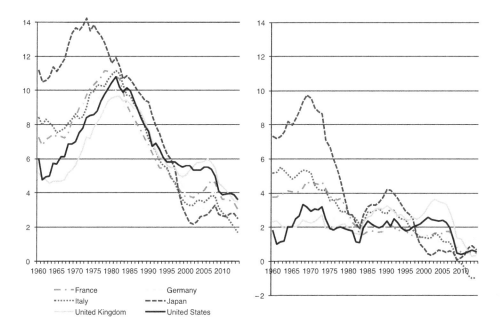

Figure 12.2 Nominal (left) and real (right) GDP percentage change on year earlier (rolling 10-year averages) in key OECD countries, 1960–2014

Source: Penn World Tables; *The Economist*.

practically the same as some of the most essential societal ambitions such as progress, wellbeing, or national power, and that growth is essentially limitless.

If the argument is taken seriously that growth is not a self-evident concept and economic desire, but rather emerged in and is intrinsically shaped by concrete historical situations, was promoted by specific interests, and has continuously been contested, this has far-reaching consequences for political debates and the research-agendas of the social sciences, in particular for economics. Not only was the discipline of economics deeply entangled in the making of the growth paradigm and partly derives its authority and influence from the prevalence of growthmanship, but more fundamentally, GDP or growth very often underlie research as seemingly unproblematic analytical categories. A historical understanding of the politically contested making and remaking of growth statistics and their underlying concepts is thus indispensable to avoid the reproduction of a geographically and temporally very narrow and particular perspective that has for many good reasons been criticized as problematic.

This analysis challenges the self-evidence and "pseudo-rationality" of the growth fetish and reopens searching questions about what societies value, what should be understood as progress, and who benefits and who bears the costs. In political discourse, this social-historical contextualization demonstrates that while the growth paradigm was convenient for the mid-twentieth century "empty" world with abundant ecological resources and sinks, it helped create drastically changed conditions. Since economic reasoning has not adjusted, the growth paradigm has increasingly become a threat to humans, the planet, and future generations.[77] Whereas a focus on economic growth was quite understandable given the circumstances of the mid-twentieth century—and the chapter has focused on these underlying reasons and on what policy experts were trying to achieve by stimulating growth—these circumstances have changed drastically today. Perhaps one could think of growth as a kind of addiction, which is pleasurable in the beginning—when the benefits outweigh the costs—and becomes impossible to shake off when the side effects increasingly emerge, not least because a world without it has become inconceivable.[78]

Ultimately, the growth paradigm is unstable and self-contradictory since the expectation it raises of continually increasing levels of material production run up to the ecological limits of a finite planet, with catastrophic consequences, noticeable today.[79] It is thus rather fortunate, as argued by Richard Wilkinson and Kate Pickett in their celebrated statistical survey of welfare over time and between societies, that

> just when the human species discovers that the environment cannot absorb further increases in emissions, we also learn that further economic growth in the developed world no longer improves health, happiness or measures of wellbeing. On top of that, we have now seen that there are ways of improving the quality of life in rich countries without further economic growth.[80]

Thus, as it is becoming more and more obvious that industrialization, economic growth, and the vast but unequal accumulation of wealth have triggered a process of global environmental degradation that will, according to current projections, fundamentally undermine the foundations of human life on this planet we also learn that the very process driving the Anthropocene is neither self-evident, natural, nor indispensable.

Acknowledgements

This chapter is a slightly updated version of a published article, Matthias Schmelzer, "The

Growth Paradigm: History, Hegemony, and the Contested Making of Economic Growth-manship," *Ecological Economics* 118 (2015), 262–71, reprinted with permission from Elsevier.

For helpful comments on the issues discussed in this article I thank Jana Flemming, Alexander Nützenadel, Hartmut Kaelble, Matthieu Leimgruber, Adam Tooze, Andrea Vetter, and two anonymous reviewers. This research was made possible through the Swiss National Fund (PP00P1_139023/1).

Notes

1   OECD, *Statistics, Knowledge and Policy: Measuring and Fostering the Progress of Societies* (Paris: OECD, 2008), cover text.

2   Tomas Sedlacek, *Economics of Good and Evil: The Quest for Economic Meaning from Gilgamesh to Wall Street* (Oxford: Oxford University Press, 2011).

3   John Robert McNeill, *Something New Under the Sun: An Environmental History of the Twentieth-Century World* (New York: W. W. Norton & Company, 2000), 236.

4   Christophe Bonneuil and Jean-Baptiste Fressoz, *L'événement anthropocène: La Terre, l'histoire et nous* (Paris: Seuil, 2013); Dipesh Chakrabarty, "The Climate of History: Four Theses," *Critical Inquiry* 35, no. 2 (2009): 197–222.

5   The differences between the older concept GNP and the newer GDP are not relevant to the argument in this chapter and will thus be neglected. For more details see Diane Coyle, *GDP: A Brief but Affectionate History* (Princeton, NJ: Princeton University Press, 2014); Lorenzo Fioramonti, *Gross Domestic Problem: The Politics Behind the World's Most Powerful Number* (London: Zed Books, 2013); Dirk Philipsen, *The Little Big Number: How GDP Came to Rule the World and What to Do about It* (Princeton: Princeton University Press, 2015); Daniel Speich Chassé, *Die Erfindung des Bruttosozialprodukts: Globale Ungleichheit in der Wissensgeschichte der Ökonomie* (Göttingen: Vandenhoeck & Ruprecht, 2013); Joseph Stiglitz, Amartya Sen, and Jean-Paul Fitoussi, *Mismeasuring Our Lives: Why GDP Doesn't Add Up* (New York: New Press, 2010).

6   The most important studies are an older study on the rise and fall of growthmanship by Arndt, and studies on growth politics in the US and Japan: Heinz W. Arndt, *The Rise and Fall of Economic Growth: A Study in Contemporary Thought* (Melbourne: Longman Cheshire, 1978); Robert M. Collins, *More: The Politics of Economic Growth in Postwar America* (Oxford: Oxford University Press, 2000); Scott O'Bryan, *The Growth Idea: Purpose and Prosperity in Postwar Japan* (Honolulu: University of Hawaii Press, 2009). See also studies on the history of GDP: Coyle, *GDP*; Fioramonti, *Gross Domestic Problem*; Philipp Lepenies, *Die Macht der einen Zahl. Eine politische Geschichte des Bruttoinlandsprodukts* (Frankfurt a.M.: Suhrkamp, 2013); Philipsen, *The Little Big Number*; Speich Chassé, *Erfindung*. And see studies based largely on these accounts, in particular Arndt, *The Rise and Fall of Economic Growth*, and also ch. 1 of Peter A. Victor, *Managing Without Growth: Slower by Design, Not Disaster*, Advances in Ecological Economics (Cheltenham: Edward Elgar, 2008). Two highly informative studies of the growth paradigm are the neo-Gramscian analysis by Stephen Purdey and an ideology-theoretical article by Gareth Dale: Stephen J. Purdey, *Economic Growth, the Environment and International Relations: The Growth Paradigm* (Abingdon: Routledge, 2009); Gareth Dale, "The Growth Paradigm: A Critique," *International Socialism* 134 (2012), retrieved from http://isj.org.uk/the-growth-paradigm-a-critique.

7   Klaus Dörre, Stephan Lessenich, and Hartmut Rosa, *Soziologie—Kapitalismus—Kritik: eine Debatte* (Frankfurt am Main: Suhrkamp, 2009); Jürgen Kocka, *Geschichte des Kapitalismus* (Munich: C.H. Beck, 2013); Karl Marx, *Das Kapital: Kritik der politischen Ökonomie. Erster Band* (Berlin: Dietz, 1962).

8   Jacques Ellul, *The Technological System* (New York: Continuum Press, 1980).

9   Wolfgang Sachs, "Introduction," in *The Development Dictionary: A Guide to Knowledge as Power*, ed. Wolfgang Sachs (London: Zed Books, 1992), 1.

10 This article builds on key results of a research project in economic history. For more details, also regarding the state of the art of historical knowledge about the growth idea, see Matthias Schmelzer, *The Hegemony of Growth: The OECD and the Making of the Economic Growth Paradigm* (Cambridge: Cambridge University Press, 2016). On the OECD see Matthieu Leimgruber and Matthias Schmelzer, eds., *The OECD and the International Political Economy Since 1948* (Basingstoke: Palgrave Macmillan, 2017).

11 Stiglitz et al., *Mismeasuring*.

12 Avner Offer, *The Challenge of Affluence: Self-Control and Well-Being in the United States and Britain since 1950* (Oxford: Oxford University Press, 2006); Marilyn Waring, *Counting for Nothing: What Men Value and What Women Are Worth* (Toronto: University of Toronto Press, 1999); Richard Wilkinson and Kate Pickett, *The Spirit Level: Why Greater Equality Makes Societies Stronger*, reprint edition (New York: Bloomsbury Press, 2011).

13 Isabelle Cassiers and Géraldine Thiry, "A High-Stakes Shift: Turning the Tide From GDP to New Prosperity Indicators," in *Redefining Prosperity*, ed. Isabelle Cassiers (New York: Routledge, 2015), 22–40; Herman E. Daly and Joshua C. Farley, *Ecological Economics: Principles and Applications*, 2nd ed. (Washington, DC: Island Press, 2011); Tim Jackson, *Prosperity without Growth: Economics for a Finite Planet* (London: Earthscan, 2009); McNeill, *Something New*.

14 Richard Baldwin and Coen Teulings, eds., *Secular Stagnation: Facts, Causes and Cures* (London: CEPR Press, 2014); Thomas Piketty, *Capital in the Twenty-First Century* (Cambridge, MA: Harvard University Press, 2014), ch. 2.

15 Robert Costanza et al., "Development: Time to Leave GDP Behind," *Nature* 505, no. 7483 (2014): 283–5; Giacomo D'Alisa, Federico Demaria, and Giorgos Kallis, eds., *Degrowth: A Vocabulary for a New Era* (Abingdon: Routledge, 2014).

16 On this debate see Jeroen C. J. M. van den Bergh and Giorgos Kallis, "Growth, A-Growth or Degrowth to Stay within Planetary Boundaries?," *Journal of Economic Issues* 46, no. 4 (2012): 909–20.

17 Offer, *The Challenge of Affluence*, 17; Jeroen C. J. M. van den Bergh, "The GDP Paradox," *Journal of Economic Psychology* 30, no. 2 (April 2009): 117–35.

18 Herman E. Daly, "In Defense of a Steady-State Economy," *American Journal of Agricultural Economics* 54, no. 5 (1972): 945–54.

19 Thomas S. Kuhn, *The Structure of Scientific Revolutions*, 3rd ed. (Chicago, IL: University of Chicago Press, 1996), 175.

20 Charles S. Maier, "The World Economy and the Cold War in the Middle of the Twentieth Century," in *The Cambridge History of the Cold War*, 3 volumes, ed. Melvyn P. Leffler and Arne Westad (Cambridge: Cambridge University Press, 2010), 48; see also Terry Eagleton, *Ideology: An Introduction* (London: Verso, 1991); Purdey, *Economic Growth*; Dale, "The Growth Paradigm: A Critique"; Iris Borowy and Matthias Schmelzer, eds., *History of the Future of Economic Growth. Historical Roots of Current Debates on Sustainable Degrowth* (Abingdon: Routledge, 2017).

21 For more details see Schmelzer, *Hegemony*.

22 For the longer-term emergence of precursors of the modern growth paradigm see the forthcoming book by Gareth Dale, but also Arndt, *Rise and Fall*; Anthony Brewer, *The Making of the Classical Theory of Economic Growth* (London: Taylor & Francis, 2010); Robert A. Nisbet, *History of the Idea of Progress* (New Brunswick: Transaction Publishers, 1994); Purdey, *Economic Growth*; James C. Scott, *Seeing Like a State: How Certain Schemes to Improve the Human Condition Have Failed* (New Haven, CT: Yale University Press, 1998).

23 François Fourquet, *Les comptes de la puissance: histoire de la comptabilité nationale et du plan* (Fontenay-sous-Bois: Recherches, 1980); Speich Chassé, *Erfindung*; Adam Tooze, "Imagining National Economies: National and International Economics Statistics 1900–1950," in *Imagining Nations*, ed. Geoffrey Cubitt (Manchester: Manchester University Press, 1998), 212–28.

24 Timothy Mitchell, "Fixing the Economy," *Cultural Studies* 12, no. 1 (1998): 82–101; Tooze, "Imagining National Economies."

25 Zoltan Kenessey, ed., *The Accounts of Nations* (Amsterdam: IOS Press, 1994), pt. 1.

26 Tomo Suzuki, "The Epistemology of Macroeconomic Reality: The Keynesian Revolution from an Accounting Point of View," *Accounting, Organizations and Society* 28, no. 5 (2003): 473.

27 Political analysts Clifford Cobb, Ted Halstead, and Jonathan Rowe, quoted in Fioramonti, *Gross Domestic Problem*, 31.

28 Coyle, *GDP*; Philipsen, *The Little Big Number*; Fioramonti, *Gross Domestic Problem*; Speich Chassé, *Erfindung*.

29 Simon Kuznets, "National Income, 1929–1932," in *Senate Document No. 124, 73rd Congress, 2nd Session* (Washington, DC: US Government Printing Office, 1934).

30 Milton Gilbert, "National Income: Concepts and Measurements," in *Measuring and Projecting National Income* (New York: National Industrial Conference Board, 1945), 5. See also Richard Stone and Colin Clark in Milton Gilbert et al., "The Measurement of National Wealth: Discussion," *Econometrica* 17 (1949): 258f.

31 Richard Stone and Kurt Hansen, "Inter-Country Comparisons of the National Accounts and the Work of the National Accounts Research Unit of the OEEC," *Review of Income and Wealth* 3, no. 1 (1953): 102–4. See also National Archives and Records Administration, College Park, Maryland, United States (NARA), RG 469, Entry UD 379, Box 93, Folder OEEC General 1955, Kaufmann, OEEC Meeting of National Accounts Experts, November 26, 1955; Angus Maddison, "Confessions of a Chiffrephile," *Banca Nazionale Del Lavoro Quarterly Review* 189 (1994): 123–85.

32 Speich Chassé, *Erfindung*; Fioramonti, *Gross Domestic Problem*.

33 Ester Boserup, *Woman's Role in Economic Development* (London: Allen & Unwin, 1970); Marianne Ferber and Julie A. Nelson, *Beyond Economic Man: Feminist Theory and Economics* (Chicago, IL: University of Chicago Press, 1993); Waring, *Counting for Nothing*.

34 Erling Joar Fløttum et al., *History of National Accounts in Norway. From Free Research to Statistics Regulated by Law* (Oslo: Statistics Norway, 2012), 24; Gilbert et al., "Measurement," 257; Einar Lie, "The 'Protestant' View: The Norwegian and Scandinavian Approach to National Accounting in the Postwar Period," *History of Political Economy* 39, no. 4 (2007): 713–34. The neglect of female labor had, of course, long historical roots, see for example Nancy Folbre, "The Unproductive Housewife: Her Evolution in Nineteenth-Century Economic Thought," *Signs* 16, no. 3 (1991): 463–84.

35 NARA, RG 56, Entry UD-UP 734-M, Box 48, Folder National Income—US, J. B. D. Derksen, The Comparability of National Income Statistics, Economic and Social Council of the United Nations, August 28, 1947; Edward F. Denison, "Report on Tripartite Discussions of National Income Measurement," *Studies in Income and Wealth* 10 (1947): 14–16.

36 Suzuki has focused on some of the less controversial but similarly powerful configurations. Suzuki, "Epistemology," 485–98. See also the discussion of Kuznets's early critique of GNP accounting in Fioramonti, *Gross Domestic Problem*, 50–68.

37 Arthur Smithies, quoted in Gilbert et al., "Measurement," 269f.

38 Schmelzer, *Hegemony*, ch. 1.

39 Arndt, *Rise and Fall*, 30; see also Collins, *More*, 14–16.

40 OECD Historical Archive Paris (OECD-HA), FAC, box 368, file 10, Agreement for OEEC, March 26, 1948. See also OECD-HA, C(48)122, August 7, 1948; OEEC, 1948, p. 17.

41 Maier, in "The World Economy and the Cold War," identifies a speech by CEA chairman Leon Keyserling in 1949 as the first public pronouncement, but Robert Collins (*More*, pp. 17–25) cites some earlier statements. See also Maier, "The Politics of Productivity," p. 177.

42 Robert Marjolin, *Architect of European Unity: Memoirs, 1911–86* (London: Weidenfeld & Nicholson, 1989), 155.

43 Eagleton, *Ideology*, 6; see also Charles S. Maier, "The Politics of Productivity: Foundations of American International Economic Policy after World War II," *International Organization* 31, no. 4 (1977): 607–33; John Gerard Ruggie, "International Regimes, Transactions, and Change: Embedded Liberalism in the Postwar Economic Order," *International Organization* 36, no. 2 (1982): 379–415.

44 Henry C. Wallich, *Newseek*, January 24, 1972. For more details see Schmelzer, *Hegemony*, ch. 2–3.

45 Quoted in Thomas Robertson, "Development," in *Encyclopedia of the Cold War, Volume 1*, ed. Ruud van Dijk (London: Taylor & Francis, 2008), 255.

46 Selwyn Lloyd used the term "competitive targetry" at the first Ministerial meeting of the OECD. OECD-HA, OECD/C/M(61)7, November 16–17, 1961. On the importance of economic targets in that period see Alan Budd, *The Politics of Economic Planning* (Manchester: Manchester University Press, 1978), 84–6; Collins, *More*, 51–67; O'Bryan, *Growth Idea*, 157–71.

47 Friedrich Haffner, "Das sowjetische Preissystem: Theorie und Praxis, Änderungsvorschläge und Reformmaßnahmen," dissertation, Freie Universität Berlin, 1968; Bundesarchiv Koblenz, Germany (BAK), B 102/77352, "Westen antwortet auf die russische Herausforderung," Deutsche Zeitung, November 16, 1961.

48 BAK, B 136/8387, Franz Blücher to Konrad Adenauer, August 20, 1951; Marjolin, *Architect of European Unity*, p. 163f.; Schmelzer, *Hegemony of Growth*.

49 Cited in: Robert Shannan Peckham, "Alexander King," *The Independent*, March 26, 2007.

50 Matthias Schmelzer, "The Crisis before the Crisis: The 'Problems of Modern Society' and the OECD, 1968–74," *European Review of History* 19, no. 6 (2012): 999–1020.

51 O'Bryan, *Growth Idea*, 157f.; James Tobin, "Economic Growth as an Objective of Government Policy," *The American Economic Review* 54, no. 3 (1964): 1.

52 OECD-HA, CPE/WP2(80)3, Factor Income Shares and Profit Rates, April 25, 1980; Communiqué of the Meeting of the OECD Council at Ministerial level in May 1982, retrieved from

www.g8.utoronto.ca/oecd/oecd82.htm; Rawi Abdelal, *Capital Rules: The Construction of Global Finance* (Cambridge: Harvard University Press, 2007); Collins, *More*, 191–7; Peter Hill, *Profits and Rates of Return* (Paris: OECD, 1979); Matthieu Leimgruber, "The Embattled Standard-Bearer of Social Insurance and Its Challenger: The ILO, the OECD and the 'Crisis of the Welfare State', 1975–1985," in *Globalizing Social Rights: The International Labor Organization and Beyond*, ed. Sandrine Kott and Joëlle Droux (Basingstoke: Palgrave Macmillan, 2013), 293–309.

53 Wendy Brown, *Undoing the Demos: Neoliberalism's Stealth Revolution* (Cambridge, MA: MIT Press, 2015), 64.

54 Max Horkheimer and Theodor W. Adorno, *Dialektik der Aufklärung: philosophische Fragmente*, 16th ed. (Frankfurt a.M.: Fischer Taschenbuch Verlag, 2006).

55 Maier, "Politics of Productivity"; O'Bryan, *Growth Idea*.

56 On this meeting see OECD/C/M(61)7, November 16–17, 1961, and the sources in NARA, RG 59, Entry 5304, Box 22; NARA, RG 56, Entry UD-UP 734-H, Box 4; NARA, RG 59, Entry A1 5605, Box 1; The National Archive, Kew, Britain (TNA), PREM 11/4228; BAK, B 102/77352, George W. Ball to Prof. Ludwig Erhard, November 7, 1961.

57 OECD-HA, OECD/C/M(61)7, November 16–17, 1961.

58 NARA, RG 59, Entry A1 5605, Box 1, Folder OECD Ministerial 1961, Copy of letter by Ball to Ministers, November 7, 1961; TNA, T 230/919, Remarks by Henry H. Fowler at the American Conference on the Atlantic Community and Economic Growth, New York, December 12, 1965; Allen Wallis, "A Philosophy of Economic Growth," *Wall Street Journal*, October 24, 1960, p. 12. See also Arndt, *Rise and Fall*, 62–5.

59 TNA, T 230/579, Vinter, Elements of a Policy for Economic Growth, February 27, 1961.

60 OECD-HA, OECD/C/M(61)7, November 16–17, 1961; BAK, B 102/77352, Draft Communiqué on Economic Growth Target, November 1961; NARA, RG 59, Entry 5304, Box 22, Folder OECD Ministerial Meeting 1961, Robert Solow, Portion of Draft Communiqué, November 2, 1961. See also Alan S. Milward, *The European Rescue of the Nation-State* (Abingdon: Routledge, 2000), 30.

61 NARA, RG 56, Entry UD-UP 734-H, Box 3, Folder OECD/5/30 EPC WP-2, Walter Heller, Economic Growth: Challenges and Opportunity, Address to the Loeb Awards Fourth Annual Presentation, New York, May 18, 1961.

62 TNA, T 299/178, Lee to Hubback, November 7, 1961. See also Allen to Lee, November 7, 1961; and attached draft Communiqué; F.J.A. to Cairncross, November 7, 1961.

63 Tobin, "Economic Growth," 1.

64 Timothy Mitchell, *Carbon Democracy: Political Power in the Age of Oil* (London: Verso, 2011), 139.

65 Steven F. Bernstein, *The Compromise of Liberal Environmentalism* (New York: Columbia University Press, 2001); OECD, *OECD at Work for the Environment* (Paris: OECD, 1973).

66 Brown, *Undoing the Demos*; Marion Fourcade, *Economists and Societies: Discipline and Profession in the United States, Britain, and France, 1890s to 1990s* (Princeton, NJ: Princeton University Press, 2009); Alexander Nützenadel, *Stunde der Ökonomen. Wissenschaft, Politik und Expertenkultur in der Bundesrepublik 1949–1974* (Göttingen: Vandenhoeck & Ruprecht, 2005); Lutz Raphael, "Die Verwissenschaftlichung des Sozialen als methodische und konzeptionelle Herausforderung für eine Sozialgeschichte des 20. Jahrhunderts," *Geschichte und Gesellschaft* 22, no. 2 (1996): 165–93.

67 OECD-HA, CES Divers 1970, Van Lennep at the Meeting of the Committee on Economic Affairs and Development of the Consultative Assembly of the Council of Europe, July 3, 1970.

68 See OECD-HA, PRESS/A(69)10, Problems of the modern society. Statement by the Secretary-General, Thorkil Kristensen, February 14, 1969; the files in OECD-HA, Box 36486 "Club of Rome"; and Matthias Schmelzer, "'Born in the Corridors of the OECD': The Forgotten Origins of the Club of Rome, Transnational Networks, and the 1970s in Global History," *Journal of Global History* 12, no. 1 (2017).

69 Collins, *More*, ch. 5; Stephen Macekura, *Of Limits and Growth: Global Environmentalism and the Rise of "Sustainable Development" in the Twentieth Century* (Cambridge: Cambridge University Press, 2015); Schmelzer, "Crisis."

70 Iris Borowy, *Defining Sustainable Development: The World Commission on Environment and Development (Brundtland Commission)* (Abingdon: Routledge, 2013); Jackson, *Prosperity without Growth*; Gareth Dale, Manu V. Mathai, and Jose Puppim De Oliveira, eds., *Green Growth: Political Ideology, Political Economy and Policy Alternatives* (London: Zed Books, 2016); Victor, *Managing Without Growth*.

71 Larry Elliot, "OECD Calls on Eurozone Finance Ministers to Take Decisive Action," *Guardian*, March 27, 2012.

72   BRAINPOoL, "Review Report on Beyond GDP Indicators: Categorisation, Intentions and Impacts," 2012, retrieved from http://wikiprogress.org/articles/initiatives/review-report-on-beyond-gdp-indicators-categorisation-intentions-and-impacts.

73   OECD-HA, Thorkil Kristensen, "Work and Policies of the OECD," Confidential Annex: Statement by Kristensen at the second Council Meeting, C/M(61)2, October 13, 1961. See, however, the recent OECD project New Approaches to Economic Challenges (NAEC), available at www.oecd.org/naec.

74   Baldwin and Teulings, *Secular Stagnation*; Piketty, *Capital*.

75   Barry Eichengreen, "Secular Stagnation: A Review of the Issues," in *Secular Stagnation: Facts, Causes and Cures*, ed. Richard Baldwin and Coen Teulings (London: CEPR Press, 2014), 41.

76   John Maynard Keynes, "The Economic Possibilities of Our Grandchildren," in *Essays in Persuasion* (New York: W.W. Norton & Company, 1963), 358–73; John Stuart Mill, *Principles of Political Economy, with Some of Their Applications to Social Philosophy* (London: Longmans, Green, Reader, and Dyer, 1909), bk. 4, ch. 6.

77   McNeill, *Something New*.

78   I would also like to thank Avner Offer for this suggestion. See also See Herman E. Daly, *Ecological Economics and Sustainable Development* (Cheltenham: Edward Elgar, 2008); Cassiers and Thiry, "A High-Stakes Shift."

79   Purdey, *Economic Growth*.

80   Wilkinson and Pickett, *Spirit Level*, 218f.

References

Abdelal, Rawi. *Capital Rules: The Construction of Global Finance*. Cambridge, MA: Harvard University Press, 2007.

Arndt, Heinz W. *The Rise and Fall of Economic Growth: A Study in Contemporary Thought*. Melbourne: Longman Cheshire, 1978.

Baldwin, Richard, and Coen Teulings, eds. *Secular Stagnation: Facts, Causes and Cures*. London: CEPR Press, 2014.

Bergh, Jeroen C.J.M. van den. "The GDP Paradox." *Journal of Economic Psychology* 30, no. 2 (April 2009): 117–35.

Bergh, Jeroen C.J.M. van den, and Giorgos Kallis. "Growth, A-Growth or Degrowth to Stay within Planetary Boundaries?" *Journal of Economic Issues* 46, no. 4 (2012): 909–20.

Bernstein, Steven F. *The Compromise of Liberal Environmentalism*. New York: Columbia University Press, 2001.

Bonneuil, Christophe, and Jean-Baptiste Fressoz. *L'événement anthropocène: La Terre, l'histoire et nous*. Paris: Seuil, 2013.

Borowy, Iris. *Defining Sustainable Development: The World Commission on Environment and Development (Brundtland Commission)*. Abingdon: Routledge, 2013.

Borowy, Iris, and Matthias Schmelzer, eds. *History of the Future of Economic Growth. Historical Roots of Current Debates on Sustainable Degrowth*. Abingdon: Routledge, 2017.

Boserup, Ester. *Woman's Role in Economic Development*. London: Allen & Unwin, 1970.

BRAINPOoL. "Review Report on Beyond GDP Indicators: Categorisation, Intentions and Impacts." 2012. Retrieved from http://wikiprogress.org/articles/initiatives/review-report-on-beyond-gdp-indicators-categorisation-intentions-and-impacts.

Brewer, Anthony. *The Making of the Classical Theory of Economic Growth*. London: Taylor & Francis, 2010.

Brown, Wendy. *Undoing the Demos: Neoliberalism's Stealth Revolution*. Cambridge, MA: MIT Press, 2015.

Budd, Alan. *The Politics of Economic Planning*. Manchester: Manchester University Press, 1978.

Cassiers, Isabelle, and Géraldine Thiry. "A High-Stakes Shift: Turning the Tide From GDP to New Prosperity Indicators." In *Redefining Prosperity*, edited by Isabelle Cassiers, 22–40. New York: Routledge, 2015.

Chakrabarty, Dipesh. "The Climate of History: Four Theses." *Critical Inquiry* 35, no. 2 (2009): 197–222.

Collins, Robert M. *More: The Politics of Economic Growth in Postwar America*. Oxford: Oxford University Press, 2000.

Costanza, Robert, Ida Kubiszewski, Enrico Giovannini, Hunter Lovins, Jacqueline McGlade, Kate E. Pickett, Kristín Vala Ragnarsdóttir, Debra Roberts, Roberto De Vogli, and Richard Wilkinson. "Development: Time to Leave GDP Behind." *Nature* 505, no. 7483 (2014): 283–5.

Coyle, Diane. *GDP: A Brief but Affectionate History*. Princeton, NJ: Princeton University Press, 2014.

D'Alisa, Giacomo, Federico Demaria, and Giorgos Kallis, eds. *Degrowth: A Vocabulary for a New Era*. Abingdon: Routledge, 2014.

Dale, Gareth. "The Growth Paradigm: A Critique." *International Socialism* 134 (2012), retrieved from http://isj.org.uk/the-growth-paradigm-a-critique.

Dale, Gareth, Manu V. Mathai, and Jose Puppim De Oliveira, eds. *Green Growth: Political Ideology, Political Economy and Policy Alternatives*. London: Zed Books, 2016.

Daly, Herman E. *Ecological Economics and Sustainable Development*. Cheltenham: Edward Elgar, 2008.

Daly, Herman E. "In Defense of a Steady-State Economy." *American Journal of Agricultural Economics* 54, no. 5 (1972): 945–54.

Daly, Herman E., and Joshua C. Farley. *Ecological Economics: Principles and Applications*. 2nd ed. Washington: Island Press, 2011.

Denison, Edward F. "Report on Tripartite Discussions of National Income Measurement." *Studies in Income and Wealth* 10 (1947): 3–22.

Dörre, Klaus, Stephan Lessenich, and Hartmut Rosa. *Soziologie—Kapitalismus—Kritik: eine Debatte*. Frankfurt am Main: Suhrkamp, 2009.

Eagleton, Terry. *Ideology: An Introduction*. London: Verso, 1991.

Eichengreen, Barry. "Secular Stagnation: A Review of the Issues." In *Secular Stagnation: Facts, Causes and Cures*, edited by Richard Baldwin and Coen Teulings, 41–6. London: CEPR Press, 2014.

Ellul, Jacques. *The Technological System*. New York: Continuum Press, 1980.

Ferber, Marianne, and Julie A. Nelson. *Beyond Economic Man: Feminist Theory and Economics*. Chicago, IL: University of Chicago Press, 1993.

Fioramonti, Lorenzo. *Gross Domestic Problem: The Politics Behind the World's Most Powerful Number*. London: Zed Books, 2013.

Fløttum, Erling Joar, Tore Halvorsen, Liv Hobbelstad Simpson, and Tor Skoglud. *History of National Accounts in Norway. From Free Research to Statistics Regulated by Law*. Oslo: Statistics Norway, 2012.

Folbre, Nancy. "The Unproductive Housewife: Her Evolution in Nineteenth-Century Economic Thought." *Signs* 16, no. 3 (1991): 463–84.

Fourcade, Marion. *Economists and Societies: Discipline and Profession in the United States, Britain, and France, 1890s to 1990s*. Princeton, NJ: Princeton University Press, 2009.

Fourquet, François. *Les comptes de la puissance: histoire de la comptabilité nationale et du plan*. Fontenay-sous-Bois: Recherches, 1980.

Gilbert, Milton. "National Income: Concepts and Measurements." In *Measuring and Projecting National Income*. New York: National Industrial Conference Board, 1945.

Gilbert, Milton, Colin Clark, J. R. N. Stone, Francois Perroux, D. K. Lieu, Evelpides, Francois Divisia, et al. "The Measurement of National Wealth: Discussion." *Econometrica* 17 (1949): 255–72.

Haffner, Friedrich. "Das sowjetische Preissystem: Theorie und Praxis, Änderungsvorschläge und Reformmaßnahmen." Dissertation, Freie Universität Berlin, 1968.

Hill, Peter. *Profits and Rates of Return*. Paris: OECD, 1979.

Horkheimer, Max, and Theodor W. Adorno. *Dialektik der Aufklärung: philosophische Fragmente*. 16th ed. Frankfurt a.M.: Fischer Taschenbuch Verlag, 2006.

Jackson, Tim. *Prosperity without Growth: Economics for a Finite Planet*. London: Earthscan, 2009.

Kenessey, Zoltan, ed. *The Accounts of Nations*. Amsterdam: IOS Press, 1994.

Keynes, John Maynard. "The Economic Possibilities of Our Grandchildren." In *Essays in Persuasion*, 358–73. New York: W. W. Norton & Company, 1963.

Kocka, Jürgen. *Geschichte des Kapitalismus*. Munich: C. H. Beck, 2013.

Kuhn, Thomas S. *The Structure of Scientific Revolutions*. 3rd ed. Chicago, IL: University of Chicago Press, 1996.

Kuznets, Simon. "National Income, 1929–1932." In *Senate Document No. 124, 73rd Congress, 2nd Session*. Washington: US Government Printing Office, 1934.

Leimgruber, Matthieu. "The Embattled Standard-Bearer of Social Insurance and Its Challenger: The ILO, the OECD and the 'Crisis of the Welfare State', 1975–1985." In *Globalizing Social Rights: The International Labor Organization and Beyond*, edited by Sandrine Kott and Joëlle Droux, 293–309. Basingstoke: Palgrave Macmillan, 2013.

Leimgruber, Matthieu, and Matthias Schmelzer, eds. *The OECD and the International Political Economy Since 1948*. Basingstoke: Palgrave Macmillan, 2017.

Lepenies, Philipp. *Die Macht der einen Zahl. Eine politische Geschichte des Bruttoinlandsprodukts.* Frankfurt a.M.: Suhrkamp, 2013.

Lie, Einar. "The 'Protestant' View: The Norwegian and Scandinavian Approach to National Accounting in the Postwar Period." *History of Political Economy* 39, no. 4 (2007): 713–34.

Macekura, Stephen. *Of Limits and Growth: Global Environmentalism and the Rise of "Sustainable Development" in the Twentieth Century.* Cambridge: Cambridge University Press, 2015.

Maddison, Angus. "Confessions of a Chiffrephile." *Banca Nazionale Del Lavoro Quarterly Review* 189 (1994): 123–85.

Maier, Charles S. "The Politics of Productivity: Foundations of American International Economic Policy after World War II." *International Organization* 31, no. 4 (1977): 607–33.

Maier, Charles S. "The World Economy and the Cold War in the Middle of the Twentieth Century." In *The Cambridge History of the Cold War, 3 Volumes,* edited by Melvyn P. Leffler and Arne Westad, 44–66. Cambridge: Cambridge University Press, 2010.

Marjolin, Robert. *Architect of European Unity: Memoirs, 1911–86.* London: Weidenfeld & Nicholson, 1989.

Marx, Karl. *Das Kapital. Kritik der politischen Ökonomie. Erster Band.* Berlin: Dietz, 1962.

McNeill, John Robert. *Something New Under the Sun: An Environmental History of the Twentieth-Century World.* New York: W.W. Norton & Company, 2000.

Mill, John Stuart. *Principles of Political Economy, with Some of Their Applications to Social Philosophy.* London: Longmans, Green, Reader, and Dyer, 1909.

Milward, Alan S. *The European Rescue of the Nation-State.* Abingdon: Routledge, 2000.

Mitchell, Timothy. *Carbon Democracy: Political Power in the Age of Oil.* London: Verso, 2011.

Mitchell, Timothy. "Fixing the Economy." *Cultural Studies* 12, no. 1 (1998): 82–101.

Nisbet, Robert. *History of the Idea of Progress.* New Brunswick: Transaction Publishers, 1994.

Nützenadel, Alexander. *Stunde der Ökonomen. Wissenschaft, Politik und Expertenkultur in der Bundesrepublik 1949–1974.* Göttingen: Vandenhoeck & Ruprecht, 2005.

O'Bryan, Scott. *The Growth Idea: Purpose and Prosperity in Postwar Japan.* Honolulu, HI: University of Hawaii Press, 2009.

OECD. *OECD at Work for the Environment.* Paris: OECD, 1973.

OECD. *Statistics, Knowledge and Policy: Measuring and Fostering the Progress of Societies.* Paris: OECD, 2008.

Offer, Avner. *The Challenge of Affluence: Self-Control and Well-Being in the United States and Britain since 1950.* Oxford: Oxford University Press, 2006.

Philipsen, Dirk. *The Little Big Number: How GDP Came to Rule the World and What to Do about It.* Princeton, NJ: Princeton University Press, 2015.

Piketty, Thomas. *Capital in the Twenty-First Century.* Cambridge, MA: Harvard University Press, 2014.

Purdey, Stephen J. *Economic Growth, the Environment and International Relations: The Growth Paradigm.* Abingdon: Routledge, 2009.

Raphael, Lutz. "Die Verwissenschaftlichung des Sozialen als methodische und konzeptionelle Herausforderung für eine Sozialgeschichte des 20. Jahrhunderts." *Geschichte und Gesellschaft* 22, no. 2 (1996): 165–93.

Robertson, Thomas. "Development." In *Encyclopedia of the Cold War, Volume 1,* edited by Ruud van Dijk, 254–6. London: Taylor & Francis, 2008.

Ruggie, John Gerard. "International Regimes, Transactions, and Change: Embedded Liberalism in the Postwar Economic Order." *International Organization* 36, no. 2 (1982): 379–415.

Sachs, Wolfgang. "Introduction." In *The Development Dictionary: A Guide to Knowledge as Power,* edited by Wolfgang Sachs, 1–5. London: Zed Books, 1992.

Schmelzer, Matthias. "'Born in the Corridors of the OECD': The Forgotten Origins of the Club of Rome, Transnational Networks, and the 1970s in Global History." *Journal of Global History* 12, no. 1 (2017).

Schmelzer, Matthias. "The Crisis before the Crisis: The 'Problems of Modern Society' and the OECD, 1968–74." *European Review of History* 19, no. 6 (2012): 999–1020.

Schmelzer, Matthias. *The Hegemony of Growth. The OECD and the Making of the Economic Growth Paradigm.* Cambridge: Cambridge University Press, 2016.

Scott, James C. *Seeing Like a State: How Certain Schemes to Improve the Human Condition Have Failed.* New Haven, CT: Yale University Press, 1998.

Sedlacek, Tomas. *Economics of Good and Evil: The Quest for Economic Meaning from Gilgamesh to Wall Street.* Oxford: Oxford University Press, 2011.

Speich Chassé, Daniel. *Die Erfindung des Bruttosozialprodukts: Globale Ungleichheit in der Wissensgeschichte der Ökonomie.* Göttingen: Vandenhoeck & Ruprecht, 2013.

Stiglitz, Joseph, Amartya Sen, and Jean-Paul Fitoussi. *Mismeasuring Our Lives: Why GDP Doesn't Add Up.* New York: New Press, 2010.

Stone, Richard, and Kurt Hansen. "Inter-Country Comparisons of the National Accounts and the Work of the National Accounts Research Unit of the OEEC." *Review of Income and Wealth* 3, no. 1 (1953): 101–41.

Suzuki, Tomo. "The Epistemology of Macroeconomic Reality: The Keynesian Revolution from an Accounting Point of View." *Accounting, Organizations and Society* 28, no. 5 (2003): 471–517.

Tobin, James. "Economic Growth as an Objective of Government Policy." *The American Economic Review* 54, no. 3 (1964): 1–20.

Tooze, Adam. "Imagining National Economies: National and International Economics Statistics 1900–1950." In *Imagining Nations*, edited by Geoffrey Cubitt, 212–28. Manchester: Manchester University Press, 1998.

Victor, Peter A. *Managing Without Growth: Slower by Design, Not Disaster.* Advances in Ecological Economics. Cheltenham: Edward Elgar, 2008.

Waring, Marilyn. *Counting for Nothing: What Men Value and What Women Are Worth.* Toronto: University of Toronto Press, 1999.

Wilkinson, Richard, and Kate Pickett. *The Spirit Level: Why Greater Equality Makes Societies Stronger.* Reprint edition. New York: Bloomsbury Press, 2011.

13

A BASIS FOR SYSTEMIC SUSTAINABILITY MEASUREMENT

An update

Simon Bell and Stephen Morse

Introduction

In our various writings during the first decade of the twenty-first century on sustainability indicators (SIs), our goal was to take a systemic approach to understanding sustainability measurement. We had read many books and papers on SIs, some of which date to the late 1980s, but we were not entirely satisfied with what we had found. Our reading spanned an eclectic range of SIs, including ecological concepts such as the maximum sustainable yield (MSY) for fisheries, which is still in use and indeed forms part of the new Sustainable Development Goals (SDGs) released by the United Nations (UN) in early 2016, as well as other, well-established indicators, such as the Shannon-Wiener index of biodiversity (H). We also covered SI frameworks, such as the AMOEBA framework, developed by Dutch researchers for assessing the sustainability of the North Sea, and wrote about it in a form that would appeal to non-specialists. We reviewed the development of SIs as part of the concept of sustainable cities and communities during the early to mid-1990s, of which there were many following the release of Local Agenda 21, which came out of the Rio 1992 conference. "Sustainable Seattle" was perhaps the best-known example at the time that this chapter was originally written, in the late 1990s, but the integration of sustainability and SIs into municipal planning has now spread throughout the world. We were also intrigued by the growing use of SIs as a means to help achieve what were called "sustainable institutions," and these were largely based on microfinance as a means to help achieve sustainable development. Microfinance gained a great deal of attention following the first Microcredit Summit held in 1997, and was being strongly promoted by its adherents, almost as a "silver bullet," as the sustainable way forward in sustainable development. Here the indicators were designed to monitor dependency of such institutions on continuing donations from aid organizations with the ultimate intention of eliminating such funding and encouraging the institution to fend for itself and live off the interest gained from microcredit. It presented an apparent win-win in development terms with the resource poor of the world gaining access to finance to help them improve the basis for their livelihood and institutions that would gradually lessen their dependence on funds from the developed world. The example we covered was the Subsidy Dependence Index (SDI) created by the World Bank. Finally, we were fascinated by the emergence and promise of the Environmental Sustainability Index (ESI) in the late 1990s; the first attempt that we were aware of to develop global rankings of nation-state environmental sustainability based upon a complex set of indicators, data transformation and aggregation methods.

It was clear to us that SIs were rapidly growing in popularity, largely because they followed the basic adage that to manage something you have to measure it. From the 1990s onward, a whole range of new SIs appeared, including Ecological Footprint (EF), Carbon Footprint (CF), Ecosystem Based Fishery Management (EBFM), the Triple Bottom Line (TBL), the Index of Sustainable Economic Welfare (ISEW), and many others. A few of the other indicators that often appear in discussions of SIs are summarized in Table 13.1 based on year of first publication.

Be it via a single index or a suite of SIs, the general motive was to create tools that set out what it is we want to see in a "system" (a fishery, city, community, corporation, institution, etc.), and, perhaps more critically, why it is we want to see it, and to assess progress towards a goal. The new "indicator epistemology" made a lot of sense and also allowed researchers to move away from rather vague definitions of sustainability, including, it has to be said, the oft-quoted Brundtland definition ("*Sustainable development is development that meets the needs of the present without compromising the ability of future generations to meet their own needs*") published by the United Nations World Commission on Environment and Development (WCED) in 1987. Such definitions provide a broad and useful sense of mission with regard to sustainability, but offer little else in terms of tangible goals. Our feeling at the turn of the century was that for SI"s to "work," a measurement needed to identify real and tangible positions of sustainability, and it needed to give a clearer sense, to the individual or institution doing the measuring, what

Table 13.1 Some common sustainability indicators

Index	Abbreviation	Starting year
Happiness Index	HI	1987
Big Mac Index	BMI	1988
Human Development Index	HDI	1990
Ecological Footprint	EF	1993
Genuine Progress Indicator	GPI	1994
Gender Empowerment Index	GEI	1995
Gender-related Development Index	GrDI	1995
Corruption Perception Index	CPI	1996
Environmental Performance Index	EPI	1996
Global Competitiveness Index	GCI	1996
Human Poverty Index	HPI	1997
Living Planet Index	LPI	1998
Bribe Payers Index	BPI	1999
Environmental Sustainability Index	ESI	2000
Carbon Footprint	CF	2001
Mothers Index	MI	2001
Press Freedom Index	PFI	2002
Commitment to Development Index	CDI	2003
Democracy Score	DS	2004
Failed States Index	FSI	2005
Climate Change Performance Index	CCPI	2006
Happy Planet Index	HaPI	2006
Global Hunger Index	GHI	2007
Global Peace Index	GlPI	2007

it is that one wanted, what this wanting might imply, and how one intended to know when one had obtained what we wanted.

As noted above, at one level the SI literature of the period made a lot of scientific sense to us, but at another level we were concerned. This chapter will seek to set out some of those concerns and how we proposed at the time to address them. We certainly did not claim to have all the answers, but we did feel that a critical but supportive voice would help. This chapter is based upon a version we first published in a 1999 book with Earthscan, with a second edition published in 2000 that allowed us to include the ESI among other developments. The book was entitled *Sustainability Indicators: Measuring the Immeasurable*. While the book, and this chapter, was published in 1999, the ideas in it had been developed over the preceding years when we were both based in the School of Development Studies at the University of East Anglia (UEA), Norwich, UK. During the early 1990s, we were constantly exposed to and engaged in participatory methods to help engage the urban and rural poor communities of Africa, Asia and elsewhere.[1] These ideas of "participation" gained a prominence within development that one simply did not see in places such as the UK. The 1992 "Earth Summit" brought the concept of sustainability gradually to the fore, and in the years immediately after the summit Norwich City Council was something of a pioneer in the development of SIs within the UK, using "Sustainable Seattle" as a role model, and was thus a major influence upon us. We were fascinated by the trade-offs that Norwich were trying to address with regard to the three pillars of sustainability (economic, social and environmental) and how they saw their SIs being put into use by communities within the city. They were obviously going through a major learning curve but the "buzz" that surrounded all of this was infectious, and we could see clear parallels with our participatory work in the developing world. It was an intriguing nexus.

The context of this chapter

One of the features of SIs that struck us in the late 1990s was that they reflected a mindset based upon assumptions that suggested a mechanistic and reductive vision of the world. This was epitomized for us in the pressure–state–response (PSR) framework that had emerged from OECD and others in the 1980s, with its inherent "cause–effect" mentality but could also be seen with concepts such as the MSY. The idea that something as rich and as complex as, for example, the sustainability of a city such as Norwich or Seattle could be reduced to a few dozen SIs was a nagging concern to us and we wondered about what was being lost in the trade-offs, while trying to use SIs as a way of achieving sustainability. Not only that but, of course, much depends on who is setting the SIs and for what purpose. We did not hear many voices from those people living and working in the system and instead it seemed to us that the SIs were being created and controlled by what we might consider to be an isolated technical elite – and yet sustainability is the issue of our age and impacts on all lives. We sought to provide what we called an "alternative system mindset" that would raise an awareness of this disparity in power and also serve to provide the basis for how it may be addressed. But it has to be stressed that we were certainly not interested in only providing criticism of the dominant mindset surrounding SIs; that is far too easy and not very constructive. It also has to be said that we took a fairly eclectic view of SIs and had no intention to focus on any one of them or indeed on one group. Our intention was to focus on the ideas behind them rather than the "nuts and bolts" of the SIs. Each of the SIs in Table 13.1 is constructed in a particular way, and some of them are relatively simple in a mathematic sense while others are more complex. Where we talked about their construction in the book this was within the context of the assumptions that were made rather than the mathematics.

While this chapter is understandably dated, despite our best attempts to update it, it does provide one of the very first attempts that we are aware of to make the case for a more systemic and participatory approach to SIs, and was followed by other chapters that set out a practical methodology for how it could be put into practice. At the time of first publication we called the methodology "Systemic Sustainability Analysis" (SSA), but this later evolved into "Imagine" and more recently (and radically) into "Triple Task." It was meant to be a generic approach that could be applied in any context, be they cities, communities or institutions, and was designed to be as participatory as possible. This chapter does not cover the details of SSA/Imagine – the reader will have to look at the book or our other writings for that – but instead sets out the epistemological foundation upon which SSA was built.

The chapters before this one in the original book set out the details of some of the SI examples we mentioned above: MSY, biodiversity, institutional sustainability, etc. The breadth of examples was purposely selected to illustrate how SIs can – and indeed do – have a wide usage in sustainability, and we knew that this choice would be contentious. We expected comments along the lines of "why did you leave out indicator X or indicator Y" and we were not disappointed. Indeed, it was nonetheless quite astonishing how people had their "favorites" when it came to SIs and the degree of passion with which they held their position. Needless to say, we could not cover every indicator, even with the limited literature on SIs at the time, and we wanted to ground our analysis within practical use of SIs, rather than just offering a theoretical treatise. In this chapter the reader will come across references to example SIs from the early chapters of the 1999/2000 book, and our intention was to show how the broader points can be observed in those examples. We ask the reader's indulgence with regard to the referencing to the other chapters.

The chapter begins with our setting of the case against a reductionist approach to SIs. To some extent we feel that this is as relevant today as it was at the time of first writing. Narrowness disguised as scientific excellence, contempt for local people's values, disguised as a lofty focus on strategic issues and denial of the right to hold a contrary view, disguised as hierarchic prioritization of perspectives, are still prevalent in contemporary approaches to sustainability, and they have not gone away. By way of contrast we follow this with our systemic perspective.

The chapter presented here is more or less as it was published in the 2000 edition. We have made a few minor changes to create this version of the chapter, including some notes in square brackets, in order to help the reader follow the argument in a stand-alone mode without the other chapters of the book, but other than that it is much the same. Hence the references are rather dated, although we have updated them with the newer editions where relevant and that explains why the reader will come across publications dated after 2000. Some of the examples we cover in this chapter (and indeed book) have been superseded. The ESI, for example, was replaced in 2006 by the Environmental Performance Index (EPI) and while microfinance is still an important tool in the development toolbox, it is now regarded as part of a mix of approaches rather than the "silver bullet" it was presented as by some in the 1990s. The chapter also misses out on some other important initiatives that were just beginning to gain prominence in the late 1990s such as the rise of the Sustainable Livelihood Approach (SLA) promoted by DfID and others, and, of course, the Millennium Development Goals (MDGs) – and now Sustainable Development Goals (SDGs) and the indicator sets and targets that were/are associated with them.

We were very happy with the impact the book made, and continues to make, within the field of SIs. It is still widely cited and occasionally we receive a request for it to be made available on sites such as ResearchGate, which, unfortunately, we cannot do for copyright reasons.

The five premises we present at the end of this chapter as the basis for our participatory approach to SI development are – we would argue – still relevant to this day. Indeed, we would argue that one of the key and still under-explored frontiers of SI research is located within the general realm of "use" and all that is associated with that. For that reason, we have always resisted invitations to select what we see as the "best" SDI or even to construct one. Such questions miss the point. There is no "best" SI in an objective sense and neither can one be made; it is a mirage. Much depends upon context and also much depends upon making an SI "usable" in the sense of facilitating a beneficial impact. It is this focus on "use" and "impact" that would allow for an evolution of SIs to take place and those doing badly should be allowed to leave the stage while others can take their place or perhaps evolve in ways that enhance their use and impact. And, of course, all of this takes place within human societies that are themselves changing in many ways, including what is regarded as "need" (in a Brundtland definition sense of the term). In our view we need to facilitate the evolution of SIs and in order to do that we need to better understand it, while accepting all the time that we live "in" and "on" a planet that is itself undergoing change. SIs have to work within that change and indeed help bring about a change for the better. Whether the reader agrees with us or not we hope they will enjoy the chapter and imagine the exciting process of discovery we were going through at the time.

Speaking of paradigms and professionals

We will start the chapter by setting out the thoughts of two authors – Chambers and Hobart – who provide some seeds of thought for us to develop. Drawing from Plato's *Republic*, Chambers powerfully sets out problems with mindsets:

> Unwitting prisoners, professionals sit chained to their central places and mistake the flat shows of figures, tables, reports, professional papers and printouts for the rounded, dynamic, multidimensional substance of the world of those others at the peripheries. But there is a twist in the analogy. Platonism is stood on its head. Plato's reality, of which the prisoners received only the shadows, was of essences, each simple, unitary, abstract and unchanging. The reality, of which core professionals perceive only the simplified shadows, is in contrast a diversity: of people, of farming systems and livelihoods, each a complex whole, concrete and changing. But professionals reconstruct that reality to make it manageable in their own alien analytic terms, seeking and selecting the universal in the diverse, the part in the whole, the simple in the complex, the controllable in the uncontrollable, the measurable in the immeasurable … For the convenience and control of normal professionals, it is not the local, complex, diverse, dynamic and unpredictable reality of those who are poor, weak and peripheral that counts, but the flat shadows of that reality that they, prisoners of their professionalism, fashion for themselves.[2]

On a similar, illustrative theme Hobart argues:

> Local knowledge often constitutes people as potential agents. For instance, in healing, the patient is widely expected to participate actively in the diagnosis and cure. By contrast, scientific knowledge as observed in development practice generally represents the superior knowing expert as an agent and the people being developed as ignorant, passive recipients or objects of his knowledge.[3]

Elsewhere [in *Sustainability Indicators*] we have reviewed the state of play with SIs generally, although not exclusively without reference to local peoples and their knowledge. ESI, MSY and AMOEBA are all the constructs of experts. But, in this chapter we renew our discussion of the value of different approaches to thinking about SIs, and we question again if SIs should be "scientifically" derived in all cases. The process of their development, for instance, may be based on science, as with the MSY, but may just as plausibly be developed by a technocratic belief process or pseudo-science such as with the ESI. We would argue that this would account for as much distortion in the final SIs as would be seen in purely subjectively gathered indicators.

In this chapter we will look at a number of topics:

- changes in thinking;
- the demise of narrow scientism;
- a systemic approach to problem solving;
- introducing a range of systems approaches;
- new definitions and new thinking – holism, eclecticism, systemism;
- emerging premises for SI development.

Building upon the layered examples of sustainability indicators set out in the previous chapters (single SI, AMOEBA, sustainable cities and communities combining SIs, institutional sustainability), and taking forward the scientific or technocratic approaches to sustainability analysis which we described there, the aim of this chapter is to introduce and discuss an alternative, systemic approach to thinking and problem solving. We will compare this with what we might call the traditional, scientific and technocratic approach. In this process we draw out the implicit problem of using SIs – by definition a reductionist technique and tool – to describe sustainability – by definition a vision of wholeness. In this chapter we justify why we are using a systems approach to developing a different way of gauging sustainability. In our view, our approach builds off and develops from a practitioner perspective the work begun by Clayton and Radcliffe.[4]

Changes in thinking – from science to systems

The value of different perceptions and the necessity for individuals involved in problem situations to learn from one another in a participatory fashion are two of the themes of this chapter. Changes in perception can involve changes in thinking, and this can be thought of as a "paradigm shift." A definition might be helpful here. A paradigm is "an outstandingly clear or typical example or archetype … a philosophical and theoretical framework of a scientific school or discipline within which theories, laws and generalizations and the experiments performed in support of them are formulated."[5]

We might say that there is a Western–Middle Eastern scientific tradition that is a paradigm of thinking. This paradigm is dominant but there are alternatives to it. One alternative might be described as a systemic approach. This is an alternative paradigm of thinking, but one which we feel does not deny the value of science; instead, it complements it and is sympathetic to its contribution while recognizing that there are other contributions which can also be made by other forms of thinking, from other individuals and groups.

Alternative views or even multiple views of reality are encouraged in a true systems approach. The unpacking of ideas relating to participation, learning and thinking in different ways requires an understanding that local people often have clear ideas of their own about what is sustainable (from their own perspective and in their own terms) without an expert's view.

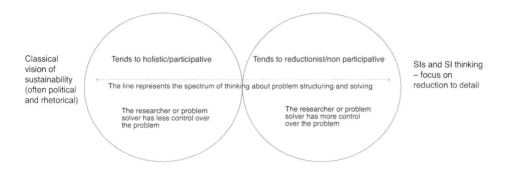

Figure 13.1   A continuum of research approaches

From one perspective the development of SIs, as set out [elsewhere in *Sustainability Indicators*] exemplify the hegemony of the technocrat. We have already reported in other aspects of the review that this hegemony is challenged by individuals within the scientific community.

There has been a dramatic change in thinking in many related areas among sections of the scientific community. The changes we are discussing here do not represent movement from a wrong way to a right way of thinking; rather it is a movement from one paradigm (and thus a set of assumptions about the world) to another. In an earlier work, Bell described this move-ment of mindset in terms of a continuum[6] (see Figure 13.1).

The horizontal line, the spectrum or continuum, provides one perspective of the range of thinking which can be undertaken in any problem-solving exercise. The range extends from the most reductionist to the most holistic. Koestler describes these two as referring to individ-uality and wholeness respectively, but again does not see either approach as being opposed to the other: "'partness' and 'wholeness' recommend themselves as a serviceable pair of compli-mentary concepts because they are derived from the ubiquitously hierarchic organization of all living matter."[7]

Whether we argue with the "hierarchic organization of all living matter" or not, the idea of complementary concepts is one which we support. However, it is possible to see them as being opposed. Therefore, before going on, we need to define these terms and understand more clearly what they include and exclude. In our work we intend to show that holism, in reality, always includes scientific and reductionistic modes of thinking. If holism were to be seen as exclusive or extreme, then it would not be holistic (by definition). To clarify the meaning of the terms, we will set them out against the background of current trends in the discussion within the academic community. An overall and rather dramatic phrase that we use to describe this stage of our description is "the demise of narrow scientism."

The demise of narrow scientism

Before we look at what reductionism and holism mean, let us get a clear idea about scientism. According to Webster, scientism is defined as follows:

> Scientism n. (1877) (1) methods and attitudes typical of or attributed to the natural scientist (2) an exaggerated trust in the efficacy of the methods of natural science applied to all areas of investigation (as in philosophy, the social sciences, and the humanities).[8]

The key word to keep in mind here is "exaggerated." There are a range of approaches to sustainability which worked on the premise that sustainability was a quantity which could be more or less defined in an absolute sense: "The measure of sustainability for wheat production, as a weighted figure, is …" This form of approach (if expressed a little facetiously here) might also be defined as an "exaggerated trust in the efficacy of the methods of natural science applied to all areas of investigation." This type of approach is exemplified in sustainability analysis that makes use of mathematical formulae to gain quantitative measure. Such formulae give the analysis a degree of respectability, but the formulae themselves colossally simplify the true complexity of the context. Unfortunately, the definition of scientism used here also raises another phrase that we need to define for clarity's sake – scientific method. What is *the* scientific method?

Here, again, is Webster:

> Scientific method (1854): principles and procedures for the systematic pursuit of knowledge involving the recognition and formulation of a problem, the collection of data through observation and experiment, and the formulation and testing of hypotheses.[9]

The method of science seems to involve observing the world in a systematic way, seeing problems (or opportunities), collecting data and testing theories about why the problems are there and rejecting hypotheses that are perceived to be "wrong." In this approach, questions arise, such as: Whose problems? Whose perception of problems? Whose justification for action? Whose idea about what data is legitimate? Who are legitimate stakeholders in the problem context? What are their views? On a similar tack, Richard Dawkins has put the essence of this issue as follows:

> If I ask an engineer how a steam engine works … I should definitely not be impressed if the engineer said it was propelled by "force locomotif." And if he started boring on about the whole being greater than the sum of its parts, I would interrupt him: "Never mind about that, tell me how it works." What I would want to hear is something about how the parts of an engine interact with each other to produce the behaviour of the whole engine.[10]

Dawkins's statement is indicative of the mindset of many scientists and also expresses the notion that within the scientific community there is an assumption that science is its own justification, that parts explain the whole and that objectivity is an accepted given truth of a well-undertaken scientific method. We will return to these issues. To get back to our definitions of reductionism and holism, it can be argued that this idea of scientific method finds its logical extreme in reductionism.

Again, Webster:

> Reductionism n. (1943) (1) the attempt to explain all biological processes by the same explanations (as by physical laws) that chemists and physicists use to interpret inanimate matter; also: the theory that complete reductionism is possible (2) a procedure or theory that reduces complex data or phenomena to simple terms.[11]

Reductionism reduces wholeness to individual parts and bits to make them understandable. Its scientific approach to understanding is to stand back, take an objective (scientific?) worldview, and seek the truth. As Bell puts it:

A reductionist approach rejects ideas about the reality and importance of unscientific aspects of life (hunches, guess-work, instincts for rightness and even in certain circumstances illogical activity, i.e. activity which is not consistent with narrow definitions of efficiency). The universe is seen through empiricism as fixed, knowable, measurable and, therefore, predictable.[12]

Developing an understanding of what we mean by reductionism, Dawkins argues that there are two forms: "reductionist" and "hierarchical reductionist." The first type, which we might refer to as the classical reductionist, is in Dawkins's words set up by "trendy intellectual magazines" as a kind of straw man: "To call oneself a reductionist will sound, in some circles, a bit like admitting to eating babies. But, just as nobody actually eats babies, so nobody is really a reductionist in any sense worth being against. The non-existent reductionist tries to explain complicated things directly in terms of the smallest part."[13]

Alternatively, the second type of hierarchical reductionist, of which he counts himself, "believes that carburettors are explained in terms of smaller units," which are "explained in terms of smaller units … which are ultimately explained in terms of the smallest of fundamental particles. Reductionism, in this sense, is just another name for an honest desire to understand how things work."[14]

The problem is that this form of analysis does not stop at carburetors but is used in all forms of social, environmental, and ecological analyses as well. In these contexts, the limitations of the approach are already evident. Something which has many units all in various states of interaction would require a substantial effort over many years using hierarchical reductionism to understand in full. In practice what happens is that a few of the key units and interactions are singled out for analysis. Reductionism as a paradigm adopted by scientific professionals, whether the baby-eating or hierarchical form, is one extreme of the continuum we set out in Figure 13.1. It is expressive of one way of thinking about the world and how we understand it. It is arguably the approach or method of understanding the world that has been the basis for much of Western and Arab science and it has been responsible for amazing and revolutionary advances in all branches of human thought and discovery. However, on the negative side, the process of dividing up the world in order to identify small parts is questionable in many areas of understanding and has led to partial analyses and the development of answers to problems which themselves cause still greater problems (a difficulty with all approaches which extrapolate from the part to the whole).

There is another problem with reductionist approaches. Dividing an entity means that the concept of wholeness is often rendered dead by the process of examination. Studying "dead" parts can be informative but can often do little to help us understand the living whole. Furthermore, the paradigm of a reductionist can be very limiting.[15] If one considers the world as disconnected parts rather than as an inclusive whole, the resulting worldview can be restricted in terms of understanding the relationships and processes that combine to make the whole. However, we are developing the argument for our approach before providing the definitions. So far we have looked at what we mean by reductionist approaches and have argued that such approaches deal with parts. Set against this is holism, which Webster defines in the following terms: "Holism n. (1926) (1) a theory that the universe and esp. living nature is correctly seen in terms of interacting wholes (as of living organisms) that are more than the mere sum of elementary particles."[16]

Another definition takes us even further into our understanding of this approach: "The theory that the fundamental principle of the universe is the creation of wholes, i.e. complete and self-contained systems from the atom and the cell by evolution to the most complex forms of life and mind."[17]

Holism deals with wholes and in this paradigm we see the universe composed of "self-contained systems." This kind of approach can be said to find a logical end-point in the notion of the world as a living system, as expressed in the work of James Lovelock and the establishment of the theory of Gaia.[18] Systems approached as wholes are fundamental and need to be understood in their entirety. To break them down into elements is to lose the point of the wholeness. Lovelock has discussed wholeness and reductionism in terms of Gaia:

> Consider Gaia as an alternative to the conventional wisdom that sees the Earth as a dead planet made of inanimate rocks, ocean and atmosphere, and merely inhabited by life. Consider it as a real system, comprising all of life and all of its environment tightly coupled so as to form a self-regulating entity.[19]

Lovelock went on to develop this theme in terms of sustainability and its corollary, health:

> Only when we think of our planetary home as if it were alive can we see, perhaps for the first time, why farming abrades the living tissue of its skin and why pollution is poisonous to it as well as to us. ... The living Earth's response to what we do will depend not merely on the extent of our land use and pollutions but also on its current state of health.[20]

To adopt an approach that deals with wholes has many implications. Possibly the first point is to recognize that the premise of the traditional, reductionist scientist – which is that the knowing process works by "a procedure or theory that reduces complex data or phenomena to simple terms" – is no longer valid for us (nor would we agree that simplicity depends on reductionism). This does not mean that the traditional scientific approaches are invalid in all cases and in all contexts. However, if we are to understand complex wholes, we will need to adopt a different paradigm or extend the old. Later in this chapter we will describe a process within the systems thinking movement, from first-order to second-order cybernetics, which attempts to explain this adoption of a different paradigm. The process can be thought of as a movement of mindset from an observer divorced from context (first order) to an observer deeply involved in the context (second order). For now, it is worth noting the comment of Buddrus upon this process: There is a parallel between first- and second-order cybernetics (which we discuss in more detail shortly) and with the movement from reductionist to holistic paradigms:

> What is needed is a transformation of awareness from cybernetics of the first order to cybernetics of the second order ... This seemingly simple transformation has fundamental impacts when applied to self-awareness and belief systems. It can cause considerable mental problems in orientation: the transition of oneself from an observer of a reality which is considered to be outside oneself, to a participant in the same reality, and then towards being a co-creator of that reality, requires fundamental cognitive and emotional reorientation.[21]

In understanding sustainability, we argue that we need to recognize and work with unities, of which we, as observers, are also a part. This is not to suggest that complex unities cannot be better understood by identifying key components, interactions and processes, but that scientific approaches need to be seen in terms of the greater whole of which the observer is a part; the observer therefore brings ideas and actions into the context.

The traditional scientific paradigm has its value and its place in our understanding, but as one view among many – and we would argue that it should not be the meta-theory that dominates all others. The benefit of the holistic approach is that we can deal with complex wholes without losing their complexity or "killing the whole,"[22] and we can ask wider questions than those which relate to individual parts. The downside for our analysis is that analysis itself becomes terribly difficult and can lose all sense of focus and organization if the practitioner is not careful. To make holism work we need to grasp the principles of systems thinking which lie at its heart.

The "seed" idea that we want the reader to take away is the value of a more holistic approach within the analysis and measurement of sustainability.

Systems approaches to problem solving

In this chapter the word "system" probably arises more often than any other. Often, the word is not used in a strict and exact fashion. In terms of daily usage, the word is almost redundant, occasionally meaning little more than "thing" or a set of related things (for instance, a dish washing system, a driving system, an office system). We now want to develop what we mean by system – but we should say at the outset that there is considerable discussion within the systems community about this definition and there are many interpretations of what a system is. There is also a vigorous and developing discussion on systems and sustainability.[23] Here we make use of widely accepted definitions. One view of the systems approach is, as the American systems thinker Peter Senge puts it, the primacy of the whole: "The primacy of the whole suggests that relationships are, in a genuine sense, more fundamental than things, and that wholes are primordial to parts. We do not have to create interrelatedness. The world is already interrelated."[24]

From this perspective, the idea of systems is a perfect foil for Senge's thinking: "A system is a perceived whole whose elements 'hang together' because they continually affect each other over time and operate toward a common purpose. The word descends from the Greek verb *sunistánai*, which originally meant 'to cause to stand together'. As this origin suggests, *the structure of a system includes the quality of perception with which you, the observer, cause it to stand together.*"[25]

This view of systems has echoes of the work of Peter Checkland in the UK, where there is great emphasis placed upon systems existing within the human mind, as perceptions, which we project into the world as a means of describing and understanding it.[26]

Systems thinking has a number of strands but is fundamentally based upon a few simple concepts. The lists of components vary with different authors, but there are substantial similarities between them.[27] For our definition of a system we make use of the one provided by Avgerou and Cornford.[28] These authors present the major features of systems as six-fold, and these are set out in Table 13.2.

Although there are different ideas about the fundamentals of systems, a systems analysis of a problem context can be undertaken. Such an analysis, whether of an information system or an ecological or a social organization, would be expected to provide an understanding of processes and relationships within a "wholeness." Emerging from this set of features and the earlier description taken from Senge, we can say some fundamental things about the basis for a systems approach:

1   System is a term that can be applied to a vast number of different things, and this application is variable depending upon the individual or shared perception of an onlooker. A system can be a physical entity (such as the carbon cycle), a social entity (a political constitution), or an abstract idea (the idea of sustainability – as we shall demonstrate).

Table 13.2 Defining features of systems

Systems feature	Description
Identification of a boundary	This defines the system as distinct from its environment
Interaction with the environment	The environment is not the system itself since it lies outside but it does affect it
Being closed or open	Concerns the interaction of the system with what lies beyond its boundary
Goal-seeking	A system is capable of changing its behaviour to produce an outcome
Being purposeful	Systems select goals
Exerting control	A true system retains its identity under changing circumstances

2   Once defined the system will have a boundary (unless it is an infinite system!), and the boundary is defined by the onlookers – or we might say stakeholders. Ison, quoting Russell, draws actor and boundary together in saying "the observer is seen as part of the system's construction and not independent of the system. Russell takes this debate further. He emphasizes that 'a system' is always a short-hand way of specifying a system environment relationship."[29]

3   The system conceived by the onlooker will take place in a larger environment that is defined by being outside the boundary agreed. The environment will have a relationship with the system but the degree to which it affects the system will largely be dependent upon the system itself.

4   Systems are changing and can be self-changing. As a purposeful wholeness, the system will be expected to seek its own optimum.

The final point is critical. If a system is purposeful then it might be expected to seek its own continuance and therefore sustainability.

Figure 13.2 provides one view of the systems approach so far described. Although it is rather artificial, let us compare this systems view of the world with an equivalent, taking the most reductionist stance possible (Figure 13.3). The difficulties that this approach raises for the study of sustainability can be juxtaposed to the advantages of systems as set out in Table 13.3.

Arising from the discussion thus far, the systems approach to understanding complex contexts is of interest for three reasons:

- The system is a construct in the mind of the onlooker(s) or stakeholder(s); the system is brought forth or created as an artificial construct by those studying it. Therefore, the system can be the result of an eclectic process, which Webster defines as elements drawn from various sources.
- The system is a whole and has the potential to change itself.
- The system is involved with its own sustainability; it can change as its environment changes in order to be sustained.

These three seed ideas, developing on the idea of wholeness set out in the previous section, will be fundamental to our thinking in later sections. So far we have described the reductionist

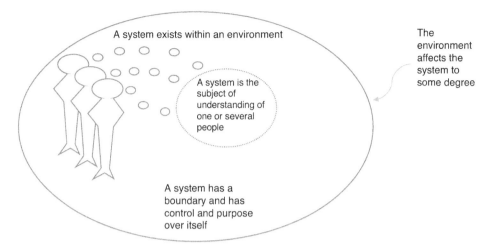

Figure 13.2   A systems view of a particular context

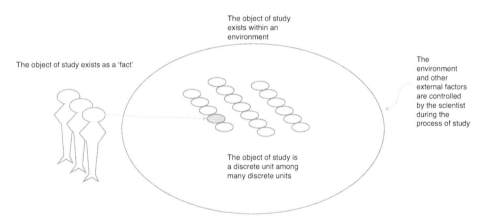

Figure 13.3   A reductionist view of a particular context

Table 13.3 Comparison of systems and reductionist approaches

Systems approach	Reductionist approach
The problem is shared by legitimate stakeholders in the problem context	The problem is in the mind of the scientist
A wholeness is reviewed	A part of a complex whole is analysed
The environment affects the system	The environment is expected to be controlled
The boundary of the system is flexible and dependent upon the perception of the stakeholder	The boundary of the part is defined by the expert

mindset, which we argue is behind much of the scientific method expressed by many of the conventional advocates and developers of SIs. We have not yet discussed systemic approaches to problem solving or SI development, but for now we want to briefly describe some forms of the systems approach to problem solving.

A range of systems approaches

As we noted in the previous section, there are numerous ways of thinking about and applying a systems approach. This is quite consistent with the systems view that the variable perceptions of different stakeholders in a problem context are legitimate but need to be justified. In this section we will quickly describe four different approaches, some analytic and some more descriptive, which are either explicitly or implicitly systems-based. We argue that they can all be understood in terms of the axis that we set out in Figure 13.4. The approaches that we illustrate here are from the fields of problem solving, problem description, project appraisal, and project planning.

The first form of systems approach is set out in Figure 13.5 and is known as the soft systems approach or soft systems method (SSM).

A problem-solving approach – the soft systems method

To describe this approach, we set out the main elements in Figure 13.5. This provides a view of all the elements of the approach and shows the manner in which they combine.

The SSM was developed, and has since been extended, by Peter Checkland and colleagues at the University of Lancaster in the UK, and has since been developed by him and others.[30]

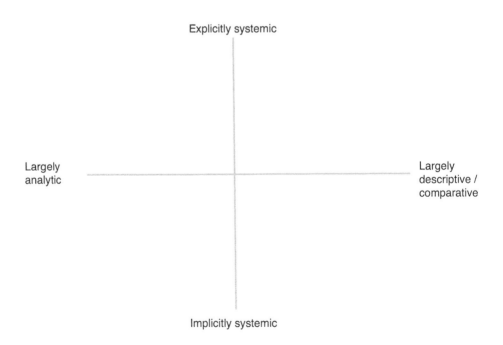

Figure 13.4 Axis for comparing systems approaches

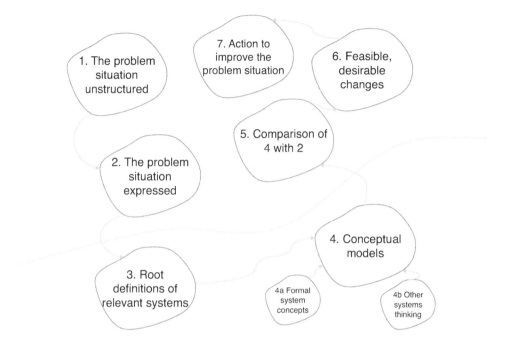

Figure 13.5 The soft system method

Today the approach is taught by universities and consultancy agencies in many locations and has taken on many nuances depending upon the requirements of the teaching and the specific aims and objectives of the practitioners. The way in which we develop our perception of the approach varies from others but is essentially related to the format set out by the Open University.[31] From our perspective, the fundamental insight of Checkland's work is that problems in the world are usually "soft." By soft we mean that objectives are unclear, purposes are muddled and solutions are not usually initially available. This contrasts to the traditional "hard" approach (of, for example, reductionist science, which sees problems as being definable and objectives as self-evident and open to empirical study), which has been the hallmark of problem investigation in much of academia. We will not go into detail about the nature of SSM, but features worth bringing out from Figure 13.5 are:

- It is often necessary to spend considerable time in perceiving the problem and exploring the tasks and issues implicit in it.[32] These are set out in elements 2 and 3 in Figure 13.5.
- There is not an assumption that the "problem" is clear. It may have many definitions.
- The next key point is that a definition of a transformation within the problem context needs to be agreed upon (element 3). It is not assumed that because, for example, I am a fish biologist looking at the problem, the solution to the problem will be maintaining production (as is often the case in MSY). We need to see that other domains may contain the "solutions" to a given problem. For example, in the Peruvian anchovy fishery, which collapsed essentially because of overfishing, the emphasis of fishery scientists was originally on setting an MSY of production, which itself became invalid because of the El Niño effect. But if the emphasis was on helping the fishing industry in more general terms,

perhaps other perceptions could have been brought to bear, such as livelihood diversification.

- The next point is that identifying a transformation is the basis for an activity plan which is then compared to the problem context as first reviewed (elements 4 and 5). It is often the case that in analyzing a given problem, one loses sight of the issues that first excited one's attention. This feedback loop requires constant recalibration and reappraisal of the original problem as first perceived.

- Stakeholders are brought together to discuss the analysis (element 6), whether that be community members affected by a specific development, or, say, an aboriginal group whose fisheries could be affected by a new approach to species management. Ideally this is not an expert-driven approach and stakeholders are performing the analysis too, but this idea of inclusivity prior to action is another strong feature of the SSM approach.

- Finally, the process is cyclical. One does not definitively "fix" problems. Rather, one achieves ways forward mutually and then works on the next issue that arises (elements 7 and 1).

The main features of the soft systems approach are that the process of thinking systemically about problems is iterative, participatory, and ongoing. The second systems approach arises from the work of Senge et al. and relates to his team's work on the learning organization (LO).[33]

Problem description – the learning organization approach

Senge has set out five "disciplines" for encouraging and developing the learning organization that is the focus for his work in making use of systems approaches. The five disciplines are: systems thinking, personal mastery, mental models, shared vision, and team learning. As with the work of Checkland and his collaborators, the five disciplines have been developed and applied by various agencies and academic institutions, in different contexts, and have produced a rich range of approaches and adaptations.[34] The five disciplines as we interpret them are set out in Table 13.4 with a brief definition of each discipline, a note on where they might be applied, and some indication of what might be the expected outcome of their application.

As with the work of Peter Checkland, the LO approach does not see problem solving as being easy or objective. Focusing heavily on dialogue and team learning, the list of outcomes shows how closely the LO approach relates to the relative merits of scientism and systemism. In defining sustainability, group consensus and insight are more vital than reductionist objectivity. In the LO approach, the systems approach is a core discipline associated with others in order to provide learning and consensus. As with SSM, processes are important and systems analysis relates to cycles of understanding. Senge makes use of what he calls "archetypes" to be compared against the real world. One such archetypal model is shown in Figure 13.6. In the snowball archetype, a situation of continuous decline or improvement is described – here demonstrated by the River Cynon example described earlier. The snowball is not a virtuous cycle in either contexts of decline or growth – it epitomizes continuous change, feeding on itself. It therefore requires a balancing and adapting action (contained in the balancing archetype described by Senge) to cause stability and equilibrium.

By using a range of archetypes such as these, situations can be considered and the consequences of actions modelled and discussed by stakeholders in the process. At first glance the approach may appear to be largely descriptive and comparative,[35] but it does allow contexts to

Table 13.4 The five disciplines

Discipline	Definition	Where applied?	Expected positive outcome?
Systems thinking	This focuses on links and loops – loops that can be reinforcing (small changes become big changes) or balancing (pushing stability, resistance and limits).	Contexts where cause and effect are unclear	Description and insight
Personal mastery	Numerous interpretations; but one threefold explanation of what this means is: (1) Articulating a personal vision. (2) Seeing reality clearly. (3) Making a commitment to the results you want.	Context where change processes threaten individuals' ability to cope	Empowerment
Mental models	We are all making mental models of the world as we experience it. The fifth discipline develops this tendency. Such models are based upon reflection and enquiry.	Any action learning situation	Clear self-analysis
Shared vision	Built around six core ideas: (1) The organization has a destiny. (2) A deep purpose is in the founders' aspirations. (3) Not all visions are equal. (4) There is a need for collective purpose. (5) There is a need to provide forums for people to speak from the heart. (6) Creative tension is useful and can be encouraged.	Contexts of dramatic change	Organization-wide clarity of purpose
Team learning	Learning through conversation, dialogue and skillful discussion – the aim is to achieve "collective mindfulness."	Contexts of team development	Group consensus

be reviewed for change processes, and therefore it appears a useful method to apply to analyzing sustainability – particularly where known forces of change are at work and their consequences need to be considered.

Appraisal – the participatory rural appraisal approach

Participatory rural appraisal (PRA), we argue, is a systemic approach to the range of issues that arise in project appraisal. Although PRA is not meant to be explicitly a "systems" approach, it contains much in common with what we described so far as central to a systems ethos in understanding complex situations – there is a shared epistemology.[36] There is no consensus as to what constitutes PRA techniques as opposed to any other set of methods for analyzing populations. As with SSM and LO, the PRA approach has been taken up and developed globally, and there is a rich literature on the various ways in which it has been applied and developed.

Working from literature produced from various sources, some of the techniques for PRA are set out in Box 13.1 with a brief description of what they involve.[37]

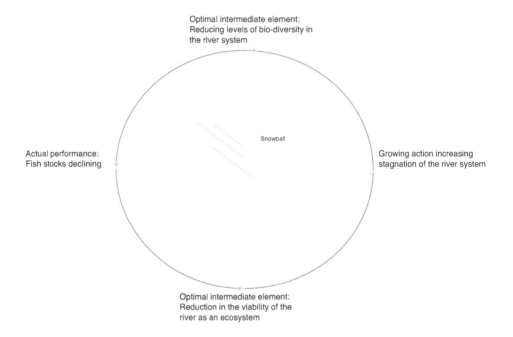

Figure 13.6   The reinforcing loop (snowball)

Box 13.1 Some of the techniques in participatory rural appraisal (PRA)

- *Participatory mapping and modelling* (all participatory diagramming). This technique encourages local people to draw and mark the ground with colors, sticks, cigarette packets and string (and anything that comes to hand, although one should be wary about bringing in pens and paper as these can block local people from expressing their views readily) in order to show variation from a local perspective of "mappable" phenomena.
- *Transect walks and participatory transect.* To gain a quick overview of local practices, the team walk a transect through the appraisal area.
- *Seasonal calendars.* This is a form of modelling or mapping where villagers are asked to show the seasonal or monthly distributions of inputs and outputs.
- *Activity profiles and daily routines.* This is used when it is important to understand how daily patterns of activity are evolving.
- *Time lines.* When there is a need to gain a view of local history. The time lines can be collected at community interview (see rapid approaches below).
- *Local histories.*
- *Venn (Chapati) diagrams.* To gain a systemic view of the overlaps between different groups, commodities, inputs and/or outputs in a village setting.
- *Wealth rankings.* To gain an insight into the distribution of wealth over time and space. Small groups can be asked to rank the wealthiest from the poorest in the village. Often piles of stones are used to indicate relative wealth. This can be done as part of the exercise to map the village social context. In this case the community as a whole might rank itself.

- *Matrices.* Communities are asked to set out a matrix for technologies and to set out attributes in the rows. Another approach might be to map the productive area of the village and then to set out problems and opportunities in the rows.
- *Inventory of local management systems and resources.* This can be used in focus group or community group interviews (see below). Local people know their management practices best. The interviews focus on how local management is undertaken. Use local classifications wherever possible.
- *Portraits, profiles, case studies and stories.* These include summaries of family histories, farm coping mechanisms, conflict resolving. Use focus group technique as described below.
- *Folklore, songs and poetry.* Sitting, listening (usually with an interpreter) and absorbing – principles of direct observation; see below.
- *Team interactions.* Evening discussion and morning brainstorming sessions with teams which can be mixed and changed but must be carefully monitored by one member of the team. The monitor should record locations of people during the interaction and draw attention to the way the team works: (i) draw a circle around the person who is talking; break the circle when they are interrupted; (ii) draw an arrow from the talker to the person being talked to with a note of duration; (iii) record each contribution in seconds.
- *The night halt.* When it is important to show that the outsiders are "with" the village: too often consultants are not in the village when people have time to talk – in mornings and evenings.
- *Survey of villagers' attitudes.*
- *Intriguing practices and beliefs.* When we have the time to try to absorb the richness of local life – taking a sideways look at expected project outcomes.
- *Key informant interview.* Interview a select group of individuals. They are preidentified as having insights and are usually owners or major stakeholders in problem areas. They are usually preidentified as being "reliable."
- *Focus group interview.* A recent addition to semi-formal techniques. The technique is historically based in market research to gauge reaction of customers to new products. The focus is on reactions to potential changes. Participants discuss among themselves.
- *Community interviews.* Focus groups are for local people to discuss their own issues and problems; in community interviews the investigator asks questions, raises issues and seeks responses. The primary response is to and from interviewer to participant.

Chambers, the major author of the approach, indicates three pillars to PRA. These are the:

- behavior and attitudes of the development professional;
- need for sharing between different actors;
- requirement for participatory methods.[38]

These three pillars are set out and developed in Figure 13.7.

SSM has many advocates and LO is now adopted by many practitioners in management science, but PRA has been adopted by the development community almost as a new orthodoxy in project practice. This has raised questions about its value and there is considerable debate around the capacity of PRA to work in context. Biggs indicates three concerns with the approach:

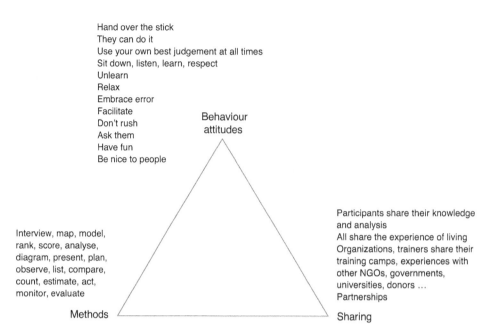

Hand over the stick
They can do it
Use your own best judgement at all times
Sit down, listen, learn, respect
Unlearn
Relax
Embrace error
Facilitate
Don't rush
Ask them
Have fun
Be nice to people

Behaviour attitudes

Participants share their knowledge and analysis
All share the experience of living
Organizations, trainers share their training camps, experiences with other NGOs, governments, universities, donors ...
Partnerships

Interview, map, model, rank, score, analyse, diagram, present, plan, observe, list, compare, count, estimate, act, monitor, evaluate

Methods

Sharing

Figure 13.7   Three pillars of PRA

Firstly, there is the risk that an exaggerated confidence in certain techniques and management tools associated, in this instance, with "participatory" approaches, can limit critical awareness of how their application proceeds in practice ... Secondly, there is a tendency to assume that simply "including" certain kinds of people (in a team process) is sufficient to affect the "participation" of the group which they are taken to represent ... Finally, it cannot be assumed that "inclusion" guarantees meaningful participation.[39]

We will return to this critique later as we develop our participatory model for measuring sustainability. PRA is widely regarded as including populations of stakeholders, and it values the insight of this population. As with both SSM and LO, all three approaches provide the stakeholders in a given context with a say in the process of understanding, as well as responsibility for the sustainability of the enterprise and a legitimate place in developing an analysis. PRA is interested in setting boundaries to appraisal, but not in narrowing the boundary to a pre-specified topic. The object of appraisal is treated as a system since it is recognized as a whole.

In this section we have been considering the PRA approach as a systemic manner of dealing with project appraisal. However, if systemism can work in appraisal, can it be applied to project development, planning, monitoring and evaluation? In the next section we examine one approach to such applications.

Project handling – the logframe approach

The logical framework can be a useful method for measuring sustainability, and incorporates cost-benefit analysis (CBA) to provide indicators of process and project impact. Much has been written about the logical framework or logframe (LF) approach to project planning and management.[40] Unlike SSM, LO and PRA, LF does not have a single point of reference or champion as provided by Peter Checkland, Peter Senge and Robert Chambers respectively. LF appears as an evolved approach with no single point of original authorship.

This approach is only implicitly systemic in that it encourages its users to think widely about their project and to represent it as a totality, with both hard and soft elements clearly demarcated. The approach can be participatory and requires a great deal of agreement within the project team to work effectively.[41] The basic LF is a four-by-four matrix and is shown in Figure 13.8.

The LF can be both descriptive and analytical. Descriptively, it allows a team or stakeholder group involved in a project to set out the formal aspects of the project (activities that lead to outputs, result in purposes and, hopefully, achieve the project goal), and also the informal or "soft" elements of the project at each level – this is shown in the "assumptions" column on the right. Therefore, the project is described in both soft and hard, formal and informal terms. Furthermore, the middle two columns allow the project to be monitored and analyzed, either qualitatively or quantitatively, in terms of the performance of the project. Performance can be measured on activities (the spending of money and the achievement of activities to date), on outputs (giving a notion of the projects, impact – has it achieved what it originally set out to do?), and at the level of purpose (evaluation – was the result as expected?).

In sum, the approach can be said to be systemic in that it sets a boundary around a complex unity and explicitly treats this unity as a whole. It should involve a range of participants in the project process (although this is not always the case in practice), and the project as a system is able to change in response to changes in the environment (it has properties of control and self-regulation). But how is the LF approach applied? When employing LF to develop or monitor a project, project activity is set out in the bottom-left cell. The activity described here can be measured and controlled by use of the related verifiable indicators and by means of verifica-

Figure 13.8   An outline of the logical framework concept

tion.[42] On the second row from the bottom, directly above activities, the verifiable indicators relating to outputs can be regarded as indicators of the project's impact. On the third row the indicators of purpose can be used as the main evaluation points for assessing the project's capacity to meet its original objectives. All indicators can, if required, be developed as indicators of sustainability. The diagram might be better understood as set out in Figure 13.9.

The LF approach might be argued to be "goal driven" and rather positivist.[43] The approach depends on the method of application for its systemic content (is it participatory; is it inclusive?) by the team involved.

Although LF was not used in the Norwich 21 example of sustainable cities discussed earlier in the book, we could apply it retrospectively to the first elements of the first column as shown in Table 13.5.

An overview of systemic approaches

The four systemic approaches are set out below in one frame (Figure 13.10) in terms of whether they are implicitly or explicitly systemic, whether they are problem solving, or descriptive or comparative. Before accepting that an approach is systemic or not, the quote from Buddrus given earlier in this chapter should be remembered as a caution (see above).

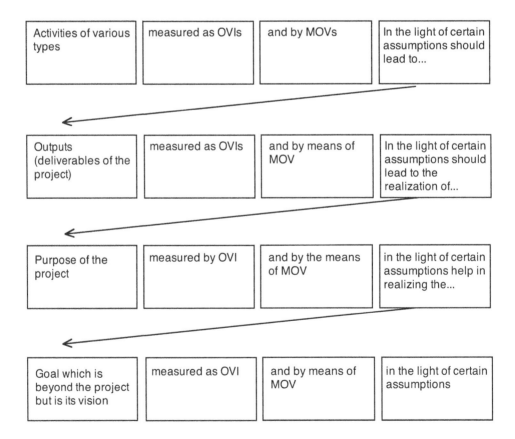

Figure 13.9 Explaining logical frameworks

Table 13.5 A partial log frame expression of Norwich 21

Goal (this would relate to the achievement of sustainability in cities at a national level)	Objectively verifiable indicators (similarly this would relate to the measurement of sustainability in cities at the national level)
Purpose	
"Promoting a prosperous and dynamic city with policies for sustainable long term growth and development that take account of the needs of the present generation of people without comprising the ability of future generations to meet their own needs" (Norwich 21, 1997)	This would relate to the achievement of the impact indicators set out below and the merging realization of sustainability which they would produce. This is an exercise for the owner of the Norwich 21 action plan
Outputs	
1. Clean air	1. 0 days poor air quality due to nitrogen oxides measured at Guildhall
2. Less domestic waste	2. Waste produced 0.36 tonnes per head, waste recycled = 0.018 tonnes per head
3. etc.	3. etc.
Activities	
Test against UK national air quality strategy standard, etc.	etc.

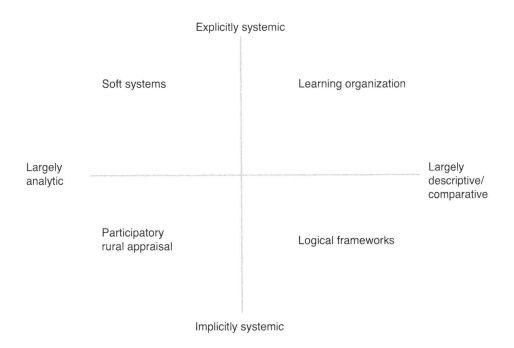

Figure 13.10 The four systemic approaches

A point made regularly among groups of managers training in the use of LF is the tendency to simplify the approach to "box-filling" in isolation, rather than exploring and describing a project context in a participatory manner. It is almost always possible to apply a systemic approach in a reductionist manner and thus lose the value of the undertaking.

In this section we have tried to demonstrate that systems approaches can vary considerably and can do quite different things, but that they retain some of the core ideas of what constitutes a systems study. We now go on to look at what implementing the systems perspective can mean to the development of a viable assessment of sustainability.

New definitions and new thinking – holism, eclecticism, systemism and future casting

We are interested in understanding the issues surrounding the measurement of the "immeasurable." It is our contention that the idea of measuring sustainability in absolute, traditional, objective, empirical, reductionist terms, as with SIs, is non-viable. It cannot be done because sustainability itself is not a single thing. Or better, it can be done but it will be done badly, oversimplifying complexity and reducing a variety of relevant and legitimate views and understandings to the dominant mindset of the scientist or analyst. A façade of objectivity can be generated, as with the ESI, but it is just that – a façade. Sustainability is, we believe, a highly complex and contested term open to a wide variety of interpretations and conceptualizations. In short, it is a concept dependent upon the various perceptions of the stakeholders residing within the problem context. Sustainability is not an absolute quantity to be measured. Sustainability changes as an idea (or as a system) in terms of the perception of the onlookers and they will also change with time. According to this approach, the view of sustainability must be developed so that it takes on-board the legitimacy of different views of the concept. When we adopt this mindset, we see that the view of a reductionist (even a mythical, "baby-eating" reductionist) may be legitimate and valuable. However, it is equally true that the view of a local inhabitant may also be legitimate and, although it may vary from that of the technical expert, may contain richness and detail that the expert does not have access to or actually loses in applying analytical tools. Narrow, expert-driven conceptions of sustainability have been problematic; the model for considering sustainability that we develop in this book is therefore developed around three premises. In measuring the immeasurable we are concerned with:

- eclectically derived, systemic wholeness; that is,
- the perception of systemic wholeness which derives from legitimate sources; and
- the sustainability of wholeness which is under observation.

In the following pages we will develop these themes in our approach and we will conduct our analysis using systems tools. It is not the purpose of this text to explore all the thinking and conceptualization behind the systems movement. However, we are aware that we are dealing with sets of concepts that require detailed analysis and justification. Such work has been undertaken elsewhere.[44] Behind the discussion of systems approaches and techniques explored in this chapter lies a theoretic discussion specifically expressed in the field of cybernetics; another definition is required. Here is Webster:

> Cybernetics ... Gk *kybernetes* pilot, governor (fr. *kybernan* to steer, govern)... (1948): the science of communication and control theory that is concerned esp. with the

comparative study of automatic control systems (as the nervous system and brain and mechanical-electrical communication systems).[45]

Developing upon this definition, the term was described by Wiener as the science of "control and communication in the animal and the machine."[46] Cybernetics is now organized into first, second, and third order categories which can be said to involve in sequence:

- the understanding of "feedback loops"[47] and control systems to explain how the world works in a scientific sense;[48]
- the understanding that individuals construct their own "reality"[49] and that this should lead to tolerance of alternative views;[50]
- reflection on the understanding of multiple realities and the means by which these multiple realities can be contained in a consensus view.

As Umpleby puts it:

> Whereas the first phase of cybernetics took an empirical approach to the nervous system, the second phase of cybernetics created a philosophy based on the findings of neurophysiological investigations. The third phase, the cybernetics of conceptual systems, looks at the community that creates and sustains ideas and the motivations of the members of that community.[51]

Our discussion in this chapter reflects thinking in the categories of first, second, and third order cybernetics.

Perhaps the issue of multiple and inclusive worldviews in the matter of sustainability is expressed most clearly in the work of Maturana and Varela.[52] Their work relates to the nature of biological systems but has implications in many related fields and is, at present, the source of much discussion among systems thinkers.[53] The core idea we wish to make use of here is that of "autopoiesis": the capacity of systems for self-making, self-renewal, or self-production. In a revolutionary departure from much of the background of systems thinking, Maturana and Varela postulate that systems are closed but there is an intimate interaction with the environment within this closure. The environment is not "out there," but as Morgan puts it, "the theory of autopoiesis accepts that systems can be recognized as having "environments" but insists that relations with any environment are internally determined."[54]

The exciting element for the sustainability debate is in working out what autopoiesis means – again, quoting Morgan, "Autopoietic systems are closed loops: self-referential systems that strive to shape themselves in their own image." Morgan gives some examples of what this means in practice. An example drawn from the fishing industry is illustrative, especially given the collapse of many fisheries in recent decades:

> the commercial fishing business … is also in the process of destroying itself because, historically, the key actors involved have seen themselves as being separate from the fish. The firms involved have enacted identities in pursuit of short-term goals, with the result that their actions have, in many parts of the world, already depleted the resource on which their business relies.[55]

The lesson seems to be that, for a truly systemic (or second or third order cybernetic) view of complex situations, the autopoietic approach explains why organizations can be progressive and inclusive or narrow and blinkered. Science, therefore, as an autopoietic system, can close itself off to factors which are not seen as central to the mindset of science itself. In this sense it can be as blinkered as the fishing industry is today. Revelations concerning the sudden explosion in cod stocks in the North Sea in 1997 suggest that the scientific analysis of sustainable fishing levels had been proved wrong – resulting in an explosion in stock rather than catastrophe.[56] This dramatic increase in stocks occurred against a background of numerous fishing boats being broken up and their crews made unemployed. In Morgan's example of a limited view, quoted above, the culprits are the fishing industry. In this example the problem is the limited under- standing and incorrect quantification of stocks as provided by scientists. A fishing industry representative said on BBC Radio 4 that the job for the scientists was now to build trust with the fishermen since the scientists were no longer believed. However, the condition of the North Sea fish stock at the time of writing appears again to be facing catastrophe. The chal- lenge for those predicting such crisis is to convince the fishermen that the future situation is more accurately understood than it is in the present. This raises the issue of futurology.

In an autopoietic sense, the systems approach to sustainability must mean that we include as much of our environment as possible in our self-referencing. As a result, the views of all involved in contentious projects are included (and their opinions valued) in the decision- making processes.

To truly engage systemically in the understanding of sustainability one final and often underutilized aspect of the wider environment needs to be "swept in" and that is the future condition of a system in question. After all, sustainability is about decisions we make now and how they impact upon this and future generations. Thus in the development of any plan, one must consider effects deep into the future. One of the great weaknesses of organizations is their resistance to change in the context in which they are embedded. For sustainability it is impor- tant to develop future plans that are capable of meeting the needs of tomorrow as well as today. Scenario Making is one means to grapple with the significant and unpredictable issue of possi- ble future outcomes.

Michel Godet, a major thinker in the "French School" of Scenario Planning, has commented that "unfortunately there are no statistics for the future".[57] Matzdorf and Ramage add:

> no-one can predict the future. Many people have tried – from prophets to mathe- maticians – but most predictions go awry. One only has to look at the divergent predictions of Global Climate Models (GCMs), created at great expense of time and money, to predict future global climate to see how uncertain all this can be … Even here it is only the natural system that is being modeled based on variables such as greenhouse gas emissions. The models have now variables for human behaviour. However, we can identify a number of possible futures, and especially the areas in which major change is likely to occur. Scenario Planning is one way of doing this.[58]

One of the founders of Scenario Making, Peter Schwartz, in an interview described the spirit of contemporary Scenario Planning as follows:

> There is a recognition that big complicated methodologies and elaborate computer models are not the optimal way. It has moved away from formal planning-like processes more toward a thinking tool. And it is not much more profound than that. So it's a methodology for contingent thinking, for thinking about different possibili- ties and asking the question "what if?"[59]

Schwartz continues, "That's why I called my book *The Art of the Long View*. The second thing that is quite important is it has moved away from a focus on the external world toward the internal world of the executive."[60] Schwartz went on to describe the application of Scenario Planning in the following terms: "This was Pierre Wack's big insight at Shell. The objective is not to get a more accurate picture of the world around us but to influence decision making inside the mind of the decision maker. The objective of good scenarios is better decisions not better predictions."[61]

Scenario Making can be seen as an element of a systemic approach to intervention – as an additional means to improve decision-making. Matzdorf and Ramage, as advocates of the Schwartz approach, have described the scenario approach as follows:

> Scenarios are alternative images: possibilities, not predictions. Scenarios are not just wild guesses or science fiction stories. However vital imagination is to the process, there are some rules that need to be followed if scenarios are to help in strategic planning. In particular, we believe it is not useful to develop just one or two scenarios. Some approaches to Scenario Planning use an optimistic one, a pessimistic one and the status quo, or two opposing scenarios. Schwartz argues, by contrast, that a range of different scenarios helps people to "think outside the box", rather than in "black-and-white" opposites, making it possible for planners to develop strategies for many different futures rather than just for one or two options. Scenarios should help managers to become aware of the mental models and frames of reference they operate in, and not leave them caught up in their "mental ruts."[62]

Originally scenario planning was developed for strategic organizational planning, but it is also highly valuable for sustainability planning.[63]

Emerging premises for SI development

In the previous four sections we have taken a wide-ranging and provocative view of the role and nature of reductionism, and we have indicated that, although this approach is useful and valid for partial understanding of many areas of analysis, it is not valid as the basis for our understanding of sustainability. We have also described some elements of a systems approach, which is concerned with wholeness and is designed to take on-board the various viewpoints of actors and stakeholders in a problem context. We have described how this approach is related to developments in the field of cybernetics and, most centrally, the autopoiesis of Maturana and Varela. Finally, we have indicated that sustainability includes the future of the system in question. Scenario making or Prospective can be one means to address this issue of futurology.

For our study, these factors were vital in helping us to set the basic premises that we used to develop our hypothesis for systemic and scenario-based sustainability analysis, which has now resulted in the development of the Imagine approach. As we go through them, the reader should be able to see how they relate back to vital aspects of the discussion so far. The premises for the development of the Imagine approach are:

- Sustainability can provide a qualitative measure of the integrality and wholeness of any given system.
- Subjectivity on the part of the stakeholders in any given system (including researchers) is unavoidable.

- Subjectively derived measures of sustainability are useful if the subjectivity is explicitly accepted and declared at the outset and if the method for deriving the measures are available to a range of stakeholders.
- Measures of sustainability can be valuable aids to future planning, forecasting and awareness-building.
- Rapid and participatory tools for developing our thinking and modelling concerning measures of sustainability are of value to a wide range of stakeholders within development policy.

Notes

1   The participatory approach is exemplified by the work of Robert Chambers. See R. Chambers, "Rapid Rural Appraisal: Rationale and Repertoire," *Public Administration and Development* 1 (1981): 95–106; R. Chambers, *Whose Reality Counts? Putting the First Last* (London: Intermediate Technology, 1997); R. Chambers, *Participatory Workshops: A Sourcebook of 21 Sets of Ideas and Activities* (London: Earthscan, 2002).
2   Chambers, *Whose Reality Counts?*, 55.
3   M. Hobart, *An Anthropological Critique of Development: The Growth of Ignorance* (London: Routledge, 1993), 5.
4   A. Clayton and N. Radcliffe, *Sustainability: A Systems Approach* (London: Earthscan, 1996).
5   Webster, *Merriam Webster's Collegiate Dictionary* (Chicago, IL: Merriam-Webster, 1995).
6   See P. Richards, *Community Environmental Knowledge in African Rural Development: IDS Bulletin* (Brighton: IDS, 1979); S. D. Biggs, "A Multiple Source of Innovation Model of Agricultural Research and Technology Promotion," *World Development* 18(11) (1990): 1481–99; S. D. Biggs and J. Farrington, "Assessing the Effects of Farming Systems Research: Time for the Re-Introduction of a Political and Institutional Perspective," Asian Farming Systems Research and Extension Symposium, Bangkok, 1990.
7   A. Koestler, *The Ghost in the Machine* (London: Arkana, 1964), 290.
8   Webster, *Merriam Webster's Collegiate Dictionary*.
9   Ibid. On scientism, see also M. Stenmark, "What is Scientism?" *Religious Studies* 33 (1997): 15–32; T. Sorrell, *Scientism: Philosophy and the Infatuation with Science* (London: Routledge, 1991).
10   R. Dawkins, *The Blind Watchmaker* (Bath: Longman, 1986), 11.
11   Webster, *Merriam Webster's Collegiate Dictionary*.
12   Bell, *Learning With Information Systems*, (London: Routledge, 1996), 63.
13   Dawkins, *Blind*, 13.
14   Ibid, 13.
15   On reductionism, see R. H. Jones, *Reductionism: Analysis and the Fullness of Reality* (Lewisburg, PA: Bucknell University Press, 1990).
16   Webster, *Merriam Webster's Collegiate Dictionary*.
17   Chambers, *Chambers Twentieth Century Dictionary* (Edinburgh: W. and R. Chambers, 1979).
18   J. Lovelock, *Gaia* (Oxford: Oxford University Press, 1979); J. Lovelock, *Healing Gaia* (New York: Harmony Books, 1991); J. Lovelock, *Gaia: A New Look at Life on Earth* (Oxford: Oxford Paperbacks, 2000); J. Lovelock, *The Revenge of Gaia: Why the Earth is Fighting Back – And How We Can Still Save Humanity* (London: Penguin Books, 2007).
19   Lovelock, *Healing*, 12.
20   Lovelock, *Revenge*, 2.
21   V. Buddrus, *East–West European Centre for Integrative Humanistic Education and Psychology: Theoretical Background and Belief System* (Morschen: East–West European Centre for Integrative Humanistic Education and Psychology, 1996), 1.
22   See P. Hardi and T. Zdan, eds., *Assessing Sustainable Development: Principles in Practice* (Winnipeg, OH: International Institute for Sustainable Development, 1997).
23   The classic work on systems is D. Meadows et al. (Club of Rome), *The Limits to Growth* (New York: Universe, 1972); see also F. Stowell et al., eds., *Systems for Sustainability: People, Organizations and Environments* (New York: Plenum, 1997); and again, Clayton and Radcliffe, *Sustainability*.
24   P. Senge et al., *The Fifth Discipline Fieldbook: Strategies and Tools for Building a Learning Organisation* (London: Nicholas Brealey, 1994), 25.

25 Ibid., 90.
26 P. B. Checkland, *Systems Thinking, Systems Practice* (Chichester: Wiley, 1981); P. B. Checkland and S. Holwell, *Information, Systems and Information Systems: Making Sense of the Field* (Chichester: Wiley, 1998); P. B. Checkland and J. Scholes, *Soft Systems Methodology in Action* (Chichester: Wiley, 1990); P. B. Checkland and J. Poulter, *Learning for Action: A Short Definitive Account of Soft Systems Methodology, and Its Use, Practitioners, Teachers and Students* (Chichester: Wiley, 2006).
27 For alternative definitions, see Checkland, *Systems Thinking*; V. Bignell and J. Fortune, *Understanding Systems Failures* (Manchester: Manchester University Press, 1984); Open University, *Complexity Management and Change: A Systems Approach* (Milton Keynes: Open University, 1987).
28 C. Avgerou and T. Cornford, *Developing Information Systems: Concepts, Issues and Practice* (London: Macmillan Information Systems Series, 1993).
29 R. Ison, "Soft Systems: A Non-Computer View of Decision Support," in J. Stuth and B. Lyons, eds., *Decision Support Systems for the Management of Grazing Lands* (Paris: UNESCO, 1993), 94, quoting D. Russell, *How We See the World Determines What We Do in the World: Preparing the Ground for Action Research* (Hawkesbury: University of Western Sydney, 1986).
30 Checkland, *Systems Thinking*; Checkland and Scholes, *Soft Systems*; Checkland and Poulter, *Learning*; D. E. Avison and A. T. Wood-Harper, *Multiview: An Exploration in Information Systems Development* (Maidenhead: McGraw-Hill, 1990).
31 Open University, *Complexity*.
32 See Hardi and Zdan, *Assessing*.
33 Senge et al., *Fifth Discipline*.
34 See www.solonline.org.
35 There is a similarity here to the "failures" approach adapted by Bignell and Fortune, *Understanding*.
36 See the systems concepts as noted in Chambers, *Whose Reality Counts?*, 138.
37 See R. Chambers, *Rural Development: Putting the Last First* (New York: John Wiley and Sons, 1991); L. E. A. Natrajan, *Comparative Study of Sample Survey and Participatory Rural Appraisal Methodologies with Special Reference to Evaluation of National Programme on Improved Chulah* (1993); P. Shah and G. E. A. Hardwaj, "Gujarat, India: Participatory Monitoring," *The Rural Extension Bulletin* 1 (1993): 34–7; S. McPherson, *Participatory Monitoring and Evaluation* Abstracts (London: Institute of Development Studies, 1994); L. Webber and R. Ison, "Participatory Rural Appraisal Design: Conceptual and Process Issues," *Agricultural Systems* 47 (1995): 107–31; also, Bell, *Learning*; Chambers, *Participatory*.
38 Chambers, *Whose Reality Counts?*, 105.
39 S. D. Biggs, *Contending Coalitions in Participatory Technology Development: Challenges to the New Orthodoxy* (London: The Limits of Participation, Intermediate Technology, 1995), 4–5.
40 G. Coleman, "Logical Framework Approach to the Monitoring and Evaluation of Agricultural and Rural Development Projects," *Project Appraisal* 2(4) (1987): 251–9; D. Cordingley, "Integrating the Logical Framework into the Management of Technical Co-operation Projects," *Project Appraisal* 10(2) (1995): 103–12; Bell, *Learning*.
41 See M. Thompson and R. Chudoba, *Case Study Municipal and Regional Planning in Northern Bohemia, Czech Republic: A Participatory Approach* (Washington, DC: World Bank, 1994); Team Technologies, *TeamUp* 2.0 (Chantilly, VA: Team Technologies, 1995). Note that this approach is not always participatory. See Chambers, *Whose Reality Counts?* 42–4, for a description of ZOPP (a version of logical frameworks).
42 See Bell, *Learning*.
43 For a discussion of this type of argument, see Checkland and Holwell, *Information*.
44 See F. Capra, *The Web of Life: A Synthesis of Mind and Matter* (London: HarperCollins, 1996).
45 Webster, *Merriam Webster's Collegiate Dictionary*.
46 See N. Wiener, *Cybernetics* (Boston, MA: MIT Press, 1948).
47 Capra, *Web*, 56.
48 See S. Umpleby, *The Cybernetics of Conceptual Systems, Department of Management Science* (Washington, DC: George Washington University, 1994).
49 H. von Foerster, "On Constructing a Reality," in his *Understanding Understanding: Essays on Cybernetics and Cognition*, 211–27 (New York: Springer, 2003).
50 Umpleby, *Cybernetics*.
51 Ibid., 13.
52 R. Macadam, I. Britton, and D. Russell, "The Use of Soft Systems Methodology to Improve the Adoption by Australian Cotton Growers of the Siratac Computer-Based Crop Management System,"

Agricultural Systems 34 (1990), 1–14; H. Maturana, *Knowing and Being* (Milton Keynes: Open University, 1997); H. R. Maturana and F. J. Varela, *The Tree of Knowledge: The Biological Roots of Human Understanding* (Boston, MA: Shambhala, 1992).

53   See J. Mingers, *Self-Producing Systems* (New York: Plenum, 1995).

54   G. Morgan, *Images of Organization: New Edition* (London: Sage, 1997), 255.

55   Ibid., 260.

56   BBC Radio 4, one o'clock news, 18 December 1997.

57   M. Godet, et al., *Scenarios and Strategies: A Toolbox for Scenario Planning. Paris, Laboratory for Investigation in Prospective and Strategy* (Paris: Conservatoire National des Arts et Metiers, 1999), 6.

58   F. Matzdorf and M. Ramage, "Out of the Box – Into the Future," *Organisations and People* 6(3) (1999), 29.

59   D. Dearlove, "Thinking the Unthinkable: An Interview with Peter Schwartz, Scenario Planning Futurist," *The Business* (September 2002), 3.

60   P. Schwartz, *The Art of the Long View: Scenario Planning – Protecting Your Company Against an Uncertain Future* (London: Century Business, 1992), 39.

61   Ibid., 39.

62   Matzdorf and Ramage, "Out of the Box," 30.

63   See A. de Geus, "Planning as Learning," *Harvard Business Review* (March–April 1988), 70–74; Schwartz, *Art of the Long View*; K. van der Heijden, Giving Scenarios a Context in the Organisation. In *"The Fifth Discipline Fieldbook"*, P. Senge, ed. (London: Nicholas Brealey, 1994).

References

Avgerou, C. and T. Cornford. *Developing Information Systems: Concepts, Issues and Practice.* London: Macmillan Information Systems Series, 1993.

Avison, D. E. and A. T. Wood-Harper. *Multiview: An Exploration in Information Systems Development.* Maidenhead: McGraw-Hill, 1990.

Bell, S. *Learning with Information Systems: Learning Cycles in Information Systems Development.* London: Routledge, 1996.

Bell, S., and S. Morse. *Sustainability Indicators: Measuring the Immeasurable.* London: Earthscan, 1999.

Biggs, S. D. "A Multiple Source of Innovation Model of Agricultural Research and Technology Promotion." *World Development* 18(11) (1990): 1481–99.

Biggs, S. D. *Contending Coalitions in Participatory Technology Development: Challenges to the New Orthodoxy.* London: The Limits of Participation, Intermediate Technology, 1995.

Biggs, S. D. and J. Farrington. "Assessing the Effects of Farming Systems Research: Time for the Re-Introduction of a Political and Institutional Perspective." Asian Farming Systems Research and Extension Symposium, Bangkok, 1990.

Bignell, V. and J. Fortune. *Understanding Systems Failures.* Manchester: Manchester University Press, 1984.

Buddrus, V. *East–West European Centre for Integrative Humanistic Education and Psychology: Theoretical background and Belief System.* Morschen: East–West European Centre for Integrative Humanistic Education and Psychology, 1996.

Capra, F. *The Web of Life: A Synthesis of Mind and Matter.* London: HarperCollins, 1996.

Chambers. *Chambers Twentieth Century Dictionary.* Edinburgh: W. and R. Chambers, 1979.

Chambers, R. "Rapid Rural Appraisal: Rationale and Repertoire." *Public Administration and Development* 1 (1981): 95–106.

Chambers, R. *Rural Development: Putting the Last First.* New York: John Wiley and Sons, 1991.

Chambers, R. *Whose Reality Counts? Putting the First Last.* London: Intermediate Technology, 1997.

Chambers, R. *Participatory Workshops: A Sourcebook of 21 Sets of Ideas and Activities.* London: Earthscan, 2002.

Checkland, P. B. *Systems Thinking, Systems Practice.* Chichester: Wiley, 1981.

Checkland, P. B. and S. Holwell. *Information, Systems and Information Systems: Making Sense of the Field.* Chichester: Wiley, 1998.

Checkland, P. B. and J. Poulter. *Learning for Action: A Short Definitive Account of Soft Systems Methodology, and Its Use, Practitioners, Teachers and Students.* Chichester: Wiley, 2006.

Checkland, P. B. and J. Scholes. *Soft Systems Methodology in Action.* Chichester: Wiley, 1990.

Clayton, A. and N. Radcliffe. *Sustainability: A Systems Approach.* London: Earthscan, 1996.

Coleman, G. "Logical Framework Approach to the Monitoring and Evaluation of Agricultural and Rural Development Projects." *Project Appraisal* 2(4) (1987): 251–9.

Cordingley, D. "Integrating the Logical Framework into the Management of Technical Co-operation Projects." *Project Appraisal* 10(2) (1995): 103–12.

Dawkins, R. *The Blind Watchmaker.* Bath: Longman, 1986.

Dearlove, D. "Thinking the Unthinkable: An Interview with Peter Schwartz, Scenario Planning Futurist." *The Business* September (2002): 22–3.

De Geus, A. "Planning as Learning." *Harvard Business Review* (March–April 1998): 70–74.

Foerster, H. von. "On Constructing a Reality." In his *Understanding Understanding: Essays on Cybernetics and Cognition,* 211–27. New York: Springer, 2003.

Godet, M. et al. *Scenarios and Strategies: A Toolbox for Scenario Planning. Paris, Laboratory for Investigation in Prospective and Strategy.* Paris: Conservatoire National des Arts et Metiers, 1999.

Hardi, P. and T. Zdan, eds. *Assessing Sustainable Development: Principles in Practice.* Winnipeg: International Institute for Sustainable Development, 1997.

Hobart, M. *An Anthropological Critique of Development: The Growth of Ignorance.* London: Routledge 1993.

Ison, R. "Soft Systems: A Non-Computer View of Decision Support." In J. Stuth and B. Lyons, eds., *Decision Support Systems for the Management of Grazing Lands,* 83–121. Paris: UNESCO, 1993.

Jones, R. H. *Reductionism: Analysis and the Fullness of Reality.* Lewisburg, PA: Bucknell University Press, 1990.

Koestler, A. *The Ghost in the Machine.* London: Arkana, 1964.

Lovelock, J. *Gaia.* Oxford: Oxford University Press, 1979.

Lovelock, J. *Healing Gaia.* New York: Harmony Books, 1991.

Lovelock, J. *Gaia: A New Look at Life on Earth.* Oxford: Oxford Paperbacks, 2000.

Lovelock, J. *The Revenge of Gaia: Why the Earth is Fighting Back – And How We Can Still Save Humanity.* London: Penguin Books, 2007.

Macadam, R., I. Britton, and D. Russell. "The Use of Soft Systems Methodology to Improve the Adoption by Australian Cotton Growers of the Siratac Computer-Based Crop Management System." *Agricultural Systems* 34 (1990): 1–14.

Maturana, H. *Knowing and Being.* Milton Keynes: Open University, 1997.

Maturana, H. R. and F. J. Varela. *The Tree of Knowledge: The Biological Roots of Human Understanding.* Boston: Shambhala, 1992.

Matzdorf, F. and M. Ramage. "Out of the Box – Into the Future." *Organisations and People* 6(3) (1999): 29–34.

McPherson, S. *Participatory Monitoring and Evaluation Abstracts.* London: Institute of Development Studies, 1994.

Meadows, D. H., D. L. Meadows, J. Randers, and W. W. Behrens III. *The Limits to Growth.* New York: Universe, 1972.

Mingers, J. *Self-Producing Systems.* New York: Plenum, 1995.

Morgan, G. *Images of Organization: New Edition.* London: Sage, 1997.

Natrajan, L. E. *A Comparative Study of Sample Survey and Participatory Rural Appraisal Methodologies with Special Reference to Evaluation of National Programme on Improved Chulah.* 1993.

Open University. *Complexity Management and Change: A Systems Approach.* Milton Keynes: Open University, 1987.

Richards, P. *Community Environmental Knowledge in African Rural Development.* Brighton: IDS, 1979.

Russell, D. *How We See the World Determines What We Do in the World: Preparing the Ground for Action Research.* Hawkesbury: University of Western Sydney, 1986.

Schwartz, P. *The Art of the Long View: Scenario Planning – Protecting Your Company Against an Uncertain Future.* London: Century Business, 1992.

Senge, P., et al. *The Fifth Discipline Fieldbook: Strategies and Tools for Building a Learning Organisation.* London: Nicholas Brealey, 1994.

Shah, P. and G. E. A. Hardwaj. "Gujerat, India: Participatory Monitoring." *The Rural Extension Bulletin* 1 (1993): 34–7.

Sorrell, T. *Scientism: Philosophy and the Infatuation with Science.* London: Routledge, 1991.

Stenmark, M. "What is Scientism?" *Religious Studies* 33 (1997): 15–32.

Stowell, F., et al., eds. *Systems for Sustainability: People, Organizations and Environments.* New York: Plenum, 1997.

Team Technologies. *TeamUp 2.0.* Chantilly, VA: Team Technologies, 1995.

Thompson, M. and R. Chudoba. *Case Study Municipal and Regional Planning in Northern Bohemia, Czech Republic: A Participatory Approach.* Washington, DC: World Bank, 1994.

Umpleby, S. *The Cybernetics of Conceptual Systems, Department of Management Science.* Washington, DC: George Washington University, 1994.

Van der Heijden, K. "Giving Scenarios a Context in the Organisation." In *The Fifth Discipline Fieldbook.* P. Senge, ed. London: Nicholas Brealey, 1994.

Webber, L. and R. Ison. "Participatory Rural Appraisal Design: Conceptual and Process Issues." *Agricultural Systems* 47 (1995): 107–31.

Webster. *Merriam-Webster's Collegiate Dictionary.* Chicago, IL: Merriam-Webster, 1995.

Wiener, N. *Cybernetics.* Boston, MA: MIT Press, 1948.

14

SUSTAINABILITY AND THE REFRAMING OF THE WORLD CITY

Stephen Zavestoski

When Klaus Toepfer, Director of the United Nations Environment Programme (UNEP), claimed in 2005 that "the battle for sustainable development, for delivering a more environmentally stable, just and healthier world, is going to be largely won and lost in our cities," he articulated a concern for sustainable urbanism that has now become a centerpiece of the sustainability movement.[1] Moreover, Toepfer recognized that growth—of economies, of populations, of urban areas—has had a mixed but largely negative impact on many of the world's city-dwellers, especially in the Global South. Although the United Nations (UN) helped establish the concept of sustainable development, which often and awkwardly paired together social and environmental wellbeing with the desire for more growth, in the twenty-first century UN representatives and others in the world of sustainability have come to see the growth imperative as deeply problematic. This enduring commitment to growth, when viewed through the prism of our global ecological crisis and a rapidly urbanizing world, has compelled Toepfer and others like him to rethink the role of cities in the vision for a sustainable society.

Since 2000, a wide range of individuals and organizations—from world leaders and mayors of so-called global cities, to urban planners, climate scientists, and even environmentalists—have made very similar assertions. At the World Summit on Sustainable Development in 2002, for example, leaders of local governments declared that "with half of the world's population now living in urban settlements ... the issue of sustainable urban management and development is one of the critical issues for the 21st century."[2] Combined with Toepfer's declaration at the halfway point of the decade, a momentum was building by 2010 that produced the following proclamations:[3]

> [T]he effects of urbanization and climate change are converging in dangerous ways. The results of this convergence threaten to have unprecedented negative impacts on quality of life, and economic and social stability. However, alongside these threats is an equally compelling set of opportunities ... urbanization will also offer many opportunities to develop cohesive mitigation and adaptation strategies to deal with climate change. The populations, enterprises and authorities of urban centres will be fundamental players in developing these strategies.[4]

Cities are at the cusp of an exciting revolution … Cities are brimming with natural advantages and economies of scale that allow them to adapt and thrive in a new low energy age … Cities, as the major source of global carbon emissions, have a responsibility to lead the way.[5]

This unprecedented urban expansion sets forth before us a once-in-a-lifetime opportunity to plan, develop, build and manage cities that are simultaneously more ecologically and economically sustainable.[6]

Even the Global Footprint Network, a partnership covering "200 cities, 23 nations, leading business, scientists, NGOs, academics and … more than 70 global partners" that aims "to enable a sustainable future where all people have the opportunity to live satisfying lives within the means of one planet" embraces the narrative: "The global effort for sustainability will be won, or lost, in the world's cities, where urban design may influence over 70 percent of people's Ecological Footprint and 80 percent of the world's population is expected to live by 2050."[7]

These pronouncements appear, on the one hand, to be motivated by demographic trends in the relative portion of human beings who live in urban areas. The tipping point was reached in 2007, when for the first time in history the world's population shifted from majority rural to majority urban. As the United Nations Population Division (UNPD) reports, 54 percent of the world lives in urban areas as of 2014, up from 30 percent in 1950. More significantly, the report forecasts that "the coming decades will bring further profound changes to the size and spatial distribution of the global population such that the world's population in 2050 is projected to be 66 per cent urban."[8] Furthermore, 90 percent of the 2.5 billion new urban-dwellers that will exist by 2050 will live in Asian and African cities.

The World Bank calls urbanization in developing countries "the most significant demographic transformation in our century as it restructures national economies and reshapes the lives of billions of people."[9] Urban land area in developing countries could triple between 2000 and 2030, according to the World Bank report. The amount of newly added urban land area over this 30 year period, an estimated 400,000 km[2], is equivalent to the entire amount of urban land area globally as of 2000.[10] Put differently, between 2000 and 2030 we could be building a whole new urban world at about ten times the speed at which the urban world of 2000 came into existence. Strikingly, all of this growth will be happening in parts of the world already constrained in terms of natural and economic resources.

On the other hand, claims like the Global Footprint Network's that "the global effort for sustainability will be won, or lost, in the world's cities" are also motivated by new understandings of planetary boundaries. The severity of human interventions into Earth systems, according to geologists, have brought about the end of the geologic epoch known as the Holocene and ushered in the Anthropocene.[11] One group of scholars highlight the post-1950 period, or "The Great Acceleration," as evidence that these recent interventions far outstrip in significance and scale any previous human impacts on earth systems. "While it is certainly true humans have always altered their environment, sometimes on a large scale, what we are now documenting since the mid-20th century is unprecedented in its rate and magnitude."[12] Furthermore, the authors add, "the Great Acceleration has, until very recently, been almost entirely driven by a small fraction of the human population, those in developed countries," whereas the predicted expansion in urbanization will take place primarily in developing countries. "In a practical sense," the authors conclude, "the future trajectory of the Anthropocene may well be determined by what development pathways urbanisation takes in the coming decades, particularly in Asia and Africa."[13]

So we have placed the Earth system at risk, yet a stable Earth system is a prerequisite for human development. The Great Acceleration, ironically, coincided with humanity's taming of "the three constant threats to all societies since the dawn of humanity—famine, disease and conflict—... yet, the backdrop to [these achievements] is one of global-scale ecological degradation," an unintended outcome that "calls for new thinking and solutions that go beyond the old model of development, beyond environmentalism and beyond traditional economic thinking."[14] Sustainable cities, given the fervor with which they are being celebrated, appear to represent precisely the "new thinking" and "new solutions" for which so many scholars and policymakers call.

How did we arrive here? Does it make sense to place such hopes on cities which, after all, consume 75 percent of the world's natural resources, 80 percent of the global energy supply and produce approximately 75 percent of the global carbon emissions while occupying just 2.6 percent of the earth's land mass?[15] What would make us believe that cities can uphold the world's commitment to economic growth as the path to human development without pushing us over the precipice of planetary ecological collapse? A critical history of the narratives of cities and sustainability is needed in order to answer this question.

A convergence of two histories

Sustainability as a concept and sustainable development as a normative goal have their own interwoven histories as elaborated elsewhere in this handbook and at great length, for example, by Ulrich Grober[16] and Jeremy Caradonna.[17] Likewise, the city as a physical form of human settlement, and "urban" as an analytical concept for describing human relations in the city, have interwoven histories as recorded by Jane Jacobs,[18] Robert Park[19] and Louis Wirth.[20] While the long histories of sustainability and cities are important, for the purposes of a critical history of the narratives of cities and sustainability, it is most useful to begin much closer to the point in time at which these two histories converge.

From "world cities" to "world-class cities"

The modern phase of globalization that began with the Industrial Revolution was institutionalized in the post-World War II Bretton Woods Conference (1944) that created the International Monetary Fund and the World Bank. Ensuing growth in the speed and breadth of global capital flows, described above as The Great Acceleration, produced a wide range of transformations over the last half of the twentieth century. Among these was a new way of conceptualizing and understanding cities in terms of their positions within global networks of capital in which a new class of cities—"world cities"—became the object of concern.

To better understand the changing views on cities and their role in a sustainable global society, it is worthwhile to consider some of the key scholarly works on cities that have shaped our views on cities, and especially world cities, since the 1960s. Modern academic usage of the term "world city" began with Peter Hall's *The World Cities* (1966). Hall's world cities were seen "primarily as national centers that channeled international forces and interests towards national interests," a conceptualization that is "thus arguably a product of a period in which cities operated primarily as nodes within national urban systems."[21] This nodal or "mosaic" view of cities shifted, according to Peter Taylor and Ben Derudder, when "economic restructuring of the world-economy ma[de] the idea of a mosaic of separate urban systems appear anachronistic and ... irrelevant" so that "new thinking about cities was required."[22]

According to Paul Knox, "during the 1970s and 1980s [there was a shift] from an international to a more global economy" characterized by "a web of global corporate networks whose operations span national boundaries but are only loosely regulated by nation-states," and this shift had profound effects on the way in which cities were characterized and valued.[23] A series of thinkers, from Manuel Castells to David Harvey "revolutionized the study of urbanization" by linking "city forming processes to the larger historical movement of industrial capitalism."[24] Although cities in the Global North were rarely centers of production and manufacturing by this time, they now became seen as central to the maintenance of financial capital. Further, there was a growing awareness in the late twentieth century that development had not affected all cities equitably, and that cities in the Global South were in general poorer and often more polluted than those in the North.

John Friedmann and Goetz Wolff's analysis of world city formation advanced this line of thinking by emphasizing how a global network of cities had formed to support "the emerging world system of production and markets" in the interest of "the great capitalist undertaking to organize the world for the efficient extraction of surplus."[25] Friedmann later put forward seven theses linking urbanization processes to global economic forces. Simplified, Friedmann's world city hypothesis asserts that structural changes in cities are related to "the form and extent of the city's integration with the world economy," an integration that can be measured in terms of the concentration of finance, multinational corporation headquarters, business services, manufacturing activity, transportation, and population.[26] A flourishing financial sector often became the measuring-stick of a successful city, whereas other factors related to social and environmental wellbeing came to be seen as of secondary importance.

By the 1990s, Friedmann's "world city hypothesis" was becoming a "world city paradigm." For example, Saskia Sassen's *The Global City* advanced earlier thinking by identifying the significance of producer services (e.g., banking, accounting, insurance and law) in making possible investment of finance capital.[27] Given its rapid rise, the world city hypothesis eventually came under fire, especially by those who saw it as privileging Western understandings of "city-ness."[28] Yet Friedmann had already noted after the first decade of world city research that one of the primary contributions of the world city paradigm was to make portions of the meta-narrative of capital visible so as to provide a basis for a critical perspective. Friedmann, in fact, urged that research not "focus only on the global cities and the space of global accumulation that they articulate. We must understand global cities in relation to their respective peripheries." Doing so, continued Friedmann, "raises questions about the possible limits to growth, not only in the ecological sense so popular today, but also in a political sense."[29] Crucially, the 1990s was also the period in which the global ecological impacts of cities became fully appreciated by scholars and the wider public. William Rees and Mathis Wackernagel created the ecological and carbon footprints and forced a new way of seeing cities as heavily reliant on "distant elsewheres" for crucial ecological services.[30]

Despite these new critical perspectives on global capital accumulation and the limits to urban growth, the seemingly never-ending impulse to rank cities based on criteria that often had little social or ecological content only amplified in this period.[31] This ranking tendency has had the perverse effect of motivating cities to compete against one another for world city status, often at the neglect of factors that would have contributed to urban sustainability, such as improved access to food, less energy consumption, lower emissions, improved transportation systems, or greater levels of wellbeing and satisfaction. Friedmann tells a story of being invited to Singapore to speak about world cities:

> In private conversations with senior government officials, it became clear to me what the government really wanted [was that] their city-state might rise to the rank of a "world city." The golden phrase had become a badge of status … But to me this

question pointed to an ongoing competitive struggle for position in the global network of capitalist cities and the inherent instability of this system.[32]

What often mattered to so-called "world-class" cities was the presence of renowned museums, the presence of high-priced sports stadium, influential architecture, or, the shiniest jewel of all, a successful Olympics bid.

The obsession with being "world class" was often linked with the ideology of economic growth. This was apparent even in 1987, when the *New York Times* described Tokyo's aim to "reshape itself as a 'World Class City.'"[33] Soon "world-class city" would "come to be something of a proxy for world cityness in popular writing and in government publications" and an idea latched onto by the corporate sector.[34] But the frenzy of world-class city rhetoric during the 1990s and early 2000s eventually produced a backlash. Within urban studies, at least, world-class cities discourse was criticized as a type of "speculative urbanism"[35] with a neoliberal agenda[36] that overemphasizes aesthetics[37] to justify everything from slum clearance[38] to suppression of political resistance.[39] Further, an avalanche of new data from the 1990s onward—in UN reports, in the growing literature on ecological footprints, and derived from a series of other methods and indicators—made it painfully clear that cities, and perhaps world class cities in particular, were quite unsustainable.

And yet the idea of the "world-class city" never went away, remaining a powerful idea shaping a belief in cities as physical places with unlimited potential for transformation and growth. The idea of sustainability over the past two decades has not challenged the world-class city so much as it has become a new lens through which to view cities and their value.

From sustainability and sustainable development to world-class sustainable cities

While some might point to Ebenezer Howard's Garden City, Patrick Geddes's "neotechnic" city, Frank Lloyd Wright's Broadacre City, or Mumford's "Biotechnic City"[40] as early attempts to conceptualize cities within nature, the starting point of a critical history of cities and sustainability falls in the latter half of the twentieth century when the concepts of sustainability and sustainable development become institutionalized.[41] Timothy Beatley roots the growing interest in the role of communities and cities in the sustainability agenda "in the activism and growing environmental awareness of the 1960s and early 1970s.[42] Mike Hodson and Simon Marvin similarly identify "contemporary sustainable cities' debates … as emerging out of the multiple crises—economic, ecological, of industrial capitalism and urbanism—particularly as they were perceived by Western nation states in the late 1960s and 1970s."[43]

Sustainability came on the radar in the 1970s with the 1972 UN Conference on the Human Environment in Stockholm, although sustainability was first discussed in the context of cities within the UN at the UN Habitat Conference on Human Settlements in Vancouver in 1976, well before the Brundtland Commission's classic definition of sustainable development appeared in its famous report of 1987.[44] The more radical critiques of earlier decades that implicated industrial capitalism in the unfolding ecological crisis (in cities and beyond) gave way by the late 1980s to "the view that the ecological crisis could be solved through an agenda developed and enacted through society's existing institutions." This view led the Brundtland Commission to produce a vague definition of sustainable development—"development that meets the needs of the present without compromising the ability of future generations to meet their own needs."[45] This vague formulation was easily integrated into the agendas of existing international institutions with minimal interest in social and environmental justice, such as the

World Bank and IMF. This approach, according to some scholars, led radical critics to view the Brundtland definition as embracing continued economic growth, thus making sustainable development "a rhetorical ploy which conceals a strategy for sustaining development rather than addressing the causes of the ecological crisis."[46] Indeed, the report itself advocates for a renewed commitment to economic growth at several points.

Cities would become more explicitly implicated in sustainable development when Agenda 21, a sustainability action plan resulting from the 1992 Rio Earth Summit, pointed to the importance of local authorities in achieving the sustainable development objectives of the Rio Declaration. What this meant was that local authorities would pick and choose policy approaches, usually involving green architecture and buildings, transportation, energy, green space, social and environmental justice, and economic development in implementing the sustainable urban development agenda. The result, according to Hodson and Marvin, was that "through Local Agenda 21, sustainability became enshrined in urban plans and policy where cities were central to the problematic of global ecological 'crisis' and sustainability."[47] Another nuance of this evolution was that cities came to be seen for their potential to solve sustainable development problems rather than part of the problem themselves.

The evolution is explained neatly by the same analyses of the "urbanization of neoliberalism" that can be applied to an understanding of world cities.[48] A new urban politics in the 1990s, according to Hodson and Marvin, "promoted an entrepreneurial and managerial role for city governing" wherein cities adapted to become outward facing in order to attract investment and tourism and consequently, in precisely the manner we see world cities research deployed, subject themselves to "comparison of their urban success vis-à-vis other cities."[49] Although Agenda 21 is not a neoliberal document, it did emerge out of the same pool of thought that viewed growth as somehow both the cause of and solution to the world's problems. Thus cities were often looked to as entities meant to grow the world out of its social and environmental problems.

This commitment to cities as the engines of economic growth required to solve the dual challenges of human development and ecological crisis has been sustained over three decades and across a wide range of international institutions. Its most recent expression can be seen at the UN Conference on Housing and Sustainable Urban Development (Habitat III) where the resulting "New Urban Agenda" acknowledges "the persistence of multiple forms of poverty, growing inequalities, and environmental degradation" despite pledges to tackle these challenges at Habitat I in 1976 and Habitat II in 1996, while doubling down on cities: "There is a need to take advantage of the opportunities of urbanization as an engine of sustained and inclusive economic growth, social and cultural development, and environmental protection, and of its potential contributions to the achievement of transformative and sustainable development."[50]

The New Urban Agenda's compatibility with the existing (often neoliberal) international institutions, not to mention the corporate sector, is perhaps best illustrated by the International Conference on World Class Sustainable Cities (WCSC). Begun in 2009, the annual conference aims to "enlighten, educate, change the mindsets, encourage community engagement and increase the knowledge of City Administrators, city stakeholders, industry players, Government agencies and the general public to deliberate on integrated urban solutions and the key challenges faced in transforming our cities into sustainable human settlements."[51] Among the conference's many corporate backers is platinum sponsor Sime Darby Property, a leading property developer in Southeast Asia. Another example of the embedding of "world-class cityness" into urban sustainability is the sustainable cities index prepared by Arcadis, a global design consultancy firm with more than €3.4 billion in revenues. In profiling the cities at the top of its rankings, the report refers to world-class "athletics, education systems, tourism, infrastructure and connectivity," not social and ecological factors.[52]

Cities and sustainability: questions and debates

Beneath the sheen created by the rhetoric of world-class sustainable cities, a range of academic disciplines and professions have been attempting to grapple pragmatically with defining and measuring sustainability in the context of cities for at least three decades. A general consensus seems to exist around the historical narrative of sustainable cities.[53] Contemporary views of the field, however, disagree as to whether there is a convergence of urban sustainability definitions and agendas[54] or a splintering[55] and whether competing conceptualizations are useful[56] or "promote confusion and cynicism."[57]

The field is plagued by a label—whether "sustainable cities" or "urban sustainability"—consisting of two words equally difficult to define. Defining "urban," explain Hilda Blanco and Daniel Mazmanian, "is almost as difficult as defining sustainability. There is no recognized, global definition of urban areas or cities."[58] Consequently, as Harriet Bulkeley and Michele Betsill state, "despite … near universal recognition that sustainable cities … are a desirable policy goal, there is less certainty about what this might mean in practice."[59] For example, different disciplines tend to emphasize different dimensions of sustainability's three Es (ecology, economy, equity).[60] As a result, explains Simon Joss, there is little agreement about what exactly the sustainable city entails or "how the relationship between social equity, economic development and environmental regeneration should be articulated and applied to various urban scales and settings," so that "the question of how to go about planning, implementing and governing sustainable urban development all too often eludes easy explanation, in theory as much as in practice."[61]

Kent Portney further problematizes the idea of urban sustainability by noting that a city's jurisdictional boundaries seldom correspond to ecosystems that are the appropriate unit from which to approach ecological sustainability.[62] In a sense, this echoes the concerns of Rees and Wackernagel with viewing cities as isolated entities independent of the broader ecological services on which they rely. Chapple proposes that "the region … can be an appropriate scale for understanding sustainability, which is much in line with the bioregionalist movement that has sprung up in different parts of the world in recent years."[63] Ultimately, underlying the various perspectives and their disagreements is the question of whether a view of cities as vehicles of economic growth can be reconciled with the goals of sustainability.

In the absence of reconciliation, variants such as "eco-cities," "low-carbon cities," "resilient cities," and "smart cities" continue to proliferate. This is particularly problematic for Hodson and Marvin whose view is that "under conditions of global ecological change, international financial and economic crisis and austerity governance new eco-logics are entering the urban sustainability lexicon—climate change, green growth, smart growth, resilience and vulnerability, ecological security."[64] Although Hodson and Marvin ask whether these eco-logics are "reinforc[ing] new strategies of economic accumulation within the existing urban hierarchy or … signal[ing] a more progressive kind of politics that may challenge existing urban hierarchies," their answer is rather pessimistic:

> Powerful social interests are seeking to ensure that particular technical fixes, socio-technical configurations, and selected trajectories of urban development are normalized, routine, rolled-out and widely replicated across many urban contexts. There is very little fundamental questioning of this new logic and why it is emerging in the particular form that it is. For us, how history and critical analysis can help us to understand what is missing from these new configurations is crucial.[65]

Surely one explanation for the attention paid to cities as possible loci for solutions to global ecological crisis is that cities appear to be the appropriate scale for intervention. It is at the scale of the city, according to Joss, "where governance structures and processes can be designed and implemented to produce significant action and effect sustainable development,"[66] but in reality, cities are dependent on national and international efforts at supporting sustainability.

How does a critical history of the narrative of cities and sustainability allow us to make sense of these present tensions? When Scott Campbell told planners in 1996 that they would face in the coming years "tough decisions about where they stand on protecting the green city, promoting the economically growing city, and advocating social justice," he was operating under the assumption that "these conflicts go to the historic core of planning, and are a leit-motif in the contemporary battles in both our cities and rural areas."[67] To grapple with the conflicts he placed environment, equity and economy at the points of a "planner's triangle," the axes of which represent the property conflict, the development conflict and the resource conflict. The goal for planning, Campbell concluded, is therefore a broader agenda: "to sustain, simultaneously and in balance, these three sometimes competing, sometimes complementary systems."[68]

In revisiting the planner's triangle 20 years later, Campbell asks "should we be surprised that planners still talk about sustainability … despite the commonplace criticisms that the idea is nebulous, imprecise, corrupted, or difficult to implement"?[69] Campbell concedes that "skeptics might argue that this persistence belies the field's slow pace of theoretical advancement and its weakness for soft, aspirational concepts" but suggests instead that the term's "endurance also reflects the power and adaptability of the concept: Sustainability is a resilient, sustainable idea."[70]

Consequently, suggests Campbell, our idea of urban sustainability may be shifting:

> If 'sustainability' once suggested achieving a balance through placing an urban-ecological system in equilibrium (i.e., a steady-state condition of resource and land uses), we may eventually grow more comfortable with the idea that sustainability is dynamic, unpredictable—even unstable—and plagued with internal contradictions.[71]

The critical history of the narrative of cities and sustainability presented here might be useful in problematizing Campbell's view of urban sustainability as inherently dynamic and unpre-dictable. When Hodson and Marvin ask "what does it mean for a particular class of world cities and associated coalitions of interests to become active in the (re-)organization of planetary ecological resources and what it means to attempt to secure 'their' ecological reproduction both in respect of particular world cities and networks of world cities?" they are reminding us that history is at least partially the outcome not of random or unpredictable events but of specific power relations.[72] What are the power relations that produced the historical progression from world cities to world-class cities, and from sustainability to sustainable development to world-class sustainable cities? Unless these power relations have changed it might be naive to expect follow-through on promises of inclusive prosperity and "just sustainabilities" made in UN-Habitat's 2016 World Cities Report and included in its New Urban Agenda.

In asking whether sustainable development is a useful concept, Campbell used the follow-ing metaphor: "Some environmentalists argue that if sustainable development is necessary, it therefore must be possible. Perhaps so, but if you are stranded at the bottom of a deep well, a ladder may be impossible even though necessary."[73] A critical perspective on the history of cities and sustainability, if nothing else, prompts us to consider whether what appears to be a sustain-able urban development ladder is an illusion—a trick played by elites hoping to retain power. If the sustainable urban development ladder is an impossibility, what are the lessons learned from the past 30 years of urban sustainability thinking and practice that can conjure a ladder capable of delivering on the aspirations of universal human development without undermining earth systems essential for survival of our species?

Notes

1   Tim Radford, "Three Decades of Environmental Degradation as Seen from Space," *The Guardian* (June 3, 2005), retrieved from www.theguardian.com/society/2005/jun/04/environment. environment.

2   "Local Government Declaration to the World Summit on Sustainable Development," 2002, p. 2, retrieved from www.dlist.org/sites/default/files/doclib/Local_Government_Declaration_to_ the_WSSD.pdf.

3   The following quotes appear in Simon Joss, *Sustainable Cities: Governing for Urban Innovation* (London: Palgrave, 2015), 14.

4   UN-Habitat, *Cities and Climate Change: Global Report on Human Settlements 2011* (London: Earthscan, 2011), 1–2.

5   Boris Johnson, Mayor of London, quoted in Embassy of Switzerland in the UK, 2010, "A Swiss-UK Dialogue: Urban Sustainability, a Contradiction in Terms?" Embassy of Switzerland in the UK, London, p. 10, retrieved from www.swisscleantech.ch/images/misc_documents/Swiss-UK_ Dialogue_1006-03_final.pdf.

6   Hiroaki Suzuki, Lead Urban Specialist, World Bank, quoted in World Bank, "Eco2 Cities: A guide for developing ecologically sustainable and economically viable cities" Press Release, November 8, 2011, World Bank, Washington DC, retrieved from www.worldbank.org/en/news/press-release/2011/11/08/eco2-cities-guide-for-developing-ecologically-sustainable-and-economically-v iable-cities.

7   Global Footprint Network, "Footprint for cities," retrieved on March 13, 2017, from www.foot-printnetwork.org/pt/index.php/GFN/page/footprint_for_cities/.

8   UN Department of Economic and Social Affairs, Population Division, *World Urbanization Prospects: The 2014 Revision—Highlights*, ST/ESA/SER.A/352 (UNDESA, 2014), 2, retrieved from https://esa.un.org/unpd/wup/Publications/Files/WUP2014-Highlights.pdf.

9   Hiroaki Suzuki, Arish Dastur, Sebastian Moffatt, Nanae Yabuki and Hinako Maruyama, *Eco2 Cities: Ecological Cities as Economic Cities* (Washington, DC: World Bank, 2010), 13.

10 Shlomo Angel, Stephen C. Sheppard and Daniel L. Civco, *The Dynamics of Global Urban Expansion* (Washington, DC: World Bank, 2005)

11 Paul J. Crutzen and Eugene F. Stoermer, "The 'Anthropocene.'" *Global Change Newsletter* 41 (2000), 12–13.

12 Will Steffen, Wendy Broadgate, Lisa Deutsch, Owen Gaffney and Cornelia Ludwig, "The Trajectory of the Anthropocene: The Great Acceleration," *The Anthropocene Review* 2, no. 1 (2015), 91.

13 Steffen et al., "The Trajectory of the Anthropocene: The Great Acceleration," 91.

14 Nebojsa Nakicenovic, Johan Rockstrom, Owen Gaffney and Caroline Zimm, 2016, "Global Commons in the Anthropocene: World Development on a Stable and Resilient Planet" (WP-16-019) IIASA Working Paper, IIASA, Laxenburg, Austria, retrieved on January 4, 2017 from http://pure.iiasa.ac.at/14003/1/WP-16-019.pdf.

15 See Alice Charles, "The New Urban Agenda Has Been Formally Adopted. So What Happens Next?" November 22, 2016, World Economic Forum, retrieved from www.weforum.org/agenda/2016/11/ last-month-a-new-global-agreement-to-drive-sustainable-urban-development-was-reached-so-what-is-it-and-happens-next.

16 Ulrich Grober, *Sustainability: A Cultural History*, trans. Ray Cunningham (Totnes: Green Books, 2012).

17 Jeremy L. Caradonna, *Sustainability: A History* (New York: Oxford University Press, 2014); and "The Historiography of Sustainability: An Emergent Subfield," *Ekonomska i ekohistorija*, 11, no. 1 (2015), 7–18.

18 Jane Jacobs, *The Death and Life of Great American Cities* (New York: Vintage, 1961).

19 Robert E. Park, "The City: Suggestions for the Investigation of Human Behavior in the City Environment," *The American Journal of Sociology* 20 no. 5 (1915), 577–612.

20 Louis Wirth, "Urbanism as a Way of Life," *The American Journal of Sociology* 44, no. 1 (1938), 1–24.

21   Neil Brenner and Roger Keil (eds.), *The Global City Reader* (London: Routledge, 2006), 20.

22   Peter J. Taylor and Ben Derudder, *World City Network: A Global Urban Analysis*, second edition (New York: Routledge, 2016), 21.

23   Paul Knox, "World Cities in a World-System" in *World Cities in a World-System* (New York: Cambridge University Press, 1995), 3.

24   John Friedmann, "The World City Hypothesis," *Development and Change* 17 (1986), 69.

25   John Friedmann and Goetz Wolff, "World City Formation: An Agenda for Research and Action," *International Journal of Urban and Regional Research* 6, no. 3 (1982), 309.

26   Friedmann, "The World City Hypothesis," 72.

27   Saskia Sassen, *The Global City: New York, London, Tokyo* (Princeton, NJ: Princeton University Press, 1991/2001).

28   See Jennifer Robinson, "Global and World Cities: A View from Off the Map," *International Journal of Urban and Regional Research* 26 no. 3 (2002), 531–54; and Colin McFarlane, "Urban Shadows: Materiality, the 'Southern City' and Urban Theory," *Geography Compass* 2, no. 2 (2008), 340–58.

29   John Friedmann, "Where We Stand: A Decade of World City Research," in *World Cities in a World System* (Cambridge: Cambridge University Press, 1995), 42.

30   See William Rees and Mathis Wackernagel, "Urban Ecological Footprints: Why Cities Cannot Be Sustainable—and Why They Are a Key to Sustainability," *Environmental Impact Assessment Review* 16, no. 4-6 (1996), 223–48; and Mathis Wackernagel and William Rees, *Our Ecological Footprint: Reducing Human Impact on the Earth* (Gabriola Island, BC: New Society Publishers, 1998).

31   See Brian J. Godfrey and Yu Zhou, "Ranking World Cities: Multinational Corporations and the Global Urban Hierarchy," *Urban Geography* 20, no. 3 (1999), 268–81.

32   Friedmann, "Where We Stand: A Decade of World City Research," 36.

33   Clyde Haberman, "Tokyo Aims to Reshape Itself as a 'World Class City.'" *The New York Times* (February 8, 1987), 4.

34   David A. McDonald, *World City Syndrome: Neo-liberalism and Inequality in Cape Town* (New York: Routledge, 2007).

35   Michael Goldman, "Speculating on the Next World City" in *Worlding Cities: Asian Experiments and the Art of Being Global* (Oxford: Wiley-Blackwell, 2011), 229–58.

36   McDonald, *World City Syndrome*.

37   D. Asher Ghertner, *Rule by Aesthetics: World-class City Making in Delhi* (New York: Oxford University Press, 2015).

38   Amita Baviskar, "Demolishing Delhi: World Class City in the Making," *Mute Magazine*, September 5, 2006, retrieved from www.metamute.org/editorial/articles/demolishing-delhi-world-class-city-making.

39   Ananya Roy, "The Blockade of the World-class City: Dialectical Images of Indian Urbanism," in *Worlding Cities: Asian Experiments and the Art of Being Global* (Oxford: Wiley-Blackwell, 2011), 259–78.

40   Ebenezer Howard, *Garden Cities of Tomorrow* (London: Faber and Faber, 1902 [1946]); Patrick Geddes, *Cities in Evolution* (London: Williams and Norgate, 1915 [1948]); Frank Lloyd Wright, "Broadacre City: A new Community Plan," *The Architectural Record* April (1935), 243–54; Lewis Mumford, *The Culture of Cities* (New York: Harcourt Brace Jovanovich, 1938).

41   Mike Hodson and Simon Marvin (eds.), *After Sustainable Cities?* (New York: Routledge, 2014).

42   Timothy Beatley, "Sustainability in Planning: The Arc and Trajectory of a Movement, and New Directions for the Twenty-first-century City," in *Planning Ideas that Matter: Livability, Territoriality, Governance, and Reflective Practice* (Cambridge, MA: MIT Press, 2012), 95.

43   Hodson and Marvin, *After Sustainable Cities?*, 1.

44   See also Richard Register, *Ecocity Berkeley: Building Cities for a Healthy Future* (Berkeley, CA: North Atlantic Books, 1987).

45   World Commission on Environment and Development, *Our Common Future: Report of the World Commission on Environment and Development* (The Brundtland Report) (New York: WCED, 1987), chapter 2, para 1.

46   See Hodson and Marvin, *After Sustainable Cities?*, 3; quote from Maarten A. Hajer *The Politics of Environmental Discourse: Ecological Modernization and the Policy Process* (Oxford: Oxford University Press, 1995), 12.

47   Hodson and Marvin, *After Sustainable Cities?*, 4

48   Neil Brenner and Nik Theodore, "Cities and the Geographies of 'Actually Existing Neoliberalism'" *Antipode* 34, no. 3 (2002), 349–79.

49 Hodson and Marvin, *After Sustainable Cities?*, 6
50 UN-Habitat, *Urbanization and Development: Emerging Futures, World Cities Report 2016* (Nairobi, Kenya: United Nations Human Settlements Programme, 2016).
51 WCSC, "About Us," WCSC 2016 International Conference on World Class Sustainable Cities, retrieved on March 13, 2017 from http://wcsckl.com/v1/index.php/the-institute/about-us.
52 Arcadis, "Sustainable Cities Index 2016: Putting People at the Heart of City Sustainability," retrieved on March 13, 2017 from https://images.arcadis.com/media/0/6/6/{06687980-3179-47AD-89FD-F6AFA76EBB73}Sustainable%20Cities%20Index%202016%20Global%20Web.pdf.
53 See Thomas L. Daniels, "A Trail across Time: American Environmental Planning from City Beautiful to Sustainability," *Journal of the American Planning Association*, 75, no. 2 (2009), 178–92.
54 Mark Roseland and Maria Spiliotopoulou, "Converging Urban Agendas: Toward Healthy and Sustainable Communities," *Social Sciences* 5, no. 3 (2016), 28.
55 Martin de Jong, Simon Joss, Dean Schraven, Changjie Zhan and Margot Weijnen, "Sustainable–Smart–Resilient–Low Carbon–Eco–Knowledge Cities: Making Sense of a Multitude of Concepts Promoting Sustainable Urbanization," *Journal of Cleaner Production* 109 (2015), 25–38.
56 Simon Guy and Simon Marvin, "Understanding Sustainable Cities: Competing Urban Futures," *European Urban and Regional Studies* 6, no. 3 (1999), 268–75.
57 Kent. E. Portney, *Taking Sustainable Cities Seriously: Economic Development, the Environment, and Quality of Life in American Cities*, 2nd edition (Cambridge, MA: MIT Press, 2013).
58 Hilda Blanco and Daniel A. Mazmanian, "The Sustainable City: Introduction and Overview," in *Elgar Companion to Sustainable Cities: Strategies, Methods and Outlook* Northampton, MA: Edward Elgar Publishing, 2014), 1–11.
59 Harriet Bulkeley and Michele Betsill, "Rethinking Sustainable Cities: Multi-level Governance and the Urban Politics of Climate Change," *Environmental Politics* 14, no. 1 (2005), 42–63.
60 Karen Chapple, *Planning Sustainable Cities and Regions: Towards More Equitable Development* (New York: Routledge, 2014).
61 Joss, *Sustainable Cities: Governing for Urban Innovation*, ix.
62 Portney, *Sustainable Cities Seriously: Economic Development, the Environment, and Quality of Life in American Cities.*
63 Chapple, *Planning Sustainable Cities and Regions: Towards More Equitable Development*, 25.
64 Hodson and Marvin, *After Sustainable Cities?*, 91.
65 Mike Hodson and Simon Marvin, *World Cities and Climate Change: Producing Urban Ecological Security* (New York: Open University Press, 2010), 3–4.
66 Joss, *Sustainable Cities: Governing for Urban Innovation*, 17.
67 Scott Campbell, "Green Cities, Growing Cities, Just Cities? Urban Planning and the Contradictions of Sustainable Development," *Journal of the American Planning Association* 62, no. 3 (1996), 296.
68 Ibid., 304.
69 Scott Campbell, "The Planner's Triangle Revisited: Sustainability and the Evolution of a Planning Ideal that Can't Stand Still," *Journal of the American Planning Association* 82, no. 4 (2016), 392.
70 Ibid., 392.
71 Ibid., 395–6.
72 Hodson and Marvin, *World Cities and Climate Change: Producing Urban Ecological Security*, 3.
73 Campbell, "Green Cities, Growing Cities, Just Cities?," 301.

References

Angel, Shlomo, Stephen C. Sheppard and Daniel L. Civco. *The Dynamics of Global Urban Expansion.* Washington, DC: World Bank, 2005.
Arcadis. "Sustainable Cities Index 2016: Putting People at the Heart of City Sustainability." Retrieved on March 13, 2017 from https://images.arcadis.com/media/0/6/6/{06687980-3179-47AD-89FD-F6AFA76EBB73}Sustainable%20Cities%20Index%202016%20Global%20Web.pdf.
Baviskar, Amita. "Demolishing Delhi: World Class City in the Making." *Mute Magazine*, 5 Sept 2006. Retrieved from www.metamute.org/editorial/articles/demolishing-delhi-world-class-city-making.
Beatley, Timothy. "Sustainability in Planning: The Arc and Trajectory of a Movement, and New Directions for the Twenty-First-Century City." In *Planning Ideas that Matter: Livability, Territoriality, Governance, and Reflective Practice*, edited by Bishwapriya Sanyal, Lawrence J. Vale and Christina D. Rosan, 91–124. Cambridge, MA: MIT Press, 2012.

Blanco, Hilda and Daniel A. Mazmanian. "The Sustainable City: Introduction and Overview." In *Elgar Companion to Sustainable Cities: Strategies, Methods and Outlook*, edited by Daniel A. Mazmanian and Hilda Blanco, 1–11. Northampton, MA: Edward Elgar Publishing, 2014.

Brenner, Neil and Roger Keil (eds.). *The Global City Reader*. London: Routledge, 2006.

Brenner, Neil and Nik Theodore. "Cities and the Geographies of 'Actually Existing Neoliberalism.'" *Antipode* 34, no. 3 (2002), 349–79.

Bulkeley, Hariet and Michele Betsill. "Rethinking Sustainable Cities: Multi-level Governance and the Urban Politics of Climate Change." *Environmental Politics* 14, no. 1 (2005), 42–63.

Campbell, Scott. "Green Cities, Growing Cities, Just Cities? Urban Planning and the Contradictions of Sustainable Development." *Journal of the American Planning Association* 62, no. 3 (1996), 296–312.

Campbell, Scott. "The Planner's Triangle Revisited: Sustainability and the Evolution of a Planning Ideal that Can't Stand Still." *Journal of the American Planning Association* 82, no. 4 (2016), 388–97.

Caradonna, Jeremy L. *Sustainability: A History*. New York: Oxford University Press, 2014.

Caradonna, Jeremy L. "The Historiography of Sustainability: An Emergent Subfield." *Ekonomska i ekohistorija* 11, no. 1 (2015), 7–18.

Chapple, Karen. *Planning Sustainable Cities and Regions: Towards More Equitable Development*. New York: Routledge, 2014.

Charles, Alice. "The New Urban Agenda has been formally adopted. So what happens next?" World Economic Forum. November 22, 2016. Retrieved from www.weforum.org/agenda/2016/11/last-month-a-new-global-agreement-to-drive-sustainable-urban-development-was-reached-so-what-is-it-and-happens-next.

Crutzen, Paul J. and Eugene F. Stoermer, "The 'Anthropocene." *Global Change Newsletter* 41, (2000), 12–13.

Daniels Thomas L. "A Trail across Time: American Environmental Planning from City Beautiful to Sustainability." *Journal of the American Planning Association* 75, no. 2 (2009), 178–92.

de Jong, Martin, Simon Joss, Dean Schraven, Changjie Zhan and Margot Weijnen. "Sustainable–Smart–Resilient–Low Carbon–Eco–Knowledge Cities: Making Sense of a Multitude of Concepts Promoting Sustainable Urbanization." *Journal of Cleaner Production* 109 (2015), 25–38.

Embassy of Switzerland in the UK, 2010. "A Swiss-UK Dialogue: Urban Sustainability, a Contradiction in Terms?" Embassy of Switzerland in the UK, London. Retrieved on March 13, 2017 from www.swiss cleantech.ch/images/misc_documents/Swiss-UK_Dialogue_1006-03_final.pdf.

Friedmann, John. "The World City Hypothesis." *Development and Change* 17 (1986), 69–83.

Friedmann, John. "Where We Stand: A Decade of World City Research." In *World Cities in a World System*, edited by Paul L. Knox and Peter J. Taylor, 21–47. Cambridge: Cambridge University Press, 1995.

Friedmann, John and Goetz Wolff "World City Formation: An Agenda for Research and Action." *International Journal of Urban and Regional Research* 6, no. 3 (1982), 309–44.

Geddes, Patrick. *Cities in Evolution*. London: Williams and Norgate, 1915 [1948].

Ghertner, D. Asher. *Rule by Aesthetics: World-class City Making in Delhi*. New York: Oxford University Press, 2015.

Global Footprint Network. "At a Glance." Retrieved on March 13, 2017 from www.footprintnetwork.org/pt/index.php/GFN/page/at_a_glance.

Global Footprint Network. "Footprint for Cities." Retrieved on March 13, 2017 from www.footprint-network.org/pt/index.php/GFN/page/footprint_for_cities.

Godfrey, Brian J. and Yu Zhou, "Ranking World Cities: Multinational Corporations and the Global Urban Hierarchy." *Urban Geography* 20, no. 3 (1999), 268–81.

Goldman, Michael. "Speculating on the Next World City." In *Worlding Cities: Asian Experiments and the Art of Being Global*, edited by Ananya Roy and Aihwa Ong, 229–58. Oxford: Wiley-Blackwell, 2011.

Grober, Ulrich. *Sustainability: A Cultural History*, trans. Ray Cunningham. Totnes: Green Books, 2012.

Guy, Simon and Simon Marvin. "Understanding Sustainable Cities: Competing Urban Futures." *European Urban and Regional Studies* 6, no. 3 (1999), 268–75.

Haberman, Clyde. "Tokyo Aims to Reshape Itself as a 'World Class City.'" *The New York Times* (February 8, 1987), 4.

Hajer Maarten A. *The Politics of Environmental Discourse: Ecological Modernization and the Policy Process*. Oxford: Oxford University Press, 1995.

Hall, Peter. *The World Cities*. London: Weidenfeld & Nicolson, 1966.

Hodson, Mike and Simon Marvin. *World Cities and Climate Change: Producing Urban Ecological Security*. New York: Open University Press, 2010.

Hodson, Mike and Simon Marvin (eds). *After Sustainable Cities?* New York: Routledge, 2014.

Howard, Ebenezer. *Garden Cities of Tomorrow.* London: Faber and Faber, 1902 [1946].

Jacobs, Jane. *The Death and Life of Great American Cities.* New York: Vintage, 1961.

Joss, Simon. *Sustainable Cities: Governing for Urban Innovation.* London: Palgrave, 2015.

Knox, Paul "World Cities in a World-System." In *World Cities in a World-System*, edited by Paul L. Knox and Peter J. Taylor, 3–20. Cambridge: Cambridge University Press, 1995.

Local Government Declaration to the World Summit on Sustainable Development. Retrieved on March 13, 2017 from www.dlist.org/sites/default/files/doclib/Local_Government_Declaration_to_the_WSSD.pdf.

McDonald, David A. *World City Syndrome: Neo-liberalism and Inequality in Cape Town.* New York: Routledge, 2007.

McFarlane, Colin. "Urban Shadows: Materiality, the 'Southern City' and Urban Theory." *Geography Compass* 2, no. 2 (2008), 340–58.

Mumford, Lewis. *The Culture of Cities.* New York: Harcourt Brace Jovanovich, 1938.

Nakicenovic, Nebojsa, Johan Rockstrom, Owen Gaffney and Caroline Zimm. "Global Commons in the Anthropocene: World Development on a Stable and Resilient Planet" (WP-16-019), IIASA Working Paper, IIASA, Laxenburg, Austria (2016). Retrieved on March 13, 2017 from http://pure.iiasa.ac.at/14003/1/WP-16-019.pdf.

Park, Robert E. "The City: Suggestions for the Investigation of Human Behavior in the City Environment." *The American Journal of Sociology* 20, no. 5 (1915), 577–612.

Portney, Kent E. *Taking Sustainable Cities Seriously: Economic Development, the Environment, and Quality of Life in American Cities*, 2nd edition. Cambridge, MA: MIT Press, 2013.

Radford, Tim. "Three Decades of Environmental Degradation as Seen from Space." *The Guardian* (June 3, 2005). Retrieved from www.theguardian.com/society/2005/jun/04/environment.environment.

Rees, William and Mathis Wackernagel. "Urban Ecological Footprints: Why Cities Cannot Be Sustainable—and Why They Are a Key to Sustainability." *Environmental Impact Assessment Review* 16, no. 4-6 (1996), 223–48.

Register, Richard. *Ecocity Berkeley: Building Cities for a Healthy Future.* Berkeley, CA: North Atlantic Books, 1987.

Robinson, Jennifer. "Global and World Cities: A View from Off the Map." *International Journal of Urban and Regional Research* 26, no. 3 (2002), 531–54.

Roseland, Mark and Maria Spiliotopoulou. "Converging Urban Agendas: Toward Healthy and Sustainable Communities." *Social Sciences* 5, no. 3 (2016), 28.

Roy, Ananya. "The Blockade of the World-class City: Dialectical Images of Indian Urbanism." In *Worlding Cities: Asian Experiments and the Art of Being Global*, edited by Ananya Roy and Aihwa Ong, 259–78. Oxford: Wiley-Blackwell, 2011.

Sassen, Saskia. *The Global City: New York, London, Tokyo.* Princeton, NJ: Princeton University Press, 1991/2001.

Steffen, Will, Wendy Broadgate, Lisa Deutsch, Owen Gaffney and Cornelia Ludwig. "The Trajectory of the Anthropocene: The Great Acceleration." *The Anthropocene Review* 2, no. 1 (2015), 81–98.

Suzuki, Hiroaki, Arish Dastur, Sebastian Moffatt, Nanae Yabuki and Hinako Maruyama. *Eco2 Cities: Ecological Cities as Economic Cities.* Washington DC: World Bank, 2010.

Taylor, Peter J. and Ben Derudder. *World City Network: A Global Urban Analysis* (second edition). New York: Routledge, 2016.

UN Department of Economic and Social Affairs, Population Division. "World Urbanization Prospects: The 2014 Revision, Highlights." ST/ESA/SER.A/352. Retrieved on March 13, 2017 from https://esa.un.org/unpd/wup/Publications/Files/WUP2014-Highlights.pdf.

UN-Habitat. *Cities and Climate Change: Global Report on Human Settlements 2011.* London: Earthscan, 2011.

UN-Habitat. *Urbanization and Development: Emerging Futures, World Cities Report 2016.* Nairobi: United Nations Human Settlements Programme (UN-Habitat), 2016.

Wackernagel, Mathis and William Rees. *Our Ecological Footprint: Reducing Human Impact on the Earth.* Gabriola Island, BC: New Society Publishers, 1998.

WCSC. "About Us." WCSC 2016 International Conference on World Class Sustainable Cities. Retrieved on March 13, 2017 from http://wcsckl.com/v1/index.php/the-institute/about-us.

Wirth, Louis. "Urbanism as a Way of Life." *The American Journal of Sociology* 44, no. 1 (1938), 1–24.

World Bank. "Eco2 Cities: A Guide for Developing Ecologically Sustainable and Economically Viable

Cities." Press Release, November 8, 2011. World Bank, Washington DC. Retrieved from www.world-bank.org/en/news/press-release/2011/11/08/eco2-cities-guide-for-developing-ecologically-sustaina ble-and-economically-viable-cities.

World Commission on Environment and Development. *Report of the World Commission on Environment and Development: Our Common Future* (The Brundtland Report). New York: WCED, 1987.

Wright, Frank Lloyd. "Broadacre City: A New Community Plan." *The Architectural Record* (April 1935), 243–54.

15

SUSTAINABLE TRANSPORTATION

From feet to wheels and machines and back to feet

Preston L. Schiller

Introduction

The history of sustainable transportation (ST) is also the history of *un*sustainable transportation—especially in the modern era. Transportation is about the movement of people, goods, and information. It is about modes and why certain ones have gained dominance—even at great environmental and social costs—while some less harmful ones have been neglected or only recently re-emphasized. It is also about the ways in which modes interacted with available technologies and infrastructure and social and cultural expectations. As with many other aspects of the history of sustainability, it is about how human society has developed in ways that are environmentally, socially, and economically unsustainable—and about the recent efforts to interrupt or reverse that damaging and destructive trajectory. More recently it is about amplifying aspects of transportation that are healthy and helpful to society and micro and macro levels. The terms "sustainable mobility" and "sustainable accessibility" are coming into use, although "sustainable transportation" appears to be the most widely used label in this field.

While several elegant and elaborate definitions of sustainable transportation are available,[1] for the purposes of this chapter it is only necessary to understand sustainable transportation in terms of:

- less polluting and more renewable energy sources for all modes, including their production;
- accessibility to key services and activities, especially employment, health, and sustenance;
- a shortening of trip distances; and
- enhancement of human safety, security, and social equity.

An important dynamic from early recorded history to the present is the tension between using transportation for the development and maximization of what is proximate, local, and easily reached with that which is distant, unknown, or little known and difficult to reach. In the modern era this tension has manifested itself in several ways: Between building the "city beautiful" versus suburbanization and exurbanization; between rootedness in locale or wandering ceaselessly in search of the novel; between shrinking the size and ecological footprint of vehicles and slowing the pace of mobility or making vehicles ever fatter and faster; and between the welcoming or acceptance of a certain level of physical exertion involved with movement and efforts to make mobility effortless—as in the extreme example of the autonomous vehicle discussed later in the chapter.

Transportation and the sustainability project

Transportation is vital to the whole of sustainability efforts because it is entwined with most aspects of existence: Personal mobility for long and short trips; the local and global movement of goods and materials; the relationship of transportation to urban development patterns; transportation's relation to landscapes; and the extremely pressing issue of transportation's relationship with the destruction of the physical environment in which it is embedded as well as the atmosphere surrounding it. This chapter will attempt to clarify several of these issues and illustrate a few of the complex ways in which transportation interacts with and shapes societal change, for good or ill, and illuminate the prospects for sustainability in this domain. While its principal focus is on the greatest "consumer" of travel, the United States, it is also informed by the situation of Canada and several other places around the planet, in historical and contemporary times.

The history of sustainable transportation invites attention for numerous reasons. Its trajectory was one of sustainability, in terms of how it fit within the confines of environmental limits and maximization of existing sustainable resources for virtually all of human history, and, then, in a short span of a little more than two centuries it managed to become dangerously unsustainable. Attention is also drawn to the ways in which technology—including infrastructure—culture and human development have interacted in the development of transportation modes from walking to space exploration. Finally the modern historical, especially the contemporary situation, is reviewed in terms of which modes and policies are or can be shaped to be sustainable.

Sustainable transportation (ST) is inherently opposed to business as usual (BAU) approaches to mobility. It questions the pressures for unlimited and unbridled mobility—whether of persons or goods. It questions whether all our trips are necessary and whether those ones most necessary could be done with minimal or no environmental burden.[2]

There are many modes of transportation, each of which has significant implications for sustainability:

- walking and non-mechanized modes, including sleds and animal-assisted transport;
- water and maritime travel, including freight movement;
- wheeled apparatus; mechanized, motorized, or not;
- air travel; and
- communications and telecommunications.

These modes interact with technology, the types of infrastructure made available and the range of culturally shaped beliefs and practices that comprise society.

The intersection of modes, infrastructure, and society

Transportation, sustainable or not, needs to be understood in terms of several factors, intersecting domains, themes, and issues which shape its realization. Key factors or domains are:

- technology and infrastructure;

- cultural, political, and economic influences;
- geographical and topological features.

Among the major themes and issues in the history of sustainable and unsustainable transportation are:

- Modal development, competition, succession: how modes developed and competed with each other; how some modes, or varieties of the same general mode, prevailed and others did not; how several modes reflected a desire for easier, speedier, more reliable and affordable travel but were not necessarily concerned about sustainability—as has been the case with those most dependent upon fossil fuels and land consumption.
- Relations between modes, infrastructure, community forms and travel: in order for certain modes to succeed, changes in infrastructure were needed. Modes, infrastructure, community design, and social and cultural factors interacted in complex and sometimes unpredictable ways. For example, choice of transportation and land-use patterns are implicated directly on a global scale in the health of people and their capacity to enjoy life and fulfill their potential.
- How modes interact with land use, infrastructure, environment, and societal and cultural factors are important considerations for planning and policy-making especially as rapid changes occur—or need to occur. Some modal, infrastructure, and urban forms are more compatible with sustainability goals than are others, and some modes work together in a sustainable manner to achieve energy efficiency, less fossil-fuel dependence, lower noise and air pollution, and more compact communities than do other modes.

This chapter will now discuss the development of specific modes and some of the ways in which their history shaped political and economic developments and perceptions about mobility as well as their prospects for sustainability.

Modes and their implications for mobility and development

Walking

Walking is foundational to sustainable transportation. The history of transportation begins with feet and only much later proceeded to ways in which humans learned to use other techniques and technologies that, in their earliest forms, still constituted environmentally sustainable ways of facilitating mobility. One could view the evolution of the human race from Homo erectus, a few million years ago, and then into Homo sapiens, which occurred over the next 500,000 of years, as the first steps in sustainable transportation. Tetrapedalism enabled early humans or proto-humans to move about quickly as they searched for sustenance or avoided danger. Bipedalism enabled humans to observe a wider panorama and to climb trees and rocks and obtain some forms of food more easily. Walking erect also spurred cognitive development and led to the transition from H. erectus to H. sapiens.

Walking was mostly a slow-paced, small-group activity as humans began to gather in very early seasonal communities. Some 20,000 years ago humans created more permanent settlements and early civilizations developed some 10,000 years ago.[3] Walking remained the primary mode even as humans developed rudimentary ways of travelling across water, used animals to lighten their loads, or developed types of sleds and sleighs.[4] For the vast majority of human settlement history, until about 1850, life proceeded at a typical speed of about 3 miles per hour

(5 km/h). Even after the introduction of mechanized and motorized urban transport in the 19th century, walking remained the dominant mobility mode. While most trips on foot were relatively short, some activities involving commerce, warfare, governance, or pilgrimage could necessitate long-distance walking or a combination of foot and water travel. Even when animals were available, either for carrying cargo or pulling wagons, people as often walked beside them as atop them.

Even as motorized forms of transport were rapidly developing and spreading in the 19th century, walking was a valued leisure activity as well as a competitive sport attracting considerable public attention. The modern Olympic sport of racewalking has its roots in the 1850s and was known as pedestrianism. Some of the historical pedestrian contests were about speed while others were about distance and endurance. Some of the contests were outdoors and cross-country, others were held in arenas filled with paying (and gambling) spectators.[5] During the same time period many women were becoming emancipated from traditional clothing, including shoes that hampered their mobility and ability to engage in athletics, and some women competed as professional "pedestriennes."[6] Newspapers from London (UK), Petersborough (ON), Canterbury (NZ), and across the US regularly reported on the contests between the 1860s and 1880s.

While walking is still the dominant mode around the planet, and pedestrians are the road-users most killed by motor vehicles, in highly motorized countries, automobility has transformed walking into a minority mode—although there has been a resurgent interest in walking and walkable communities in recent years. These will be addressed more in the latter part of this chapter.

Maritime and water travel

One early way of getting around sustainably that either extended the range of feet in certain terrains, or offered faster mobility, was travel along or across waterways: rivers, lakes, coasts, and much later, canals. Devices such as logs, later lashed together with vines to create rafts, and early dugout canoes could traverse bodies of water and manage their flows and currents. Eventually wind power was harnessed for water-borne mobility and ever-larger vessels were developed. Travel between coastal communities, along river systems and across large bodies of water resulted in increased trade, population exchanges, and early forms of warfare. Improvements in the technology of water travel, which expanded its range, efficiency, and hauling capacities began as far back as 5000 years ago and facilitated the transatlantic crossings in the centuries of conquest and colonization of the Americas—although mortality remained high, especially among those transported as slaves.[7] New routes and the exploitation of existing indigenous waterways, portages, and canoe technology played a paramount role in the development of the Canadian fur trade—and the early history of Canada.[8] By the beginning of the 19th century steam-driven boats were successfully used for passengers and freight; mostly for river and coastal routes where they could frequently stop for fuel—wood in the early days—often contributing to unsustainable forestry practices, and coal later in the 19th century. Transoceanic steamships were slower in development due to their inefficient engines and the bulkiness of the wood or coal needed. Beginning in the Early Modern Period, the development of canals and locks systems also played an important role in the growth of continental and intra-state shipping. In England and France, canals were also common transportation paths. The more recent development of super-tankers[9] and cargo containerization beginning in 1966 radically transformed shipping and was an important factor in the rapid growth of world trade from the 1970s onward.[10]

In modern times Europe is doing a good job of keeping its inland waterways useful for goods shipments but the US much less so. While there is discussion in the US of the need for inland waterways infrastructure improvements, or at least good maintenance, most assessments indicate their poor repair. Ferries, whether passenger-only or for vehicles and passengers, are very inefficient ways to move people in comparison to transit or trains. While there is discussion of using wind and photoelectric powered batteries to reduce shipping's ecological footprint, there has been little success in developing these.[11]

From wheelbarrows, wagons, railways, and bicycles to autonomous vehicles

The wheel was first developed about 7000 years ago for milling and only later applied to mobility when the wheelbarrow was invented in Classical Greece between the sixth and fourth centuries BCE. Civilization was not dependent upon the wheel for mobility: several societies with complex systems of mobility developed without the wheel, as in the pre-Columbian Americas. Nor was wheeled transport suited for all cultures and terrains. After adopting the wheel much of the Middle East abandoned it in favor of the camel for several centuries. As the wheel improved, the devices it supported became more differentiated: hand carts, animal-drawn wagons, and chariots, and these, in turn, created traffic congestion and conflicts with the pedestrian majority of early towns and cities.[12] Travel remained very difficult even after the domestication of animals, wheeled transportation, and boats and navigation improved.

Wheeled coaches were pulled by horses along tracks in the earliest form of rail transit, called the "omnibus."[13] George Stephenson's "Rocket" locomotive won the competition for the first modern railway line that linked Liverpool and Manchester in 1829.[14] Incidentally, it killed a Member of Parliament on its maiden voyage. Railroads, too, made the transition from wood to coal, but not before decimating many of the forests of England, New England, and wherever else early railroading spread. Railways made possible the joining of east and west across the vast continent of North America, and became the final item of bargain that made the Confederation of Canada into a reality. The settler societies of the US and Canada were able to populate the countryside, move agricultural products, and exploit natural resources principally because of railways.[15]

Nineteenth-century motorization, along with engineering improvements to waterways[16] and roads, bridges and tunnels, made possible a major expansion of transportation's capabilities. Modern imperialism and growing world trade were fueled by fossils, and the global logistics of military and cargo transportation were shaped by the strategic location of coal and, later, oil depots. In many ways, transportation became less sustainable as it became more dependent upon non-renewable resources, and especially fossil fuels.

A bicycle by any other name

The bicycle, an invention of the 19th century, illustrates the complex relationship between modal innovation, competition and succession, infrastructure, transportation politics, and how a culture of mobility can develop. The bicycle began as a wheel-assisted walking or running machine, the "velocipede" (Latin for "fast foot"; still called "*vélo*" in France) or "Draisine" after its German founder, Karl (von) Drais; a forester with a humanitarian concern for replacing horses with a mechanical device. It was simply an inline two-wheeled device with a crossbar for sitting and a steering bar, and is somewhat reminiscent of the "runner bikes" that have become popular with two- and three-year-olds in North America.[17]

After many decades and a bumpy road of many technical improvements, with some societal resistance to overcome, a significant portion of Western society adopted the bicycle. In places such as France, it created unprecedented access to cities and the countryside, especially for the working classes.[18] It soon spread to popular use in Africa and Asia. It led to the articulation of a sizeable "bicycle culture" involving garb, clubs, and competitions. In Europe it became well institutionalized and accepted as "normal" for use by both sexes and all ages, in town and country. Bicycling was seen as part of the hygienic movement's healthy lifestyle, and the emancipation of women—where it influenced early feminist practices around mobility and clothing.[19]

Conflicts of sustainable infrastructure: from animal paths to "Good Roads"

In the late 19th century, British, French, and North American cyclists led political movements that aimed to improve streets and roads. The street and road improvement movements begun by bicyclists, especially the "Good Roads" movement in the US and the Road Improvements Association in England, spearheaded efforts that unintentionally paved the way for the rise of automobility and the dynamic of "annihilation of space and time," as bicycle culture and infrastructure entered into a crash course with the automobile and its expansive needs—a conflict that has played itself out in many ways and in many parts of the world.[20] Before exploring the rise of the automobile it is necessary to examine the development of roads and streets—the infrastructural basis for walking, animal-assisted transport (wagons and carriages), bicycling, and motor vehicles.[21]

The path to roadways

The earliest routes that bipeds trod made use of animal tracks. Improving paths and trails probably began after humans formed permanent settlements and needed better connections with other settlements. Road improvements in and beyond cities began as early as 3000 to 4000 BCE in India and parts of Mesopotamia. However, caring for roads was more common within settlements than between settlements. Road-building between settlements increased as commerce and the use of wheeled and animal-assisted vehicles developed. Road-building began as a large-scale endeavor with the peak of the Roman Empire, roughly between 400 BCE and 400 AD. Roman roads and bridges, many of which survive today, were extremely well built and stretched for a total of 53,000 miles (85,000 km). In comparison, today's vast US Interstate Highway System totals less than 50,000 linear miles (80,000 km).[22] Extensive road projects in other parts of the world included the Great Wall of China—actually a 5500-mile (8800 km) network of elevated roadways for military movement to prevent invasions, keep nomadic peoples and their herds out of protected agricultural lands, and to protect the Silk Road running alongside part of it.[23] The vast and well-engineered Inca road network (approximately 25,000 miles/40,000 km) was constructed mostly in the 15th and 16th centuries, prior to the Spanish conquest. Many parts of it are still in use today by pedestrians and motor vehicles.[24]

Roads outside cities in Europe and North America from colonial days until the 20th century were a hodgepodge of generally poor-quality government routes, often with fees or mandatory in-kind labor for their use, as well as many private and tolled routes (on roads and waterways), which left travelers with an aversion to tolls and road privatization that has survived to modern times. Overall, there was generally a lack of government interest in centrally directed systems of roads.[25] Some cities even had toll gates at their entries.[26] France had begun improving the status of road-building and civil engineering with the establishment of the École des

Ponts et Chaussées (School of Bridges and Roads) in 1747. Napoleon Bonaparte's extensive road-building program was remarkably well-planned and designed for troop movements; following a French tradition that began in the 17th century, roads were lined by trees to shade marching soldiers and horses. The early enhancement of trees had the important consequence of aestheticizing road travel for later users.

Most transportation attention and investment in England went into developing an extensive railroad system; yet roads were good enough to allow some bicycle travel and competitions between cities, thanks, in part, to the work of the British pavement innovator John McAdam, after whom the type of pavement known as macadam was named.[27] While support for road improvements grew among bicycle advocates in France and Britain during the 1880s and 1890s, the strongest pro-roads movement developed in the rapidly growing cities of the US, whose streets were generally in poor condition.[28]

Bicycle boom and the Good Roads movement pave the way for early automobility

The 1890s Good Roads movement in the US began in the 1880s as an outgrowth of the League of American Wheelmen (LAW; now The League of American Bicyclists or LAB), aided by bicycle (and later electric car) manufacturer Alfred A. Pope. During the 1890s, its advocacy joined together urban bicyclists with the growing largely rural and small town populist movement. Farmers exploited by monopolistic railways were willing to join them because good roads were needed to move goods from farm to market, and railways also joined in the expectation that better roads would extend the reach of railroads into rural areas still dependent upon rail shipment.[29]

The launching of the Good Roads movement preceded the introduction of the automobile into the US; but by the time the movement became influential in the early 20th century, the automobile was fast becoming a major force in American life. By the end of World War I, most industrial nations, either at federal or state–provincial levels, had resolved to establish government funding streams to improve and expand road networks. No good deed goes unpunished; the bicyclists were driven off the road by the more powerful motor-vehicle interests.

The rise of automobility: the challenge of sustainable transportation

The first successful automobile was patented in 1885 by Karl Benz in Germany, produced as the "Velo,"[30] and was reinvented many times across Europe and the US over the next two decades. Many, perhaps most, of the automobile's early inventors and producers had roots in the bicycle industry, again illustrating an ironical connection between the bicycle and subsequent developments, such as good roads, that challenged its viability.[31] Electric automobiles were introduced in France and Great Britain in the 1880s, and in the 1890s in the US. Many of the electric car's earliest adapters were women—although some feminist scholars, including Scharf, have criticized its gender-oriented marketing. It could travel 50 to 80 miles (84–134 km) on a charge. The scarcity of charging facilities between towns, the time needed to recharge, and the high purchase price limited its application as a touring vehicle. The emerging automobile culture valued speed and distance over less noise and pollution. In 1912 the electronic starter was introduced and interest in electric vehicles dampened, although they never disappeared in the 20th century.[32]

The automobile's popularity grew rapidly in the late 19th and early 20th centuries, and an international "car culture" emerged that engaged in a bitter contest with other users of streets:

pedestrians, children at play and vendors.[33] Automobile clubs, the precursors of present-day automobile associations found around the globe, emerged on both sides of the Atlantic and eventually across the Pacific. Their original focus was organizing automobile racing and touring. Eventually the clubs merged their efforts with those of the manufacturing and road construction interests into "highway lobbies" that campaigned successfully for road funding, standards, and regulations. Some automobile clubs developed into insurance companies offering various services to motorists.[34]

By the eve of World War I the development of mechanized and motorized transportation modes was well under way, with the automobile in the lead and the truck quickly replacing horse-drawn wagons.[35] Demands for improved and expanded roads increased and Fordism led to the availability of relatively affordable and useful automobiles for rural, urban, professional, and working folk alike. Transit improvements, especially the streetcar, improved urban livability as well as making suburbanization possible.[36]

After World War I motor vehicle use and highway building increased rapidly. Consolidation and merger in the automobile sector occurred as it expanded and became a major employer in many industrial nations. The Fordist assembly line mode of production became pre-eminent and Sloanism influenced the design and marketing of automobiles. Political leadership on both sides of the Atlantic used expanding roads and increasing motorization as ways of appealing to the citizenry. The creation of road, street, and traffic standards began.[37]

Motoring for the masses: depression parkways, superhighways, and Hoovervilles

While Herbert Hoover's 1928 presidential campaign was promising "a chicken in every pot. And a car in every backyard, to boot," the specter of the Great Depression and its Joads families[38] and "Hoovervilles"[39] was looming. During the 1930s, Hitler accelerated the expansion of the ambitious Autobahn network, which had begun more democratically in the 1920s, and began to develop the "people's car," or Volkswagen, as a way of helping the masses experience the value of the superhighways. The Italian Autostrada network of high-speed roads, begun in the more democratic time of 1921, was expanded upon by the Fascist regime in 1930s, but never achieved the scale of the Autobahn.[40]

In the US, motor parkways became important expressions of the new highway culture. Some, like the 1930s federal Blue Ridge Parkway, served recreational travel for many citizens.[41] Others, such as the many built under Robert Moses, served recreational, commuting, and social exclusion purposes; their design excluded buses —thus many poor and people of color could not use them.[42]

The automobile and the emerging road network that supported it became a major driver of industrialization, fossil fuel consumption, economic growth, and air pollution. If the 19th century, with its coal-powered trains, began the transition away from a low-impact, non-mechanized, and pre-industrial era of human transportation and toward one that was based on non-renewable fuels and transportation-related pollution, it was the 20th century that completed this transition. According to the environmental historian J. R. McNeill, in the 20th century, the world economy grew by factor 14, industrial output by factor 40, and air pollution by factor 5, and the automobile contributed directly to all of these changes.[43] Thus only in the 20th century did transportation become a serious problem, in terms of environmental sustainability, and here the new corporate-backed, privatized car culture played a major role.

In keeping with this new car culture public transportation systems were under siege by automobility and automobile interests. In the US, many, if not most, streetcar systems had been built by developers or utilities developers or utilities had built many, if not most, streetcar systems in order to promote their realty or energy source. They had promised a perpetual five cent ride and regulatory agencies held them to it long after it was feasible to operate a quality service at such a low fare. Quality suffered greatly as services became more crowded and less well maintained, becoming easy targets for closure or replacement by buses. A combination of strategic neglect and hostile private interests eventually sunk streetcars and damaged other forms of public transportation, which were tarred with the brush of being used mainly by the lower classes.[44]

During World War II, the industrial power of automobile manufacturing in several nations was retooled to produce trucks, troop carriers, tanks, cannons, and even aircraft. Transit services achieved record ridership as automobile manufacture was suspended, fuel was rationed and massive numbers of newly recruited women commuted by public transportation to defense-related production jobs.

Post-World War II: automobility dominates, the fly-drive culture takes hold, and sustainability becomes a concern

The period from the end of World War II to the year 2000 presents a complex, somewhat contradictory, and often confusing, array of transportation and land-use trends and counter-trends, movements, and counter-movements.[45] But what is clear is that automobility created unprecedented land-use changes and increased the dangers associated with transportation, thus affecting both social wellbeing and ecosystems. In the United States alone, an area equivalent to the arable land of Ohio, Indiana, and Pennsylvania was paved over in the twentieth century, mostly for the automobile.[46] By some estimates, as many as 25 million people died in automobile-related accidents in the twentieth century. But in order to best understand the impacts of automobility, it is helpful to divide this period into four approximate phases that vary somewhat from country to country and continent to continent:

1. 1945 to 1960

The first period is one of post-war expansion of cities; rebuilding of bombed European and Asian cities; restoration and expansion of transportation infrastructure; a rush to create more housing—which encouraged suburbanization and, in the US and Canada, a lowering of urban and suburban densities and the creation of suburban mono-cultures; a weakening of transit and the expansion of automobility; from the expansion of the automobile industry to highway expansions which also imposed highway engineering standards unnecessarily on residential streets.[47] Urban regions' transportation planning, especially in the US, became captive to questionable models, posing as scientific, which predicted great increases in traffic and then provided roadway capacity in the form of new roads and freeway and arterial expansions as the solution. The cultural desire for private transportation, "freedom," and "modernity" further undermined interest in trains and public transportation, which now seemed shabby, lower class, and old-fashioned.

2. 1960 to 1980

The second period is one of large-scale transportation infrastructure expansion: a heavy emphasis on highways in many nations and a mixture of highways, railways, and public transportation emphases in others. Resistance to highway expansion, especially in cities, grew considerably and a fair amount of fresh thinking about transportation came to light.[48] Interest in pedestrianization and traffic calming began in several cities. Some abandonment or neglect of transportation infrastructure occurred under decolonization. Increasingly severe air pollution prompted government regulatory responses, mostly aimed at reducing certain "criteria pollutants" from emissions (especially lead, which was a fuel additive that contributed significantly to "smog," a term popularized by Barry Commoner), increasing fuel efficiency, and some renewed interest in public transportation. The 1973 and 1979 energy crises, which caused fossil fuel shortages in the West, brought to light the problems of the transportation sector's complete reliance on foreign oil. In the US and Canada, private railways withdrew from inter-city passenger services and Amtrak and VIA Rail were created as federal entities in their place. Railway technology greatly improved the speed of trains as well as the ease and rapidity of laying tracks. There was wholesale abandonment of short freight, rail lines in North America increased, and around the planet the amount of freight transported by trucks grew significantly. In highly motorized countries the car culture flourished and began to influence all aspects of societal life.[49]

3. 1980 to 2000

The third period is one of growing recognition of problems associated with motor vehicle dependencies and road expansion emphases as unsustainable throughout the planet. In most of western and northern Europe there was further refinement and expansion of passenger rail and urban transit systems. Resistance to traffic expansion in cities grew, as did concern over pollu-tion and growth in truck traffic. Climate change came on the radar in the 1980s, mainly via the United Nations, and transportation-related greenhouse gas (GHG) emissions, which constituted at least a quarter of all emissions, began to paint the automobile in a new and nega-tive light. Several countries and many cities increased their investment in bicycle facilities, with the Netherlands, Scandinavia, and Germany in the lead. Different styles of privatization and deregulation emerged; some countries engaged in increased partnerships between public and private transportation entities and refocused regulatory efforts, while others, such as the UK, engaged in sweeping deregulation and transfer of public transportation assets to private inter-ests. Pedestrianization and traffic calming were undertaken on a broad scale in many parts of Europe and Asia and attracted interest in North America.[50] In Latin America the city of Curitiba began its sustainable transportation and development reforms that have led it to become a world model.[51] During the early 1990s a short-lived effort to reform US federal policy was initiated with the passage of the Intermodal Surface Transportation Efficiency Act (ISTEA) although subsequent iterations, including the most recent "FAST Act" show that reform has been slowed or even reversed.[52]

4. 2000 to the present

The fourth period is marked by a rapidly growing awareness of transportation's impact on atmospheric pollution and greenhouse gas accumulation, as well as increasing concern about the various social and environmental problems associated with the globalization of commerce. A spirited scholarly critique of automobility also emerged.[53] There are many cities on several continents where efforts to achieve a greater degree of sustainability have succeeded, with asso-ciations of cities and mayors playing an active role in this matter.[54] That said, the automobile remains a major barrier to sustainability, and we are yet to reach "peak cars" on the world's roads. There has been an increasing tension between the embedded car culture and the desire to improve the quality of urban life, with less emphasis on automobility and more emphasis on green modes and urban design that encourage accessibility. A few cities are removing central city freeways, while other suburban areas experience ongoing expansion of roads. Cities such as Copenhagen have made major strides in decreasing reliance on cars and increasing urban-core walkability and bicycle ridership, all of which have increased social wellbeing.

Several unproven or even untested technologies that promise faster, less impactful, less stressful daily commutes and long distance travel have attracted mostly uncritical attention. These include the autonomous vehicle (AV) and connected car, and Elon Musk's proposed hyperloop alternative to high-speed rail—albeit existing mostly as a "back of an envelope" concept.[55] On the horizon, one can find billionaires competing to develop commercial rocketry and space tourism—or even emigration to some other part of the solar system or galaxy; in short, fanciful ideas that neither reduce reliance on high-impact fuels nor increase overall social wellbeing.

In some heavily populated countries, such as India and China, there is heavy pressure from government and the automobile industry to create a transportation regime that serves only a small sliver of the population, while making other, viable modes, such as walking and cycling, difficult if not sometimes illegal. In many cases there simply has not been much thought or analysis given to the spatial limitations of crowded cities, which simply do not have room for more motor vehicles. The epic traffic jams, in places such as China and Bangladesh, of recent years will thus likely continue to hamper urban sustainability.

Is there a future for sustainable transportation?

This chapter has reviewed key aspects of the history of sustainable transportation from early times to the present. Whether transportation becomes more or less sustainable in the future depends on whether and how societies around the planet, but especially the most highly motorized ones, respond to the necessity of addressing climate change, land-use changes, and social wellbeing in ways that are beneficial to the environment and attractive to citizens. There are formidable challenges, as well as possible solutions, at local as well as global levels of society. In this section we will describe these challenges and possible solutions.

Institutionalizing sustainable transportation in policy and planning

Institutionalizing sustainable transportation in governmental policy and planning, at all levels, is probably the most important set of actions that can be taken to reduce unsustainable transportation's environmental and social impacts. National governmental policy and planning directly shapes its expenditures and regulatory functions. These, in turn, leverage related matters at all other levels of government and, when properly crafted, can also shape corporate behavior in the transportation realm.

In the US the principal response to the energy crisis of 1973, as well as subsequent nodding in the direction of actions in response to climate change, has been to focus on increasing the fuel efficiency of motor vehicles—also known as CAFE (corporate average fuel economy). This effort was effectively derailed by loopholes and the granting of tax breaks for sports utility vehicles (SUVs) and light trucks, which led to a stalling of motor vehicle fuel efficiency for two decades. Even after such standards became strengthened under President Obama, careful analysis has shown that CAFE itself is woefully insufficient to meet US greenhouse gases (GHGs) reductions; driving itself, as measured in vehicle miles travelled (VMTs) must also be reduced.[56] This underlines the need for action at several levels of government.

Perhaps one of the most influential transportation issue that government can change is the way that transportation is funded. Although lip service is paid to the "polluter pays" principle by actors across the political spectrum, it is rarely or only lightly applied to surface or air transportation. In the US, the principal transportation funding mechanism has been an excise tax on fuel at the pump, which is then deposited in the Highway Trust Fund from which projects and regulatory enforcement are funded. There are many problems with this tax: It has not been substantially increased over the past two decades; it is not indexed to inflation or rising costs; it has proven politically very difficult—almost impossible—to increase. Consequently the Highway Trust Fund is drained and Congressional stalemate has resulted in inaction and no alternative-funding source identified or enacted. Most transportation economists agree that this tax has been crafted to address only the direct costs of transportation and does not address indirect costs or externalities such as pollution, traffic safety, etc. While important for its leveraging effects, the federal expenditures are only a relatively small fraction of total transportation expenditures—governmental or personal. Expenditures by state and local governments far surpass those of the federal government and personal investments and expenditures far surpass those of all government. The citizens of the US pay a very steep price for their over-dependence on the personal motor vehicle for most mobility. That said, carbon taxes in some places, such as the Province of British Columbia (2008), have had tangible impacts on automobile reliance. But the overall impact of this and other proposed carbon taxes on the sustainability of transportation is yet to be determined.

Another way that government, at various levels, could help reverse the course of unsustainable transportation would be to enact a number of measures that have already proven effective in reducing driving and promoting green modes. These include: workplace-based programs to raise the costs of parking for drive-alone employee commuters—often referred to as "parking cash-out"—while providing those who avail themselves of alternatives such as transit or cycling rewards in the form of cash or transit passes;[57] the promotion of similar incentives for college students; the promotion and funding of walking or biking to school among younger student populations; and improving access to necessary goods and services through measures such as influencing land use planning away from current exclusionary forms of zoning to more inclusionary ones where green modes could play a role in access. Some programs that simply offer citizens good information about alternatives to driving have been shown to reap considerable reductions in discretionary driving.[58]

Government at all levels could also set good examples by locating public meetings only in transit accessible locations and encouraging less travel, especially by unsustainable modes, by its employees and contractors. A huge amount of travel is generated and funded by government itself. Public officials, including educators, could also work on being role models and using green modes more themselves.

Making telecommunications work for the environment

The rise of telecommunications, greatly assisted by public initiative and funding, has advanced many aspects of communications, research, and information seeking. Political campaigns of recent years, from the "Arab Spring" to the 2016 US presidential election, replete with its "fake news," have demonstrated the power and reach of internet-based media, for better or worse. Recent efforts to further deregulate this industry forget its origin in the intertwined sectors of the public sectors of government and academia. These sectors envisioned a broader public benefit than making it ever more easy for a handful of monopolistic high-tech robber barons and their corporations to enrich themselves while shirking public responsibilities.

Promoters have often claimed that advanced telecommunications will substitute for a fair amount of travel through teleconferencing and information transfers that will render face-to-face contact unnecessary. There are several problems with these assumptions: First such substitution has yet to be empirically verified; greater intensity of telecommunications may, in fact, lead to more travel as the work of Patricia Mokhtarian and her colleagues suggests.[59] Second, telecommunication may do little to dampen discretionary travel or it may actually facilitate such travel by making it easier.

There are ways that advanced telecommunications, properly directed, could help in sustainable transportation efforts.[60] One example would be an expansion and formalization of the way that episodic conclaves, "transit camps," of internet experts have helped agencies use available mapping resources, such as Google maps and publicly available repositories of geographic information systems (GIS) data, to develop online trip planners for government at all levels, including public transportation providers.[61] Governments could be encouraged to apply these widely in numerous ways, from bus stop or transit center information provision, to providing a range of incentives for public employees and government consultants to use these as part of a concerted effort to reduce travel and shift necessary travel, local and long distance, to greener modes.

Can less be made into more? Creating a culture of green modes

There is demonstrably a considerable environmental and social burden stemming from automobility and excessive long distance travel.[62] This points to the need for more localization of travel, less gratuitous mobility as just another form of excess consumption—as in the transatlantic weekend trip or the 5000-mile Caesar salad, and improved accessibility to necessities. One positive role that government can play is to work with citizenry, NGOs, and sustainable transportation advocacy groups to create a culture of green modes at all levels of society. A good example is the way that Portland, Oregon transformed itself from an average city for transit and land use planning to a national leader over several decades.[63] More recently, in the early 2000s, Portland made a conscious policy and planning decision to improve upon its lackluster 1 percent mode share for bicycling to a 6 percent share by 2014 (very high for the US) by improving the situation for cycling in several ways:

1   The most important was to improve the conditions on the several bridges connecting its CBD with adjacent communities—which dramatically increased riding in those corridors.
2   An accelerated program of bicycle lanes and boulevards which have come to be known as neighborhood greenways.
3   Launching of a very successful bike-share program.
4   Several annual summer closures of about six miles of streets in different neighborhoods to motor vehicle traffic and the creation of a festival atmosphere around these Sunday Parkways.

This effort has resulted in the surpassing of that goal in less than a decade and the creation of a "bicycle culture" that has made cycling a socially valued activity and cyclists proud of their efforts and of their city.[64]

Similarly there is a great need for the US and Canada to increase efforts aimed at promoting and restoring rail travel as an environmentally sound mode and one that could, with proper planning, funding and programming, become a viable alternative to driving or flying for a range of travel needs—especially for trips in the range of a few hundred miles. Several European and Asian countries are developing high-speed or higher speed rail systems (HSR, HrSR) or increments thereof and many have improved and expanded slower but highly comprehensive networks.[65] One of the policy goals for the promotion of HSR has been to create a viable means for short (100–300 miles/150–500 km) and medium distance (300–600 miles/500–1000 km) travel. Improving passenger rail links between major cities in several US regions and in the Quebec–Windsor corridor in Canada could help each nation with climate change goals and reduce automobile dependence and pressures to expand regional airports.[66]

Active transportation; healthy communities by design

One of the positive elements which grew out of the US transportation reform movement and sustainable transportation activism of the 1990s and beyond has been the ability to connect transportation to health, education, and other interests, and to promote a "new paradigm of active transportation"[67] and "healthy communities by design," or "active living by design."[68] As defined by the Centers for Disease Control (CDC, US), "active transportation is any self-propelled, human-powered mode of transportation, such as walking or bicycling."[69] The Public Health Agency of Canada offers a broader, more inclusive and fun-oriented list that includes non-mechanized wheelchairing, in-line skating and, of course, snowshoeing and skiing.[70] The daily walk to and from the bus stop or transit station is also shown to have positive health effects, especially in comparison with commuter driving.[71] Other attention has come from sources more closely associated with urban planning[72] and educators, parents and child advocates have been promoting Safe Routes to School programs to enable children to walk and bike to school safely and more physical activities, including longer recesses, at school.[73]

A parallel effort has been the movement towards "complete streets," which have infrastructure for the mobility spectrum that is defined well in a publication of the Federal Highway Administration of the US:

> According to the National Complete Streets Coalition, established in 2005, complete streets are those designed and operated to enable safe access and travel for all users. Pedestrians, bicyclists, motorists, transit users, and travelers of all ages and abilities will be able to move along the street network safely. Typical elements that make up a complete street include sidewalks, bicycle lanes (or wide, paved shoulders), shared-use paths, designated bus lanes, safe and accessible transit stops, and frequent and safe crossings for pedestrians.[74]

It is heartening that this is seen as a necessity by the formerly "motor vehicles first" transportation establishment, which historically has paid far more attention to the needs automobility than of the mobility needs and safety of other transportation infrastructure users.

Conclusion: back to feet

Historically, most travel, even locally, was done out of necessity; mainly for sustenance, commerce, and military purposes. Communities were located and shaped to minimize travel and maximize accessibility. The history of transportation can be seen as a long arc from feet and sustainable wheels to unsustainable over motorization and hypermobility and now back to a renewed interest in travel by foot and green modes. At present this dynamic faces the daunting challenge of pacifying or reversing the forced march led by denial or insufficient addressing of climate change, as well as the spread of near-magical beliefs in salvatory fixes—such as the hype surrounding autonomous vehicles. Similarly, in the center of the automobilized world, the weakening and near-gutting of the ISTEA reforms,[75] and widespread citizen antipathy towards tax measures and reforms that could direct society away from its current trajectory all call into question the direction that the motorized world is heading. This offers advocates for sustainable transportation planning and policy-making many interesting challenges and opportunities that must be addressed for their cause to advance.

It may be that the catastrophic effects of climate change will leave humans with the necessity of walking and cycling much more and living in compact communities. Or perhaps an enlightened transformation will take hold and a voluntary movement towards sustainability in transportation will occur. Either way it might be wise to invest in sturdy walking shoes, keep bicycle chains well lubricated, and a transit schedule handy.

Notes

1   Banister (2008: 73–80); OECD (1998).
2   For more about the opposition between ST and BAU see Schiller and Kenworthy (2017).
3   These dates are, of course, subject to debate. See WikiBooks (2014); see also Solnit (2000) and Amato (2004).
4   Lay (1992: 25, *passim*).
5   Duane (2010).
6   Park (2012: 730–49).
7   For more on the development of marine technology, see Derry and Williams (1993: 14, 19, 190–211); Falola and Warnock (2007: 277–8).
8   See Glazebrook (1964).
9   Approximately one-third of oceanic cargo is the transport of fuels.
10   See World Shipping Council (undated).
11   For a discussion of the concept of "ecological footprint," see Wackernagel and Rees (1996). On the subject of the state of inland waterways infrastructure in the US, see Kelley (2016).
12   For more on the development of the wheel, see Lay (1992: 26–41); Derry and Williams (1993: ch. 6); Hindley (1971). For more information on the Inca, see Hyslop (1984); Gambino (2009); Schiller and Kenworthy (2017: ch. 3).
13   For more on the omnibus, see Lay (1992: 129); Derry and Williams (1993: 385–8); Bellis (undated).
14   For more on steam vehicles, see Lay (1992: 137–8); Derry and Williams (1993: ch. 6); Goddard (1994: 6–7); see also Steamindex (2015).
15   See Goddard (1994); Berton (1972).
16   For more on canals and locks, see Derry and Williams (1993: 436–46).
17   For more about the rich history of bicycling and its implications for sustainability see Schiller and Kenworthy (2017).
18   Robb (2008).
19   See Schiller and Kenworthy (2017).
20   On this term, see Marx (1857); see also Christensen (2013); Schivelbusch (1986); Sachs (1983, 1992); on "Good Roads," see Hamer (1987: 23–4); Waller (1983: 243–54); also Goddard (1994); Gutfreund (2004); Dobb (undated).
21   See illustrations of several early bicycles and velocipedes on Wikipedia's "velocipede" entry (Wikipedia undated).
22   This is the figure for the interstates' linear mileage, as opposed to "lane mileage," which is the sum of all parallel lanes, which would equal over 200,000 miles (334,000 km). In the year 2000, the total lane mileage of US intercity highways, federal, state, and local lanes was 8,239,625 (13,760,173 km). See BTS and US Government (2002). For more on ancient roads, see Lay (1992); NCHRP (2006: 35).
23   The Wall was travelled and chronicled by the merchant Marco Polo and the scholar Ibn Battuta in the late 13th and early 14th centuries. For more on the various routes and histories of the great Silk Road, see Wild (1992); Wilford (1993); Whitfield (2015); Wood (2002); Manchester (2007).
24   Especially remarkable because the Inca did not use the wheel for mobility; for its extent, see the map in Hyslop (1984: 4).
25   France and Germany, as exceptions, each had centrally directed interests in roads, although Germany's central direction began considerably later than that of France. For more information, see Hindley (1971: 74–7); Lay (1992: 60, 70, 101–2, 117); McShane (1994: 103–4).

26 See City of Toronto (2009).

27 Hindley (1971).

28 For road history in France, see Hindley (1971: 74–7). For roads and the popularity of bicycling in France and the Roads Improvement Association see Flink (1988: 2–4, 169–71). For England, see Hamer (1987: 23–4); see also Herlihy (2004). For a different perspective, see McShane (1994), which emphasizes the role played by improved roads in generating interest in both bicycles and automobility, and in creating conflicts between vehicles and pedestrians.

29 For more about the history, revivals, and renaming of LAW to the League of American Bicyclists, see League of American Bicyclists (undated). For more about the Good Roads movement and the "Gospel of Good Roads," see Flink (1988); Goddard (1994); Gutfreund (2004).

30 As noted, this is the French term for bicycle.

31 For Fordism and the GM response to it, the marketing strategy of "Sloanism," see Lay (1992: 160–61); Rothschild (1973). For the relationship between bicycles and early automobile development, see Flink (1988); Barker and Gerhold (1995: 52–5); Rae (1965).

32 For more on electric cars, see Flink (1988); Scharf (1991: 35–50); Goddard (1994); Kirsch (2000); Didik (undated).

33 See Norton (2008).

34 For early automobile clubs and how they shaped automobile culture and politics, see Flink (1988); Scharf (1991); McShane (1994); Barker and Gerhold (1995: 55); Goddard (1994); Sachs (1992); Gutfreund (2004); Rae (1965).

35 For a discussion of the importance of the truck in cities as a replacement for horse-drawn wagons and the role that it played in reshaping urban transportation and road needs, see McShane (1994).

36 See Jackson (1985).

37 For more on standards, see Eno Transportation Foundation (undated); Blanchard (1919); Rothschild (1973).

38 At the center of Steinbeck's *The Grapes of Wrath* (1939).

39 The pejorative term for the car-filled encampments that accompanied the displaced yet motorized population of the 1930s US; named after the president who had promised them much but led them into the Great Depression.

40 For the Autostrada, see Hindley (1971: 98). For the Autobahn and Volkswagen, see Sachs (1992); Schivelbusch (2007).

41 Literally a linear park, 469 miles (780 km) in its final form; for more on it, see Schivelbusch (2007).

42 See Caro (1975); Schivelbusch (2007).

43 McNeill (2000: 360–61).

44 For more, see Jackson (1985: 168–71); Klein and Olson (1996). The extent to which US streetcar lines were "killed" by conspiracy or by neglect is subject to debate. For more, see Flink (1988); McShane (1994); Goddard (1994); Gutfreund (2004); Bottles (1987).

45 For a more detailed examination of these periods as well as their implications for policy and planning see Schiller and Kenworthy (2017: chs. 3, 7, 8).

46 Hawken et al. (1999: 22).

47 For critiques of US suburban development and its relation to street and road design, see Newman and Kenworthy (1999: 150); ITE (2006); Kunstler (1996); Kay (1997).

48 For a sample of critiques, reactions, and rejections of highway expansion, see Jacobs (1961); Nowlan and Nowlan (1970); Plowden (1972); Schneider (1972); Rothschild (1973); Davies (1975); Schaeffer and Scalar (1975); Illich (1974); Sachs (1983). For a sample of fresh ideas and approaches, see Stone (1971); Richards (1976); Stringer and Wenzel (1976); Gakenheimer (1978).

49 Flink (1988).

50 For European transportation policy changes during this time, see Pucher (1997, 2004); Pucher and Lefevre (1996). For the spread of pedestrianization and traffic calming, see Wynne (1980); Appleyard et al. (1981); Untermann (1984); Moudon (1987); Roberts (1988); Engwicht (1989, 1993); CNU (undated).

51 See Vasconcellos (2001).

52 See discussions of ISTEA in Schiller and Kenworthy (2017); Dilger (2015). FAST is an acronym for Fixing America's Surface Transportation.

53 See Sheller and Urry (2000); Whitelegg (2015).

54 See Schiller and Kenworthy (2017: chs. 9–10).

55 See Schiller (2016).

56　See Winkelman (2009).
57　See Shoup (2011).
58　See Roberts (1988).
59　For example, Mokhtarian (2002).
60　See FHWA and USDOT (undated).
61　Schiller et al. (2010).
62　See Freund and Martin (1993); Whitelegg (2015).
63　Schiller and Kenworthy (2017).
64　See Portland Bureau of Transportation (2006); see also Portland Bureau of Transportation (2014); City Clock (2014); and bikeportland.org.
65　See Schiller and Kenworthy (2017: chs. 5 and 7) for more discussion of deregulation, privatization, etc., especially for the EU and UK.
66　With a fair amount of variance between regions, approximately 25 percent to 40 percent of commercial aviation flights are within the distances easily served by high and higher speed rail (HSR, HrSR). See Schiller and Kenworthy (2017), esp. ch. 4.
67　Killingsworth et al. (2003).
68　See Jackson and Sinclair 2012.
69　See Centers for Disease Control (2011).
70　See Public Health Agency of Canada (2014).
71　See Besser and Dannenberg (2005).
72　See Handy et al. (2002).
73　See APHA (2012).
74　Smith et al. (2010: n.p.).
75　See Dilger (2015).

References

Amato, J. *On Foot: A History of Walking*. New York: New York University Press, 2004.

APHA. *Promoting Active Transportation: An Opportunity for Public Health*. Washington, DC: American Public Health Association, 2012.

Appleyard, D., Gerson, M., and Lintell, M. *Livable Streets*. Berkeley, CA: University of California Press, 1981.

Banister, D. "The Sustainable Mobility Paradigm." *Transport Policy* 15 (2008): 73–80.

Barker, T. C., and Gerhold, D. *The Rise and Rise of Road Transport, 1700–1990*. Cambridge: Cambridge University Press, 1995.

Bellis, M. "Streetcars–Cable Cars." Undated. Retrieved on July 14, 2009 from http://inventors.about. com/library/inventors/blstreetcars.htm.

Berton, P. *The Great Railway, Illustrated*. Toronto: McClelland and Stewart, 1972.

Besser, L. M., and Dannenberg, A. L. "Walking to Public Transit: Steps to Help Meet Physical Activity Recommendations." *American Journal of Preventive Medicine* 29, no. 4 (2005): 273–80.

Blanchard, A. (ed.) *American Highway Engineer's Handbook*. New York: John Wiley, 1919.

Bottles, S. L. *Los Angeles and the Automobile: The Making of the Modern City*. Berkeley, CA: University of California Press, 1987.

BTS and US Government. "System Mileage within the United States (Statute Miles)." 2002. Retrieved on July 15, 2009 from www.bts.gov/publications/national_transportation_statistics/2002/html/table_01_01.html.

Caro, R. A. *The Power Broker: Robert Moses and the Fall of New York*. New York: Vintage Books, 1975.

Centers for Disease Control. "Strategies for Health-Oriented Transportation Projects and Policies Promote Active Transportation." 2011. Retrieved on November 22, 2016 from www.cdc.gov/healthyplaces/transportation/promote_strategy.htm.

Christensen, J. "The Hyperloop and the Annihilation of Space and Time." *The New Yorker* (August 20, 2013). Retrieved on May 21, 2016 from www.newyorker.com/tech/elements/the-hyperloop-and-the-annihilation-of-space-and-time.

City Clock. "Cycling Mode Share Data for 700 Cities." *City Clock Magazine* (August 8, 2014). Retrieved on February 4, 2017 from www.cityclock.org/urban-cycling-mode-share/#.WJZIALSdJ0e.

City of Toronto. "Did Toronto Have Toll Gates?" 2009. Retrieved on July 10, 2009 from www.toronto. ca/archives/toronto_history_faqs. htm#tollgates.

CNU. *Urban Thoroughfares Manual.* Undated. Washington, DC: Congress for the New Urbanism.

Davies, R. O. *The Age of Asphalt: The Automobile, the Freeway, and the Condition of Metropolitan America.* The America's Alternatives Series. Philadelphia, PA: Lippincott.

Derry, T. K., and Williams, T. I. *A Short History of Technology: From the Earliest times to AD 1900.* New York: Dover Publications, 1993.

Didik, F. "History and Directory of Electric Cars from 1834–1987." Undated. Retrieved on July 6, 2009 from www.didik.com/ev_hist.htm.

Dilger, R. J. "Federalism Issues in Surface Transportation Policy: A Historical Perspective." 2015. *Congressional Research Service 7-5700.* Retrieved on August 22, 2016 from www.crs.gov, 8 December, www.fas.org/sgp/crs/misc/R40431.pdf.

Dobb, K. "An Alternative Form of Long Distance Cycling: The British Roads Records Association." Undated. Retrieved on July 10, 2009 from www.randonneurs.bc.ca/history/an-alternative-form-of-long-distance-cycling_part-4.html.

Duane, C. "Weston & Pedestrian Era Walking Contest Rules." 2010. Retrieved on February 3, 2017 from http://walkapedia.org/walking/reading/research/136-weston-a-pedestrian-era-walking-contest-rules.html.

Engwicht, D. (ed.). *Traffic Calming: The Solution to Urban Traffic and a New Vision for Neighborhood Livability.* Ashgrove, Australia: Citizens Advocating Responsible Transportation, 1989.

Engwicht, D. *Reclaiming Our Cities and Towns: Better Living with Less Traffic.* Philadelphia, PA: New Society Publishers, 1993.

Eno Transportation Foundation. "William Phelps Eno." Undated. Eno Transportation Foundation. Retrieved on February 3, 2009 from www.enotrans.org/about-eno.

Falola, T., and Warnock, A. *Encyclopedia of the Middle Passage.* Greenwood Milestones in African American History. Westport, CT: Greenwood Press, 2007.

FHWA and USDOT. "Teleconferencing." Undated. Retrieved on July 4, 2009 from www.fhwa.dot.gov/REPORTS/PITTD/teleconf.htm.

Flink, J. J. *The Automobile Age.* Cambridge, MA: MIT Press, 1988.

Freund, P. E. S., and Martin, G. T. *The Ecology of the Automobile.* Montréal: Black Rose Books, 1993.

Gakenheimer, R. A. *The Automobile and the Environment: An International Perspective.* MIT Press Series in Transportation Studies; 1. Cambridge, MA: MIT Press, 1978.

Gambino, M. "A Salute to the Wheel." June 17, 2009. Retrieved on July 17, 2016 from www.smithsonianmag.com/science-nature/a-salute-to-the-wheel-31805121/?no-ist.

Glazebrook, G. P. de T. *A History of Transportation in Canada,* 2 vols. Ottawa: Carleton University Press, 1964.

Goddard, S. B. *Getting There: The Epic Struggle between Road and Rail in the American Century.* New York: Basic Books, 1994.

Gutfreund, O. D. *Twentieth Century Sprawl: Highways and the Reshaping of the American Landscape.* New York: Oxford University Press, 2004.

Hamer, M. *Wheels within Wheels: A Study of the Road Lobby.* London: Routledge & Kegan Paul, 1987.

Handy, S. L., Boarnet, M. G., Ewing, R., and Killingsworth, R. E. "How the Built Environment Affects Physical Activity: Views from Urban Planning." *American Journal of Preventive Medicine* 23, no. 2 (2002): 64–73.

Hawken, P., Lovins, A. B., and Lovins, L. H. *Natural Capitalism: Creating the Next Industrial Revolution,* 1st edition. Boston: Little, Brown, 1999.

Herlihy, D. V. *Bicycle: The History.* New Haven, CT: Yale University Press, 2004.

Hindley, G. *A History of Roads.* London: Peter Davies, 1971.

Hyslop, J. *The Inka Road System.* Orlando, FL: Academic Press, 1984.

Illich, I. *Energy and Equity.* New York: Harper & Row, 1974.

ITE. *Context Sensitive Solutions in Designing Major Urban Thoroughfares for Walkable Communities: An ITE Proposed Recommended Practice.* Washington, DC: Institute of Transportation Engineers, 2006.

Jackson, K. T. *Crabgrass Frontier: The Suburbanization of the United States.* New York: Oxford University Press, 1985.

Jackson, R., and Sinclair, S. *Designing Healthy Communities,* 1st edition. San Francisco, CA: Jossey-Bass, 2012.

Jacobs, J. (1961) *The Death and Life of Great American Cities.* New York: Random House.

Kay, J. H. *Asphalt Nation: How the Automobile Took Over America, and How We Can Take It Back.* New York: Crown Publishers, 1997.

Kelley, T. J. "Choke Point of a Nation: The High Cost of an Aging River Lock." *The New York Times* (November 23, 2016). Retrieved on November 27, 2016 from http://nyti.ms/2ggbzj7.

Killingsworth, R. E. E., De Nazelle, A. H., and Bell, R. H. "Building a New Paradigm: Improving Public Health through Transportation." *ITE Journal (Institute of Transportation Engineers)* 73, no. 6 (2003): 28–32.

Kirsch, D. *The Electric Vehicle and the Burden of History*. New Brunswick, NJ: Rutgers University Press, 2000.

Klein, J., and Olson, M. *Taken for a Ride*. Hohokus, NJ: New Day Films, 1996.

Kunstler, J. H. *Home from Nowhere: Remaking Our Everyday World for the 21st Century*. New York: Simon & Schuster, 1996.

Lay, M. *Ways of the World: A History of the World's Roads and of the Vehicles That Used Them*. New Brunswick, NJ: Rutgers University Press, 1992.

League of American Bicyclists. "Detailed History." Undated. Retrieved on February 3, 2017 from www.bikeleague.org/content/mission-and-history.

Manchester, K. "The Silk Road and Beyond—Travel, Trade, and Transformation—Introduction." *Art Institute Of Chicago Museum Studies* 33, no. 1 (2007): 5.

Marx, K. "Grundrisse: Notebook V—The Chapter on Capital." 1857. Marxists Internet Archive. Retrieved on July 17, 2016 from www.marxists.org/archive/marx/works/1857/grundrisse/ch10.htm.

McNeill, J. R. *Something New under the Sun: An Environmental History of the Twentieth-century World*, 1st edition. Global Century Series. New York: W.W. Norton & Company, 2000.

McShane, C. *Down the Asphalt Path: The Automobile and the American City*. Columbia History of Urban Life. New York: Columbia University Press, 1994.

Mokhtarian, P. L. "Telecommunications and Travel: The Case for Complementarity." *Journal of Industrial Ecology* 6, no. 2 (2002): 43–57.

Moudon, A. V. *Public Streets for Public Use*. New York: Van Nostrand Reinhold, 1987.

NCHRP. *Guide to Transportation's Role in Public Health Disasters*. Report 525. Washington, DC: Transportation Research Board, 2006. Retrieved on July 15, 2009www.trb.org/publications/nchrp/nchrp_rpt_525v10.pdf.

Newman, P. W. G., and Kenworthy, J. R. *Sustainability and Cities: Overcoming Automobile Dependence*. Washington, DC: Island Press, 1999.

Norton, P. D. *Fighting Traffic: The Dawn of the Motor Age in the American City*. Inside Technology. Cambridge, MA: MIT Press, 2008.

Nowlan, D., and Nowlan, N. *The Bad Trip: The Untold Story of the Spadina Expressway*. Toronto: New Press, House of Anansi, 1970.

OECD. "Economic and Social Implications of Sustainable Transportation." In *Proceedings of the Ottawa Workshop*, October 20–21. Paris: OECD Publications, 1998.

Park, R. J. "Contesting the Norm: Women and Professional Sports in Late Nineteenth-Century America." *The International Journal of the History of Sport* 29, no. 5 (2012): 730–49.

Plowden, S. *Towns against Traffic*. London: Deutsch, 1972.

Portland Bureau of Transportation. "2006 Bicycle Counts." 2006. Retrieved on February 4, 2017 from www.portlandoregon.gov/transportation/article/156490.

Portland Bureau of Transportation. "Portland Bicycle Count Report 2013–2014." 2014. Retrieved on February 4, 2017 from www.portlandoregon.gov/transportation/article/545858.

Public Health Agency of Canada. "What is Active Transportation?" 2014. Retrieved on November 22, 2016 from www.phac-aspc.gc.ca/hp-ps/hl-mvs/pa-ap/at-ta-eng.php.

Pucher, J. "Bicycling Boom in Germany: A Revival Engineered by Public Policy," *Transportation Quarterly*, vol 51, no 4 (1997): 31–46.

Pucher, J. "Public Transportation," in S. Hanson and G. Giuliano, *The Geography of Urban Transportation*, 3rd edition. New York: Guilford Press, 2004.

Pucher, J., and Lefevre, C. *The Urban Transport Crisis in Europe and North America*. Basingstoke: Macmillan, 1996.

Rae, J. B. *The American Automobile: A Brief History*. Chicago, IL: University of Chicago Press, 1965.

Richards, B. *Moving in Cities*. Boulder, CO: Westview, 1976.

Robb, G. *The Discovery of France: A Historical Geography from the Revolution to the First World War*. New York: W. W. Norton & Co., 2008.

Roberts, J. *Quality Streets: How Traditional Urban Centers Benefit from Traffic-Calming*. TEST report no. 75. London: Transport and Environment Studies, 1988.

Rothschild, E. *Paradise Lost: The Decline of the Auto-Industrial Age.* New York: Random House, 1973.

Sachs, W. "Are Energy Intensive Life-Images Fading? The Cultural Meaning of the Automobile in Transition." *Journal of Economic Psychology* 3 (1983): 347–65.

Sachs, W. *For Love of the Automobile: Looking Back into the History of Our Desires.* Berkeley, CA: University of California Press, 1992.

Schaeffer, K. M and Scalar, E. *Access for All: Transportation and Urban Growth.* Harmondsworth: Penguin, 1975.

Scharf, V. *Taking the Wheel: Women and the Coming of the Motor Age.* New York: Free Press, 1991.

Schiller, P. L. "Automated and Connected Vehicles: High Tech Hope or Hype?" *World Transport Policy and Practice* 22, no. 3 (2016): 28–44.

Schiller, P. L. and Kenworthy, J. R. *An Introduction to Sustainable Transportation: Policy, Planning and Implementation*, 2nd edition. London: Taylor & Francis, 2017.

Schiller, P. L., Bruun, E. C., and Kenworthy, J. R. *An Introduction to Sustainable Transportation: Policy, Planning and Implementation.* London: Earthscan, 2010.

Schivelbusch, W. *The Railway Journey.* Berkeley, CA: University of California Press, 1986.

Schivelbusch, W. *Three New Deals: Reflections on Roosevelt's America, Mussolini's Italy, and Hitler's Germany, 1933–1939.* New York: Picador, 2007.

Schneider, K. *Autokind vs Mankind: An Analysis of Tyranny, a Proposal for Rebellion, a Plan for Reconstruction.* New York: Schocken, 1972.

Sheller, M., and Urry, J. "The City and the Car." *International Journal of Urban and Regional Research* 24, no. 4, December (2000): 737–57.

Shoup, D. *The High Cost of Free Parking*, updated edition. Chicago, IL: Planners Press, American Planning Association, 2011.

Smith, R., Reed, S., and Baker, S. "Complete STREETS." *Public Roads* 74, no. 1 (2010): 12–17.

Solnit, R. *Wanderlust: A History of Walking.* New York: Penguin, 2000.

Steamindex. "George Stephenson." 2015. Retrieved on February 4, 2017 from www.steamindex.com/people/stephen.htm.

Steinbeck, J. *The Grapes of Wrath.* New York: Viking Compass, 1939.

Stone, T. *Beyond the Automobile: Reshaping the Transportation Environment.* Englewood Cliffs, NJ: Prentice Hall, 1971.

Stringer, P., and Wenzel, H. *Transportation Planning for a Better Environment.* NATO Conference series II, Systems Science, vol. 1. New York: NATO Scientific Affairs Division by Plenum Press, 1976.

Untermann, R. K., and Lewicki, L. *Accommodating the Pedestrian: Adapting Towns and Neighborhoods for Walking and Bicycling.* New York: Van Nostrand Reinhold, 1984.

Vasconcellos, Eduardo A. De. *Urban Transport, Environment, and Equity: The Case for Developing Countries.* London: Earthscan, 2001.

Wackernagel, M., and Rees, W. E. *Our Ecological Footprint: Reducing Human Impact on the Earth.* Gabriola Island, BC: New Society Publishers, 1996.

Waller, P. J. *Town, City, and Nation: England, 1850–1914.* Oxford: Oxford University Press, 1983.

Whitelegg, J. *Mobility: A New Urban Design and Transport Planning Philosophy for a Sustainable Future.* Church Stretton: Straw Barnes Press, 2015.

Whitfield, Susan. *Life along the Silk Road*, 2nd edition. Oakland, CA: University of California Press, 2015.

WikiBooks. *Introduction to Paleoanthropology.* 2014. Retrieved on February 4, 2017 from http://en.wikibooks.org/wiki/Introduction_to_Paleoanthropology.

Wikipedia. "Velocipede." Undated. Retrieved on July 9, 2009 from http://en.wikipedia.org/wiki/Velocipede.

Wild, O. *The Silk Road.* 1992. Retrieved on February 3, 2017 from www.cais-soas.com/CAIS/Geography/silk_road.htm.

Wilford, J. "New Finds Suggest Even Earlier Trade on Fabled Silk Road." *New York Times* (March 16, 1993). Retrieved on June 25, 2009 from www.nytimes. com/1993/03/16/science/new-finds-suggest-even-earlier-trade-on-fabled-silk-road.html.

Winkelman, S. "Transportation's Role in Climate Change and Reducing Greenhouse Gases: Testimony Before (US) Senate Committee on Environment and Public Works." 2009. Retrieved from www.epw.senate.gov/public/_cache/files/8a92a671-848a-4eef-b75d-79b150d3e944/winkelmanepwtestimony71409.pdf.

Wood, Frances. *The Silk Road: Two Thousand Years in the Heart of Asia.* Berkeley, CA: University of California Press, 2002.

World Shipping Council. "History of Containerization." Undated. Retrieved on February 4, 2017 from www.worldshipping.org/about-the-industry/history-of-containerization.

Wynne, George G. *Traffic Restraints in Residential Neighborhoods.* New Brunswick, NJ: Transaction Books, 1980.

16

FROM HYDROLOGY TO HYDROSOCIALITY

Historiography of waters in India

Jenia Mukherjee

Introduction

This chapter introduces readers to the history and historiography of waters in India, shedding light on the development and recent trends within the field. The purpose however is not narrowly confined towards an exploration of the state of the art but also an evolution of a framework that might be helpful to critically interrogate unilinear and accepted perceptions, visions, and social agendas since the advent of "modernity." The present statistics on the unsustainable use of water in India is shocking. This chapter will capture the complex, multi-layered dimensions of water challenges within the Indian context and will address the issue via political economy, history, and the ecology of water management.

The per capita water availability in India in 1951 was 5177 m³ per year when the total population was only 361 million. In 2001, as the population increased to 1027 million, the per capita water availability reduced drastically to 1820 m³ per year. As per the estimate of scientists, the per capita water availability will further drop to 1341 m³ in 2025 and 1140 m³ in 2050, when the population is estimated to increase to 1640 million.[1] Although 5 percent of the total water is used for domestic use, 27 percent of villages and 4–6 percent of the urban population in India do not have access to drinking water. The irrigation sector also suffers from water availability, and the ground water level is dropping drastically, culminating in catastrophic droughts, especially in the western states in recent years.

The eminent journalist and the author of *Everybody Loves a Good Drought*, P. Sainath explains that India's mega-water crisis is an outcome of the growing inequality and lop-sided, environmentally disastrous policies that had swept the scene since the last 25 to 30 years.[2] Citing NSSO and other data and survey reports on inequality of wealth distribution and water crisis, Sainath argues that the crisis is taking place due to the predominance of global corporate capitalist projects, including large-scale concretization of pilgrim towns, deforestation, construction of hotels and resorts in the ecologically-sensitive belts, housing complexes with large numbers of swimming pools, diversion to cash crop, and tremendous misuse and abuse of water.[3]

Apart from quantity, the quality of water in India is also a major area of concern. Seventy percent of the water consumed by the rural population in India does not meet the WHO –standards. It has been reported that 80 percent of rural illnesses, 21 percent of transmissible diseases and 20 percent of deaths among children in the age group of 5 years are directly linked to consumption of unsafe water. Only 31 percent of the sewage water generated in 23 major cities is treated and the rest is polluting 18 major rivers in the country. Only 30 percent of the rural population and 65 percent of the urban population use toilets. The chapter is a timely contribution against the present Indian water scenario, which is beset by many issues: inter-state water disputes, disputes over corporate control of limited water resources, severe inequity to water access and entitlements across class, caste, gender, and spatial lines, and frequent floods (often manipulated) and droughts, which tend to have catastrophic impacts on the poorest sections of the society. It will also address the reemergence of the "grand" Inter-linking of Rivers (ILR) Project and different aspects of urban river rejuvenation and beautification plans.[4]

The historiography of water in India reveals the deep-rootedness of these challenges, schemes, and initiatives, which have been largely dictated by and limited to "hydrological" (technical) understandings of water. Although this chapter addresses the usefulness of "colonial hydrology" as a significant theoretical lens, the emerging "hydrosocial" framework and its tropical complexities surrounding "muddyscapes" offers radical re-conceptualizations of water usage, even though it requires more detailed, case-specific, and rigorous empirical research before it can inform policy recommendations and actions. Given the often unsustainable use of water in India, combined with increased conflicts over water rights and access, it is crucial not only to understand the historical background of water issues, but also to present a helpful framework for understanding how water and social wellbeing are indelibly linked.

Historiography of waters in colonial India

The inception

Reflecting on the origins and institutionalization of environmental history, J. R. McNeill mentions that it has countless, tangled roots. Concern for nature and accounts of environmental change generated by human actions find reflection in extant texts such as the Epic of Gilgamesh, the writings of Herodotus, Ibn Khaldun, Montesquieu, Annales School historians, such as Fernand Braudel, Le Roy Ladurie, and some historical geographers. But environmental history as a self-conscious undertaking dates only to about 1970, when the world witnessed waves of environmental activism or popular environmentalism.[5] The United States is the homeland of this field; American historians came together both intellectually and institutionally to launch environmental history as a self-conscious discipline and crafted methods and methodologies to pursue the same.[6] The institutionalization of the discipline came later elsewhere and gradually it flourished in several corners of the world: Europe, Latin America, Africa, and South Asia. South Asia, and especially India, evolved as a rich ground. Indeed, in India, environmental historians attracted the attention of fellow historians, as successfully as in the US.[7]

Though environmental activism propagated and propelled environmental history as a distinct field of inquiry in both the US and South Asia,[8] the nature of these movements, and hence the thrust of analysis and description, differed widely between the North and the South. This is evident in Ramchandra Guha and Joan Martinez-Alier's conceptualization of "ecology of affluence" vs. "environmentalism of the poor," which are terms they use to illustrate the divergent American and South Asian contexts and pathways.[9] In contradistinction to the wilderness cult that emphasizes preservation of nature for nature's sake, Indian environmentalism emerged out of access to habitat and ecosystem resource that was directly linked to the question of livelihood and distributive justice. The divergence in the two approaches is evident, for instance, in the different forms of environmental activism that one finds in the US and India. Guha and Martinez-Alier offer two examples in which activists offered their lives to defend the environment. In the US, Mark Dubois tried to stop dam construction in the Stanislaus River in California by chaining himself to a boulder, whereas, in India, Medha Patkar decided to drown herself in the Narmada River. Though the strategy of protest was similar, the goal was not absolutely the same. Dubois's protest was to preserve the Stanislaus canyon as the last remaining example and icon of virgin and unspoilt American wilderness. Patkar, by contrast, was fighting not only to save the Narmada River but also for the huge number of inhabitants of the area who would be displaced with the construction of the Sardar Sarovar dam project.

During 1980s and 1990s, numerous works on environmental history flooded the Indian scene of social science. However, the core component of it remained restricted to forest history, the origins of which could be traced back to the Chipko movement in the Garhwal Himalaya in early 1970s.[10] Between 1973 and 1980, countrywide resistance in defense of community rights pitted the rich against the poor; the state against the common people; the trans-national corporations against the sons of the soil, and prompted a thorough critique of forest policy of modern India. This in turn led historians to delve deep into the relation between forest, forest products and local communities, and the colonial and post-colonial forestry perception of the state.

Eventually, environmental activism surrounding anti-dam resistances brought water issues into environmental studies. Environmental sociologists played a significant role and numerous studies were conducted on social and ecological consequences of displacement due to the construction of dams and other development projects in post-independence India.[11] However, the major limitation within the field of environmental sociology was that, to an extent, it failed "to locate these projects as part of a wider, more gradual, historical process of change."[12] The contribution of environmental history and more specifically the history of waters became significant within this context.

Pre-colonial equilibrium vs. colonial hydrology

A strong feature of Indian environmental history is the presumed existence of a pre-colonial equilibrium that was destroyed by the "metabolic rift"[13] of the colonial period (fifteenth to mid-twentieth centuries). In forest history, the majority of scholars assume a more or less peaceful pre-colonial (wo)man-nature equilibrium, in sharp contrast to the violent politics and the destructive effects of the colonial regime.[14] Similar currents of thought also flow through narratives and discourses on the history of water management and governance in India.

Numerous works exist to portray the negative effects of colonial irrigation systems and efforts of flood control and management through interventionist policies, such as the construction of big, permanent embankments on tropical rivers, such as the Kosi, the Gandak, the Mahanadi, the Damodar, and the Godavari, which replaced the environmentally benign pre-colonial water-harvesting structures and agrarian regime that evolved in tune to the "moods of rivers."[15] Elizabeth Whitcombe's *Agrarian Conditions in Northern India* could be considered the first attempt to highlight the profound differences between pre-colonial and colonial water regimes. Drawing from a vast bibliography of government records on agriculture, famine, finance, railways, revenue, public works, trade, law, and especially the comprehensive land-revenue settlement reports for each of the nearly 40 districts covering United Provinces of Agra and Oudh (present day Uttar Pradesh), Whitcombe explored how the introduction of perennial irrigation had an adverse impact on the environment and the peasantry. The new hydraulic experiments, heavily underwritten by commercial principles, led to saline deserts, waterlogged swamps, and decreased soil fertility. She examined how more reliance on large-scale water storage systems and lengthy diversion channels replaced and uprooted the traditional well irrigation systems, which was more dependable in the semi-arid region with low rainfall. Her work also studied power relations and the nexus involved in distribution and access to irrigation water and collection of water tax; rich landlords controlled the layout of distribution channels and local agents who collected water charges followed extractive means burdening peasants who were already suffering from ecological problems. Whitcombe illustrated in detail how a "depressed peasantry laboured in a distorted environment."[16]

Other studies on colonial canal irrigation in the south and the eastern deltas of the sub-continent have shown how hydraulics along with socio-political structures of spheres of annexation and control were manipulated, re-ordered, and realigned to suit colonial capitalist motives and interests of making profits from revenue collection.[17] Studies of colonial experiments with flood control through the construction of large embankments are mainly limited to eastern part of the sub-continent.[18] Flooding came to be perceived as an obstacle restraining routine and regular revenue collection, especially after the Permanent Settlement in eastern India (Bengal, Bihar, and Orissa).[19] The unusual volatility of the deltaic stretch of tropical rivers, such as the Mahanadi and the Damodar, became an immediate source of anxiety to the British as their attempt to consolidate and rule in the fertile (eastern) region became "critically dependent on their ability to subdue and train the volatility of the delta's hydrology."[20] The colonial policy of land rent through a series of regulations replacing the pre-colonial tax on gross produce led to the onset of an embankment regime through the constitution of embankment committees required to prepare annual estimates for construction, maintenance, and repairs.[21] Praveen Singh's chapter on "Flood Control in North Bihar" draws attention to the construction of private marginal embankments by European planters, indigo factories, and big landowners or zamindars, as well as publicly funded embankments.[22] According to the stipulations of the Permanent Settlement, zamindars who became beneficiaries of any increase in crop output were made responsible to maintain certain stretches and bear the losses if floods washed them away.[23] Singh argues that "the colonial officers believed that a mix of compulsion and incentives for the zamindars would protect existing agricultural lands and bring new areas under cultivation."[24] Weaving complex arrangements with sections of prosperous peasantry, railway officials, irrigation engineers, district administrators, colonial zamindars crafted an embankment-driven, flood-control regime.[25] Dinesh Mishra's study on the Kosi and the Gandak Rivers also demonstrate administrative optimism surrounding embankments, which were considered pivotal to bring huge acres of fallow land under the plough.[26]

These studies also reveal the acute environmental and social consequences of interventions to rivers. Rohan D'Souza's research on the Mahanadi River in Orissa shows how embankments, designed to insulate lands from inundation, resulted in the silting up of the channels and aggravated floods by raising flood lines to dangerous levels.[27] Embankments, by confining floods within the narrow stretch of the river channel, did not allow the silt to be deposited uniformly over the flood plains, resulting instead in the deposit of silt in the riverbed itself. It also deprived flood plains of nutrient deposits essential for continued fertility. Low-lying areas remained waterlogged and the pre-existing drainage network was destroyed. Moreover, the constant breaching of embankments leading to more floods caused by the concentrated discharge of water proved to be disastrous.[28] This was a complete departure from the pre-colonial agrarian relations to flood, floodwaters, and sediment in the region, which drew the attention of British civil engineers, such as William Wilcocks, who wrote extensively about the beneficial side of what he understood and categorized as "overflow irrigation."[29] Daniel Klingensmith investigates the debate among civil engineers, public health officials, academicians, and politicians on the dissipation of large sections of the Bengal delta and the deleterious effects of replacing "overflow irrigation" through the construction of embankments (including railway embankments) on the Damodar.[30] D'Souza details the transformation of the "flood dependent agrarian regime" in the Orissa delta into a "flood vulnerable landscape," when the nature of floods in the "protected" (embanked) areas changed with floods of higher levels and longer duration.[31] These studies also shed light on the changing social structure and hierarchies with the introduction of interventionist hydraulic technologies that crafted new distributive dynamics. The case of North Bihar captures local dynamics of how British-induced embankment regulations acquired "fatal competitive momentum" and resulted in "overall aggravation of the flood line as it tended to clog drainage and deteriorate the bed of the river."[32]

Embankment failures led to the imposition of a more severe interventionist regime on rivers as the empire's "triad of bourgeois property, the Company zamindari system and an extensive revenue collection strategy" was used "to secure its social and economic foundations."[33] Eswara Rao's study on colonial irrigational works in the Madras Presidency and more specifically the Godavari Anicut is a clear reflection of this system.[34] Rao explores the mindset of Arthur Cotton, the chief architect of the project, to unveil "economic and political rhetoric that proved decisive for initiating the construction of the Godavari scheme."[35] Perennial irrigation and navigation schemes superseded upon the phase of embankment, leading to serious socio-ecological ramifications. Canal irrigation not only upset traditional water access and distribution mechanisms by introducing water rates and embankment rates, but replaced all non-canal irrigation sources and traditional harvesting structures. As early as 1980, Nirmal Sengupta explored the decay of the traditional network of *ahar* (tank) and *pyne* (channel) irrigation system in South Bihar following the perennial canal irrigation system of the British to introduce new revenue streams.[36] Indu Agnihotri's article depicts how colonial canal systems in Punjab overran the existing inundation canal system, affecting pastoralists and eroding the vibrant pastoral economy of the region.[37] Arun Agrawal and Sunita Narain's report reveals how traditional water harvesting systems in India declined through a series of colonial hydraulic actions for capitalist profits.[38] Reflecting on the Indus Basin, David Gilmartin and Benjamin Weilargue that the perennial canal system between the late nineteenth and early twentieth century was loaded with technochauvinism of "imperial science," leading to the replacement of benefits of environmentally benign, community-managed, small-scale structures with socially and environmentally disruptive interventions centrally designed and engineered by scientists and technocrats under the aegis of the Irrigation Department.[39]

The final move towards total river control strategies by damming rivers was made during the last phase of the colonial rule.[40] One hydraulic phase was superseded by another through the adoption of newer technologies, leading to an acute "metabolic rift" between (wo)man and nature, river and society. On his case study on the Mahanadi River, D'Souza postulates, "in the instance of our example, the recasting of the delta as a flood vulnerable landscape provoked an upward spiral in technological choices that over the span of a century and a half moved from embankments to a canal system and finally the construction of the Hirakud dam on the Mahanadi River."[41] Everywhere the British created problems where none formerly existed, and disrupted and privatized an essentially stable water system. The Damodar River in Bengal and the Kosi in Bihar underwent similar phases of interventions among others.[42] Multi-purpose river valley development (MPRVD) schemes were introduced in India in a political context where Indian capital and the colonial state constituted a new rhetoric and paradigm for rule.[43]

The binary of pre-colonial environmentally benign and socially accommodative water management system versus environmentally malign and socially disruptive hydraulic interventions induced by distinct phases of colonial capitalism is a very strong component of the historiography of waters in India. Scholars have also moved further to argue that the conceptual notion of "colonial hydrology" might provide scientific and theoretical traction by attempting to characterize the British experience as "comprising an altogether distinct paradigm for hydraulic interventions."[44]

Beyond reductionist dualisms

Paul Greenough's "standard environmental narrative" shows that scholars, from Marxists to New Age Utopianists, have insisted upon the transition from harmony to disruption, justice to inequity, and prosperity to misery.[45] This interpretation makes sense for understanding colonial rule in the Indian subcontinent and its effects on water. However, there are few studies that go beyond these reductionist dualisms and trace continuities between the pre-colonial and the colonial state in South Asia, thus making the distinction between the phases partly irrelevant. Some studies have taken an altogether different stance to reveal the advantages of colonial water management system. Ian Stone countered Whitcombe's findings to argue that canals in the northwest became a source of economic dynamism and constant innovation. Similarly, the pessimism versus optimism debate flows through South India, as well.[46] Peter Schmitthenner contradicts Rao by arguing that hydraulic engineering projects in the deltas of Cauvery and Godavari Rivers were "less environmentally disruptive or destructive than colonial riparian works of the north and blended more into the environmental and cultural landscape of the respective delta regions."[47] However, these are very few in number and capture only a partial glimpse of the big picture.

The departure from pre-colonial/colonial binary and optimism/pessimism debate becomes profound in nuanced narratives of historians who seem to be convinced about the continuities and continuous development along historical trajectories and conjectures. Questioning whether colonialism did indeed have a major impact on traditional water structures, David Hardiman's study on indigenous water systems in Gujarat showed that commercialization and peasant indebtedness were processes that predated the colonial regime in the region, and which were integral to the expansion of well irrigation in the west.[48] On a different note, Rosin's work on western India (Rajasthan) reflects on how a series of complex groundwater irrigation and drinking water devices (such as silt-ponds, step wells, reservoirs, and L-shaped embankments) remained in operation well into the colonial and post-colonial periods.[49] These systems survived by overlapping with other hydrological regimes, and thus, rather than being replaced or displaced, co-existed with the "modern" hydraulic technologies introduced by the British.

Exploring the South Indian terrain, David Mosse's seminal work interrogates the notion of a "pre-colonial equilibrium" and sketches a picture of stable and sustained water management practices by "organic" and "autonomous" villages.[50] Studying larger processes of statecraft and regional politics in the pre-colonial period, Mosse shows that village communities were unstable entities driven and shaped by hierarchies. Mosse argues in favor of continuities, since the construction and maintenance of tanks in South India underwent phases of efflorescence, expansion, intensification as well as decadence and decline prior to colonial rule. Moreover, "pre-colonial tank irrigation systems … were not resourced, maintained and operated by autonomous village institutions, but by a wider set of political relations of the decentralized, or "segmentary" pre-colonial state."[51] In a critique of the pre-colonial equity and harmony framework, Mosse unpacked the inherent social and political inequities within the tank management system.[52] His ethno-historical work (1997) in Sivagangai and Ramnad districts of Tamil Nadu establishes "a persistent link between systems of caste honour and tank irrigation works in the articulation of authority at the level of village, micro-region and kingdom."[53]

Esha Shah's arguments are in line with those of Mosse and add detail and nuance to our understanding of tank irrigation practices in pre-modern South India (mainly Karnataka).[54] Shah draws mainly on historical ethnography, and especially on "subaltern" evidence, including oral narratives and folklores in the form of stories, legends, and songs. The folksongs and stories inscribed in popular memory bear testimony to hydrological irregularity, technological vulnerability, and social anxiety. According to Shah, this historical memory does not even "remotely resemble an idyllic picture of a blissful community living on the banks of a bountiful tank in the pre-modern era."[55] "Seeing like a subaltern" and drawing a continuous line between post-modern and modern times, Shah explores how in each epoch the reproduction of technology implied reproduction of radically different social and cultural spaces and, most significantly, social and power relations.[56]

Towards political ecology, cultural politics, and hydrosociality

Cultural politics and social dynamics

Recently, an emerging literature redefining cultural politics of natural (water) resources has gone beyond the dull rigor of economic determinism to capture lived experiences and sedimented histories of situated cultural practices and spaces.[57] Scholars critically interrogate reductionistic dualisms while enlarging the canvas to cover post-colonial struggles surrounding water resources. Reviewing policy debates on agricultural water use in India, Peter Mollinga claims that it is caught between strategic essentialism and analytical reductionism.[58] Strategic essentialism assumes a binary opposition between state–village and state–community, and the resulting analytical reductionism overemphasizes the village and community institutions and pays scant attention to the ways in which the state functions at the local level.[59] Mollinga asserts the importance of focusing on material conditions, including the impact of more recent processes of economic liberalization and globalization on water use, as multinational corporations have increased their consumption of water in many parts of India, creating a whole range of conflicts.[60] The context was already laid by scholars like Vandana Shiva and Lyla Mehta, the most notable among others. Shiva apprehended 'water wars' in the near future and alarmed against the vicissitudes of corporate culture and the historical erosion of communal water rights.[61] Mehta drew our consciousness on the construction of scarcity as a meta-narrative to justify and allow simplistic portrayals of property rights by ignoring cultural and symbolic dimensions of water.[62] Anathakrishnan Aiyer's article discusses in detail the "corporate theft" of water resources and opposition and resistances against transnational corporations like Suez, Coca-Cola, etc. and focuses on the struggle for water rights in Plachimada, Kerala.[63] Beyond the conventional literature on water privatization in India, Aiyer's study concludes that corporate control of resources in India must be located and analyzed within a framework that is not necessarily restricted to neoliberal globalization and transnational corporations. The Plachimada case could be analyzed, for instance, as part of the unfolding agrarian crisis in India. He suggests that "old" and "new" conceptual or theoretical and political concerns should be combined with contemporary encounters with neoliberal globalization.[64] The literature on corporatization and privatization of water has another important dimension in Vandana Asthana's work on urban Delhi.[65] Asthana explores emerging dynamics of water policy processes in India in the post-economic reform era. She draws attention to how water management is undergoing a dramatic transformation through the process of privatization, liberalization, and deregulation. Her work theorizes the overlap across local, subnational, and global scales, expressed by a variety of voices and "the dynamics of inclusion and exclusion that surround the widely diverse actors involved in the processes of producing water policy.[66] These works provoke us to understand ways in which multiple fields of power manifest in complex ways. The political ecology framework, edifying explorations of asymmetric workings of power is embedded in these researches.

Some scholars have combined "Third World" dynamics of colonial legacy, national regulations, and the politics of international aid agencies in India with extensive research on forests and water, using a political ecology framework that focuses on state-led extractive natural resource management.[67] Recent studies on water focus on the imaginings, appropriations, and contestations that make way to the unfolding of "waterscapes."[68] The political ecology framework has been enlarged to accommodate cultural politics that "treats identities, interests and resources, not as pre-determined givens, but as emergent products of the practices of cultural production."[69] By emphasizing power, process, and practice, this approach treats "culture itself as a site of political struggle."[70] In the words of Baviskar, "while cultural politics shares political ecology's commitment to understanding the asymmetric workings of power, it has a greater appreciation of the complex and contingent conditions under which people make history."[71]

Studies in the Indian urban water scene in recent years have moved beyond the Anglophonic urban political ecology of water perspective and its southern manifestation.[72] Although the urban perspective is important, in that it explores the fractured nature of drinking water provision in cities and the ways in which this reflects historical and contemporary power differentials among multiple stakeholders,[73] what is needed is a more comprehensive political, social, and cultural approach across the "many waters" of urban and rural India.[74] Nonetheless, the focus on "blue infrastructures" within urban spaces and their peripheries is a significant aspect of the cultural politics of water management.[75] While Baviskar shows how the Yamuna riverbed has changed from being a neglected "non-place" to prized real estate for private and public corporations, tracing the shifting visibility of the river in the social and ecological imagination of Delhi, Jenia Mukherjee brings out the significance of sustainable flows between Kolkata and its peri-urban wetlands, and how it has transformed from a mutually reinforcing system to a truncated one through the commoditization of the commons.[76] Considering primary historical sources, including official documents, reports, letters, extracts, maps, and plans and using a long temporal scale, Mukherjee narrates the co-evolution of the city and its peri-urban interface (in the form of around 264 sewage-fed fish ponds at present) as consciously constructed spaces when the British tamed natural creeks, inlets from rivers, and swamps and marshes into waterscapes (i.e. extensive excavated canals connecting the city with its hinterland) to accomplish colonial capitalist motives of revenue generation, unintentionally giving birth to the eastern sewage-fed wetlands that in turn emerged as the space for informal, untamed practices of marginal peri-urban communities.[77] It further captures how post-colonial urban planning narrative, tuned to the neo-liberal rhetoric of "sustainable urbanization," has been strategically constructed to tame the rapid urban sprawl at the cost of the peri-urban wetlands, for seeking financial gains.[78] Going beyond the Baviskarian paradigm of "bourgeois environmentalism" (where state-led and middle-class supported ecological (riverfront) conservation excludes the aspirations of the poor), Mukherjee unpacks varieties of urban environmentalism surrounding the conservation of canals and wetlands of Kolkata.[79]

Karan Coelho and Nithya V. Raman's work shows how the complex and changing landscapes of urban land and water are closely interwoven with the complex and changing landscapes of eviction and relocation in Chennai.[80] It narrates stories of the making and remaking of the city's physical form through historical processes of land reclamation, and through the state's urban expansion programs for housing and institutional development. It traces large-scale relocations of the urban poor to flood-prone marshlands, a process through which the boundaries between land and water in urban peripheries were reshaped (sometimes un-consciously blurred and sometimes consciously and irreversibly re-inscribed). These stories center on a common theme: the filling of water bodies to house people, followed by displacement of people to restore water bodies and relocating them onto yet other water bodies. This story encapsulates historical shifts in the political and economic rationalities that connect municipal governance to paradigms of planning and social engineering. It records the complicated journey of urban water bodies perceived as land-in-the-making, to be filled and reclaimed for housing projects, dump yards, resettlement sites, etc. to lakes-in-the-making, to be cleared, dredged, desilted, and beautified, all at the expense of the vast majority of the local inhabitants.

New works explore everyday negotiations over access to composite water resources, such as lakes and ponds, which are simultaneously water, land, and public space. Examining the transformation of a lake in Bangalore, Jayaraj Sundaresan argues that the making and unmaking of "the commons" involves the making and unmaking of communities and vice versa that occur at the interface between democratic struggles and bureaucratic systems.[81] While Neha Singh (in a forthcoming work) studies the making and unmaking of waterscapes (mainly lakes) in Udaipur, mapping the contours of contestations embedded in circulation of water, Natasha Cornea, Anna Zimmer, and René Véron demonstrate that in a context of ambiguity of the statutory governance regime and fragmented control in a small city like Burdwan (West Bengal), the (re)production of the pondscape is embedded within complex relationships of power whereby social marginalization can be offset at least momentarily by local institutions such as neighborhood clubs and political parties.[82]

Thus, our understanding of waters has moved from unilinear, singular narratives to comprehensive portrayals, capturing complicated interwoven aspects across temporal scales. The development and incorporation of the notion of "hydrosocial" for undertaking critical political ecologies of water has been a significant breakthrough in this regard. It captures socio-natural processes "by which water and society make and remake each other over space and time."[83] The hydrosocial cycle, in contrast to the hydrologic cycle, highlights the "dialectical and relational processes through which water and society interrelate."[84] Within the tropical context, this assumes greater significance and complexity, when water-society dialogue gets transformed into water-sediment-society dialectics.[85]

Treading the muddyscapes

What happens when hydrosociality meets its sediments? "Hydrosociality" implies the manifold ways in which humans and non-humans mesh in relations that are simultaneously social and hydrological, in the sense of involving or impinging on the circulation, distribution, and quality of water, as well as on the materials that water gathers: muddy sediments. The muddy terrain of human engagements with water for understanding current and past water predicaments assumes significance and relevance within the tropical context of India.

D'Souza reveals the "great hydraulic transition" that took place in eastern India when European hydraulic knowledge and rationality loaded with new economic, legal, and quantitative calculations was imposed on the colonized territory.[86] The work sheds light on the European environmental imagination in the early modern period that considered soil–liquid hybrids (in the form of marshes, swamps, fens, etc.) as uninhabitable and treacherous places and crafted innumerable drainage, reclamation, and embankment campaigns to deploy pumps, dredging devices, locks, and sluices to transform these "once soluble and precarious waterscapes into firm and durable landscapes."[87] The European modernist interventions effected "separations" between "water and land, between fresh and salt waters, between clear and turbid flows, between individual channels, between lagoon water and river water and between city and terraferma."[88] Furthermore:

> Land exorcised of water was transformed into property, to be then elaborated as socio-economic-legal objects. Flowing waters telescoped into contained channels, on the other hand, were revealed principally as engineering visions … rationalised chiefly as communication, transport and movement rather than as a site for production.[89]

Within the context of South Asia and more specifically eastern India, the study demonstrates how hydraulic modernity, by rendering land and water into discrete domains, deployed new water technologies to constitute a "modern ecological and productive regime."[90] By the 1940s, the natural soil-water erasure caused havoc and Bengal's rivers were declared to be an indisputable water problem. Colonial suggestions for "balancing" the Bengal Rivers were through a slew of large dams and barrages across the Ganges, Teesta, Damodar, and so on.

Introducing the "wet" theory and focusing on the floodplains of Bengal, Kuntala Lahiri-Dutt also argues that here, hybrid environments "potentially destabilize the conventional water-land binary characterised by their uncertain existence, their indeterminacy, and their fluid liminal presence as ambiguous temporal, cultural, and political geographies."[91]

Building upon these notional tractions, and concentrating on the Lower Gangetic Basin "that is part land, part water, but is neither in its entirety,"[92] Mukherjee and Flore Lafaye De Micheaux capture how intervention on this tropical river altered deposition patterns of its alluvial sediments, affecting changing perceptions towards riverine islands (*chars*) among multiple social actors across temporal trajectories.[93] Traversing local hydrosocial dialectics through the coproduction of mud, waters, and social order, the study examines the impact of profound hydrological transformations that cross colonial river management and post-colonial dam construction (the Farakka Barrage) initiatives. *Char* formation was a natural phenomenon on the meandering stretch of the river. However, the construction of the Farakka Barrage in the 1960s, intended to manipulate flow in one of the major distributaries of the Ganges, intercepted the normal sediment transport-deposition pattern of the river, causing increased flood intensity and leading to the rise of running/temporary *chars* in upstream and downstream of the Barrage. The continuous emergence, submergence, re-emergence, and re-submergence of *chars* have influenced social processes of settlement, displacement, re-settlement, and re-displacement among *choruas* (people inhabiting *chars*). While *chars* were also looked into as revenue assets during the pre-Farakka period,[94] the post-colonial state considers running *chars* as *sikasti* (water) to legitimize its strategy of not providing infrastructural provisions in these fragile spaces in turn affecting the livelihood of *choruas* who keep on migrating from one island to the other. These theoretical tractions such as hydrosociality and muddyscapes await rigorous empirical researches to critically interrogate accepted frameworks and binaries that mask conventional disciplines.

Conclusion

Within the Indian water scenario, there seems to be a strong contradiction between policy frameworks and planning driven by hydrology-oriented solutions[95] and the real need of the hour towards an integrated, comprehensive, and hydrosocial understanding of water resources for addressing water sustainability. The state of the literature on water captures snapshots of major hydraulic transitions in India. Though the pre-colonial/colonial binary has been challenged as reductionist, the "modern" hydraulic transition and interventions, and their ramifications cannot be underestimated. In a radio broadcast in 1945, the Governor of Bengal, R.G. Casey, suggested that flow of rivers (in Bengal) between summer and winter should be controlled and equalized to avoid disastrous flooding during the monsoon and drought during winter. The same old colonial logic is resurrected in the contemporary ILR project plan with aggravated fatal outcomes due to more powerful technologies and much greater capital investments. Recent studies have focused on the urban predicament and its blue infrastructures, capturing multi-layered, dialectical-relational aspects of the water issues that India faces. This literature offers a critical lens to identify and explore severe challenges within contemporary externally funded and state-led urban environmental planning surrounding riverfront development projects, water parks, and other urban features. This literature also shows the dire social and ecological costs of "externalization," which occurs when social and environmental concerns are ignored in the pursuit of financial gain. An exploration of the various theories, empirical findings, and frameworks on water history (past and present) is a timely contribution to question unilinear, reductionist hydrological solutions. The incorporation of these multi-layered, yet integrated approaches in policy formulations could be the next major transformative context to address water sustainability within the Indian context.

Acknowledgements

I would like to acknowledge the Indian Council of Social Science Research (ICSSR) and University of Lausanne for supporting this research through the Scholars Exchange Grant (SEG) on "Political Ecology of the River: Indo-Swiss Exploration of Hydrosocial Dynamics in the Lower Stretch of the Ganges." Additionally, I would like to thank the editor for providing comments that were extremely useful to revise the chapter. Finally, I would also like to thank my PhD scholar Archita Chatterjee for copy-editing support.

Notes

1 S. K. Gupta and R. D. Deshpande, "Water for India in 2050: First-Order Assessment of Available Options," *Current Science* 86 (2004): 1216.
2 P. Sainath, *Everybody Loves a Good Drought: Stories from India's Poorest Districts* (New Delhi: Penguin Books, 1996).
3 "Delhi Assembly Lecture Series: Basic Reason for Water Crisis in India is Inequality, says P. Sainath," Firstpost, retrieved on January 18, 2017 from www.firstpost.com/india/delhi-assembly-lecture-series-basic-reason-for-water-crisis-in-india-is-inequality-says-p-sainath-2859070.html.
4 The project is planned to build 30 links and approximately 3,000 storages to connect 37 Himalayan and Peninsular rivers to form a gigantic South Asian water grid. The canals, planned to be 50 to 100 meters wide and more than 6 meters deep, will facilitate the navigation of water. The estimates of key project variables, still in the nature of "back-of-the-envelope calculations," suggest that it will cost a staggering US$123 billion (or Indian Rs. 560,000 crore at 2002 prices), handle 178 km³ of inter-basin water transfer/per year, build 12,500 km of canals, create 35 gigawatts in hydropower capacity, add 35 million hectares to India's irrigated areas, and generate an unknown volume of navigation and fishery benefits.
5 J. R. McNeill, "The State of the Field of Environmental History," *Annual Review of Environment and Resources* 35 (2010): 345–74.
6 Ibid.
7 Ibid.
8 It is to be noted that South Asia is used synonymously with India in Indian environmental history. Though this is problematic and needs further inquiry, however, one of the possible reasons might be the colonial context as the major center of analysis across undivided India (including present Pakistan and Bangladesh).

9   R. Guha and J. Martinez-Alier, *Varieties of Environmentalism: Essays North and South* (London: Earthscan, 1997).

10   R. Chakrabarti, *Situating Environmental History* (New Delhi: Manohar, 2007).

11   P. McCully, *Silenced Rivers: The Ecology and Politics of Large Dams* (London: Zed Books, 1996). S. Sangvai, *The River and Life* (Mumbai: Earthcare Books, 2000). A. Baviskar, *In the Belly of the River: Tribal Conflicts over Development in the Narmada Valley* (Delhi: Oxford University Press, 1997).

12   A. Baviskar, "Ecology and Development in India: A Field and its Future," *Sociological Bulletin* 46/2 (1997): 193–207.

13   J.B. Foster, "Marx's Theory of Metabolic Rift: Classical Foundations for Environmental Sociology 1," *American Journal of Sociology* 105/2 (1999): 366–405.

14   M. Gadgil and R. Guha, *This Fissured Land: An Ecological History of India* (Delhi: Oxford University Press, 1992). R. Guha, *The Unquiet Woods: Ecological Change and Peasant Resistance in the Himalayas* (Delhi: Oxford University Press, 1991). V. Damodaran, *Broken Promises: Popular Protest, Indian Nationalism and the Congress Party in Bihar, 1935–46* (Delhi: Oxford University Press, 1992). M. Rangarajan, *Fencing the Forests: Conservation and Ecological Change in India's Central Provinces, 1860– 1914* (Oxford: Oxford University Press, 1996). V. Saberwal, *Pastoral Politics: Shepherds, Bureaucrats, and Conservation in the Western Himalaya* (Delhi: Oxford University Press, 1999).

15   K. Lahiri-Dutt and G. Samanta, *Dancing with the River: People and Life on the Chars of South Asia* (New Haven, CT: Yale University Press, 2013).

16   E. Whitcombe, *Agrarian Conditions in Northern India: The United Provinces under the British Rule, 1860– 1900*, vol. 1 (Berkeley, CA: California University Press, 1972), xi.

17   G. N. Rao, "Transition from Subsistence to Commercialised Agriculture: A Study of Krishna District in Andhra, 1850–90," *Economic and Political Weekly* 20/25–26 (1985): 60–69. G. N. Rao, "Canal Irrigation and Agrarian Change in Colonial Andhra: A Study of Godavari District, c. 1850–90," *The Indian Economic and Social History Review* 25/1 (1988): 25–60. M. A. Reddy, "Travails of an Irrigation Canal Company in South India, 1857–1882," *Economic and Political Weekly* 25/12 (1990): 619–28. P. Chaudhuri, "Peasants and British Rule in Orissa," *Social Scientist* 19/7 (1991): 36–42. R. D'Souza, "Colonialism, Capitalism and Nature: Debating the Origins of Mahanadi Delta's Hydraulic Crisis (1803–1928)," *Economic and Political Weekly* 37/13 (2002): 1261–72. R. D'Souza, "Damming the Mahanadi River: The Emergence of Multi-Purpose River Valley Development in India (1943–46)," *The Indian Economic and Social History Review* 40/1 (2003): 81–105.

18   There are very few exceptions, they include Benjamin Weil's study on flood control measures in the Indus Basin between the 1840s and 1930s and its socio-environmental impacts. Weil illustrates how embankments and hurdle dykes built upon the western stretch of the Indus to protect the town of Dera Ghazi Khan led to the eventual relocation of the same. He elucidates how technochauvinism through complex hydraulic engineering works replaced traditional warning systems and mobility, undermining alluvial farming systems as well as a precautionary approach to environmental management. B. Weil, "The Rivers Come: Colonial Flood Control and Knowledge Systems in the Indus Basin, 1840s–1930s," *Environment and History* 12/1 (2006): 3–29. Weil laments "It might have been cheaper and easier to have moved the town first and let the river have its way" (ibid., 17).

19   The Permanent Settlement was introduced by Lord Cornwallis in 1793. It was an agreement between the British East India Company and the Landlords of the Bengal Province (Bengal, Bihar and Orissa) to settle the Land Revenue to be raised.

20   D'Souza, "Colonialism, Capitalism and Nature," 1264.

21   D. K. Mishra, "The Bihar Flood Story," *Economic and Political Weekly* 32/1 (1997): 2206–17. Mishra, *Trapped Between The Devil and the Deep Waters—Story of Bihar's Kosi River* (Delhi: People's Science Institute and SANDRP, 2008). R. D'Souza, "Colonialism, Capitalism and Nature," 1261–72. R. D'Souza, *Drowned and Dammed: Colonial Capitalism and Flood Control in Eastern India* (New Delhi: Oxford University Press, 2006). P. Singh, "The Colonial State, Zamindars and the Politics of Flood Control in North Bihar (1850–1945)," *The Indian Economic and Social History Review* 45/2 (2008): 239–59. P. Singh, "Flood Control for North Bihar: An Environmental History from the 'Ground-Level' (1850–1954)," in *The British Empire and the Natural World: Environmental Encounters in South Asia*, ed. D. Kumar et al. (Oxford: Oxford University Press, 2011).

22   Singh, "Flood Control for North Bihar."

23   H. L. Harrison, *The Bengal Embankment Manual: Containing an Account of the Action of the Government in Dealing with the Embankments and Water-Courses since the Permanent Settlement: A Discussion of the Principles of the Act of 1873* (Calcutta: Bengal Secretariat Press, 1875).

24   Singh, "Flood Control for North Bihar," 162.

25   Ibid.

26   Mishra, *Trapped Between The Devil and the Deep Waters.*

27   D'Souza, "Colonialism, Capitalism and Nature." D'Souza, *Drowned and Dammed.*

28   The scale of devastation could be understood with historical instances including the breaching of the Gandak embankment in Bihar and the inundation of around 30 villages in 1872. Mishra, "The Bihar Flood Story," 15.

29   It was a system by which the nutrient-rich, silt-laden monsoon floodwaters from the upper regions of various rivers flowing into Bengal were distributed evenly over the delta, watering and more importantly fertilizing fields, spreading fish over the countryside and sweeping away the mosquito populations that spread malaria. Floodwater was not allowed to go where it would; rather, it was directed through a system of wide, shallow canals with minimal embankments, in which, at an appropriate time during the yearly monsoon, breaches were made to allow more even flooding. The local water boards which worked with cultivators were in charge of the proper distribution of this water. The canals also drained the land, preventing water-logging. W. Willcocks, *Lectures on the Ancient System of Irrigation in Bengal* (Calcutta: University of Calcutta, 1930). D. Klingensmith, *One Valley and a Thousand: Dams, Nationalism, and Development* (New York: Oxford University Press, 2007).

30   Ibid.

31   D'Souza, "Colonialism, Capitalism and Nature." P. Singh, "Flood Control for North Bihar."

32   Singh, "Flood Control for North Bihar," 174.

33   D'Souza, "Colonialism, Capitalism and Nature," 1267.

34   E. Rao, "Taming 'Liquid Gold' and Dam Technology: A Study of the Godavari Anicut," in *The British Empire and the Natural World: Environmental Encounters in South Asia*, ed. D. Kumar et al. (Oxford: Oxford University Press, 2011).

35   Sir Arthur Cotton wanted to convert the "water of the Godavari into money instead of letting it run into the sea." Cotton went to the extent of representing liquid flows in financial units. He calculated that 420,000 cubic yards of water per hour flowed into the sea and at the rate of Rs. 500 per hour or 12,000 per day, for 240 days it gave Rs. 2,880,000; flowing water was liquid gold. He was the first irrigation engineer who envisioned a plan to link the rivers in South India for inland navigation. See Rao, "Taming 'Liquid Gold' and Dam Technology," 146–9. See also H. Morris, *A Description and Historical Account of Godavery District in the Presidency of Madras* (London: Truber and Company, 1878).

36   N. Sengupta, "The Indigenous Irrigation Organisation in South Bihar," *The Indian Economic and Social History Review* 37/2 (1980): 157–87.

37   I. Agnihotri, "Ecology, Land Use and Colonisation: The Canal Colonies of Punjab," *The Indian Economic and Social History Review* 33/1 (1996): 37–58.

38   A. Agrawal and S. Narain ed., *Dying Wisdom: Rise, Fall and Potential of India's Traditional Water Harvesting Systems* (New Delhi: Centre for Science and Environment, 1997).

39   D. Gilmartin, "Scientific Empire and Imperial Science: Colonialism and Irrigation Technology in the Indus Basin," *The Journal of Asian Studies* 53/4(1994): 1127–48. Weil, "The Rivers Come."

40   D'Souza, "Colonialism, Capitalism and Nature." D'Souza, "Damming the Mahanadi River." D'Souza, *Drowned and Dammed.* Singh, "Flood Control for North Bihar." Mishra, *Trapped Between The Devil and the Deep Waters.* Klingensmith, *One Valley and a Thousand.*

41   D'Souza, "Colonialism, Capitalism and Nature," 1270.

42   Klingensmith, *One Valley and a Thousand.* Mishra, *Trapped Between The Devil and the Deep Waters.* Singh, "Flood Control for North Bihar."

43   D'Souza, "Damming the Mahanadi River."

44   R. D'Souza, "Water in British India: The Making of a 'Colonial Hydrology'," *History Compass* 4/4 (2006): 621–8.

45   P. Greenough, "Naturae Ferae: Wild Animals in South Asia and the Standard Environmental Narrative," in *Agrarian Societies: Synthetic Work at the Cutting Edge*, ed. J. Schott and N. Bhatt (New Haven, CT: Yale University Press, 2001), 141–85.

46   I. Stone, *Canal Irrigation in British India: Perspectives on Technological Change in a Peasant Economy* (Cambridge: Cambridge University Press, 1885).

47   Rao, "Canal Irrigation and Agrarian Change in Colonial Andhra." P. Schmitthenner,"Colonial Hydraulic Projects in South India: Environmental and Cultural Legacy," in *British Empire and the Natural World: Environmental Encounters in South Asia*, 181.

48   D. Hardiman, "Well Irrigation in Gujarat: Systems of Use, Hierarchies and Control," *Economic and*

Political Weekly 33/25(1998): 1533–44.

49   R. T. Rosin, "The Tradition of Groundwater Irrigation in Northwestern India," *Human Ecology* 21/1(1993): 51–83.

50   D. Mosse, *The Rule of Water: Statecraft, Ecology, and Collective Action in South India* (New Delhi: Oxford University Press, 2003).

51   B. Stein, *Peasant State and Society in Medieval South India* (Delhi: Oxford University Press, 1980). D. Mosse, "Colonial and Contemporary Ideologies of 'Community Management': The Case of Tank Irrigation Development in South India," *Modern Asian Studies* 33/2 (1999): 314.

52   Mosse, "Colonial and Contemporary Ideologies of 'Community Management'." Mosse, *The Rule of Water*.

53   Mosse, "Colonial and Contemporary Ideologies of 'Community Management'," 314.

54   E. Shah, "Telling Otherwise: A Historical Anthropology of Tank Irrigation Technology in South India," *Technology and Culture* 49/3(2008): 652–74. E. Shah, "Seeing Like a Subaltern: Historical Ethnography of Pre-modern and Modern Tank Irrigation Technology in Karnataka, India," *Water Alternatives* 5/2 (2012): 507–38.

55   Shah traces from folklores how the upper class used to appropriate the labor of the lower Vodda caste, and also events of life sacrifice of Dalit women in tanks when there were droughts or natural calamities. Shah, "Seeing Like a Subaltern," 519.

56   Ibid.

57   D. S. Moore, "The Crucible of Cultural Politics: Reworking "Development" in Zimbabwe's Eastern Highlands," *American Ethnologist* 26/3 (1999): 654–89. A. Baviskar, "For a Cultural Politics of Natural Resources," *Economic and Political Weekly* 38/48 (2003): 5051–5.

58   P. Mollinga, "The Material Conditions of a Polarized Discourse: Clamours and Silences in Critical Analysis of Agricultural Water Use in India," *Journal of Agrarian Change* 10/3 (2010): 414–36.

59   Mollinga, "Material Conditions of a Polarized Discourse." Shah, "Seeing Like a Subaltern."

60   Mollinga, "Material Conditions of a Polarized Discourse."

61   V. Shiva, *Water Wars: Privatization, Pollution and Profit* (London: Pluto Press, 2002). L. Mehta, "Whose Scarcity? Whose Property? The Case of Water in Western India," *Land Use Policy* 24 (2007): 654–63.

62   Mehta, "Whose Scarcity? Whose Property?"

63   A. Aiyer, "The Allure of the Transnational: Notes on Some Aspects of the Political Economy of Water in India," *Cultural Anthropology* 22 (2007): 640–58.

64   Ibid.

65   V. Asthana, *Water Policy Processes in India: Discourses of Power and Resistance* (London: Routledge, 2009).

66   Ibid., 7.

67   R. L. Bryant and S. Bailey, *Third World Political Ecology* (London: Routledge, 1997). M. Gadgil and R. Guha, *Ecology and Equity: The Use and Abuse of Nature in Contemporary India* (London: Routledge, 1995). Gadgil and Guha, *This Fissured Land*.

68   A. Baviskar, *Waterscapes* (Delhi: Permanent Black, 2007).

69   Baviskar, "For a Cultural Politics of Natural Resources," 5052.

70   D. S. Moore, J. Kosek, and A. Pandian, *Race, Nature, and the Politics of Difference* (Durham, NC: Duke University Press, 2003), 2.

71   Baviskar, "For a Cultural Politics of Natural Resources," 5052–3.

72   K. Bakker, "Privatizing Water, Producing Scarcity: The Yorkshire Drought of 1995," *Economic Geography* 76/1 (2000): 4–27. Bakker "Paying for Water: Water Pricing and Equity in England and Wales," *Transactions of the Institute of British Geographers* 26/2 (2001): 143–64. E. Castro, E. Swyngedouw and M. Kaika, "Metropolitan Areas and Sustainable Use of Water: The Case of London," unpublished report (Oxford: School of Geography and the Environment, University of Oxford, 2000).

73   E. Swyngedouw, "Power, Nature, and the City: The Conquest of Water and the Political Ecology of Urbanization in Guayaquil, Ecuador," *Environment and Planning* A 29/2 (1997): 311–32. K. Bakker, "Archipelagos and Networks: Urbanization and Water Privatization in the South," 169/4 (2003): 328–41. M. Gandy, "Rethinking Urban Metabolism: Water, Space and the Modern City," *City* 8/3(2004): 363–79. Gandy, "Landscapes of Disaster: Water, Modernity, and Urban Fragmentation in Mumbai," *Environment and Planning* A 40/1(2008): 108–30. A. Loftus and F. Lumsden, "Reworking Hegemony in the Urban Waterscape," *Transactions of the Institute of British Geographers* 33/1(2008): 109–26. N. Anand, "Pressure: The PoliTechnics of Water Supply in Mumbai," *Cultural Anthropology* 26/4 (2011): 542–64. A. A. Ioris, "The Geography of Multiple Scarcities: Urban Development and

Water Problems in Lima, Peru," *Geoforum* 43/3 (2012): 612–22. M. Ranganathan, "Paying for Pipes, Claiming Citizenship: Political Agency and Water Reforms at the Urban Periphery," *International Journal of Urban and Regional Research* 38/2 (2014): 590–608.

74 J. Mukherjee and F. Lafaye de Micheaux, "Intervened River, Transformed Muddyscapes: Exploring Clashing Perceptions across State–Society Interface in the 'Chars' of Lower Gangetic Bengal, India," paper presented at the American Sociological Association conference on *Footprints and Futures: The Time of Anthropology* in the panel on "Muddy Footsteps and Hydrosocial Futures: Understanding Relationality with, through and about Water," Durham, July 4–7, 2016.

75 J. Mukherjee, "Ecological Utopia or Radical (Utopian) Imagining? Rethinking Urban Sustainability across Blue Infrastructures of Kolkata," paper presented at the international conference on Urban Futures and Urban Utopia in South-Asian Megacities: Narrative, Play, Planning, Kolkata, July 25, 2016.

76 A. Baviskar, "What the Eye Does Not See: The Yamuna in the Imagination of Delhi," *Economic and Political Weekly* 46/50 (2011): 45–53. J. Mukherjee, "Sustainable Flows between Kolkata and its Peri-urban Interface" in *Untamed Urbanisms*, ed. A. Allen et al. (London: Routledge, 2015). J. Mukherjee, "Beyond the Urban: Rethinking Urban Ecology using Kolkata as a Case Study," *International Journal of Urban Sustainable Development* 7/2 (2015): 131–46. J. Mukherjee and G. Chakraborty, "Commons vs Commodity: Urban Environmentalisms and the Transforming Tale of the East Kolkata Wetlands," *Urbanities* 6/2 (2016): 78–91.

77 Mukherjee, "Sustainable Flows between Kolkata and its Peri-urban Interface."

78 Mukherjee, "Sustainable Flows between Kolkata and its Peri-urban Interface." Mukherjee and Chakraborty, "Commons vs Commodity."

79 Baviskar, "What the Eye Does Not See." Mukherjee, "Sustainable Flows between Kolkata and its Peri-urban Interface." Mukherjee, "Ecological Utopia or Radical (Utopian) Imagining?"

80 K. Coelho and N.V. Raman, "From the Frying-Pan to the Floodplain: Negotiating Land, Water and Fire in Chennai's Development," in *Ecologies of Urbanism in India: Metropolitan Civility and Sustainability*, ed. A. Rademacher et al. (Hong Kong: Hong Kong University Press, 2013), 145–68.

81 J. Sundaresan, "Planning as Commoning: Transformation of a Bangalore Lake," *Economic and Political Weekly* 46/50 (2011): 71–9.

82 N. Singh, "Contested Urban Waterscape of Udaipur," in *Sustainable Urbanization in India: Challenges and Opportunities*, ed. J. Mukherjee (Singapore: Springer, forthcoming). N. Cornea, A. Zimmer and R. Veron, "Ponds, Power and Institutions: The Everyday Governance of Accessing Urban Waterbodies in a Small Bengali City," *International Journal of Urban and Regional Research* 40/2 (2016): 395–409.

83 J. Linton and J. Budds, "The Hydrosocial Cycle: Defining and Mobilizing a Relational-Dialectical Approach to Water," *Geoforum* 57 (2014): 170.

84 Ibid., 170.

85 Mukherjee and Lafaye de Micheaux, "Intervened River, Transformed Muddyscapes."

86 R. D'Souza, "River as Resource and Land to Own: The Great Hydraulic Transition in Eastern India," paper presented at the conference on Asian Environments Shaping the World: Conceptions of Nature and Environmental Practices, Singapore, March 20–21, 2009.

87 D'Souza, "River as Resource and Land to Own," 3.

88 Ibid.

89 Ibid., 3–4.

90 Ibid., 6.

91 K. Lahiri-Dutt, "Beyond the Water-land Binary in Geography: Water/Lands of Bengal Re-visioning Hybridity," *ACME: An International E-Journal for Critical Geographies* 13/3 (2014): 505.

92 Ibid., 507.

93 Mukherjee and Lafaye de Micheaux, "Intervened River, Transformed Muddyscapes."

94 This is evident in the introduction of the Bengal Alluvion and Diluvion Act (BADA) of 1825; the key to establishing land rights in the court of law was the payment of rent, even on diluviated land. Massive colonial survey operations were also initiated to produce cadastral maps for revenue survey lists (or *khatians*).

95 The provisions in the Draft National Water Framework Bill 2016 is also a reflection towards this direction (see www.prsindia.org/uploads/media/draft/Draft%20National%20Water%20Framework%20Bill%202016.pdf., accessed November 11, 2016).

References

Agnihotri, I. "Ecology, Land Use and Colonisation: The Canal Colonies of Punjab." *The Indian Economic and Social History Review* 33/1 (1996): 37–58.

Agrawal, A., and S. Narain, eds. *Dying Wisdom: Rise, Fall and Potential of India's Traditional Water Harvesting Systems.* New Delhi: Centre for Science and Environment, 1997.

Aiyer, A. "The Allure of the Transnational: Notes on Some Aspects of the Political Economy of Water in India." *Cultural Anthropology* 22 (2007): 640–58.

Anand, N. "Pressure: The PoliTechnics of Water Supply in Mumbai." *Cultural Anthropology* 26/4 (2011): 542–64.

Asthana, V. *Water Policy Processes in India: Discourses of Power and Resistance.* London: Routledge, 2009.

Bakker, K. "Privatizing Water, Producing Scarcity: The Yorkshire Drought of 1995." *Economic Geography* 76/1 (2000): 4–27.

Bakker, K. "Paying for Water: Water Pricing and Equity in England and Wales." *Transactions of the Institute of British Geographers* 26/2(2001): 143–64.

Bakker, K. "Archipelagos and Networks: Urbanization and Water Privatization in the South." *The Geographical Journal* 169/4(2003): 328–41.

Baviskar, A. *In the Belly of the River: Tribal Conflicts over Development in the Narmada Valley.* Delhi: Oxford University Press, 1997.

Baviskar, A. "Ecology and Development in India: A Field and its Future." *Sociological Bulletin* 46/2 (1997): 193–207.

Baviskar, A. "For a Cultural Politics of Natural Resources." *Economic and Political Weekly* 38/48 (2003): 5051–5.

Baviskar, A. *Waterscapes.* Delhi: Permanent Black, 2007.

Baviskar, A. "What the Eye Does Not See: The Yamuna in the Imagination of Delhi." *Economic and Political Weekly* 46/50 (2011): 45–53.

Bryant, R. L., and S. Bailey. *Third World Political Ecology.* London: Routledge, 1997.

Castro, E., Swyngedouw, E. and Kaïka, M. "Metropolitan Areas and Sustainable Use of Water: The Case of London". University of Oxford: School of Geography and the Environment, 2000. Unpublished report.

Chakrabarti, R. *Situating Environmental History.* New Delhi: Manohar, 2007.

Chaudhuri, P. "Peasants and British Rule in Orissa." *Social Scientist* 9/7(1991): 36–42.

Coelho, K. and N. Raman. "From the Frying-Pan to the Floodplain: Negotiating Land, Water and Fire in Chennai's Development." In *Ecologies of Urbanism in India: Metropolitan Civility and Sustainability,* edited by A. Rademacher and K. Sivaramakrishnan, 145–68. Hong Kong: Hong Kong University Press, 2013.

Cornea, N., A. Zimmer, and R. Veron. "Ponds, Power and Institutions: The Everyday Governance of Accessing Urban Waterbodies in a Small Bengali City." *International Journal of Urban and Regional Research* 40/2 (2016): 395–409.

Damodaran, V. *Broken Promises: Popular Protest, Indian Nationalism and the Congress Party in Bihar, 1935–46.* Delhi: Oxford University Press, 1992.

D'Souza, R. "Colonialism, Capitalism and Nature: Debating the Origins of Mahanadi Delta's Hydraulic Crisis (1803–1928)." *Economic and Political Weekly* 37/13(2002): 1261–72.

D'Souza, R. "Damming the Mahanadi River: The Emergence of Multi-Purpose River Valley Development in India (1943-46)." *The Indian Economic and Social History Review.* 40/1 (2003): 81–105.

D'Souza, R. *Drowned and Dammed: Colonial Capitalism and Flood Control in Eastern India.* New Delhi: Oxford University Press, 2006.

D'Souza, R. "Water in British India: The Making of a 'Colonial Hydrology'." *History Compass* 4/4 (2006a): 621–8.

D'Souza, R. "River as Resource and Land to Own: The Great Hydraulic Transition in Eastern India." Paper presented at the conference on Asian Environments Shaping the World: Conceptions of Nature and Environmental Practices, Singapore, March 20–21, 2009.

Foster, J. B. "Marx's Theory of Metabolic Rift: Classical Foundations for Environmental Sociology 1." *American Journal of Sociology* 105/2 (1999): 366–405.

Gadgil, M., and R. Guha. *This Fissured Land: An Ecological History of India.* Delhi: Oxford University Press, 1992.

Gadgil, M., and R. Guha. *Ecology and Equity: The Use and Abuse of Nature in Contemporary India.* London: Routledge, 1995.

Gandy, M. "Rethinking Urban Metabolism: Water, Space and the Modern City." *City* 8/3 (2004): 363–79.

Gandy, M. "Landscapes of Disaster: Water, Modernity, and Urban Fragmentation in Mumbai." *Environment and Planning* A 40/1 (2008): 108–30.

Gilmartin, D. "Scientific Empire and Imperial Science: Colonialism and Irrigation Technology in the Indus Basin." *The Journal of Asian Studies* 53/4 (1994): 1127–48.

Greenough, P. (2001). "Naturae Ferae: Wild Animals in South Asia and the Standard Environmental Narrative." In *Agrarian Societies: Synthetic Work at the Cutting Edge,* edited by J. Schott and N. Bhatt, 141–85. New Haven, CT: Yale University Press.

Guha, R. *The Unquiet Woods: Ecological Change and Peasant Resistance in the Himalayas.* Delhi: Oxford University Press, 1991.

Guha, R., and J. Martinez-Alier. *Varieties of Environmentalism: Essays North and South.* London: Earthscan, 1997.

Gupta, S. K. and R. D. Deshpande, "Water for India in 2050: First-Order Assessment of Available Options." *Current Science* 86 (2004): 1216.

Hardiman, D. "Well Irrigation in Gujarat: Systems of Use, Hierarchies and Control." *Economic and Political Weekly* 33/25(1998): 1533–44.

Harrison, H. L. *The Bengal Embankment Manual: Containing an Account of the Action of the Government in Dealing with the Embankments and Water-Courses since the Permanent Settlement: A Discussion of the Principles of the Act of 1873.* Calcutta: Bengal Secretariat Press, 1875.

Ioris, A. A. "The Geography of Multiple Scarcities: Urban Development and Water Problems in Lima, Peru." *Geoforum* 43/3(2012): 612–22.

Klingensmith, D. *One Valley and a Thousand: Dams, Nationalism, and Development.* New York: Oxford University Press, 2007.

Lahiri-Dutt, K., and G. Samanta. *Dancing with the River: People and Life on the Chars of South Asia.* New Haven, CT: Yale University Press, 2013.

Lahiri-Dutt, K. "Beyond the Water-Land Binary in Geography: Water/Lands of Bengal Re-visioning Hybridity." *ACME: An International E-Journal for Critical Geographies* 13/3(2014): 505–29.

Linton, J., and J. Budds. "The Hydrosocial Cycle: Defining and Mobilizing a Relational-Dialectical Approach to Water." *Geoforum* 57 (2014): 170–80.

Loftus, A., and F. Lumsden. "Reworking Hegemony in the Urban Waterscape." *Transactions of the Institute of British Geographers* 33/1(2008): 109–26.

McCully, P. *Silenced Rivers: The Ecology and Politics of Large Dams.* London: Zed Books, 1996.

McNeill, J. R. "The State of the Field of Environmental History." *Annual Review of Environment and Resources* 35(2010): 345–74.

Mehta, L. "Whose Scarcity? Whose Property? The Case of Water in Western India." *Land Use Policy* 24 (2007): 654–63.

Mishra, D. K. "The Bihar Flood Story." *Economic and Political Weekly,* 32/1(1997): 2206–17.

Mishra, D. K. *Trapped Between the Devil and the Deep Waters—Story of Bihar's Kosi River.* Delhi: People's Science Institute and SANDRP, 2008.

Mollinga, P. P. "The Material Conditions of a Polarized Discourse: Clamours and Silences in Critical Analysis of Agricultural Water Use in India." *Journal of Agrarian Change* 10/3 (2010): 414–36.

Moore, D. S. "The Crucible of Cultural Politics: Reworking 'development'" in Zimbabwe's Eastern Highlands." *American Ethnologist* 26/3(1999): 654–89.

Moore, D.S., J. Kosek, and A. Pandian. *Race, Nature, and the Politics of Difference.* Durham, NC: Duke University Press, 2003.

Morris, H. *A Description and Historical Account of Godavery District in the Presidency of Madras.* London: Truber and Company, 1878.

Mosse, D. "The Symbolic Making of a Common Property Resource: History, Ecology and Locality in a Tank Irrigated Landscape in South India." *Development and Change* 28/3 (1997): 467–504.

Mosse, D. "Colonial and Contemporary Ideologies of 'Community Management': The Case of Tank Irrigation Development in South India." *Modern Asian Studies* 33/2 (1999): 303–38.

Mosse, D. *The Rule of Water: Statecraft, Ecology, and Collective Action in South India.* New Delhi: Oxford University Press, 2003.

Mukherjee, J. "Sustainable Flows between Kolkata and its Peri-urban Interface." In *Untamed Urbanisms,* edited by A. Allen, A. Lampis and M. Swilling. New York: Routledge, 2015.

Mukherjee, J. "Beyond the Urban: Rethinking Urban Ecology using Kolkata as a Case Study." *International Journal of Urban Sustainable Development* 7/2 (2015a): 131–46.

Mukherjee, J. "Ecological Utopia or Radical (Utopian) Imagining? Rethinking Urban Sustainability across Blue Infrastructures of Kolkata." Paper presented at the international conference on Urban Futures and Urban Utopia in South-Asian Megacities: Narrative, Play, Planning, Kolkata, July 25, 2016.

Mukherjee, J., and F. Lafaye de Micheaux. "Intervened River, Transformed Muddyscapes: Exploring Clashing Perceptions across State–Society Interface in the 'Chars' of Lower Gangetic Bengal, India." Paper presented at the American Sociological Association conference on Footprints and Futures: The Time of Anthropology in the panel on "Muddy Footsteps and Hydrosocial Futures: Understanding Relationality with, through and about Water," Durham, July 4–7, 2016.

Mukherjee, J., and G. Chakraborty. "Commons vs Commodity: Urban Environmentalisms and the Transforming Tale of the East Kolkata Wetlands." *Urbanities* 6/2 (2016): 78–91.

Rademacher, A., and K. Sivaramakrishnan, eds. *Ecologies of Urbanism in India: Metropolitan Civility and Sustainability.* Hong Kong: Hong Kong University Press, 2013.

Ranganathan, M., Paying for Pipes, Claiming Citizenship: Political Agency and Water Reforms at the Urban Periphery. *International Journal of Urban and Regional Research* 38/2 (2014): 590–608.

Rangarajan, M. *Fencing the Forests: Conservation and Ecological Change in India's Central Provinces, 1860–1914.* Oxford: Oxford University Press, 1996.

Rao, G. N. "Transition from Subsistence to Commercialised Agriculture: A Study of Krishna District in Andhra, 1850–90." *Economic and Political Weekly* 20/25–6 (1985): 60–69.

Rao, G. N. "Canal Irrigation and Agrarian Change in Colonial Andhra: A Study of Godavari District, c. 1850-90." *The Indian Economic and Social History Review* 25/1(1988): 25–60.

Rao, E. "Taming 'Liquid Gold' and Dam Technology: A Study of the Godavari Anicut." In *The British Empire and the Natural World: Environmental Encounters in South Asia,* edited by D. Kumar, V. Damodaran and R. D'Souza. New York: Oxford University Press, 2011.

Reddy, M. A. "Travails of an Irrigation Canal Company in South India, 1857–1882." *Economic and Political Weekly* 25/12 (1990): 619–28.

Rosin, R. T. "The Tradition of Groundwater Irrigation in Northwestern India." *Human Ecology* 21/1 (1993): 51–83.

Saberwal, V. *Pastoral Politics: Shepherds, Bureaucrats, and Conservation in the Western Himalaya.* Delhi: Oxford University Press, 1999.

Sainath, P. *Everybody Loves a Good Drought: Stories from India's Poorest Districts.* New Delhi: Penguin Books, 1996.

Sangvai, S. *The River and Life.* Mumbai: Earthcare Books, 2000.

Schmitthenner, P. "Colonial Hydraulic Projects in South India: Environmental and Cultural Legacy." In *The British Empire and the Natural World: Environmental Encounters in South Asia,* edited by D. Kumar, V. Damodaran and R. D'Souza. Oxford: Oxford University Press, 2011.

Sengupta, N. "The Indigenous Irrigation Organisation in South Bihar." *The Indian Economic and Social History Review* 37/2 (1980): 157–87.

Shah, E. "Telling Otherwise: A Historical Anthropology of Tank Irrigation Technology in South India." *Technology and Culture* 49/3 (2008): 652–74.

Shah, E. "Seeing Like a Subaltern: Historical Ethnography of Pre-modern and Modern Tank Irrigation Technology in Karnataka, India." *Water Alternatives* 5/2 (2012): 507–38.

Shiva, V. *Water Wars: Privatization, Pollution and Profit.* London: Pluto Press, 2002.

Singh, P. "The Colonial State, Zamindars and the Politics of Flood Control in North Bihar (1850–1945)." *The Indian Economic and Social History Review* 45/2 (2008): 239–59.

Singh, P. "Flood Control for North Bihar: An Environmental History from the 'Ground-Level' (1850–1954)." In *The British Empire and the Natural World: Environmental Encounters in South Asia,* edited by D. Kumar, V. Damodaran and R. D'Souza. Oxford: Oxford University Press, 2011.

Singh, N. "Contested Urban Waterscape of Udaipur." In *Sustainable Urbanization in India: Challenges and Opportunities,* edited by J. Mukherjee. Singapore: Springer, forthcoming.

Stein, B. *Peasant State and Society in Medieval South India.* Delhi: Oxford University Press, 1980.

Stone, I. *Canal Irrigation in British India: Perspectives on Technological Change in a Peasant Economy.* Cambridge: Cambridge University Press, 1885.

Sundaresan, J. "Planning as Commoning: Transformation of a Bangalore Lake." *Economic and Political Weekly* 46/50 (2011): 71–9.

Swyngedouw, E. "Power, Nature, and the City: The Conquest of Water and the Political Ecology of Urbanization in Guayaquil, Ecuador." *Environment and Planning* A 29/2 (1997): 311–32.

Weil, B. "The Rivers Come: Colonial Flood Control and Knowledge Systems in the Indus Basin, 1840s–1930s." *Environment and History* 12/1 (2006): 3–29.

Whitcombe, E. *Agrarian Conditions in Northern India: The United Provinces under the British Rule, 1860–1900*, vol. 1. Berkeley, CA: California University Press, 1972.

Willcocks, W. *Lectures on the Ancient System of Irrigation in Bengal.* Calcutta: University of Calcutta, 1930.

17

SUSTAINABLE ARCHITECTURE

A short history

Vandana Baweja

Climate change and its metrics—energy consumption and the carbon cycle—have come to dominate contemporary discourses on sustainable architecture and design. Competing and overlapping design paradigms and environmental assessment methods such as—Cradle to Cradle, Bioclimatic Architecture, Biomimicry, Passive and Low Energy Architecture (PLEA), Ecological Design, Net Zero buildings, and Zero-carbon building, Leadership in Energy and Environmental Design (LEED), Building Research Establishment Environmental Assessment Method (BREEAM), and Passivhaus—promise sustainability. These design paradigms are targeted towards sustainable development through a reduction in greenhouse gas emissions and accomplishing efficiencies in the use of energy and materials. The larger goal is to attain an ecological balance between consuming the earth's finite resources and its regenerative capacity. Sustainable development was first defined in the Brundtland Report, titled *Our Common Future*, as development that "meets the needs of the present without compromising the ability of future generations to meet their own needs."[1] Since the 1990s, as sustainable development emerged as the new paradigm of economic growth based on the carrying capacity of the earth, the term "sustainability" entered the academic discourse and has had an enduring impact on several disciplines in academia. Although, urbanism was amongst the core sectors that sought sustainable development, but the Brundtland Report did not address extensively the role of architecture and urbanism in building a sustainable society. The report stressed minimizing pollution, achieving clean air, conserving water, reliance on renewable energy, and poverty reduction in cities of the Global South as the overarching goals for accomplishing sustainable development.[2] Although the Brundtland Report and the blossoming of the sustainability movement helped to bring awareness to many sectors of society, including architecture and design, the concern for environmental building dates back to the 1960s and 1970s. The growth of the sustainability movement, combined with the realization that humans were affecting the climate through the use of fossil fuels (including those used in the built environment) further pushed the architectural world toward sustainable design. Thus, since the late 1980s and early 1990s, sustainable architecture has become an articulated value, and is now regularly associated with the carbon cycle, global ecology, and various facets of sustainability (urbanism, transportation, consumption, and so on). To assess the paradigmatic genealogy of sustainable architecture, let us examine how anthropogenic climate change became a global architectural concern and how architects have responded to shifting environmental concerns.

Architecture and response to environmental problems

The United Nations Environment Programme (UNEP), the environmental agency of the United Nations, was formed out of concerns raised at the United Nations Conference on the Human Environment (also known as the Stockholm Conference) in June of 1972, which focused on the rising population and its pressure on agriculture and human settlements in the developing world, improving the agricultural sector, finding suitable technologies of modernization for the developing world, regulating chemical and radioactive pollution, and the rationalization and integration of resource management for energy.[3] Although the Stockholm Conference was meant to create a global forum to address environmental concerns, there was a clear split between the environmental concerns of the developed and the developing world. At the global level, environmental concerns in the 1960s and 1970s were informed and shaped by events and phenomena such as the decolonization and modernization of the tropics, the Cold War, the threat of nuclear holocaust, the Vietnam War, space exploration, the counter-cultural movement of the 1960s, the civil rights movement, the feminist movement, the OPEC oil embargo 1973–4, rising population, and poverty. The environmental components of international crises—population induced pressures on earth's resources, inadequate agricultural production, inadequate energy production in the underdeveloped world, pollution, radioactive waste, and diminishing and deteriorating natural resources such as water, forests, land, and ecosystems—became, for the first time in the 1970s, an abiding concern of the international community.

These concerns were articulated through a series of publications, such as Rachel Carson's *Silent Spring* (1962), Garrett Hardin's journal paper in *Science*, "The Tragedy of the Commons" (1968), Paul R. Ehrlich's *The Population Bomb* (1971), the Club of Rome's *The Limits to Growth: A Report for the Club of Rome's Project on the Predicament of Mankind* (1972), and E. F. Schumacher's *Small is Beautiful: Economics as if People Mattered* (1973).[4] Carson's *Silent Spring* presented an ecological critique of chemical pesticides, exposing their toxic hazards to a mass audience, and therefore, transformed public consciousness about the human impact on the environment. Garrett Hardin's "The Tragedy of the Commons" blamed the deterioration and pollution of the commons on population growth. The Club of Rome's *The Limits to Growth* report projected that given the current rate growth of world population, industrialization, pollution, food production, and resource depletion were not sustainable. E. F. Schumacher's *Small is Beautiful: A Study of Economics as If People Mattered* was a manifesto on appropriate technology that expressed extreme skepticism of conventional economics, large-scale modernization projects, and the global dissemination of development based on Western industrialization. 'Small is Beautiful' builds on Gandhi's critique of industrialization and modernization to argue for simpler technologies that are less capital intensive and gentler on the environment, but which employ large populations in the developing world, and gentle on the environment.

In response to these environmental problems—pollution, energy scarcity, social injustice, poverty, agricultural deficit, ecological catastrophe—that dominated the public consciousness in the 1960s and 1970s, architects responded with a range of paradigms that depended on their specific cultural, ideological, and technological context. These paradigms of the 1960s and 1970s centered on:

- architecture that followed the material and climatic logic of vernacular architecture as a means to provide social justice;
- bio-climatic architecture;

- tropical architecture;
- technocentrist high-technology solutions to maximize structural and material efficiency;
- back-to-nature and off-the-grid architectural solutions proposed through countercultural architecture;
- ecological design;
- solar architecture; and
- architecture-based appropriate technology.

These design solutions identified a key problem of architecture's reliance on energy grids that were fossil-fuel-dependent, and therefore contingent on a high-risk, centralized geopolitical supply chain, which was non-renewable and could be disrupted in the case of war or an ecological collapse. In the case of the developing world, there simply was not enough supply of energy to meet the demands of large populations, and therefore material and energy efficiencies were also geared toward distributive justice. In the developed world, autonomy and sustaining life in the case of an ecological catastrophe was one of the key environmental themes in these works. In all of these proposed solutions, the relationship between architecture and sustainability, including climate concerns, relied on architecture's capacity to mediate the external climate through optimization of the building envelope. Before architecture was informed by climate change, architects viewed climate as a generator of design attributes and metrics that included form, building envelope, orientation, energy consumption, and active and passive conditioning systems. That is, climate has always played a role in building science and philosophy, but climate *change* and resource concerns forced many architects to rethink the ways in which structures could reduce environmental impacts and use resources and technologies in new ways.

The precursors to sustainable architecture since the 1960s, with key actors and publications, can be summarized as follows.

1. Architecture that followed the material and climatic logic of vernacular architecture and primarily aimed at the poor

Key actors:
- Hassan Fathy (1900–89), Egyptian architect
- Laurie Baker (1917–2007), British–Indian architect

Key publication:
- Hassan Fathy, *Architecture for the Poor: An Experiment in Rural Egypt* (Chicago, IL: University of Chicago Press, 1973)

2. Bio-climatic architecture: architecture that was based on regionalism as a critique of the modernist universal architecture

Key actors:
- Victor Olgyay, Hungarian–American
- Aladar Olgyay, Hungarian–American

Key publication:
- Victor Olgyay and Aladar Olgyay, *Design with Climate: Bioclimatic Approach to Architectural Regionalism* (Princeton, NJ: Princeton University Press, 1963)

3. Tropical architecture

Key actors:
- Fello Atkinson (1919–82), British architect
- Jane Drew (1911–96) and Maxwell Fry (1899–1987), British wife–husband architectural team
- Otto Koenigsberger (1908–99), German–Indian

Key publications:
- Maxwell Fry and Jane Drew, *Tropical Architecture: In the Dry and Humid Zone* (London: B. T. Batsford, 1956)
- Maxwell Fry and Jane Drew, *Tropical Architecture in the Humid Zone* (London: B. T. Batsford, 1964)
- Otto H. Koenigsberger, T. G. Ingersoll, Alan Mayhew, and S. V. Szokolay, *Manual of Tropical Housing and Building* (London: Longman, 1974)

4. Technocentrist high-technology architecture that sought to maximize structural and material efficiency

Key actors:
- Buckminster Fuller (1895–1983)
- Frei Otto (1925–2015)

Key publications:
- Richard Buckminster Fuller, *Operating Manual for Spaceship Earth* (New York: Pocket books, 1970)
- Frei Otto, Rudolf Trostel, and Friedrich Karl Schleyer, *Tensile Structures; Design, Structure, and Calculation of Buildings of Cables, Nets, and Membranes* (Cambridge, MA: MIT Press, 1974)

5. Countercultural architecture

Key actors:
- Steve Baer (1938–), architect
- Stewart Brand (1938–), American counterculture author
- Hippie inhabitants of Drop City, Colorado

Key publications:
- Steve Baer, *Zome Primer: Elements of Zonohedra Geometry: Two and Three Dimensional Growths of Stars with Five-Fold Symmetry* (Albuquerque, NM: Zomeworks, 1973)
- *Whole Earth Catalog*
- David Kruschke, *Dome Cookbook of Geodesic Geometry* (Wild Rose, WI: Kruschke, 1975)

6. Ecological design

Key actors:
- Ian L. McHarg (1920–2001)
- John Todd (1939–), Nancy Todd, and the New Alchemists

- Alexander Pike and John Frazer
- Brenda and Robert Vale
- Ken Yeang
- Phil Hawes
- Sim Van der Ryn

Key publications:
- Ian L. McHarg, *Design with Nature* (Garden City, NY: Doubleday, 1969)
- Helga Olkowski, William Olkowski, Tom Javits, Farallones Institute, *The Integral Urban House: Self-Reliant Living in the City* (San Francisco, CA: Sierra Club Books, 1979)
- Nancy Jack Todd, *The Book of the New Alchemists* (New York: Dutton, 1977)
- Brenda Vale and Robert Vale, *The Autonomous House: Design and Planning for Self-Sufficiency* (London, Thames & Hudson, 1975)

7. Solar architecture

Key actors:
- Jeffery Cook
- Peter van Dresser
- Steve Baer
- Keith Haggard
- George Löf

Key publications:
- American Institute of Architects et al., *Proceedings of the ASC/AIA Forum 75, Solar Architecture, November 26–29, 1975, Arizona State University, Tempe, Arizona* (Washington, DC: Govt. Printing Office, 1976)
- Norma Skurka and Jon Naar, *Design for a Limited Planet: Living with Natural Energy* (New York: McGraw-Hill, 1978)
- AIA Research Corporation, *Solar Dwelling Design Concepts* (Washington: US Dept. of Housing and Urban Development, 1978)

8. Appropriate technology

Key actors:
- E. F. Schumacher

Key publication:
- E. F. Schumacher, *Small Is Beautiful: A Study of Economics as If People Really Mattered* (London: Blond & Briggs, 1973)

Architecture that followed the material and climatic logic of vernacular architecture and primarily aimed at the poor

Architects such as the Egyptian Hassan Fathy (1900–89) and the British–Indian Laurie Baker (1917–) embraced the material and climatic logic of vernacular architecture to design buildings for poor people with limited capital to build and even less to pay for operating costs of the buildings. The early work of Fathy was a clear departure from the universalist tendencies of modernism and the modernization paradigm. His most well-known project is New Gourna (1945–49)—a settlement that Fathy designed to resettle the residents of a village in the archae-ological zone of Gourna al-Jadida on the west bank of the Nile River. The residents of Gourna al-Jadida served as the labor pool on the archeological site and in the process sold archeologi-cal objects that belonged to the state.[5] This development created tensions between the state and the Gournies over the theft of historic relics. The village was designed to rehabilitate and socialize them into good citizens. Fathy built the village using sun-dried mud bricks and synthesized a range of indigenous precedents in structural, spatial, and organizational realms, such as Nubian vaults for construction, and the mud brick screens that were based on the *mashrabiyya* wooden screens in the Cairo mansions.[6] Fathy published his New Gourna project in a University of Chicago Press publication titled *Architecture for the Poor* in the 1970s, which was acclaimed as one of the most robust critiques of modernist architecture and the universal-ity of the modernization paradigm that informed architectural modernism.[7] This book was published at the same time as E. F. Schumacher's 1973 publication *Small is Beautiful: Economics as if People Mattered*, which also questioned the universality of modernization. Although the aim of the work is more "social" than "environmental," it had a strong influence on architects work-ing with low-impact materials and developments.

Bio-climatic architecture

Victor and Aladar Olgyay, who were twin brothers and Hungarian immigrants, pioneered the idea of Bio-climatic architecture. Their key text *Design with Climate: Bioclimatic Approach to Architectural Regionalism* was published in 1963 as a critique of universal modernism. As a basis for Bio-climatic architecture, the Olgyays categorized climatic zones as defined by the Köppen classification system: tropical-rainy, dry, warm-temperate, cool-snow-forest, and Polar. These zones were predicated on the relationship between regional vegetation to the climate. The Olgyays simplified and reduced the Köppen classification so that the climatic zones could be better related to architectural types.[8] These four zone were: hot and humid, hot and arid, cool, and temperate. Based on these climatic zones, the Olgyays corroborated Jean Dollfus's global climatic architectural typologies to conclude that vernacular architecture has a causal relation-ship with the climate and the environment. The Olgyays used examples of Native American architecture to argue that ethnically similar groups developed diverse housing forms when they encountered different climatic regions. They associated their climatic classifications of cool, temperate, hot and arid, and hot and humid climate to Native American shelter types, includ-ing the Igloo (cool), wigwam (temperate), adobe structures (hot and arid), and steep pitched roof structures with no walls (hot and humid) to reinforce the thesis of a causal relationship between architectural form and the climate.[9] On this basis, the Olgyay brothers concluded that in designing modern houses, regional thermal stresses should be the primary determinant of form. They developed a four-step design method based on sequential variables: climate, biol-ogy, technology, and architecture. The process involved collecting climatic data, assessing the impact of climate on humans in the given context, determining the best technology to achieve comfort, and finally coalescing the climatic requirements with architectural technology to provide an optimal solution. Bio-climatic architecture questions the universality of both modernism and modernization. Its audience is not the developing world architect, but an archi-tect in the developed world. Although it came about prior to the awareness of anthropogenic climate change, it forced architects in subsequent decades to take climate conditions more seriously.

Tropical architecture

As a region, the tropics are defined as zones between the Tropic of Cancer in the northern hemisphere and the Tropic of Capricorn in the southern hemisphere. The tropics, also known as the Torrid Zone, include the adjacent areas on either side of the equator, encompassing South East Asia, most of South Asia, North Australia, Central America, regions of South America, parts of the Middle East, and a large part of Africa. The tropical regions were also largely colonized by British, French, Portuguese, and Dutch imperial powers. In the nineteenth and early twentieth centuries, the design of tropical buildings developed in hygiene manuals, which were circulated across the British Empire. As the discipline of sanitary engineering emerged from hygiene, the design of tropical buildings and urbanism became a field of scientific enquiry for sanitary engineers. According to the nineteenth-century sanitation manuals, tropical spatial practices comprised a series of prescriptive measures to ensure good ventilation, lighting, effective sewage disposal, clean-water supply, and sanitation. These practices were designed to counter tropical diseases that were believed to be the consequence of the hot and humid climate. From the 1930s onward, as European modernist architects looked for opportunities in the colonial tropics, their work emerged as a syncretic practice that was informed by modernist architectural tenets and colonial discourses from the fields of climatology, sanitation, and architecture. A new Tropical Architecture produced by modernist architects in the tropics began to emerge in the 1950s and was disseminated through conferences, books, journal articles, and manuals. Consequently, the disciplinary home of Tropical Architecture was established in its natural home with the inauguration of the department of Tropical Architecture at the Architectural Association School of Architecture in London in 1954. Tropical Architecture as an architectural professional practice in the colonies developed somewhat belatedly as the British Empire declined after the Second World War. While sanitation and hygiene were focused on disease prevention, the architects of the 1950s stressed comfort, which was defined comprehensively in terms of thermal, hygrometric, ergonomic, acoustic, and psychological wellbeing. Tropical Architecture emerged as a practice that was designed to create physiological comfort with minimum or no mechanical conditioning, to ensure minimal reliance on fossil fuels. By the 1960s Tropical Architecture developed into an environmental design field that relied on quantifying climate-responsive design practices into an empirical domain, relying on a causal relationship between architectural form and climate.

Technocentrist approach: high-technology solutions that sought to maximize structural and material efficiency

Frei Otto (1925–2015) and Buckminster Fuller (1895–1983) were technocentrists, who believed in technology as a powerful tool to optimize material use. Both Fuller and Otto developed unconventional structures that were inspired by geometry in nature to streamline material use. Frei Otto is famous for lightweight pneumatic structures, cable nets, tents, and shell structures, especially the German Pavilion in Montreal (1967) and the roofs for the Olympic Games in Munich (1972). He founded the Institute for Lightweight Structures in Stuttgart in 1964, which aimed at research on creating lightweight structures to achieve material and structural efficiencies. The central question addressed in his research was the following: "Which shape should an object take in order to be lightweight?"[10] Frei Otto collaborated with the German biologist and anthropologist Johann-Gerhard Helmcke (1908–93). This collaboration led to an investigation for animate and inanimate forms such as hair, bones, spider webs, and seashells to learn structural principles in nature.[11] By researching the architecture that is not built through the agency of human beings, Frei Otto turned to the agency of non-human actors in creating optimal structures that founded on principles of material efficiency. Otto authored a book in 1962 titled *Zugbeanspruchte Konstruktionen* (Tensile Structures) in which he suggested how pneumatic structures followed the structural logic found in nature—actors in the plant and animal world such as fruit, air bubbles, blood vessels, and skin.[12]

Buckminster Fuller (1895–1983) was a Cold War designer, who became globally famous for his patent of the geodesic dome as structural system in 1954. The American pavilion at the Expo '67—the world's fair held in Montreal, Canada in the summer of 1967—was a three-quarter geodesic structure. Buckminster Fuller, with Shoji Sadao and Geometrics Inc., designed the American Expo '67 pavilion, making the geodesic dome a robust symbol of Cold-War American technological advances, free market capitalism, and consumerism.[13] The term geodesic refers to great circles of a sphere and segments of such circles. The geodesic dome is an icosahedron—polyhedron of twenty faces—inscribed on the surface of a sphere to form a geodesic grid. Diamonds or hexagons or equilateral triangles comprise the geodesic grid, which is formed when the geodesic lines intersect. The geodesic dome achieves extremely high material efficiencies in a dome structure that can cover large spaces without columns.[14] Fuller's geodesic dome patent was preceded by a series of technological innovations in the 1930s such as the Dymaxion Dwelling Machine, Dymaxion Car, Dymaxion Trailer, Dymaxion A-Frame Carrier, and Dymaxion Bathroom, which replicated nature's processes through design to set examples of how technology can solve ecological problems through design.[15] Countless geodesic domes have been built since the 1970s, and for many years were the most iconic symbol of environmental design. The intellectual rationale of material, structural, and energy efficiencies, which informs Fuller's designs, is theorized in his manifesto—*Operating Manual for Spaceship Earth* (1970)—which takes a managerial approach to the environment. Through this manifesto, Fuller advanced the notion that spaceship earth had limited resources and carrying capacity. The increasing population pressure on earth's finite resources could only be balanced through an unrelenting emphasis on material and energy efficiency, achievable through better technology.[16] Fuller's ideas expressed in the *Operating Manual for Spaceship Earth* inform several contemporary design paradigms such as cradle-to-cradle, industrial ecology, eco-managerialism, and eco-city planning.[17] Both Otto and Fuller's legacy of high-end technology as a mechanism to solve environmental problems is carried forward in works of present-day technocentrist architects such as Norman Foster, Richard Rogers, and Renzo Piano.

Countercultural architecture

Although the sociologist J. Yinger Milton first used the term "counterculture" in the early 1960s, the ideological leanings and political ramifications of the term counterculture were subsequently elucidated in 1969 in Theodore Roszak's book *The Making of a Counterculture: Reflections on the Technocratic Society and its Youthful Opposition*.[18] Roszak explained how the 1960s student activists and hippie dropouts dissented against conventional society through their rejection of postwar regimes of consumerism, capitalism, corporate power, industrial production, technological expertise, and endless growth, all deeply entrenched in war and social injustice.[19] In 1960s United States, the conviction in technology as a catalyst of progress became highly contested due to increasing belief in industrialization as a destructive force, as suggested by Rachel Carson's *Silent Spring*, loss of faith in the governmental establishment due to political malfeasance, as exemplified by the Vietnam War, and a general disillusionment with materialism and consumerism.[20] The rejection of mainstream sexuality, nuclear familial structures, patriarchal gender roles, social hierarchies, established career trajectories, capitalistic commerce, and social norms translated into a disengagement with prevailing architectural and urban practices as well. Countercultural anarchists viewed the postwar suburban boom of tract houses with chemically saturated lawns as products of the property, banking, industrial, and technological regimes that they sought to rebel against.[21] Countercultural communities rejected the material, constructional, architectural, and urban practices of conventional life to seek anti-urban, back-to-nature, and off-the-grid living. They built rudimentary structures out of trash and discarded materials with their bare hands, using simple tools to establish a corporeal relationship with materials, the process of building, and the environment, often facilitated by hallucinogenic drugs.[22] Drop City—a countercultural commune established near Trinidad, Colorado, in 1965, and inhabited until 1973—was the first commune in the United States where countercultural values were translated into everyday lived experience, especially in the realm of architecture and appropriate technologies.[23] Drop City's inhabitants, known as droppers, experimented with self-built architecture—especially the geodesic dome. In the process of building domes at Drop City, the hippies transformed Buckminster Fuller's high-tech, industrially-manufactured geodesic dome into a form of countercultural vernacular architecture, which was hand-made without architectural drawings, using discarded materials, shared communal labor, narcotic creativity, and sheer ingenuity.[24] The Drop City dwellers also used rear-view mirrors from abandoned cars to make a solar heat collector. Although Drop City was hooked to the water and electricity grids, the use of passive solar energy through DIY solar collectors was symbolic of liberation from infrastructural grids associated with conventional society. Steve Baer, the countercultural architect, visited Drop City in April 1966, which led to the publication of his counterculture treatise on domes—the *Dome Cookbook*.[25] Droppers used conventional society's trash—bottle tops, scavenged automobile glass, rummaged metal panels from junk cars, chicken wire, discarded lumber poles, and tar to make domes and its geometric polyhedral avatar the *Zome*.[26] In the process, the droppers transformed the ideological meaning of the geodesic dome from a symbol of cold-war American military-industrial complex—the *very* ideology that the droppers were protesting against—into a DIY grassroots countercultural icon.[27] This ideological and technological morphing of the geodesic dome was a paradox that permeated countercultural environmentalism through co-existing modern and anti-modern tendencies, which were best illustrated in the *Whole Earth Catalogue*—the countercultural catalogue published by Stewart Brand—where rudimentary back-to-nature technologies co-existed with DIY recipes for geodesics domes.[28] Countercultural environmentalists would form a symbiotic intellectual relationship with advocates of ecological design in their pursuit of liberation from infrastructural grids, especially the energy grid, which was dependent on an politically contingent global supply chain of fossil fuels, as the OPEC oil embargo clearly demonstrated.

Ecological design

The advocates of ecological design—John Todd, Nancy Todd, and the New Alchemists; Alexander Pike and John Frazer at the University of Cambridge, UK and their student Brenda Vale and her partner Robert Vale; Grahame Caine, at the Architectural Association in London; and Sim Van der Ryn at the University of California Berkeley all experimented with autonomous or semi-autonomous shelters that were designed to survive without conventional food supply chains and infrastructure grids of energy, water, and waste disposal. The autonomous house projects, based on ecological design, represented a paradigmatic shift in the relationship between architecture and the environment. Prior to the autonomous house concept, the emphasis of Solar, Bio-climatic, and Tropical Architecture paradigms was to save energy. The autonomous projects integrated numerous infrastructural systems and supply chains in the house to operationalize the architecture in nearly closed loops of material and energy inputs and outputs.[29] The intellectual genealogy of Autonomous architecture can be traced to the space research, which led to creation of artificial ecosystems in space that were created through quantifying energy and resource flows as input and output closed loops.[30] The impact of space research and thinking of the earth as one single spaceship with a finite carrying capacity motivated ecological designers to the self-contained and self-sustaining architecture. Not only architects, but professionals and educators working on the urban scale found the idea of earth as a one closed loop ecosystem compelling. Ian L. McHarg (1920–2001) was a professor at the Department of Landscape Design at the University of Pennsylvania and authored a book titled *Design with Nature* (1969). This book advocates for treating the earth as a finite closed loop and suggests that planners adopt an ecological approach to urban growth.[31]

The key autonomous and semi-autonomous structures were:

- The Autonomous House (1971 model only), Cambridge University, UK; designed and model built by Alexander Pike, assisted by James Thring, Gerry Smith, John Littler and students Christina Freeman and Randall Thomas.
- Eco-House (1972–5), London, UK; designed and built by Grahame Caine, a diploma student at the Architectural Association School of Architecture in London.
- Integral Urban House (1972), Berkeley, California; designed and built by Helga Olkowski, Bill Olkowski, and Tom Javits—founding members of the Farallones Institute at the University of California, Berkeley.
- Bioshelters—Cape Cod Ark (1976), architects: Solsearch; Prince Edward Island Ark (1976), architects: BGHJ; for the New Alchemists— Jack Todd, Nancy Todd, and Bill McLarney.
- Ouroboros House (1973–6), Rosemont, Minnesota; designed and built by Professor Dennis Holloway and his students at the University of Minnesota's School of Architecture and Landscape Architecture.

Alexander Pike led "The Autonomous House Research Programme" in the Department of Architecture, in the Technical Research Division at the University of Cambridge. Since the inception of the Autonomous House project, several technocrats lay claim to "autonomous structures" by actually building semi-autonomous structures, which were designed to improve material and energy consumption through achieving efficiencies over conventional structures. These semi-autonomous structures utilized ambient energy and partial recycling to lighten the load on existing infrastructure grids of water and energy supply and waste disposal. However, these structures could not function off the grids. The Cambridge Autonomous House was designed to function without conventional infrastructural grids. The project was conceived as a totally autonomous project as an experiment in decentralized infrastructure delivery with a fully integrated servicing system.[32]

Grahame Caine, a diploma student at the Architectural Association School of Architecture in London, built an autonomous house called the Eco-House in 1972 as a diploma thesis project. The Environmental Council of London sponsored the Eco-House as an ecological experiment on land borrowed from the Thames Polytechnic with a seed fund of two thousand pounds from Alvin Boyarsky, the chairman of the Architectural Association. The Eco-House was self-sufficient with a bio-digester that converted human waste into methane for fuel and produced food through a hydroponic greenhouse, thus signifying freedom from the infrastructural and agricultural grids as an ideological shift towards ecological, anarchist, and libertarian lifestyle.[33]

The Integral Urban House (1972), Berkeley, California was a demonstration project of the Farallones Institute at the University of California. The Farallones Institute is a non-profit coalition of scientists, designers, horticulturists, and technicians dedicated to teaching and research on ecological design through establishing an interchange between the fields of architecture, agriculture, waste recycling, water conservation, and renewable energy. The Integral Urban House was built as a self-sustaining spaceship capsule in anticipation of a total collapse of infrastructural grids and food supply chains in the event of natural or man-made disasters, including famine, poverty-induced human conflict, large-scale labor strikes, uncontrollable diseases, disruption of fossil fuel supplies, and war.[34] The house was based on the synthesis of biology and architecture to redefine architecture as *biotecture* or *ecotecture*.[35] The house, liberated from energy supply grids, signified freedom from dependence on powerful international fossil fuel corporations, which sourced energy from distant sources mired in geopolitical conflicts. The advocates of the Integral Urban House did not view self-sufficiency as a solution to environmental problems, but the self-contained house was designed as a building block of a self-reliant neighborhood and thus, a self-sustaining community. The Integral Urban House offered the promise of gaining greater control over an individual's ecological relationship with the earth through meeting basic needs, namely food, which was nutritious, free from pathogens and toxins, uncontaminated water, clean air, a low-impact waste management system to close ecological loops, protection from the extremes of weather, and freedom from pests without the use of toxic chemicals.[36]

Jack Todd, Nancy Todd, and Bill McLarney founded the New Alchemy Institute (1971–91). Their solution to the energy crises, looming nuclear disaster, and pollution was a living unit that they called Bioshelter, which was a large solar greenhouse that provided optimum conditions conducive to agriculture all year-round in a temperate climate. The Bioshelter was a self-sustaining architectural unit that established a positive relationship with the environment through renewable energy, local self-sustaining agriculture and aquaculture, low-impact housing, and landscapes. The arks provided heating, cooling, and electricity for lighting and appliances from renewable energy sources, recycled wastes in some instances were installed with water collection, and finally, they sustained their habitants through growing a large portion of their food requirements.[37]

The Ouroboros (1973–6) was a semi-autonomous house, designed and built by the University of Minnesota's School of Architecture and Landscape Architecture with the aid of solar-house plans from Harry E. Thomason of Washington, D.C. Under the guidance of Professor Dennis Holloway, the Ouroboros project was built in Rosemount on land that belonged to the University of Minnesota. This was essentially a solar house that included recycling programs and as such represented the transition from energy conservation to the new paradigm of autonomous structures of the 1970s. The house experimented with energy conservation, energy production, and recycling. The house was heated through heat amassed from a solar collector, had a heavily insulated sod roof to prevent heat losses, relied on a windmill for generating electricity, and had a composting toilet and enclosed sewage system for recycling wastes and conserving water.[38] The house was named Ouroboros after an ancient Egyptian and Greek mythical serpent that is depicted with its tail in its mouth, signifying its continual self-devoured existence in a state of complete self-sufficiency, neither taking or adding anything to the environment.[39]

Solar architecture

The vicissitudes of the popularity and acceptance of solar energy in architecture in the United States were closely tied to the prices and availability of fossil fuels. Prior to the Second World

War, there were sporadic experiments in active and passive solar technologies, which were interrupted by the war. After the war, engineers and entrepreneurs viewed solar energy as a new frontier, but apart from solar water heaters, there were no commercially available solar energy products till the establishment of federal programs in solar energy in the early 1970s.[40] Photovoltaic cells were produced for highly specialized functions in the late 1950s and throughout the 1960s to provide electrical power for satellites orbiting the earth, but photo-voltaic technology did not become commercially available till the 1990s. In the 1950s, the United States government agencies directed their research towards developing nuclear energy —and with the abundance of cheap fossil fuels—research on solar energy diminished signifi-cantly by the 1960s.[41] However, the OPEC oil embargo of 1973 resulted in a sharp rise in fossil fuel prices and a sudden breakdown of the supply chain. This was later exacerbated by the Iranian Revolution in 1979, and the situation peaked with the Iran-Iraq War in 1981. The oil embargo restored the research and investment in solar energy. Consequently, architects, engi-neers, entrepreneurs, and government agencies joined forces to develop new technologies. Homeowners were given state subsidies to install solar devices. However, the decline oil prices in the 1980s and the expiration of solar tax credits led to another phase of market decline in solar energy, which ended in the early 1990s with concerns over greenhouse gas emission, energy security, and pollution.[42]

Prior to the Second World War, the earliest experiments in the use of solar energy were with passive techniques, a simple low-end conditioning method in which the natural renew-able energy source being used must come in direct contact with the building surface or envelop, such as sunlight falling on a glass wall, or breeze flowing through a building envelop. The early interwar pioneers of passive solar houses were the Keck brothers, George Fred Keck and William Keck. George Fred Keck shot to national fame with his design of two houses for Chicago's "Century of Progress" World's Fair in 1933–4. In their subsequent projects, they experimented with solar passive technologies—solar orientation, using earth berms as ther-mal mass, exterior Venetian blind, roof ponds for cooling, and black slate floors in sun rooms as heat sinks.[43]

Just before the Second World War, Professor Hoyt Hottel at MIT launched a solar house program to experiment with active solar energy. His team created a house with fourteen flat-plate collectors installed on the south side of the roof. Each collector comprised a shallow plywood box with three glass sheet covers separated by air spaces on top, a five-inch layer of rock wool insulation on the bottom, and copper tubes inside it soldered to a blackened copper plate in between the top and bottom layers.[44] The collectors heated up water, which was used for heat storage in a tank in the basement. The air in the building was drawn from the rooms by fans and heated by being blown over the hot tank in the basement.[45]

Prior to the OPEC oil embargo, the key solar buildings were by and large experiments by engineers—the Desert Grassland station, 1954, Amado, Arizona; and the Tucson Laboratory, 1959, Tucson, Arizona (designed by engineers Ray Bliss and Mary Donovan); the Boulder Bungalow, 1944, Boulder, Colorado; and the Denver House, 1956, Denver, Colorado (designed by Dr. George Löf, who was a Professor of Civil Engineering at Colorado State University and Director of the Solar Energy Applications Laboratory); the Saunders House, 1960, Weston, Massachusetts (designed by inventor Norm Saunders); the Zome House, 1972, Corrales, New Mexico (designed by Steve Baer, a countercultural self-taught architect); the Bridgers and Paxton Office Building, 1956, Albuquerque, New Mexico (designed by Bridgers and Paxton, an engineering firm); and the Odeillo Houses, 1956, Odeillo, France (the Centre Nationale de la Recherche Scientifique: an energy laboratory).[46] These buildings used only passive or a combination of active and passive solar systems.

Sustainable architecture

A Green building is a building that is sustainable or close to being sustainable, that is it does not exceed the carrying capacity of the earth in terms of material, energy, and water use.

Climate change and its metrics—energy consumption and the carbon cycle—have come to dominate contemporary discourses on Sustainable Architecture. This development has signified a major shift in the relationship between climate and architecture. The architectural environmental paradigms of the 1960s, 70s, and 80s were predicated on architecture as mediator between the human body and the outdoor climate. Thus, energy-efficient designs relied on optimization of the architecture in response to the climate. Climate was viewed as a stable environmental actor, which determined architecture. As it became apparent that buildings, as one of the key consumers of fossil fuels, contribute significantly to climate change, the relationship between architecture and climate went through a paradigmatic shift—from one in which climate was a determinant of architectural metrics, to one in which architecture became an active agent in the transformation of global climatic systems. Numerous paradigms, strategies, and technologies promise to minimize architecture's impact on the fragile ecosystems and climate systems of the earth.

Paradigms: Passive and Low Energy Architecture

PLEA (Passive and Low Energy Architecture) International is a worldwide non-profit network of professionals committed to solving environmental issues through gentle, low-impact technologies. PLEA, founded in 1981, fosters ecological and environmental responsibility in architecture and planning. Passive and low energy systems arose out of energy scarcity and rely on low-cost, low-tech, and simple technologies aimed at the poor. As a paradigm, PLEA is the logical successor to the 1960s and 1970s paradigms of Bio-climatic architecture, Tropical Architecture, and appropriate technology. PLEA systems of heating and cooling—like Tropical Architecture, Bioclimatic Architecture, and vernacular architecture—are passive technologies that operate with natural mechanisms of conduction, convection, radiation, and evaporation, and therefore consume zero or very little energy.[47] Like Appropriate Technology, they rely on buildings that are built with low-capital investment and simple and easy to operate technologies. PLEA systems, due to their low cost and easy operation, can work in the underdeveloped world to serve the needs of the poor.

Paradigms: biomimicry and biotechnology

The current biomimicry movement materialized with Janine M. Benyus's book *Biomimicry: Innovation Inspired by Nature* in 1997. The term biomimicry comes from the Greek bios, meaning life; and mimesis, which implies mimicking. According to Benyus, the intellectual basis of biomimicry is threefold—to treat nature as a model, measure, and mentor.[48] Benyus proposes that by viewing nature as methodological model and a yardstick against which the design efficacies of energy and material loops are measured, the field of design can go through a paradigmatic shift by emulating ecological relationships and processes that create performance efficiencies in nature.[49] Biomimetic research, as part of the growing transdisciplinary field of biotechnology, has received intellectual and financial support from the US Department of Energy, the US Department of Defense, the Defense Advanced Research Projects Agency, and several corporations. This has led to interdisciplinary biology and engineering programs at prominent universities such as: the Centers for Biologically Inspired Design at the University of

California Berkeley and the Wyss Institute for Biologically Inspired Engineering at Harvard University.[50] As a paradigm, biomimicry is predicated on the intellectual premise that nature comprises a self-sustaining, self-governing, dynamic set of interactions that are held in a ecosystemic equilibrium through evolutionary systems of checks and balances, which render nature as an ethical design prototype.[51] Biomimicry operates within the framework of the ecological modernization theory, which advocates that both economic growth and environmental safeguard are not mutually exclusive, advanced through manifestoes such as Miriam Horn and Fred Krupp's *Earth: The Sequel: The Race to Reinvent Energy and Stop Global Warming* (2008) and Paul Hawken, Amory B. Lovins, and L. Hunter Lovins's *Natural Capitalism: Creating the Next Industrial Revolution* (2010). Horn and Krupp's *Earth: The Sequel* argues that the conventional view of the relationship between economics and environmental protection as a zero sum game is deeply flawed; because the future growth of the economy lies in environmental innovations that will transition society to a new energy regime, rendering the current fossil fuel economy obsolete, and in the process stabilize greenhouse gas emissions and eliminate resource waste.[52] Hawken et al.'s *Natural Capitalism* proposes biomimicry as a key strategy to obviate the fossil fuel economy, which has disrupted nature's carbon cycle due to excessive use of fossil fuels.[53]

Architectural examples of biomimicry include: bricks grown from bacteria, developed by bioMason, a biotechnology start-up company; Victoria and Albert Museum's Elytra Filament Pavilion, which is structurally inspired by the lightweight construction of the forewings of flying beetles known as elytra; and biology student turned architect Doris Kim Sung's thermobimetals, smart materials that respond dynamically to temperature changes by self-shading and self-ventilating.[54] The most influential biomimetic manifesto in the field of architecture to date has been William McDonough and Michael Braungart's *Cradle to Cradle: Remaking the Way We Make Things* (2002). McDonough and Braungart argue that our present manufacturing system follows a cradle to grave trajectory in which we extract precious resources from earth at a great human and environmental cost, which are then manufactured into objects of consumption through industrial processes that "put billions of pounds of toxic material into the air, water, and soil every year."[55] These objects, once used and discarded, find their way into a landfill where they decompose as waste to generate toxic byproducts that further pollute air, land, and water.[56] McDonough and Braungart critique the cradle to grave model of manufacturing and consumption as a failure because:

- it results in huge amounts of waste, which means that this system puts large of amounts of valuable materials in landfills;
- it needs to be constantly monitored and regulated, which often results in complex regulations that transcend borders of nation-states and require a complicated legal system to implement;
- it measures productivity through the least input of human labor, rendering vast populations without work; and
- it severely reduces biodiversity and homogenizes different cultural paradigms of environmentalism.[57]

McDonough and Braungart propose a paradigmatic shift in manufacturing based on nature's economy to a model they call cradle to cradle, which operates through closed loops of energy and materials flows, so that there is zero waste.[58] The cradle to cradle paradigm originated in the autonomous architecture projects that were built within the ecological design paradigm of the 1970s. The idea of creating almost closed energy and material loops was experimented through structures such as The Autonomous House (1971 model only), Cambridge UK, model built by Alexander Pike; the Eco-House (1972–5), London, UK, designed and built by Grahame Caine; the Integral Urban House (1972), Berkeley, California, designed and built by Helga Olkowski, Bill Olkowski, and Tom Javits; Bioshelter Arks, such as Cape Cod Ark (1976), designed by Solsearch architects and Prince Edward Island Ark (1976), designed by BGHJ architect for the New Alchemists— Jack Todd, Nancy Todd, and Bill McLarney; and the Ouroboros House (1973–6), Rosemont, Minnesota, designed and built by Professor Dennis Holloway and his students at the University of Minnesota's School of Architecture and Landscape Architecture. Biomimicry is not new to the field of modern architecture and engineering. Buckminster Fuller's geodesic dome was inspired by the geometry of marine planktons, Radiolaria (holoplanktonic protozoa). Frei Otto's pneumatic structures were inspired by nature's geometry occurring in fruit, air bubbles, blood vessels, and skin.

Green strategies: greening infrastructure

Green roofs and green walls are design strategies based on the premise that buildings consume large amounts of energy and materials and destroy the natural habitat that existed on land on which the building was built. These negative consequences of building can partly be remedied through the rendering of building surfaces as living vegetated surfaces that promote biodiversity. As part of the building envelop, roofs represent up to 32 percent of the horizontal surface, which renders the roof surface a key determinant in the building's energy and water cycle.[59] Green roofs have several advantages over conventional roofs: they reduce a building's energy consumption through insulation, provide better acoustic protection from urban noise, prolong the life of the roof membrane, make the building more fire resistant, mitigate storm-water runoff to attenuate the load on storm-water infrastructure and urban waterways, improve air quality through absorption of air pollutants and dust, and moderate the urban heat-island effect.[60] Green roofs cannot replicate the ground level biodiversity, but they can promote rooftop biodiversity by providing an urban habitat for birds, native plants, reptiles, mammals, bees, butterflies, moths, spiders, beetles, grasshoppers, and flies.[61] Green roofs consists of three layers: the top layer is the soil layer which is separated by a fabric filter from the middle layer which is a drainage layer, and finally under these two strata is a protective layer to guard the underlying roof structure from vegetation roots.[62] Green roofs are categorized as extensive and intensive. Extensive green roofs are simplest and most inexpensive due to minimal construction, low maintenance costs, shallow soil stratum depth, lack of irrigation, and therefore tend to be self-sustaining green patches of native species.[63] Consequently, these tend to have minimal benefit with reducing energy loads on buildings, but they do not require expensive building structure for support and can be an urban habitat for birds. Intensive roofs are expensive as they require a thick layer of soil, which adds to the weight of the building and therefore necessitates a stronger building structure. In addition, intensive roofs require regular maintenance and irrigation. However they provide the best returns in terms of reducing energy loads on the building, proving an urban habitat for biodiversity, and a recreational space for building occupants.[64] Instead of being seen solely as a recreational horticultural space, green roofs are now viewed as a bioengineering technology along with other constructed green ecosystem services like sewage-treatment wetlands, bioswales, and green walls, because of their potential to create unique rooftop ecosystem.[65] Like green roofs, green walls provide similar benefits in terms of building envelop insulation, protection from solar radiation intake, attenuation of noise pollution, cleaning the air, and sustenance of habitats for urban wildlife, especially birds. Green walls require a less elaborate structure than a green roof and therefore can be significantly cheaper. They can also easily be added to existing buildings without major structural interventions and costs, for the green wall in its simplest form is a vertical surface for climbing plants. Green walls can play a significant role in reversing the decline of urban avian species that are rapidly declining.[66]

Green strategies: closed energy loops

A net-zero building (NZEB) is one that is connected to centralized electricity supply grids, and yet meets all its energy requirements from low-cost, locally available, nonpolluting, renewable sources, over a period time that can be one month to one year.[67] For a building to be net-zero it must have a two-way supplier-consumer relationship with the grid and must generate renewable energy on site that is equal to or more than its annual energy drawn from the grid. When a net-zero building requires more operational energy than it can generate on-site, it consumes energy from the grid, which is often fossil-fuel based; and when the building produces more energy from renewable sources on-site than it consumes, it feeds the excess electricity into the grid.[68] Net-zero buildings are an effective solution to overcome an aging infrastructure that is completely or partially dependent on non-renewable sources. For example, the US power grid draws 70 percent of its electrical energy from fossil fuel sources, 20 percent from nuclear energy, 7 percent from hydropower and 3 percent from renewable resources.[69] Thus, grid connectivity is crucial to net-zero buildings, because it aids supplying the grid with green energy and secondly, off-grid buildings often rely on extraneous fossil-fuel sources such as propane for cooking, space heating, water heating, and backup generators.[70] The very first step in achieving a net-zero status for a building is to begin with reductions in energy consumption through highly efficient building envelop design, orientation, passive heating and cooling, energy-efficient mechanical systems, natural ventilation, daylight illumination, and high-efficiency lighting and electric appliances. The next step is to generate renewable energy from power sources that will last through the life of the building, either within the building footprint (rooftop photovoltaics, solar hot water, windmills located on the building), or on the site (parking lot photovoltaics, low-impact hydroelectric plants, geothermal plants, and windmills located on-site, but not within the boiling footprint).[71] Less desirable are buildings that rely on biofuels from waste produced onsite or biofuels imported from offsite, to achieve the net-zero status, for these generate carbon emissions. Lastly, buildings can also purchase energy from off-site renewable energy sources.[72]

One of the key limitations of the net-zero building paradigm is that the energy calculations only account for operational energy, which is energy consumed for lighting, HVAC, water heating, fans, cooking, or in other words the energy required to run the building on a day-to-day basis. Buildings also consume other forms of energy: embodied energy, which is energy required to construct the building in terms of production and transport of construction materials, building maintenance, and energy consumed in demolition and safe disposition of the building or end of life energy requirements.[73] To overcome this limitation, a more rigorous concept of the life cycle zero-energy building (LC-ZEB) can be used, which accounts for both embodied and operational energy.[74]

The net-zero paradigm is based on biomimetic and ecological design paradigms, which advocate for managing the built environment in closed material and energy loops, mimicking the energy cycles of natural systems.[75] As the net-zero compliance concept expands its scope from building energy budget to carbon emissions, waste generation, and water consumption, the idea of net-zero water buildings that relies on closed loop cycles of water within the building boundary is developing to conserve fresh water supplies.[76]

Green strategies: closed water cycles

By the time the Brundtland report was published there was a growing consensus that water problems—water pollution, water scarcity, and water-related land degradation—were no longer localized and their impact was global. By the 1990s it was clear that due to the uninterrupted nature of the global water cycle, anthropogenic environmental disruptions—such as greenhouse gas emissions, reduced rates of groundwater recharge, polluted runoff, and atmospheric pollution result in changed precipitation patterns, reduced groundwater tables, depleted water sources, and nonpoint source pollution of water ecosystems.[77] Conventional buildings, which are connected to a centralized municipal water supply and discharged waste water grids, rely on a linear sequence of water consumption. These centralized grids supply potable water, which is sourced from groundwater, surface water, desalination, rivers, and springs. This water is consumed and discarded as waste. To create on-site water recycling systems, domestic wastewaters need to be separated into grey wastewater (from faucets, showers, kitchen) and black wastewater (from toilets and urinals). Treated greywater can be recycled for non-potable indoor use (flushing toilets, laundry, bathing), outdoor use (irrigation, car wash, window cleaning, pressure washing, fire extinguishing) and ecosystem services (developing and preserving wetlands and recharging groundwater). Grey water treatment technologies include three processes: physical, biological, and chemical.[78] The most effective grey water treatment is a combination of these processes.[79]

Environmental assessment methods: LEED and BREEAM

The Building Research Establishment Environmental Assessment Method (BREEAM) was introduced in the UK in 1990. The UK Building Research Establishment's BRE Global implements the program. Its American counterpart: Leadership in Energy and Environmental Design (LEED) is managed by the US Green Building Council (USGBC), which was founded in 1993 as a nonprofit organization. The purpose of BREEAM and LEED assessment methods is to standardize and quantify sustainable architecture at a national level. The LEED program began in 1998 with the objective of creating metrics to measure a building's environmental performance. There is no precise definition of a green building, but there is general consensus that a green building will help reduce greenhouse gas emissions and be sustainable to exist in a net ecological balance with the carrying capacity of earth. LEED sets a shared set of goals towards sustainability that designers, builders, and engineers can target and enables their occupants to lead lifestyles that can achieve sustainability as a goal. LEED is a voluntary certification program which operates though points and credits. LEED allocates its rating system to promote sustainability in six credit categories—Location and Transport; Sustainable Sites; Water Efficiency; Energy and Atmosphere; Materials and Resources; and Indoor Environmental Quality.[80] These credit categories operate through mandatory and optional goals. The mandatory goals must be met, while the optional goals, if achieved, result in credits. To get LEED certification, the building owner or architect must comply with prerequisites and credits to get at least 40 out of 100 points. Higher points result in superior certification levels. Thus, a building must score 40 points to be LEED certified, 50 points to achieve LEED Silver status, 60 points to get to LEED Gold certification, and the highest is 80 points which get awarded with a LEED platinum rating.[81] LEED quantification of sustainability indices comprises broad areas called impact categories such as Reverse Contribution to Global Climate Change; Enhance Individual Human Health and Well-Being; Protect and Restore Water Resources; Protect, Enhance and Restore Biodiversity and Ecosystem Services; Promote Sustainable and Regenerative Material

Resources Cycles; Build a Greener Economy; and Enhance Social Equity, Environmental Justice, Community Health and Quality of Life.[82] The categories are broken into key indicators which are comprehensive sustainability indices for each category. While LEED does push the industry towards higher efficiencies of energy and material cycles, it does not yet have a robust program to enforce environmental equity and justice, which can often be difficult to quantify with each building project.

Environmental assessment methods: Passivhaus

Passivhaus or Passive House is a building type, a paradigm, a strategy to significantly reduce energy consumption, and a certification standard. Passivhaus is a building that has a highly-reduced heating and cooling load and yet provides comfortable indoor conditions, with peak daily average heating and cooling loads typically below 10 W/m² and annual useful energy demands below 15 kWh/(m²a).[83] Thus, Passivhaus can function at such low energy loads by saving on operational energy costs. Heat losses or gains in a Passivhaus are extremely low and therefore the only heating or cooling demand is from conditioning the mechanical ventilation supply air to cover the entire peak heating or cooling load.[84] Thus, the building can be heated, cooled, and dehumidified simply by conditioning the small amount of supply air. The current Passivhaus concept was first tested in a four-townhouse development in Kranichstein in Darmstadt, Germany in the early 1990s. The Kranichstein townhouses were part of a research project that had the sole objective of minimizing the total energy demand through passive strategies such as extremely high levels of insulation; high-quality windows; compact building form; hermetically sealed building envelop with no thermal bridges; mechanical ventilation heat recovery; and maximization of passive solar gains through orientation, glazing ratios, and natural daylighting.[85] The intellectual origins of the current Passivhaus paradigm are in the 1970s investigations in architectural energy efficiency that followed the OPEC embargo, chiefly the North American passive solar architecture and Swedish super insulated houses. In 1977, Canadian researchers (Robert Besant, Oliver Drerup, Rob Dumont, David Eyre, and Harold Orr) built a superinsulated house in Regina, called the Saskatchewan Conservation House, which was almost an airtight building, with triple-glazed windows, insulating night-time window shutters, south-facing windows for passive solar gains, waste-water heat exchanger to recover heat from the laundry and bath water, and heat-recovery ventilators.[86] Numerous people working towards energy efficiency contributed to the development of the Passivhaus such as Swiss building researcher Conrad Brunner, who researched low energy buildings through seasonal energy balances. In 1988 Wolfgang Feist of the Institut Wohnen und Umwelt (Institute for Housing and Environment), Germany and Bo Adamson at the Lund University, Sweden collaborated to develop research that eventually resulted in the Passivhaus as a paradigm, technology, and standard.[87] The Passivhaus Institut (PHI) is the German agency that supervises and implements the Passivhaus certification.

Conclusion

The architectural environmental paradigms that developed after the Second World War—architecture that defies global universalism and follows the material and climatic logic of vernacular architecture and which is primarily aimed at the poor—Bio-climatic Architecture; Tropical Architecture; High-Technology Architecture; Countercultural Architecture; Solar Architecture; Ecological Design; and Appropriate Technology—emerged as regional responses to specific environmental problems that were viewed as localized. Until the 1990s buildings were not see as global agents of environmental degradation. With the understanding of the global scale and impact of environmental problems—such as greenhouse gas emissions and transboundary water and air pollution—the built environment is no longer understood as a local actor, but seen as a global agent in protecting fragile climate systems and scarce water resources. The paradigm of not exceeding the carrying capacity of earth is not new, and indeed sustainable architecture often draws intellectually from various 1960s and 1970s paradigms.

Notes

1   World Commission on Environment and Development, *Our Common Future* (Oxford: Oxford University Press, 1987).
2   Ibid.
3   United Nations Conference on the Human Environment, ed., *Report of the United Nations Conference on the Human Environment: Held at Stockholm, 5–16 June 1972* (New York: United Nations, 1975).
4   Rachel Carson, *Silent Spring* (Boston, MA: Houghton Mifflin Co., 1962); Garrett Hardin, "The Tragedy of the Commons," *Science* 162, no. 3859 (December 13, 1968): 1243–8; Paul R. Ehrlich, *The Population Bomb.* (London: Ballantine in association with Pan, 1971); Donella H. Meadows and Club of Rome, *The Limits to Growth: A Report for the Club of Rome's Project on the Predicament of Mankind* (New York: Universe Books, 1972); E. F. Schumacher, *Small is Beautiful: A Study of Economics as If People Really Mattered* (London: Blond & Briggs, 1973); for summaries, see Jeremy L. Caradonna, *Sustainability: A History* (Oxford: Oxford University Press, 2014).
5   Malcolm Miles, "Utopias of Mud? Hassan Fathy and Alternative Modernisms," *Space and Culture* 9, no. 2 (2006): 115–39.
6   Panayiota Pyla, "The Many Lives of New Gourna: Alternative Histories of a Model Community and Their Current Significance," *Journal of Architecture* 14, no. 6 (2009): 715–30.
7   Hassan Fathy, *Architecture for the Poor: An Experiment in Rural Egypt* (Chicago, IL: University of Chicago Press, 1973).
8   Victor Olgyay and Aladar Olgyay, *Design with Climate: Bioclimatic Approach to Architectural Regionalism* (Princeton, NJ: Princeton University Press, 1963).
9   Ibid.
10 Berthold Burkhardt, "Natural Structures—the Research of Frei Otto in Natural Sciences," *International Journal of Space Structures,* 31, no. 1 (2016): 9–15.
11 Daniela Fabricius, "Architecture before Architecture: Frei Otto's 'Deep History,'" *The Journal of Architecture* 21, no. 8 (2016): 1253–73.
12 Ibid.
13 Jonathan Massey, "Buckminster Fuller's Cybernetic Pastoral: The United States Pavilion at Expo 67," *The Journal of Architecture* 11, no. 4 (2006): 463–83.
14 Timothy Luke, "Ephemeralization as Environmentalism: Rereading R. Buckminster Fuller's Operating Manual for Spaceship Earth," *Organization & Environment* 23, no. 3 (2010): 354–62.
15 Peder Anker, "Buckminster Fuller as Captain of Spaceship Earth," *Minerva* 45, no. 4 (2007): 417–34.
16 Ibid.
17 Luke, "Ephemeralization as Environmentalism."
18 Caroline Maniaque Benton, *Alternative Architecture: French Encounters with the American Counterculture 1960–1980* (Farnham: Ashgate, 2011).
19 Theodore Roszak, *The Making of a Counter Culture: Reflections on the Technocratic Society and Its Youthful Opposition* (New York: Doubleday & Company, 1969).
20 Benton, *Alternative Architecture.*
21 Erin Elder, "How to Build a Commune: Drop City's Influence on the Southwestern Commune Movement," in *West of Center Art and the Counterculture Experiment in America, 1965–1977*, eds. Elissa Auther and Adam Lerner (Minnesota, MN: University of Minnesota Press, 2011): 3–22.
22 Ibid.
23 Timothy Miller, *The 60s Communes: Hippies and Beyond* (Syracuse, NY: Syracuse University Press, 2002).
24 Simon Sadler, "Drop City Revisited," *Journal of Architectural Education* 59, no. 3 (2006): 5–14.
25 Felicity Dale Elliston Scott, *Architecture or Techno-Utopia: Politics after Modernism* (Cambridge, Massachusetts: MIT Press, 2010): 151–84.

26 Ibid. The use of radically recycled materials is also used by Michael Reynolds in his off-grid Earthships, many of which are found outside of Taos, New Mexico.

27 Eva Díaz, "Dome Culture in the Twenty-First Century," *Grey Room* 42 (2011): 80–105.

28 Andrew Kirk, "Appropriating Technology: The Whole Earth Catalog and Counterculture Environmental Politics," *Environmental History* 6, no. 3 (2001): 374–94.

29 Lee Stickells, "Exiting the Grid: Autonomous House Design in the 1970s," in *Proceedings of the Society of Architectural Historians, Australia and New Zealand: 32, Architecture, Institutions and Change,* eds. Paul Hogben and Judith O'Callaghan (Sydney: Society of Architectural Historians, Australia and New Zealand, 2015): 652–62.

30 Peder Anker, "The Closed World of Ecological Architecture," *Journal of Architecture* 10, no. 5 (2005): 527–52.

31 Ibid.

32 "The Alexander Pike Autonomous House, Cambridge," *Architectural Design* 44, no. 11 (1974): 681–9.

33 Lydia Kallipoliti, "From Shit to Food: Graham Caine's Eco-House in South London, 1972-1975," *Buildings & Landscapes: Journal of the Vernacular Architecture Forum* 19, no. 1 (Spring 2012): 87–106.

34 Helga Olkowski, William Olkowski, Tom Javits, Farallones Institute, *The Integral Urban House: Self-Reliant Living in the City* (San Francisco, CA: Sierra Club Books, 1979).

35 Ibid.

36 Ibid.

37 Nancy Jack Todd, "Bioshelters and Their Implications for Lifestyle," *Habitat International* 2, no. 1 (1977): 87–100.

38 Sharon J. Marcovich, "Autonomous Living in the Ouroboros House," *Popular Science Magazine* (December 1975): 80–82, 111.

39 Dennis Holloway, "Project Ouroboros," *Solar Age* 1:9 (1976): 14–17, 30–33.

40 George O. G. Löf, "Solar Energy: An Infinite Source of Clean Energy," *The Annals of the American Academy of Political and Social Science* 410, no. 1 (1973): 52–64.

41 Adam Rome, *The Bulldozer in the Countryside: Suburban Sprawl and the Rise of American Environmentalism* (Cambridge: Cambridge University Press, 2005): 15–44.

42 J. Douglas Balcomb, *Passive Solar Buildings* (Cambridge, MA: MIT Press, 2008).

43 Allen Freeman, "Passive Solar Concepts," *AIA Journal* 68, no. 14 (1979): 48–59.

44 Ken Butti and John Perlin, *A Golden Thread: 2500 Years of Solar Architecture and Technology* (London, Boston: M. Boyars, 1981).

45 Ibid.

46 Bruce Anderson, *Solar Building Architecture* (Cambridge, MA: MIT Press, 1990): 1–35.

47 J. Douglas Balcomb, "Passive and Low Energy Research and Development: A Global View," paper presented at Conference on Passive and Low Energy Alternatives, Mexico City, Mexico, 1984.

48 Janine M. Benyus, *Biomimicry Innovation Inspired by Nature* (New York: Harper Perennial, 1997).

49 Ibid.

50 Jesse Goldstein and Elizabeth Johnson, "Biomimicry: New Natures, New Enclosures," *Theory, Culture and Society* 32, no. 1 (2015): 61–81.

51 Michael Fisch, "The Nature of Biomimicry: Toward a Novel Technological Culture," *Science, Technology, and Human Values* (January 23, 2017): 1–27.

52 Miriam Horn and Fred Krupp, *Earth: The Sequel: The Race to Reinvent Energy and Stop Global Warming* (New York: W. W. Norton & Company, 2008).

53 Paul Hawken, Amory B Lovins, and L. Hunter Lovins, *Natural Capitalism: Creating the Next Industrial Revolution* (London: Earthscan, 2010).

54 "Bricks Grown from Bacteria," *ArchDaily,* www.archdaily.com/472905/bricks-grown-from-bacteria; "Elytra Filament Pavilion Explores Biomimicry at London's Victoria and Albert Museum," *ArchDaily,* www.archdaily.com/787943/elytra-filament-pavilion-explores-biomimicry-in-london; "Bloom / DO|SU Studio Architecture," *ArchDaily,* www.archdaily.com/215280/bloom-dosu-studio-architecture.

55 William McDonough and Michael Braungart, *Cradle to Cradle: Remaking the Way We Make Things* (New York: North Point Press, 2002).

56 Ibid.

57 Ibid.

58 Ibid.

59 Erica Oberndorfer, Jeremy Lundholm, Brad Bass, Reid R. Coffman, Hitesh Doshi, Nigel Dunnett,

Stuart Gaffin, Manfred Köhler, Karen K. Y. Liu, and Bradley Rowe, "Green Roofs as Urban Ecosystems: Ecological Structures, Functions, and Services," *Bioscience* 57, no. 10 (2007): 823–34.

60 Ibid.

61 Nicholas S. G. Williams, Jeremy Lundholm, and J. Scott MacIvor, "Do Green Roofs Help Urban Biodiversity Conservation?" *Journal of Applied Ecology* 51, no. 6 (2014): 1643–9.

62 Theodore Theodosiou, "Green Roofs in Buildings: Thermal and Environmental Behaviour," *Advances in Building Energy Research* 3, no. 1 (2009): 271–88.

63 Ibid.

64 Ibid.

65 Oberndorfer et al., "Green Roofs as Urban Ecosystems: Ecological Structures, Functions, and Services."

66 Caroline Chiquet, John W. Dover, and Paul Mitchell, "Birds and the Urban Environment: The Value of Green Walls," *Urban Ecosystems* 16, no. 3 (2013): 453–62.

67 Paul A. Torcellini et al., *Zero Energy Buildings: A Critical Look at the Definition* (Golden, CO: National Renewable Energy Laboratory, 2006), http://purl.access.gpo.gov/GPO/LPS82978.

68 Ibid.

69 Charles J. Kibert and Maryam Mirhadi Fard, "Differentiating among Low-Energy, Low-Carbon and Net-Zero-Energy Building Strategies for Policy Formulation," *Building Research and Information* 40, no. 5 (2012): 625–37.

70 Torcellini et al., *Zero Energy Buildings.*

71 Ibid.

72 Ibid.

73 Ayman Mohamed and Ala Hasan, "Energy Matching Analysis for Net-Zero Energy Buildings," *Science and Technology for the Built Environment* 22, no. 7 (2016): 885–901.

74 Kibert and Fard, "Differentiating among Low-Energy, Low-Carbon and Net-Zero-Energy."

75 Ibid.

76 Caryssa M. Joustra and Daniel H. Yeh, "Framework for Net-Zero and Net-Positive Building Water Cycle Management," *Building Research and Information* 43, no. 1 (2015): 121–32.

77 Malin Falkenmark, "Global Water Issues Confronting Humanity," *Journal of Peace Research* 27, no. 2 (1990): 177–90.

78 For a detailed review of grey water treatment technologies, see Lina Abu Ghunmi, Grietje Zeeman, Manar Fayyad, and Jules B. van Lier, "Grey Water Treatment Systems: A Review," *Critical Reviews in Environmental Science and Technology* 41, no. 7 (2011): 657–98.

79 Fangyue Li, Knut Wichmann, and Ralf Otterpohl, "Review of the Technological Approaches for Grey Water Treatment and Reuses," *Science of the Total Environment* 407, no. 11 (2009): 3439–49.

80 Brendan Owens et al., *LEED v4: Impact Category and Point Allocation Development Process* (Washington, DC: USGBC, 2013), www.usgbc.org/leed-v4.

81 Ibid.

82 Ibid.

83 Jürgen Schnieders, Wolfgang Feist, and Ludwig Rongen, "Passive Houses for Different Climate Zones," *Energy and Buildings* 105 (2015): 71–87.

84 Christina J Hopfe and Robert S. McLeod, *The Passivhaus Designer's Manual: A Technical Guide to Low and Zero Energy Buildings* (New York: Routledge, 2015).

85 Schnieders, Feist, and Rongen, "Passive Houses for Different Climate Zones."

86 Robert W. Besant, Robert S. Dumont, and Greg Schoenau, "The Saskatchewan Conservation House: Some Preliminary Performance Results," *Energy and Buildings* 2, no. 2 (1979): 163–74.

87 Hopfe and McLeod, *The Passivhaus Designer's Manual.*

References

Anderson, Bruce. *Solar Building Architecture.* Cambridge, MA: MIT Press, 1990.

Anker, Peder. "Buckminster Fuller as Captain of Spaceship Earth." *Minerva* 45, no. 4 (2007): 417–34.

Anker, Peder. "The Closed World of Ecological Architecture." *Journal of Architecture* 10, no. 5 (2005): 527–52.

Balcomb, J. Douglas. "Passive and Low Energy Research and Development: A Global View." Paper presented at Conference on Passive and Low Energy Alternatives, Mexico City, Mexico, 1984.

Balcomb, J. Douglas. *Passive Solar Buildings.* Cambridge, MA: MIT Press, 2008.

Benton, Caroline Maniaque. *Alternative Architecture: French Encounters with the American Counterculture 1960–1980*. Farnham: Ashgate, 2011.

Benyus, Janine M. *Biomimicry Innovation Inspired by Nature*. New York: Harper Perennial, 1997.

Besant, Robert W., Robert S. Dumont, and Greg Schoenau. "The Saskatchewan Conservation House: Some Preliminary Performance Results." *Energy and Buildings* 2, no. 2 (1979): 163–74.

Burkhardt, Berthold. "Natural Structures—the Research of Frei Otto in Natural Sciences." *International Journal of Space Structures* 31, no. 1 (2016): 9–15.

Butti, Ken, and John Perlin. *A Golden Thread: 2500 Years of Solar Architecture and Technology*. London: M. Boyars, 1981.

Caradonna, Jeremy L. *Sustainability: A History*. Oxford: Oxford University Press, 2014.

Carson, Rachel. *Silent Spring*. Boston, MA: Houghton Mifflin Co., 1962.

Chiquet, Caroline, John W. Dover, and Paul Mitchell, "Birds and the Urban Environment: The Value of Green Walls." *Urban Ecosystems* 16, no. 3 (2013): 453–62.

Díaz, Eva. "Dome Culture in the Twenty-First Century." *Grey Room* 42 (2011): 80–105.

Ehrlich, Paul R. *The Population Bomb*. London: Ballantine/Pan, 1971.

Elder, Erin. "How to Build a Commune: Drop City's Influence on the Southwestern Commune Movement." In *West of Center Art and the Counterculture Experiment in America, 1965–1977*, eds. Elissa Auther and Adam Lerner. Minnesota, MN: University of Minnesota Press, 2011: 3–22.

Fabricius, Daniela. "Architecture before Architecture: Frei Otto's 'Deep History.'" *The Journal of Architecture* 21, no. 8 (2016): 1253–73.

Falkenmark, Malin. "Global Water Issues Confronting Humanity." *Journal of Peace Research* 27, no. 2 (1990): 177–90.

Fathy, Hassan. *Architecture for the Poor: An Experiment in Rural Egypt*. Chicago, IL: University of Chicago Press, 1973.

Fisch, Michael. "The Nature of Biomimicry: Toward a Novel Technological Culture." *Science, Technology, and Human Values* (January 23, 2017): 1–27.

Freeman, Allen. "Passive Solar Concepts." *AIA Journal* 68, no. 14 (1979): 48–59.

Ghunmi, Lina Abu, Grietje Zeeman, Manar Fayyad, and Jules B. van Lier. "Grey Water Treatment Systems: A Review." *Critical Reviews in Environmental Science and Technology* 41, no. 7 (2011): 657–98.

Goldstein, Jesse, and Elizabeth Johnson. "Biomimicry: New Natures, New Enclosures." *Theory, Culture and Society* 32, no. 1 (2015): 61–81.

Hardin, Garrett. "The Tragedy of the Commons." *Science* 162, no. 3859 (December 13, 1968): 1243–8.

Hawken, Paul, Amory B. Lovins, and L. Hunter Lovins. *Natural Capitalism: Creating the Next Industrial Revolution*. London: Earthscan, 2010.

Holloway Dennis. "Project Ouroboros." *Solar Age* 1, no. 9 (1976): 14–17, 30–33.

Hopfe, Christina J., and Robert S. McLeod. *The Passivhaus Designer's Manual: A Technical Guide to Low and Zero Energy Buildings*. New York: Routledge, 2015.

Horn, Miriam, and Fred Krupp. *Earth: The Sequel: The Race to Reinvent Energy and Stop Global Warming*. New York: W. W. Norton & Company, 2008.

Joustra, Caryssa M., and Daniel H. Yeh. "Framework for Net-Zero and Net-Positive Building Water Cycle Management." *Building Research and Information* 43, no. 1 (2015): 121–32.

Kallipoliti Lydia. "From Shit to Food: Graham Caine's Eco-House in South London, 1972–1975." *Buildings and Landscapes: Journal of the Vernacular Architecture Forum* 19, no. 1 (Spring 2012): 87–106.

Kibert, Charles J., and Maryam Mirhadi Fard. "Differentiating among Low-Energy, Low-Carbon and Net-Zero-Energy Building Strategies for Policy Formulation." *Building Research and Information* 40, no. 5 (2012): 625–37.

Kirk, Andrew. "Appropriating Technology: The Whole Earth Catalog and Counterculture Environmental Politics." *Environmental History* 6, no. 3 (2001): 374–94.

Li, Fangyue, Knut Wichmann, and Ralf Otterpohl. "Review of the Technological Approaches for Grey Water Treatment and Reuses." *Science of the Total Environment* 407, no. 11 (2009): 3439–49.

Löf, George O. G. "Solar Energy: An Infinite Source of Clean Energy." *The Annals of the American Academy of Political and Social Science* 410, no. 1 (1973): 52–64.

Luke, Timothy. "Ephemeralization as Environmentalism: Rereading R. Buckminster Fuller's Operating Manual for Spaceship Earth." *Organization and Environment* 23, no. 3 (2010): 354–62.

Marcovich, Sharon J. "Autonomous Living in the Ouroboros House." *Popular Science Magazine* (December 1975): 80–82, 111.

Massey, Jonathan. "Buckminster Fuller's Cybernetic Pastoral: The United States Pavilion at Expo 67." *The Journal of Architecture* 11, no. 4 (2006): 463–83.

McDonough, William, and Michael Braungart. *Cradle to Cradle: Remaking the Way We Make Things.* New York: North Point Press, 2002.

Meadows, Donella H. and Club of Rome. *The Limits to Growth: A Report for the Club of Rome's Project on the Predicament of Mankind.* New York: Universe Books, 1972.

Miles, Malcolm. "Utopias of Mud? Hassan Fathy and Alternative Modernisms." *Space and Culture* 9, no. 2 (2006): 115–39.

Miller, Timothy. *The 60s Communes: Hippies and Beyond.* Syracuse, NY: Syracuse University Press, 2002.

Mohamed, Ayman, and Ala Hasan. "Energy Matching Analysis for Net-Zero Energy Buildings." *Science and Technology for the Built Environment* 22, no. 7 (2016): 885–901.

Oberndorfer, Erica, Jeremy Lundholm, Brad Bass, Reid R. Coffman, Hitesh Doshi, Nigel Dunnett, Stuart Gaffin, Manfred Köhler, Karen K. Y. Liu, and Bradley Rowe. "Green Roofs as Urban Ecosystems: Ecological Structures, Functions, and Services." *Bioscience* 57, no. 10 (2007): 823–34.

Olgyay, Victor, and Aladar Olgyay. *Design with Climate: Bioclimatic Approach to Architectural Regionalism.* Princeton, NJ: Princeton University Press, 1963.

Olkowski, Helga, William Olkowski, Tom Javits, Farallones Institute. *The Integral Urban House: Self-Reliant Living in the City.* San Francisco, CA: Sierra Club Books, 1979.

Owens, Brendan, et al. *LEED v4: Impact Category and Point Allocation Development Process.* Washington, DC: USGBC, 2013, www.usgbc.org/leed-v4.

Pyla, Panayiota. "The Many Lives of New Gourna: Alternative Histories of a Model Community and Their Current Significance." *Journal of Architecture* 14, no. 6 (2009): 715–30.

Rome, Adam. *The Bulldozer in the Countryside: Suburban Sprawl and the Rise of American Environmentalism.* Cambridge: Cambridge University Press, 2005.

Roszak Theodore. *The Making of a Counter Culture: Reflections on the Technocratic Society and Its Youthful Opposition.* New York: Doubleday & Company, 1969.

Sadler, Simon. "Drop City Revisited." *Journal of Architectural Education* 59, no. 3 (2006): 5–14.

Schnieders, Jürgen, Wolfgang Feist, and Ludwig Rongen, "Passive Houses for Different Climate Zones." *Energy and Buildings* 105 (2015): 71–87.

Schumacher, E. F. *Small is Beautiful: A Study of Economics as If People Really Mattered.* London: Blond & Briggs, 1973.

Scott, Felicity Dale Elliston. *Architecture or Techno-Utopia: Politics after Modernism.* Cambridge, MA: MIT Press, 2010.

Stickells, Lee. "Exiting the Grid: Autonomous House Design in the 1970s." In *Proceedings of the Society of Architectural Historians, Australia and New Zealand: 32, Architecture, Institutions and Change*, eds. Paul Hogben and Judith O'Callaghan. Sydney: Society of Architectural Historians, Australia and New Zealand, 2015: 652–62.

Theodosiou, Theodore. "Green Roofs in Buildings: Thermal and Environmental Behaviour." *Advances in Building Energy Research* 3, no. 1 (2009): 271–88.

Todd, Nancy Jack. "Bioshelters and Their Implications for Lifestyle." *Habitat International* 2, no. 1 (1977): 87–100.

Torcellini, Paul A., et al. *Zero Energy Buildings: A Critical Look at the Definition.* Golden, CO: National Renewable Energy Laboratory, 2006, http://purl.access.gpo.gov/GPO/LPS82978.

United Nations Conference on the Human Environment, ed. *Report of the United Nations Conference on the Human Environment: Held at Stockholm, 5–16 June 1972.* New York: United Nations, 1975.

Williams, Nicholas S. G., Jeremy Lundholm, and J. Scott MacIvor. "Do Green Roofs Help Urban Biodiversity Conservation?" *Journal of Applied Ecology* 51, no. 6 (2014): 1643–9.

World Commission on Environment and Development. *Our Common Future.* Oxford: Oxford University Press, 1987.

18

SUSTAINABILITY STUDIES IN HIGHER EDUCATION

Teresa Sabol Spezio

The study and institutionalization of sustainability in colleges and universities evolved out of environmental studies (ES) programs, increasing corporate thought in academic decision-making, and the desire for colleges to meet sustainability goals that minimize costs, promote social justice, and attract students.[1] There are many other factors that drove the study of sustainability, but these three factors assist in explaining the reasons for sustainability's sometime incoherent definition and presence on campus. Beginning with the Brundtland Report's (1987) definition of sustainable development, which empowered non-profit organizations and other organizations to consider their own impacts and the ability to incorporate non-monetary methods for considering development successes, to John Elkington's Triple Bottom Line (TBL), which gave corporations and other managed entities a way to integrate sustainability into their operations, the term sustainability gave institutions of higher education a subject that most academic disciplines could use as a framework for curriculum.[2] Sustainability related to just about everything and therefore found a home in diverse faculties and departments. Moreover, just as many researchers have adopted sustainability as a prism through which to view their own research, the bureaucracies within universities also turned to sustainability, hence the near-universal presence of "offices of sustainability" on university campuses. But these linkages complicate the means and methods used by universities and colleges to incorporate sustainability into their respective curricula. The complicated concept has not slowed down the growth of sustainability as a field of study; if anything it has increased the spread of sustainability, as "sustainability studies" has now become, in some places, its own interdisciplinary department or program. Many colleges and universities have created schools, programs, minors, and certificates in sustainability studies. Sustainability certificates, minors, and degrees can be earned in disciplines as disparate as gender and women studies and engineering programs.

The inter- and multi-disciplinary nature of sustainability studies allows many educational disciplines, methods, and pedagogies to address the concept, but it is clear that the diversity of applications of sustainability has diluted its meaning and mission. Unlike ES programs that attempt to educate students to become environmentally literate and responsible citizens who can analyze environmental concerns using an interdisciplinary approach, sustainability studies does not have a "straightforward" mission.[3] The desire to incorporate sustainability into college and university facility operations only further problematizes the issue, especially given that on-campus offices of sustainability can play an active role in generating curricula and shaping the research orientation of scholars. Also, the proliferation of sustainable or "green" rankings, especially in the United States, has caused colleges and universities to embrace sustainability and publicize it to their students without an understanding of its definition or objective. In this chapter, I attempt to explore the role of the study of sustainability in university and college classrooms and operations. Further, I will put forward a more unified idea of the role that sustainability studies could play in the academy.

Although UN reports, such as the Brundtland Report, and sustainability metrics, including TBL, have played a big role in the growth of sustainability as a field, universities and colleges have done more than any other entity to transform sustainability into an object of study. The creation of ES programs, beginning in the early 1970s, paved the way for a proliferation of interdisciplinary studies program centered on the environment. These programs were started largely by professors concerned with the effects of pollution, such as the pollution from chemicals, pesticides and oil spills, and sought to expand students' understanding of the environment by incorporating into curricula critical approaches to chemistry, toxicology, economics, policy, and law. As academics began to consider and attempt to understand environmental concerns, they understood that environmental concerns interconnected with social and economic ones. Thus, many scholars wanted ES to expand beyond a narrow focus on the environment. Sustainability studies became the framework for this increasingly interdisciplinary work, whether integrated into ES curricula or as stand-alone interdisciplinary programs.

The formal origins of sustainability in higher education began with the Talloires Declaration, launched in Talloires, France in 1990. The document signed by twenty university leaders merged the study of sustainability with the practice of sustainability in the university setting.[4] In the ten-point plan, signatories pledged to increase awareness by educating students, collaborating with communities, creating sustainable operations, and maintaining the sustainability movement. The document inaugurated the idea of integrating sustainability into every aspect of university life, including operations. Its far-reaching statements challenged the conventional practices of most colleges and universities, and thus few institutions chose to sign the statement. But a lot has changed since 1990, and as of 2017, 367 institutions have signed the declaration, and its influence has spread throughout academia.[5]

The outcome of the declaration mirrored the trend of ES programs. These programs began to look outside the classroom for formal education opportunities. By the late 1990s, most ES programs required some type of non-classroom learning, and this trend passed to the new domain of sustainability studies. Initially, these programs used internships and laboratory research to expand students' involvement in the community. Embracing the challenge to look outside the classroom as part of the course curriculum, professors and programs began offering service-based learning opportunities. The opportunities engaged students, professors, and researchers with organizations and people outside the classroom and outside the institution, and indeed most sustainability studies programs now offer a service-learning component or capstone project as part of the curriculum requirement, which requires outreach to the larger community. Thus students can now apply theoretical and interdisciplinary knowledge to identify, measure, and solve environmental problems—especially in local communities—under the supervision of professors *and* non-academic practitioners most internships, which often lack direct academic oversight. The "real world" activities also gave (and continue to give) students with ES degrees the ability to demonstrate critical thinking and community engagement to prospective employers and graduate schools.[6]

The interdisciplinary structure of ES programs allowed educators in myriad traditional disciplines to create new curricula tailored toward the emerging interest in the environment. Additionally, the popularity of ES programs at the undergraduate level, along with the lack of

ES doctoral candidates who could teach in ES programs, forced colleges and universities to have professors from traditional disciplines teach in the newly formed programs. These cross-listed courses filled up with eager students and ES programs grew exponentially. The Association for Environmental Studies and Sciences reports that, in 2016, there are approximately 1,500 ES or environmental science programs and departments in the United States alone, and it is based upon the success of these programs that sustainability emerged as an area of focus in higher education.

With an understanding that programs with an environmental focus drew students to colleges and classrooms, professors and universities created new and innovative programs. In addition to its popularity with students, concern over global climate change, large-scale habitat destruction, water shortages, global health issues relating to chemical pollution and resource exploitation and myriad other environmental and health issues drove the traditional disciplines to explore more deeply the interconnections of these issues. As they evolved, many of these new courses and their service learning components did not fit into established ES programs, many of which already possessed tracks or areas of focus that could not accommodate the far-reaching domains of sustainability. As a response, traditional disciplines first implemented new curriculum and courses to address the growing popularity and concerns. This move required departments or programs to build upon ideas of environmental concern and led them to create programs that embraced sustainability as its core framework. With sustainability, courses that strayed away from the now traditional ES programs could continue to recruit *and* interest students, and by the early 2000s many programs had appeared that centered on sustainability and energy, sustainable food systems, the human dimensions of climate change, and much more. Often, these programs retain a link to ES departments, but possess their own institutional structure and identity.

The scope of sustainability studies programs exploded, thus enticing a more expansive audience than traditional ES programs. At the same time, aspects of sustainability have entered into more traditional departments and schools, too. For instance, management schools introduced sustainable business and supply chain programs, as Corporate Social Responsibility (CSR) gained a more solid footing in business education. Many engineering programs created innovate programs that concentrated on alternative energy technologies, such as fuel cells, photovoltaics, various forms of green design. Gender and cultural studies programs tied sustainability to social justice issues and neocolonialism. These programs provide more sustainable solutions to issues related to fuel, trade, and war. These programs do not operate with a single definition of sustainability, and often adapt to fit the needs of a particular discipline.

This flexibility can be considered one of the strengths of the academic study of sustainability, but it also created conflicts between departments, which often take oppositional views on subjects such as energy and economics. For instance, very few economics departments now teach ecological economics, degrowth, or other aspects of the economics of sustainability. But many campuses offer such courses in newer and more flexible departments, including public health, ES, resource economics, forestry, and sociology. Many undergraduates learn about neoclassical economics in one course and then learn about ecological economics in another, and are left to navigate these different approaches on their own. The choices about what subjects are taught are ultimately left up to schools, departments, and programs. As Bill Rees, the founder of the ecological footprint and an advocate for the economics of sustainability, has argued, the result is that many university departments work at cross purposes with one another, and may actually "impede sustainability."[7]

Some colleges and universities have created a more systematic program on sustainability, embedding sustainability into the curriculum. At the University of California, Riverside, for instance, a student can earn a BA in Sustainability Studies from the Department of Gender & Sexuality Studies. At the University of Michigan, all engineering students are educated with sustainable engineering principles. Comprehensive programs in sustainability also exist at the University of British Columbia, Arizona State University, and Dalhousie University to name only a few, and new programs are in the works.[8]

At other schools, faculty members work together to create a matrix of courses that have aspects of sustainability, often cobbling together programs from pre-existing courses across campus. To facilitate this outcome, professors and administrators create focused learning communities, develope funding mechanisms to facilitate the creation of sustainability-based courses, and set up educational seminars to teach faculty about sustainability. These programs concentrated on expanding faculty's understanding of sustainability as it related to their respective disciplines. For colleges and universities that did not want a specific sustainability department, program, or school, educating faculty about sustainability and its relationship to each discipline allowed the institution to infuse sustainability into the entire school's learning objectives. In some cases, colleges and universities developed a first-year sustainability seminar and in other cases they required all students regardless of major to enroll in at least one course that addressed sustainability. With this growth in sustainability across the curriculum, faculty from institutions created formal and informal associations to meet at conferences. Additionally, they worked with publishers to create journals that addressed interdisciplinary research in sustainability and methods and pedagogies of teaching sustainability. The number of journals with the term sustainability in the title has increased exponentially since the first decade of the twenty-first century.

Elsewhere on campus, college and university administrators embraced sustainability. They understood that the increased interest, along with the service learning activities, allowed students and professors to assist colleges and universities with their environmental responsibilities and operating costs, while at the same time teaching undergraduate and graduate students.[9] The idea of "campus as a living lab" has caught on in North America, and many schools have seen collaborations between academics and facility and operations to improve the environmental performance of a university. The sustainability-based projects and programs combined student and faculty interests with the institutions need to plan, develop, and implement sustainable business practices that minimized costs and lessened environmental and health impacts. In addition, universities and colleges became the leaders in green building certifications. Since 2003, colleges and universities have built or renovated approximately 3,000 US Green Building Council-certified buildings.[10] Universities and colleges are frontrunners in constructing Leadership in Energy and Environmental Design (LEED) buildings, including Belmont College's Wedgewood Academic Center, which was the first building to achieve LEED Platinum certification in Tennessee. Many of these developments were championed by progressive administrators and trustees influenced by TBL thinking, and in many places the greening of campuses has dovetailed with corporate interests, which play an increasingly prominent role in higher education.

For many academic institutions, sustainability proved to be a useful set of concepts for campus operations, and also fruitfully tied together existing classroom and non-classroom programs. As sustainability became a more prominent feature on campuses after the year 2000, university researchers and students began to collaborate more regularly with surrounding communities, working towards economic, environmental, and social wellbeing. Common sustainability initiatives have included programs on recycling, grey water, xeriscaping, stormwater features, prison education, energy conservation, organic farming, farm-to-table cafeteria food, vegan selections, grade school tutoring, public transportation, bicycle paths, solar power, waste minimization and countless other programs. University administrators and trustees began to see academic sustainability programs as compatible with administrative and business practices and used sustainability actions to showcase their responsibilities to the community, the larger academy, and to non-government and corporate funding organizations. As government funding for public education has decreased over the past two decades, administrators and faculty began to draw on sustainability actions as a way of bringing new funding streams to academic institutions. Funding agencies paid for buildings, professorships, speaking forums and myriad other activities surrounding sustainability. Like curriculum advances, these funds poured into varied departments. For example, Arizona State University received $27.5 million from Rob and Melani Walton to create The Walton Sustainability Solutions Initiative to assist the university in developing, engaging, and educating students on complex issues surrounding sustainability.

The embrace of sustainability by many corporations, NGOs, and governments at all levels has had a significant impact on higher education since universities have close ties to these entities. Corporations created sustainability reports and scorecards, governments looked to fund and create sustainability projects, and non-governmental agencies promoted their ability to develop programs for economic, environmental, and social justice. With sustainability becoming a buzzword, universities began to care about how they ranked in terms of both on-campus sustainability and sustainability curricula, and many schools began to incorporate sustainability into recruitment efforts. At first, schools touted green buildings, organic, local, and vegetarian food options, and service-learning opportunities. As competition increased for sustainability-oriented students, higher education rankings began to include sustainable or green measures. For example, the Association for the Advancement of Sustainability in Higher Education (AASHE) founded in 2003, created the Sustainability Tracking Assessment and Rating System (STARS) that gives colleges and universities the ability to measure their own performance and be recognized as a bronze, silver, gold, or platinum institution. Close to 400 institutions in nine countries have submitted reports for review. The rankings include "diversity and affordability," "investment and finance," and other categories. In 2016, only one institution, Colorado State University, has met the platinum rating.[11] Other organizations, including the Sierra Club, Universitas Indonesia, and the Sustainable Endowments Institute also have their own sustainability rankings.

With the popularity of college rankings, universities and colleges have embraced these rankings, and institution web pages often tout their rank and performance. For example, the University of California Davis was ranked the most sustainable university in the world in December 2016. The ranking featured prominently on their web page. Clearly sustainability has become an important value in higher education, but also another means of competing with other schools. This desire to compare and compete can create more sustainable institutions but at the same time, it dilutes and problematizes how sustainability is taught, communicated, and measured. The concern at many schools, as with many corporations, is that sustainability, which is suddenly and quickly ranked and valorized, is used and manipulated to paper over conventional practices. For instance, many universities support less sustainable practices such as petroleum engineering and conventional agriculture as opposed to environmental engineering and organic agriculture.

Moving forward, the challenge for institutions of higher education in terms of sustainability, is threefold. First, sustainability studies needs a clearer vision, mandate, and identity so that the concept of sustainability is not abused or watered down to reinforce the status quo. As sustainability becomes a discipline of sorts, it needs to figure out what its core identity will be, what methods it uses, and how it will relate to other areas of study. Developing a canon of texts could go a long way to clarifying this identity. The aforementioned Brundtland Report, the *World Conservation Strategy* (1980), and *Cannibals with Forks: The Triple Bottom Line of 21st Century Business* (1997) are a good start, but one could add many more texts.[12] Further, the many methods that have been developed to measure sustainability, including Ecological Footprint Analysis, the Genuine Progress Indicator, the Happy Planet Index, Energy Return on Investment, and Life-Cycle Analysis, should be reviewed and critiqued in campus courses.[13] But ultimately, the question remains: How can such disparate fields as engineering, African-American studies, sociology, supply-chain management, law, planning, and environmental studies work together to create a curriculum to educate students on the value of sustainability? How do we create a common language so that students can leave the academy and work together for economic, environmental, and social justice? The solution can be found only if the disciplines work together to consider the foundations of sustainability and create robust programs that foster its diversity. The jury is still out on whether this will be achieved via an interdisciplinary program called sustainability studies or through sustainability-based courses in separate disciplines.

Second, universities need to figure out how to deal with the various value conflicts that arise as sustainability enters academia. In a sense, there are two conflicts. The first is between professors and/or departments that teach sustainability and those wedded to conventional practices that often conflict with the values of sustainability. As mentioned above, few economics departments have embraced alternative approaches to economics and essentially still teach neoclassical economics—the same business-as-usual economics that externalizes the natural world and promotes growth. As a result, the economics of sustainability has had to find a home in more marginal and less prestigious departments. The effect is often confusion and puzzlement among students who learn very different bodies of knowledge. How will universities address this issue? The other conflict is between university administrators, and their desire to create programs, minimize costs, and build new buildings. Universities have seen conflicts between students and researchers who want to use scarce university land to grow food and study sustainable food systems, on the one hand, with university administrators who would much prefer to use the land for new parking structures or other buildings that can be considered less sustainable.

Third, universities need to be careful to avoid greenwashing, which is the abuse of sustainability language and the subtle maintenance of the status quo. Some universities tout their STARS ranking instead of enacting substantive change. While the piecemeal efforts to improve campus sustainability, and the wellbeing of students and staff, is important, it should not be allowed to overshadow the other ways in which universities reduce sustainability efforts, via certain departments or activities.[14]

In sum, the academy needs to create a comprehensive and coherent curriculum so that sustainability studies can be defined and implemented on campuses. The challenge is to create a curriculum that harmonizes the approaches of different disciplines, to educate students on the value of sustainability studies, and to develop programs that meet on-campus sustainability objectives. We need to create a common language so the students can leave the academy and work together for economic, environmental, and social justice. The challenge will be to help the disciplines work together and create robust programs that support sustainability's diversity, values, and wider importance while engaging the administration's sustainability measures.

Notes

1   I will use environmental studies as a catchall for environmental studies and environmental science programs.

2   See World Commission on Environment and Development (WCED), *Our Common Future* ("the Brundtland Report") (Oxford: Oxford University Press, 1987); John Elkington, *Cannibals with Forks: The Triple Bottom Line of 21st Century Business* (Oxford: Capstone, 1997).

3   One could argue that this is also the case for environmental studies although as a discipline that extends back to the 1970s, it has had much more time to establish an identity.

4   Jeremy L. Caradonna, *Sustainability: A History* (Oxford: Oxford University Press, 2014), 206.

5   For a current list of Talloires Network members, see: http://talloiresnetwork.tufts.edu/who-we-ar/talloires-network-members.

6   See the Association for Environmental Studies and Sciences website: https://aessonline.org/about-aess/history.

7   William Rees, "Impeding Sustainability? The Ecological Footprint of Higher Education," *Planning for Higher Education* 31, 3 (2003): 88–98.

8   See each school's website for more information. Also, visit AASHE's website: www.aashe.org.

9   While at the University of Oregon, I co-taught courses that performed environmental and energy audits for the university. These courses led to the creation of the University of Oregon's Environmental Service Learning Program. Many college and universities created programs using the same model.

10 USGBC has a database of LEED certified projects. A search of university and colleges produced the list. See www.usgbc.org/projects.

11 The STARS report has undoubtedly helped many schools improve their operations and curricular offerings, but it is often criticized for being a self-assessment without serious critical examination.

12 WCED, *Our Common Future*; International Union for Conservation of Nature and Natural Resources, *World Conservation Strategy* (Gland: IUCN, 1980); Elkington, *Cannibals with Forks*.

13 See Caradonna, *Sustainability*, and also Chapter 13 in the present volume, by Simon Bell and Steve Morse.

14 Corporations are major donors to universities, and corporations primarily direct funding to finance research that will benefit their industrial sector. These investments and the endowment portfolios allow colleges and universities to profit from less sustainable corporate practices. In response to criticism from students, alumni and faculty, some universities in the US and Canada have divested from fossil fuel corporations. Divestiture and developing a goal of zero carbon fuel use can be seen as the only responsible action to mitigate climate change.

References

Caradonna, Jeremy L. *Sustainability: A History*. Oxford: Oxford University Press, 2014.

Elkington, John. *Cannibals with Forks: the Triple Bottom Line of 21st Century Business*. Oxford: Capstone, 1997.

Rees, William. "Impeding Sustainability: The Ecological Footprint of Higher Education." *Planning for Higher Education* 31, 3 (2003): 88–98.

World Commission on Environment and Development (WCED). *Our Common Future* ("the Brundtland Report"). Oxford: Oxford University Press, 1987.

PART VI

Core issues and key debates on sustainability

19

CLIMATE CHANGE AND ITS HISTORIES

Hervé Le Treut and Claire Weill

Introduction

Studying the development of the "climate change" issue requires an analysis of different histories, factors, and time scales. From the very outset, one needs to distinguish between three interlocking histories: the history of the climate system itself, the history of human impacts on the climate (and its subsequent changes), and the history of climate science. But addressing climate change policies also requires studies concerning a huge array of different environmental and societal risks, studies that have too often been developed in separate contexts. This complexity is itself the distinctive mark of all the debates and research on climate, and it must be reflected in the present chapter. That said, it is impossible to cover comprehensively in only a few pages such a large subject. The approach has therefore been to focus on a short period of time, mostly the decades after World War II. This post-war period was marked by the dramatic increase of human impacts on our environment, as well as unprecedented scientific attention on climate, all with the intention of evaluating and raising cautions about the risks of a rapidly changing climate. In the last decades, the search for solutions has become much more pressing, and has led to the development of new areas of research, expertise, and debates.

Elements of context

The notion of "climate" is a complex one. It is derived from a Greek word meaning "inclination" and refers to the inclination of the Sun's radiation when it reaches the surface of the Earth: a small inclination results in cold conditions (because the radiation is dispersed over a wider surface) whereas, by opposition, a larger inclination (when the Sun is high in the sky) creates warmer conditions. The current definition of climate takes into account many more parameters: it is defined by reference to a complex system, the "climate system." The very diverse elements which compose this climate system were already identified at the beginning of the twentieth century, and named as different "spheres": the atmosphere, the hydrosphere (mainly the ocean, but also the whole water cycle), the cryosphere (the glaciers, the sea-ice and the snow-covered land surfaces), the geosphere, and the lithosphere (the upper layer of the solid earth). The biosphere was added in 1926 by the Russian scientist Vladimir Vernadsky to summarize the many dimensions of the actions of life on the Earth's environment.[1] The climate is then defined as the sum of the statistical information which characterizes the evolution of the climate system over a given time period. We can describe in this way a global climate, characteristic of the planet as a whole, or more regional climates. Over the last centuries, geography has been the science studying the multiple and multidisciplinary dimensions of climate fluctuations over given areas. But after World War II, the strong development of observational and modelling tools using physical, chemical, or biochemical methods also modified in a decisive manner the understanding and predictive capacity of the climate system.

For a planet that is more than 4 billion years old, there are many ways to choose the "given period" defining climate, which offers widely different perspectives over the Earth's history. We may, for example, define climate and climate changes through statistics and statistic evolutions covering hundreds of millions of years: these changes will describe how the magma that formed the initial planet was able to cool, how the oceans were formed, how the chemical composition of the atmosphere established itself, how it was affected by the development of life, or the role of the constantly but slowly increasing strength of the Sun's radiation intensity. At the scale of tens of millions of years, the continental drift, or such accidents as huge volcanic eruptions, left a strong mark on the planet. This was also the case 65 million years before present (a date which marks the transition between the Secondary and Tertiary eras) when a collision of the Earth with an asteroid caused the extinction of many living species, including all the largest dinosaurs. Nearer to us, in the Quaternary era (during the last two million years), we are able to document the climate fluctuations that result from the oscillations of the astronomical parameters governing the rotation of the Earth and its rotation around the Sun. Those Quaternary fluctuations cover changes from glacial maxima, when North America and parts of Eurasia were covered by ice shelves that were a few kilometers thick, to interglacial periods which were generally shorter, where large ice caps were found in Greenland and Antarctica only. As human beings, we can consider the Quaternary climate as *our* climate, because it is the climate of a planet that has not changed its global geography since then. Also, although there were constant changes between glacial and interglacial conditions, the range of those changes, in terms of Earth's surface temperature and chemical composition of the atmosphere, remained constant until the beginning of the industrial era. Since around 1750, humans have exerted a much stronger influence on the climate than at any time in the past, which can be measured, for instance, in the amount of carbon dioxide that exists in the atmosphere. In pre-industrial times, the parts per million (ppm) of CO_2 was approximately 280; today the ppm is over 400.

The scientific history that is addressed in this chapter concerns only the last phase of this Quaternary era, a warm and anomalously long interglacial period that is called the Holocene. The Holocene extends back about 10,000 years, to the end of the last ice age. Its long duration may be explained by the astronomical theory, and it is likely to last some thousand years more. Of course, the Holocene covers only a small fraction of mankind's history: human beings were already present well before it started. Mankind had to confront multiple ice ages, and our Neanderthal cousins had already disappeared when the Holocene began. But it is only during the Holocene that human demography began to increase massively and our civilizations to develop.

The Holocene has witnessed some large-scale climate changes, in which humans played little or no role, such as the desertification of the Sahara 5,000 years ago, in accordance with the astronomical forcing already mentioned, or the Little Ice Age which marked the seventeenth, eighteenth and nineteenth centuries, and appeared as a combined consequence of changes in the solar irradiation and volcanism.[2] But, globally, the Holocene is defined mostly as a stable period up to the end of the nineteenth century: the atmospheric CO_2 concentrations, for example, remained throughout this period within a very narrow range between 270 and 300 ppm approximately. What is often called "climate change" is the rupture of this equilibrium, or quasi-equilibrium, through human activities.

The first developments of scientific diagnosis and concern

Two main factors have been demonstrated to carry out a direct human influence on the climate of the Earth: the modifications of the soil conditions and those of the atmospheric composition—mainly the "greenhouse gases." These two factors act together, but their manifestations are very different, and they affect the scientific history of climate sciences in very different ways. The soil modifications (deforestation, agriculture, irrigation, soil degradation, fires, urban development, and so on) are "local," a word which can describe regions of up to a continent. Their manifestation is easy to perceive: the very large destruction of the European forest since the middle ages was easy to perceive, as was the quick destruction, over the last century, of many types of landscapes that were modified by human societies over millennia. The most immediate impacts on climate are easy to diagnose, and their importance is huge because they condition water and food availability, and are a direct determinant of poverty. In the 1960s and 1970s, in the context of decolonization, of new agricultural practices ("Green Revolution"), of demographic rise and continuous poverty in many parts of the world, the emerging climate science was extremely attentive to these issues, which were also present in the first UN Conference on the Human Environment in Stockholm (1972) and later in the notion of sustainable development. One example of this emerging climate science is the work of Jule Charney, who provided an explanation for the two-decade-long Sahel drought through soil damage.[3] Although these explanations now appear insufficient to explain complex climatic processes, they have retained a huge impact.

The introduction of persistent greenhouse gases into the atmosphere (i.e. gases whose additional atmospheric content may last a few decades after their emission) has offered very different perspectives, and required very different scientific studies. The first distinctive characteristic of those gases is that they stay long enough in the atmosphere to be mixed by the winds: they constitute a warming agent which, far from being local, is on the contrary distributed almost uniformly over the whole globe. Their second characteristic is that they affect the future climate, as their persistence causes a warming effect that we have no possibility to curve down in the short term. The strong increase in greenhouse gases emissions has therefore created a new vision of the planet, as a complex and solidary body, which requires study and management at the global scale. Although evidence exists that humans have had an impact on local climates, in various ways, it is only with the Industrial Revolution and the widespread use of fossil fuels (in combination with other human practices) that humans began to change the climate at the global scale.

This understanding of the "greenhouse effect" and its impact on climate is, in fact, of fairly recent origin. The work published in 1896 by the Swedish chemist Svante Arrhenius (Nobel Prize in 1903), is often cited as an early warning about the risk of CO_2 increase. Arrhenius established that a doubling of the atmospheric CO_2 could cause a 5°C increase of the Earth temperature, an estimate which is more or less in line with modern estimates. But at the beginning of the twentieth century, CO_2 emissions were many times smaller than now and climate change due to greenhouses gases appeared as a remote process and a nice opportunity to save mankind from the next glacial era.

The real concern about greenhouse gases came much later, in 1957, when Roger Revelle and Hans Suess argued that the ocean would not quickly absorb the huge mass of CO_2 emitted by the use of fossil fuels—about 1 billion tons of carbon yearly, at that time, 10 billion tons now—and that it might be stored in the atmosphere for long periods of time.[4] The creation of an observatory on the slopes of the Mauna Loa (in 1957–8, in connection with the International Polar Year), the highest Hawaiian volcano, was intended to produce precise measurements of CO_2 in the atmosphere. The laboratory at Mauna Loa provided incontrovertible evidence that carbon dioxide levels were rising steadily—a finding continuously confirmed over the years since then. Indeed, the atmospheric CO_2 content measured at Mauna Loa (and in many other places) is now over 400 ppm, which means there is nearly a third more CO_2 in the atmosphere today than there was in pre-industrial times. In the 1970s, the measured figures were less threatening (the CO_2 had reached a value of around 325 ppm) but the findings immediately raised another question: what would be the danger of such a trend? Answering this question required the capacity to model the future, which remains an important, albeit controversial aspect of climate science.

At this point, it is useful to address "The Myth of the 1970s Global Cooling Scientific Consensus."[5] There is an erroneous belief, still widely shared, that in the 1970s there was a scientific consensus that the global climate was cooling. At that time, the scientific community was not organized as it is nowadays, and climate science was still in its infancy; important data was lacking and it was difficult to balance different effects. The natural variability of climate, the warming from newly increasing greenhouse gases, and the cooling from newly increasing aerosols led some climate scientists to argue that the global climate was cooling. Yet in a comprehensive study focusing on peer-reviewed papers published between 1965 and 1979, a group of researchers found that 44 of them presented arguments toward a warming climate, 20 were neutral, and 7 argued that there was a cooling climate.[6] The total number of published papers was increasing steadily over the whole period, and so did the yearly number of "warming" papers. Those figures describe a community beginning to organize itself and slowly reaching a consensus as CO_2 emissions became bigger and began to dominate other climate drivers. Further, climate fluctuations are complex and often non-linear, so very short-term dips in temperature were sometimes misinterpreted as constituting a cooling trend. We now know that, although the climate has fluctuations, the clear overall trend is toward a warming planet.

In spite of this growing consensus, the scientific community was confronted with a difficulty: direct observations of CO_2 increase in the atmosphere could not be immediately followed by direct observation of surface temperature increase, due to the lag generated by the long warming time of the ocean, and the necessity to wait for a signal larger than the natural variability of climate. For several decades, conjectures on future climate evolution had to rely on models, and indirect indications from past climates.

The 1979 report to the US National Academy of Sciences, coordinated by Jule Charney, marked a decisive turning point.[7] Relying on scientific understanding, but also on the results of the first climate numerical models, this report predicted that a doubling of the atmospheric CO_2 content would induce an equilibrium warming of the Earth's surface by between 1.5°C and 4.5°C—very large values when compared with the approximately 5°C separating glacial and interglacial conditions. The main conclusions of the Charney report have all been confirmed by the more advanced models developed since then, and by the evolution of the climate system itself. This is a tribute to the fundamental scientific advances that were developed since World War II in this field.

Predicting climate evolution

The science of climate changes has benefited from research that developed throughout the twentieth century, and entered a fruitful era after World War II, when the benefits of new tools, such as computers, satellites, and a large variety of isotopic measurements, made the work of climate science easier and more reliable. A complete chapter would be necessary to pay tribute to this scientific progress and the pioneering role of the Bergen school at the beginning of the twentieth century. Here we will emphasize two important issues only.

The first one concerns models. In spite of, or because of their crucial role in anticipating climate risks, they have suffered heavy criticisms. It is therefore necessary to explain the development of climate models and modeling science. Lewis Fry Richardson, from the British Met Office, is rightly credited for the first numerical weather predictions based on the equations of fluid mechanics, in 1920–22, but his work was concluded by inaccurate results. These problematic findings sparked a reflection on the formulation of the equations, which had to be consistent with the large geographical scale of the meteorological patterns that needed to be predicted. The quasi-geostrophic theory of Jule Charney (1948) offered a first elegant solution to this problem.[8] The amount of time and effort that Richardson necessitated to realize his prediction required new calculation tools. The first computers that appeared at the end of World War II offered a revolutionary solution, as they were able to carry out one thousand operations per second. This revolution is still ongoing, as petascale computers are one trillion times more powerful. Models were first designed for weather prediction, but in 1963, Edward Lorenz showed that there was an intrinsic limit of about ten days to those predictions, due to the chaotic component of the atmospheric circulation. Therefore the design and use of these emerging atmospheric models for climate purposes, had to be redefined. Rather than aiming at direct predictions, they are now to use the laws of physics, chemistry, or quantitative ecology to create some "sister planet," whose statistical behavior would be similar to that of the real Earth. The development of those models has constituted a big challenge. By the time of the Charney report, in 1979, there were only two models relevant for climate change studies. Today, there are more than 20 of them (although it may be only half, if we consider only those which were developed independently). The development of each model has required the work of about 100 scientists during one or several decades. Their growing accuracy, their capacity to take into account all the climate spheres described above, their successful use for different applications, in marine meteorology, air quality forecast, or ENSO prediction, and their capacity to reproduce ancient climate conditions, such as those of the glacial maximum, have all contributed to their relevance in diagnosing future climate change. Although uncertainties will always remain when it comes to modeling future scenarios, the accuracy of climate models has clearly increased over time. Further, the many models created by the community are not used in isolation: there is a need to consider hierarchies of models, from the conceptually simplest ones to the most complex. There is also a need to use simultaneously the models developed in different laboratories.

This brings us to a second characteristic of the work carried out by meteorologists, oceanographers, and climatologists, and that is the long-standing reliance on international cooperation to gather data worldwide. Diverse data sets are necessary to initialize meteorological models, or operational oceanic predictions. But when it comes to climate issues, the situation is different. The need for observations derives mainly from the necessity to formulate equations systems that describe in mixed statistical/empirical mode the role of small-scale processes such as clouds, marine biochemistry, vegetation, and so on. These processes affect climate at scales ranging from days to decades to centuries. These process-oriented data are often collected by large field campaigns involving ships, planes, balloons, and other equipment, and hundreds of scientists. Other independent data sets are needed to verify the quality and relevance of climate models, which includes satellite data, past data from the last century, or data from much older paleo-reconstructions. All of this requires a large and dedicated international organization.

The birth of new institutions

The progress of science has never been independent from the design and evolution of the institutions that organize it. The growing concern about climate change was one of the drivers that favored new institutions, which have gained a considerable importance over time. Without referring to these institutions, it is impossible to understand how the issue of climate change, which was virtually absent from the UN summit of Stockholm (1972) could became, 20 years later, one of the major concern of the Earth Summit at Rio (1992) that was attended by most countries and many NGOs, and which gave birth to the United Nations Framework Convention on Climate Change (UNFCC).

The World Meteorological Organization (WMO)[9] was and remains one of the main institutions which helped shape this landscape. It was founded in 1873 as the International Meteorological Organization, following the first European network of meteorological stations linked by telegraph, which was organized from 1853 onwards by the director of Paris observatory, Urbain Le Verrier.[10] The security of marine transportation was a major incentive to develop this international cooperation: Le Verrier's initiative followed the loss of many military ships during the Crimean War. WMO acquired its present name in 1947, and is placed under the authority of the United Nations. It continues the tradition of free and transparent sharing of data between nations.

The First World Climate Conference in 1979, which was organized by WMO, played a decisive role in developing climate science. A major international forum devoted exclusively to climate change, it resulted in the creation (a year later, in 1980) of the World Climate Programme (WCP) and, as part of it, the World Climate Research Programme (WCRP), which is still a major organizer of climatic research. The WCP and WCRP are placed under the umbrella of WMO, the International Council of Scientific Unions (ICSU), and UNESCO. Their creation followed decades of a major work to organize the exchange of meteorological and oceanographic data, as well as the development of numerical models. The Global Atmospheric Research Programme (GARP), set up in 1962, under the chairmanship of Jule Charney, managed to gather large-scale cooperative experiments.[11] This work got strong support from the young and quickly growing space agencies.[12]

In the wake of WCP creation, a number of meetings were set up in Villach, Austria, under the auspices of WMO and ICSU, in 1980, 1983, and 1985. They also drew on a United Nations Environment Programme (UNEP) funded research effort under way at International Meteorological Institute, Stockholm, between 1982 and 1985. It was at Villach, in 1985, that a consensus was reached by an international group of scientists, participating in their personal capacities, that "in the first half of the next century a rise of global mean temperature would occur which is greater than any in man's history." These experts also recommended that "scientists and policymakers should begin active collaboration to explore the effectiveness of alternative policies and adjustment."[13] Meanwhile, the Director of UNEP, Mostafa Tolba, a key actor in the negotiations on the protection of the ozone layer, urged the US to take appropriate actions, in a letter to the US Secretary of State, George Schultz.[14] An Ad Hoc Working Group on Greenhouse Gases (AGGG) was also created in July 1986. It comprised six experts designated by UNEP, ICSU, and WMO, including the Swedish chemist and meteorologist Bert Bolin, who became later the first director of IPCC, and the chemist and botanist Gordon Goodman, founding Director of the Beijer Institute in the Royal Swedish Academy of Sciences, The first AGGG report in four volumes formed a basis for the first report of IPCC in 1990.[15]

In 1985, the scientific knowledge was already sufficient to consider action. But the period was also marked by uncertainties that were large enough to stir debates. The US played a central role, since the country was the largest emitter of greenhouse gases, and since it contained the largest cumulated expertise in the area of climate change research and assessment. The Republican government, as well as powerful fossil fuel lobbies, was reluctant or opposed to any action on climate change. In addition, conflicting positions existed inside the US administration, between the Department of Energy (DOE), on the one hand, and the Environmental Protection Agency (EPA) and the Department of State, on the other hand. The response to Mustafa Tolba was the proposal to create an Intergovernmental Panel on Climate Change (IPCC). The creation of IPCC was able to satisfy these different parties, considering either that it would delay action, or that the creation of the IPCC would be a first step to engage the negotiation of an international convention.[16]

As a result of this and many other actions, the IPCC was created in 1998 under the auspices of WMO and UNEP. A large number of scientists participated in this creation. In addition to Bert Bolin, director from the creation to 1997, one should cite the name of N. Sundaraman, a US scientist on deputation at WMO, a former member of the Climate Assessment Programme (CIAP), the first integrated scientific assessment aimed to study the atmospheric impacts of supersonic transport aircraft in the atmosphere in the early 1970s,[17] and personalities such as Sir John Houghton, from the British Met Office.

The mission of IPCC was to provide a comprehensive assessment on the status of climate science, on the impacts of climate change and on the responses required to mitigate climate risks. Three separate working groups were set up to work on each of those three tasks. The IPCC construction is a complex and subtle mix. It is an independent panel commissioned by governments and follows a very strict procedure. The "I" of IPCC stands for "intergovernmental" and not "international," as the command of the reports, and part of the review process, is in the hands of experts named by the governments. But the text is signed by lead-authors and coordinating lead authors, co-signed by review editors, who all act "*intuitu personae*." This is a process that engages government experts, while the reports are written by independent experts. The first IPCC report was published in 1990, in phase with the Second World Climate Conference. Although the conference was not immediately seen as a success, it helped the process leading to the establishment of the UNFCCC at the occasion of the Rio 1992 Earth Summit. Since then the role of IPCC has certainly grown beyond any expectation, and its reports are seen as the "state of the art" in the field of climate science. Its most recent report, the Fifth Assessment Report, was published in 2014.

The establishment of an international political framework for action

The recommendations made at Villach in 1985 had a long filiation. They were in particular integrated into the report *Our Common Future*[18] presented at the UN General Assembly in 1987, and they provoked a significant discussion about climate change. The report was also largely disseminated, and the climate change issue became higher on the political agenda. Gordon Goodman, who had been instrumental in shaping science policy exchanges on global environmental issues,[19] aimed to provide a link between the Villach conference and the elaboration of specific measures to limit—or adapt to—warming. Therefore he proposed the organization of two interlocking international conferences in 1987 under the auspices of AGGG—one in Villach on the impact of greenhouse gases concentration increases on regional parts of the world, and another in Bellagio, Italy, on policy responses to climate change, with a mixed audience. This led to the Bellagio proposal that policymakers should set long-term quantitative targets rates of sea level rise—between 20mm and 50mm per decade—and of mean global temperature increase—a maximum of 0.1°C per decade. Sixth months later in Toronto the

"Changing Atmosphere" conference called for an international convention, and for a reduction of CO_2 concentrations of 20 percent by 2005, compared to 1988. A strong heat wave in the US contributed to a large coverage of the conference in the media and to put climate change even higher on the political agenda.[20]

In 1990, the negotiations on the United Nations Framework Convention on Climate Change (UNFCCC) formally began in the Intergovernmental Negotiating Committee (INC), under the auspices of the UN General Assembly. The Convention was then signed at the Earth Summit in Rio in 1992 and entered into force in 1994. If the creation of IPCC can be considered as the first political answer to the scientific alarm, the elaboration and signing of the convention would "definitively not" have been possible without the IPCC.[21] The IPCC work is carried out by three groups: Working Group (WG) I deals with the physical basis of the problem, WG II with the impacts of climate change, and WG III with the solutions. In the first IPCC assessment report, the scientific consensus reached by a credible, international group of scientists in WG I, as well as the work of WG III experts to define some of the terms of a climate convention, were key to push the negotiations. The UNFCCC objectives have served as a reference for more than two decades. They are clearly defined as the "stabilization of greenhouse gas concentrations in the atmosphere at a level that would prevent dangerous anthropogenic interference with the climate system. Such a level should be achieved within a time frame sufficient to allow ecosystems to adapt naturally to climate change, to ensure that food production is not threatened and to enable economic development to proceed in a sustainable manner."[22] This text is balanced between the necessities of the planet, of biodiversity, and sustainable development. The definition of action was left to the yearly Conferences of the Parties (COP).

The COP met for the first time in 1995 in Berlin, and began the discussion on a protocol, unanimously adopted in Kyoto two years later. The Kyoto Protocol fixed a global objective of GES reduction to industrialized countries, which was set to 5 percent for the first commitment period, from 2008 to 2012. The reference year to establish the reduction level was 1990. Each country that was cited in the "Annex I" to the convention (e.g. developed countries) agreed on a "burden sharing" which took the form of national quantitative objectives. The developing countries and emerging countries, members of the same group of negotiation, the "G77," had no quantitative commitments. The Kyoto Protocol eventually entered into force in 2005 after a very long process of ratification. The Kyoto Protocol was a very important first step in the international action, it was a laboratory to design new policies, but it had structural deficiencies. Its architecture was very rigid, with a deep North–South division. After China became the largest greenhouse gases emitter in 2006, this inability to engage large emerging countries into any form of meaningful action, became clearly unbearable. This may be partly the reason why it was never ratified by the United States. Soon after Canada left, in 2011, the developed countries which were still Parties to the Protocol, represented less than 15 percent of the global emissions. At COP15 in 2009 in Copenhagen, the United Nations failed to negotiate a new climate agreement. Nevertheless, a G20 initiative involving countries which were not part of Annex I, such as China, Brazil, and India, agreed on a quantitative target: to maintain the average surface temperature of the Earth under a 2°C increase, compared to preindustrial values. The Copenhagen conference was therefore not the complete failure that is often called. It also showed that a future agreement had to be universally validated in a transparent way. Thereafter, a new reflection on such an agreement progressively started. It had to be built on new foundations, which were already set up in Durban in 2011.[23]

The COP designated for the final definition of this agreement was COP21, in 2015, which the French government proposed to host in Paris. What occurred in Paris was a universal agreement, legally binding, transparent, equitable, and sustainable. It retained the objective to stabilize climate warming under 2°C, and expressed the will of staying well under this target, in response to the many countries asking for a 1.5°C commitment, in particular small island countries. One of the major characteristics of the agreement is probably to include a process of periodic review of the nationally determined contributions (NDCs), every five years. Each country has to set up its own objectives, and, if needed, to revise them with increasing ambitions, in order to reach collectively the global necessary commitment, in a bottom-up process. The role of non-state actors is also recognized, and will be key to any success of this agreement. This new climate governance regime is in fact built on four pillars: the negotiated text, the intended nationally determined contributions (INDCs) of the countries for the period 2020–2030, the initiatives of non-state actors in the Paris–Lima Action Plan, and the financial contribution of developed countries to support action on adaptation and on mitigation in developing countries.

COP21 ended up as a full success, the Paris Agreement entered into force in November, 2016, although, at the time of writing this chapter, many countries are still in the process of ratifying the accord. The whole process was extremely and unusually short (less than one year as compared with the seven years necessary for the Kyoto protocol). A credit for this success should be given to the negotiators. Another driving factor was certainly the fact that the impacts of climate change are increasingly observed and measured worldwide. In 30 years, the status of climate change risks has changed for many policymakers: from a potential danger, it has become a real and informed risk. Climate change perception is progressively entering our everyday life, culture, and history.

The context in which climate change has to be considered is also evolving very rapidly. In September 2015, the adoption by the UN Assembly in New York of the 2030 Agenda for Sustainable Development, and of the Sustainable Development Goals (SDGs) therein, was also a major milestone. Succeeding the Millennium Development Goals (MDGs), with the intention to reduce poverty by 2015, the 17 SDGs are inclusive and universal: each country is involved—either developing, emerging, or developed—and has to set an action plan to reach the targets which it will determine internally, depending on its own priorities. Here also the nature and ambition of the actions taken at the national level will determine the overall capacity to reach global targets. The agenda of the SDGs is very broad, and encompasses *inter alia* health, gender equality, water, education, food security and climate change: all those problems need to be considered collectively.

From the alert to the solutions

The Paris Agreement, as the SDGs, constitutes an important and necessary framework for action, but it is obviously far from being sufficient.

We have left the development and the results of the "physical science" of climate in 1990, at the time of the first IPCC report. There is a widespread belief that since then the only change in the scientific message has been the growing insistence with which it has been issued by the scientific community, as the first signs of the changes anticipated by the models were appearing in the real world. This is of course an important part of what happened. There are now many observations whose careful analysis show that climate change is already starting: the increase of the mean average surface temperature (around 0.8°C since the 1970s, with larger values in the Arctic regions and over the continents), the regular increase of the mean sea level (more than 3mm per year, with larger trends expected in the future). More generally many parameters that describe the planet in its mineral and life-related aspects have undergone changes that are now more difficult to explain through the effects of natural variability, whereas they match what can be expected from the impact of greenhouse gases. The Fourth Assessment Report of the IPCC, issued in 2007, was the first one to state that "climate change is unequivocal."[24]

But the sense of urgency has also changed with the realization that climate risks should not only be estimated through what already happened, but also through the very strong engagement of the future which results from past GES emissions. This emerged clearly from the Fifth Assessment Report of the IPCC (2014).[25] Roughly half of the CO_2 mass that is emitted at a given moment is retained by the atmosphere, rather than taken back by the continental soil or the oceans, and this atmospheric CO_2 surplus disappears very slowly: half of it will still be there after a century. The decay time for N_2O (nitrous oxide) is also about a century, although it is shorter—about twelve years—for CH_4 (methane), at which point it converts to CO_2, and longer for CFC (chlorofluorocarbons), and also indirectly produces O_3 (ozone). In any case, past emissions will continue to act on climate during the coming decades, whereas the rate of emissions has increased continuously. A useful diagnosis of the current situation is therefore the amount of greenhouse gas emissions aggregated since the beginning of the industrial era: if it goes over a limit of approximately 3,000 billion tons of equivalent CO_2, the chances of staying under the 2°C warming level during the coming century will be less than 66 percent. We are now rather close to this situation: we have emitted so far a bit less than 2,000 billion tons of equivalent CO_2, and at the current rate of about 50 billion tons yearly, little more than 20 years will be enough to consume our total allowance. A simple image may summarize the situation: the atmosphere is like a basin that leaks very slowly, and which we fill with greenhouse gases at an increasing rate. The yearly rate of global CO_2 emissions, for example, was multiplied by a factor of 2 between the early 1980s and today.

Models have tried to add a calendar to these figures. As the models are affected by quantitative uncertainties, and therefore collectively produce a range of results, we may retain here only orders of magnitude. To maintain a global temperature below 2°C at the end of the century, societies have to reach "carbon neutrality" in the second half of the century. Models show that after reaching their peak value (as soon as possible in the two forthcoming decades), greenhouse gas emissions have to decrease to zero before the end of the century: this is necessary because of the strong impact by past and current emissions already underlined. This perspective, which was approved by all nations through the Paris agreement, and ratified by the largest emitters of greenhouse gases, implies a huge transformation of our societies, because it requires us to find in a matter of decades an alternative energy mix to replace fossil fuels, which account for around 80 percent of our energy needs.

IPCC and the many manifestations of scientific expertise have been crucial to bring this diagnostic into the Paris agreement, to define objectives that take it into account, and establish a mechanism for the periodical review of the nationally determined contributions.[26] But it is not enough to design, implement, assess, and monitor the strategies and action plans necessary to face a global situation that is much more pressing than anticipated. Real change is needed. The current contributions of the nations are too low by at least a factor 2 to limit the warming under 2°C.[27] And yet, the most vulnerable nations—such as small islands and intertropical countries potentially facing stronger droughts or extreme events—ask for more stringent measures that would limit the warming to 1.5°C. They also ask for a financial compensation that would enable them to develop in a more sustainable manner, and to adapt, as much as possible to the component of climate changes which cannot be avoided: these so-called "green funds" have been one of the key issues of the climate negotiations.

All this defines a new, enlarged role for science, scientific expertise, and scientists in our societies. In addition, the adoption of the SDGs mobilizes far beyond the climate science. This necessary engagement of scientists in the policy process is already visible through the succession of conferences that were organized prior to major negotiations on global changes, such as "Climate Change—Global Risks, Challenges and Decisions" in Copenhagen in March 2009, before the COP15[28] and "Planet under Pressure" in 2012 in London, before the "Rio + 20" conference.

Ahead of COP21, an important effort has also been devoted to mobilize the scientific community. Building on the findings of the IPCC Fifth Assessment Report published between September 2013 and November 2014, two months before the adoption of the SDGs in New York, and four months before COP21, a large scientific conference was held in Paris on July 7–10, 2015, called "Our Common Future under Climate Change" (CFCC15). More than 2,000 scientists from 95 countries attended the 160 sessions of the conference held at UNESCO and University Pierre and Marie Curie. They included also stakeholders, national delegates, and non-governmental representatives. Climate scientists representing the three IPCC working groups, together with experts of other global challenges, such as biodiversity loss, food security, soil, poverty, social justice, and equity were present. It was perhaps the first time that such a diverse scientific assembly came together to discuss all aspects of climate change, in connection with other global challenges and in view of achieving sustainable forms of development. In particular, a significant participation of social and human scientists allowed a reflexive thinking on the evolution of the role of science.

We may use the definition and results of this conference to illustrate the new paradigms that science has to confront. Its aim was presented by the conference's organizers as follows:

> The new climate governance regime is supposed to strengthen confidence, support implementation, maximize benefits of international cooperation, and bring all stakeholders to the realization that a new development model (low carbon, resilient) is actually emerging.
>
> For science, the question has progressively shifted from consolidating the scientific basis for assessing risks and options for action, to defining the form that action has to take in order to engage in a necessary transition to low-carbon and adapted economies and societies. For stakeholders, the question has shifted from reasons for action to the form action has to take. Today, the scientific community, in partnerships with a variety of stakeholders, plays a major role for shaping our future under climate change, by identifying potential sustainable futures and innovations at different spatial and time scales, by designing and assessing relevant and coherent solutions, policies and measures, and therefore increasing the credibility of the Paris agreement.[29]

Further, a statement was presented at the end of the conference, which was eventually signed by more than 100 research and higher education organizations and institutions worldwide. It contains two sets of recommendations for the "Solution Space" and the "Problem Space," and states in its preamble:

> Science is a foundation for smart decisions at COP21 and beyond. Solving the challenge of climate change requires ambition, dedication, and leadership from governments, the private sector, and civil society, in addition to the scientific community.
>
> We in the scientific community are thoroughly committed to understanding all dimensions of the challenge, aligning the research agenda with options for solutions, informing the public, and supporting the policy process.[30]

The conversation between conference participants, which represented a large diversity of approaches, disciplines, and geographic origins—including a representation of indigenous knowledge—and also countries with different levels of development, offers a model for future scientific cooperation. The hope is that the actions in and since 2015 will determine a new interdisciplinary role for science, and increase the responsibility of scientists in facing global changes.

Environmental sciences and society: an interface with growing complexity

As emphasized above, changing the role of science from sounding alarms to contributing to solutions has been a recurrent issue debated at the CFCC15, at COP21, COP22, and in the preparation phase of new IPPC Reports.

IPCC is obviously a major actor in this perspective. It was conceived as a very subtle process associating the diagnosis of science and the necessities of decision-making. It is a rare and maybe only established example of such a structure. It is now almost 30 years old, was reviewed in 2010 by an InterAcademy Panel and served as a model for further initiatives, such as the Intergovernmental Science-Policy Platform on Biodiversity and Ecosystem Services (IPBES). One of the reasons for this success is certainly the prior existence of a truly international scientific community. We have mentioned WMO, an effort of sharing data, models and practices which is more than 140 years old, and WCRP, which has set new infrastructures to monitor the climate of the Earth since 1980: it has structured the work of thousands of researchers. As the diversity of scientific approaches grew, other international projects, placed under the auspices of the International Council of Scientific Unions (ICSU) or the International Council of Social Sciences (ISCC), have also played a major role: the International Geosphere Biosphere Programme (IGBP, founded in 1987) dealing with global biogeochemical cycles and past climates; Diversitas (founded in 1991) dealing with biodiversity, the International Human Dimension Programme (IHDP, founded in 1990) addressing the coupled human–natural system, have also organized their respective research communities. They have helped organize observatories, field work, polar or maritime expeditions, and much more. The community of scientists working in all those programmes, their results, their debates, their analysis, have fed the different Working Groups of IPCC.

IPCC remains a pillar of crucial importance. Its stated goal is to establish the message of climate sciences, in a way that is "relevant" to decision-makers but not "prescriptive." Indeed IPCC has always been cautious to maintain a clear distinction between its mission in assessing factual climate issues, and the use of science by democratically elected governments. But the central importance given to national action plans by the Paris agreement has reaffirmed the necessity of other forms of expertise that may complement the mandate of IPCC. The negotiation process taking place every five years and starting in 2018 will require quick reactions, and a capacity to judge the relative engagement of the different nations. These important debates will be difficult to carry out at the level of UN organizations, without a prior investigation at the level of each nation. Many decisions at a more local level also need to be informed by science. Finally, a new factor is that the prime decision-makers will increasingly be the citizens themselves: it is very difficult to carry out transitions at the scale that is foreseen without their involvement and consent.

Science and scientific expertise have already begun to evolve. In particular the need to develop multidisciplinary research has motivated the creation of "Future Earth," a global platform which aims to develop research for sustainability objectives, ensuring that the corresponding knowledge is generated in partnership with society and users of science. "Future

Earth" emerged in 2012 at the "Rio + 20" Conference, after a long maturation process.[31] It merges IGBP, Diversitas, and IHDP, acts as a partner of WCRP, and develops new instruments of transverse science.

A few examples may illuminate the new scientific avenues that have begun to emerge through such partnerships. Cities now constitute the largest human settlements on the planet, and among the most vulnerable to climate change. Scientific work needs to be developed to determine the best strategies to ensure their sustainable development. Each related debate (on the question of increased versus decreased density, for example) requires a vast expertise extending from meteorology or air quality, to urban planning, transportation issues, protection of the specific urban biodiversity, health, social and cultural requirements, and so on. To develop a careful analysis of these issues, a large number of actors are necessary, and the technical texts and analysis produced by local authorities need to rely on comprehensive research—it is also an obligation to meet the necessary criteria to be taken into account and assessed by IPCC assessments. Some real academic interdisciplinary work is therefore necessary. It is complex to organize, but it is happening. A forthcoming international "Climate Change and Cities" conference, for example, will be held in 2018, and will be co-organized by a long list of institutions: Future Earth, IPCC, but also Atelier Cities Alliance, C-40, ICLEI-Local Governments for Sustainability, SDSN, United Cities and Local Governments (UCLG), UN-Habitat, UNEP and WCRP. Science as it relates to cities and climate change is in an active process of formation.

Another promising avenue is the development of common scientific methodologies to provide *ex ante* and *ex post* assessments of national strategies and policies. New tools and analysis are also needed to help decision-makers and stakeholders to build sustainable pathways, and initiatives are currently being developed along these lines. For instance, the Deep Decarbonization Pathways Project,[32] and the Agricultural Transformation Pathways Initiative,[33] building on backcasting scenarios, address the energy sector and the land use sector respectively.

There are many lessons which have already been learned from such initiatives, and many more which cannot be cited here. The first one is that interdisciplinary approaches need to be built on strong disciplinary pillars. A motion that would just shift the research effort and neglect fundamental work would end in failure. Another one is that building "relevant" science is difficult. The procedures and processes to engage productive interactions between scientists, policymakers, and citizens can sometimes play a role as important as the production of scientific knowledge itself. In some situations it also appears that, even if the relevant knowledge and data already exist, the demand for scientific expertise to determine action is very low. Scientists have therefore to be proactive and bring a strong attention to the societal and political debates, and let the different actors of the society know what science can produce and offer. Developing "Climate Services" along these lines was precisely the aim of the World Climate Conference-3 organized by WMO in 2009 in Geneva. It has induced the development worldwide of a large number of projects, such as the ambitious Copernicus project of the European Commission.

More generally, building a constructive interface between science and society is a delicate task that requires meeting a wide range of conditions. For example: a common basis in terms of acknowledged scientific results; clear and transparent processes of interaction between –scientists and decision-makers; scientists already engaged in the debate with policymakers, stakeholders or different societal partners; a clear and shared formulation of what is expected from science; research tools and methodologies which actors can easily understand, use, and make their own.

Conclusion

This chapter on climate change history deliberately stresses the importance of the complex interface between science and society, at the expense of many other issues. Adopting such a point of view means that we have developed an historical perspective that is very focused on the present and the future. We have good reasons to believe that this is necessary. The diagnosis, the perception, the modes of actions, and the risks associated with climate change have been constantly evolving during the last half-century, and the most important changes are very often the more recent.

In a matter of decades, climate issues have gone from a remote, academic concern about long-term changes at the end of the century, to an immediate concern about current and future consequences in the real world. What is required to stabilize climate under a low or moderate level of warming is a change in energy production and use, as well as land use and forestry changes, which amounts to nothing less than a total transformation of our societies worldwide. Other sectors of activity need to evolve, too: agriculture, for example is emitting nitrous oxide and methane, but may also constitute a solution, since agricultural soils could store (or sequester) much more carbon than they do currently.

The important decisions to come will require a way of balancing different risks: climate risks, because the attenuation of the changes will necessarily remain partial, but also social risks, or other environmental risks, such as those linked with biodiversity. Climate science is therefore no longer the only source of guidance, but precisely because there is a need to balance climate risks and other risks, the monitoring and anticipation of climate changes has taken a new importance. And considering actions raises many pending questions that have so far received only partial or uncertain answers. The first one is of course how to define a timeline for the reduction of greenhouse gas emissions. But there are many others. How should we define what is a fair contribution for a given country? Can we try to artificially reduce the solar heating of the planet and how dangerous would this be? Is it already necessary to capture atmospheric carbon and sequestrate it underground to stay under a warming of 1.5°C (or under a warming of 2°C)? Can agriculture help us in this process? What are the damages that may result from insufficient greenhouse reduction? How far can adaptation help us? What are the criteria to adequately finance adaptation measures through the Green Funds?

The uncertainty that affects climate predictions limits the capacity of science to answer these questions. This may require some clarification, and for that purpose it is possible to split the consequences of climate change into two parts. One concerns the processes that are a direct consequence of the greenhouse warming. This is the case of the melting of glaciers and sea-ice, of the sea-level rise, of certain changes in the ecosystems, and so on. If the amplitude and therefore the timing of these risks may be uncertain, the direction of the evolution is generally unambiguous. But climate warming may also have more indirect consequences, associated with modifications of the atmospheric circulation: this is how it will affect precipitation, cyclone occurrence, severe droughts, cloud and water vapor radiative effects, and so on. As the evolution of atmospheric circulation has an unpredictable and chaotic component, the range of potential changes is more wide open, and regional features are especially difficult to predict. This process is often described as a "deregulation" of climate resulting from the greenhouse effect.

Even in these mostly physical approaches of climate changes, it is impossible to forget about society. Climate risks cannot be defined only on the basis of the severity and frequency of meteorological hazards. They also depend on the vulnerability of the socio-ecosystems that are affected. The notion of vulnerability (and that of resilience) have opened the way to a wide range of new studies, many of them at the local or regional scale (i.e. at the scale where important infrastructure is being constructed). Vulnerability studies also reveal a deep gap between intertropical and mid-latitude countries. In the former, the climate hazards may be much more important and take more easily an extreme and dangerous form. This gap is reinforced by the situation of many of those countries that face poverty, war, or civil unrest, malnutrition, and uncontrolled demographic increase, while they have not contributed meaningfully to greenhouse gases emissions. In that sense, the design of adaptation policies is a good manner to discuss these issues, provide a quick remedy, and anticipate potential conflicts, within a nation or between nations.

To summarize, we are at a crucial moment: the choices facing our societies have never been so complex and intricately linked. Tackling climate change and engaging in sustainable pathways encompasses unprecedented challenges for societies, science, and democracy. This creates for scientists, institutions, and organizations supporting science, a new obligation to bring the necessary information to the citizens and the policy process, while listening carefully to the expectations of the society.

Notes

1   The notion of "biosphere" is linked with the Gaia theory proposed by James Lovelock. Vernadsky also proposed a "technosphere" in relations with human action, which is related with the notion of Anthropocene, and a "noosphere," the sphere of human consciousness. This latter idea was also developed by Pierre Teilhard de Chardin (1922), and the Internet may be seen as one its realizations.
2   There are theories that human ecosystem change in Europe and North America contributed to the Little Ice Age, but these theories are yet unproven.
3   J. G. Charney, "Dynamics of Deserts and Drought in the Sahel," *Quarterly Journal of the Royal Meteorological Society* 101, no. 428 (1975): 193–202.
4   R. Revelle and H. Suess, "Carbon Dioxide Exchange Between Atmosphere and Ocean and the Question of an Increase of Atmospheric CO_2 During the Past Decades," *Tellus* 9 (1957): 18–27.
5   T. C. Peterson, W. M. Connolley, and J. Fleck, "The Myth of the 1970s Global Cooling Scientific Consensus," *Bulletin of the American Meteorological Society* (2008), retrieved from http://journals.ametsoc.org/doi/abs/10.1175/2008BAMS2370.1.
6   H. Le Treut et al., "Historical Overview of Climate Change," in IPCC, *Climate Change 2007: The Physical Science Basis*, Contribution of Working Group I to the Fourth Assessment Report of the Intergovernmental Panel on Climate Change (Cambridge: Cambridge University Press, 2007).
7   Ad Hoc Study Group on Carbon Dioxide and Climate, *Carbon Dioxide and Climate: A Scientific Assessment* (Washington, DC: National Academy of Science, 1979).
8   J. G. Charney, "On the Scale of Atmospheric Motions," *Geofys Publikasjoner*, 17 (1948): 1–17.
9   WMO is an intergovernmental organization that was reorganized in 1947. It is the specialized agency of the United Nations for meteorology (weather and climate), operational hydrology, and related geophysical sciences.
10 Le Verrier is better known for the discovery of planet Neptune, almost simultaneously with the British scientist John Adams, in 1846. This may serve a reminder that climate sciences never developed as a separate entity, but within the general context of science at large.
11 Examples include the GARP Atlantic Tropical Experiment (GATE) in 1974, the First GARP Global Experiment in 1979, or the Alpine Experiment (ALPEX) in 1982.
12 The first director of WCRP, was Pierre Morel, previously at CNES (French Space Research Center) and later on at NASA. See P. Morel, L. Charles, and H. Le Treut, "Regard historique sur la recherche climatique, entre observations et modèles: Climate Sciences, Observation and Modelling: An Historical Perspective," *Pollution Atmosphérique* special issue (June 2013), retrieved from www.appa.asso.fr/_docs/1/fckeditor/file/Revues/PollutionAtmospherique/Hors-serie-climat-juin-2013/Morel.pdf.
13 World Climate Programme, International Council of Scientific Unions, United Nations Environment Programme, and World Meteorological Organization, *Report of the International Conference on the Assessment of the Role of Carbon Dioxide and of Other Greenhouse Gases in Climate*

Variations and Associated Impacts, Villach, Austria, 9–15 October 1985 (Paris: International Council of Scientific Unions, 1986).

14 A. D. Hecht and D. Tirpak, "Framework Agreement on Climate Change: A Scientific and Policy History," *Climate Change* 29 (1995): 371–402.

15 Wendy E. Franz, *The Development of an International Agenda for Climate Change: Connecting Science to Policy*, IIASA Interim Report, IR-97-034 (Laxenburg, Austria: IIASA, 1997).

16 Shardul Agrawala, "Context and Early Origins of the Intergovernmental Panel on Climate Change," *Climatic Change* 39, no. 4 (1998): 605–20.

17 Ibid.

18 WCED, *Our Common Future* (Oxford University Press, 1987): 383. The report is also known as the Brundtland Report, from the name of the former Norwegian prime minister Mrs. Gro Harlem Brundtland, who chaired its redaction.

19 Gordon Goodman wrote the terms of reference for SCOPE (the Scientific Committee on Problems of the Environment) and presented the scientific submission that persuaded governments to accept the establishment of a Global Environmental Monitoring System (GEMS).

20 Franz, The Development of an International Agenda for Climate Change.

21 Jean Ripert, interview by Shardul Agrawala, Paris, France, March 14, 1997, cited in Agrawala, "Context and Early Origins of the Intergovernmental Panel on Climate Change." Jean Ripert was a French diplomat closely involved in the IPCC process and then elected as chair of the INC.

22 United Nations Framework Convention on Climate Change, 1771 UNTS 107; S. Treaty Doc No. 102-38; UN Doc. A/AC.237/18 (Part II)/Add.1; 31 ILM 849 (1992).

23 Michel Colombier, *COP21: Building an Unprecedented and Sustainable Agreement*, Working Paper no. 13/2015 (Paris: IDDRI, 2015), retrieved from www.iddri.org/publications/cop21-building-an-unprecedented-and-sustainable-agreement.

24 "Summary for Policymakers," in IPCC, *Climate Change 2007*.

25 "Summary for Policymakers," in *Climate Change 2013: The Physical Science Basis*, Contribution of Working Group I to the Fifth Assessment Report of the Intergovernmental Panel on Climate Change (Cambridge: Cambridge University Press, 2013).

26 Laurence Tubiana, First Future Earth Days Conference, Paris, December 1, 2016. Laurence Tubiana is Chief Executive Officer of the European Climate Foundation and has been Special Representative for the 2015 Paris Climate Conference.

27 From the "Groupe Interdisciplinaire sur les Contributions Nationales" established during the COP21. See also Olivier Boucher, Valentin Bellassen, Hélène Benveniste, Philippe Ciais, Patrick Criqui, Céline Guivarch, Hervé Le Treut, Sandrine Mathy, and Roland Séférian, "Opinion: In the Wake of Paris Agreement, Scientists Must Embrace New Directions for Climate Change Research," *PNAS* 113, no. 27 (2016): 7287–90.

28 K. Richardson, W. Steffen, and D. Liverman, *Climate Change: Global Risks, Challenges and Decisions* (Cambridge: Cambridge University Press, 2011): 501.

29 Our Common Future Under Climate Change, "Why This Conference?," 2015, retrieved from www.commonfuture-paris2015.org/The-Conference/Objectives.htm.

30 Our Common Future under Climate Change, "Outcome Statement," 2015, retrieved from www.commonfuture-paris2015.org/The-Conference/Outcome-Statement.htm.

31 Future Earth, "Who We Are," undated, retrieved from www. futureearth.org/who-we-are.

32 Deep Decarbonization Pathways Project, "Publication of a Climate Policy Special Issue on the DDPP," July 4, 2016, retrieved from http://deepdecarbonization.org/2016/07/publication-of-a-climate-policy-special-issue-on-the-ddpp.

33 Schwoob, M.-H., et al., *Agricultural Transformation Pathways Initiative: Summary* (Paris: IDDRI/Harpenden: Rothamsted Research, 2016), retrieved from www.iddri.org/Publications/Rapports-and-briefing-papers/ATPi%20-%20Summary%202016.pdf.

References

Ad Hoc Study Group on Carbon Dioxide and Climate. *Carbon Dioxide and Climate: A Scientific Assessment*. Washington, DC: National Academy of Science, 1979.

Agrawala, Shardul. "Context and Early Origins of the Intergovernmental Panel on Climate Change." *Climatic Change* 39, no. 4 (1998): 605–20.

Boucher, Olivier, Valentin Bellassen, Hélène Benveniste, Philippe Ciais, Patrick Criqui, Céline Guivarch, Hervé Le Treut, Sandrine Mathy, and Roland Séférian. "Opinion: In the Wake of Paris Agreement, Scientists Must Embrace New Directions for Climate Change Research." *PNAS* 113, no. 27 (2016): 7287–90.

Charney, J. G. "On the Scale of Atmospheric Motions." *Geofys Publikasjoner*, 17 (1948): 1–17.

Charney, J. G. "Dynamics of Deserts and Drought in the Sahel." *Quarterly Journal of the Royal Meteorological Society* 101, no. 428 (1975): 193–202.

Colombier, Michel. *COP21: Building an Unprecedented and Sustainable Agreement.* Working Paper no. 13/2015. Paris: IDDRI, 2015. Retrieved from www.iddri.org/publications/cop21-building-an-unprecedented-and-sustainable-agreement.

Deep Decarbonization Pathways Project. "Publication of a Climate Policy Special Issue on the DDPP." July 4, 2016. Retrieved from http://deepdecarbonization.org/2016/07/publication-of-a-climate-policy-special-issue-on-the-ddpp.

Franz, Wendy E. *The Development of an International Agenda for Climate Change: Connecting Science to Policy.* IIASA Interim Report, IR-97-034. Laxenburg, Austria: IIASA, 1997.

Future Earth. "Who We Are." Undated. Retrieved from www. futureearth.org/who-we-are.

Hecht, A. D., and D. Tirpak, "Framework Agreement on Climate Change: A Scientific and Policy History." *Climatic Change* 29 (1995): 371–402.

IPCC. *Climate Change 2007: The Physical Science Basis.* Contribution of Working Group I to the Fourth Assessment Report of the Intergovernmental Panel on Climate Change. Cambridge: Cambridge University Press, 2007.

IPCC. *Climate Change 2013: The Physical Science Basis.* Contribution of Working Group I to the Fifth Assessment Report of the Intergovernmental Panel on Climate Change. Cambridge: Cambridge University Press, 2013.

Morel, P., L. Charles, and H. Le Treut. "Regard historique sur la recherche climatique, entre observations et modèles: Climate Sciences, Observation and Modelling: An Historical Perspective." *Pollution Atmosphérique* special issue (June 2013). Retrieved from www.appa.asso.fr/_docs/1/fckeditor/file/Revues/PollutionAtmospherique/Hors-serie-climat-juin-2013/Morel.pdf.

Our Common Future under Climate Change. "Outcome Statement." 2015. Retrieved from www.commonfuture-paris2015.org/The-Conference/Outcome-Statement.htm.

Our Common Future Under Climate Change. "Why This Conference?" 2015. Retrieved from www.commonfuture-paris2015.org/The-Conference/Objectives.htm.

Peterson, T. C., W. M. Connolley, and J. Fleck. "The Myth of the 1970s Global Cooling Scientific Consensus." *Bulletin of the American Meteorological Society* (2008). Retrieved from http://journals.ametsoc.org/doi/abs/10.1175/2008BAMS2370.1.

Revelle, R., and H. Suess. "Carbon Dioxide Exchange Between Atmosphere and Ocean and the Question of an Increase of Atmospheric CO_2 During the Past Decades." *Tellus* 9 (1957): 18–27.

Richardson, K., W. Steffen, and D. Liverman. *Climate Change: Global Risks, Challenges and Decisions.* Cambridge: Cambridge University Press, 2011.

Schwoob, M.-H., et al. *Agricultural Transformation Pathways Initiative: Summary.* (Paris: IDDRI/Harpenden: Rothamsted Research, 2016. Retrieved from www.iddri.org/Publications/Rapports-and-briefing-papers/ATPi%20-%20Summary%202016.pdf.

WCED. *Our Common Future.* Oxford University Press, 1987.

World Climate Programme, International Council of Scientific Unions, United Nations Environment Programme, and World Meteorological Organization. *Report of the International Conference on the Assessment of the Role of Carbon Dioxide and of Other Greenhouse Gases in Climate Variations and Associated Impacts, Villach, Austria, 9–15 October 1985.* Paris: International Council of Scientific Unions, 1986.

20

THE PROBLEM OF ECONOMIC GROWTH

Richard Heinberg

Introduction

Economic growth is both medicine and poison. Nations have come to rely on ever-expanding economic activity, measured in terms of gross domestic product (GDP), in order to create jobs, generate higher incomes for workers, fund government services, and provide returns to investors. Virtually every politician, regardless of party affiliation, promises growth. It is the universal prescription for all economic ills.

However, economic growth is tied (whether loosely or tightly is debated, as we will see) to the expansion of overall energy use and materials consumption. In a world of finite resources, long periods of growth are therefore likely to end in an eventual crash. Indeed, efforts to address climate change and other pressing environmental problems are already confronting the political imperative to keep economies growing. Every new increment of growth makes it harder to reduce greenhouse gas emissions, or to slow resource depletion, biodiversity loss, topsoil degradation, environmental pollution, and other rapidly spiraling crises. It may not be an exaggeration to say that the success or failure of the entire human project during the current century may hinge upon the problem of growth, and whether societies can come to terms with it.

In this chapter, we will explore reasons for the unprecedented economic growth of recent decades, especially during the latter half of the twentieth century. We will identify some of the ingredients needed for substantial growth to occur in an economy. We will see how GDP became the world's single most important measure of economic success. We will examine criticisms of continued reliance on growth. We will explore whether growth can be maintained, perhaps in other and less destructive ways than is currently the case. And finally, we will see why the insistence on unending economic growth is making climate change mitigation so difficult.

A short history of economic growth

An economy is a system whereby human beings use energy to extract and transform Earth's renewable and non-renewable resources; distribute and use goods; and dispose of wastes. The growth of an economy can be described in terms of scope (the number of people in the system) or intensity (the quantities of resources—and hence energy as well—used per capita). Further, economic systems can be organized in a variety of very different ways.

Historically, the simplest economies were based on the direct harvesting of wild foods, fuels (mostly in the form of firewood), and other materials from the environment, and the sharing of goods without an intermediary token of exchange. Human muscle provided virtually all motive power. The vast majority of human economies, throughout tens of thousands of years, were organized this way. Because humans were spreading out to inhabit new territories, total global population and consumption slowly increased, but per capita rates of consumption were small and not prone to continual growth.

The agricultural revolution in Mesopotamia (roughly 10,000 years ago) and the creation of cities in the present-day Middle East (about 7,000 years ago) led to full-time division of labor, the introduction of money, and the harnessing of animal muscle; these developments in turn enabled expanded trade and increased wealth accumulation. Per capita consumption grew, along with wealth inequality. Civilizations emerged and some overall economic growth occurred—though societies still had access only to whatever amounts of wealth could be produced through the engagement of human and animal muscle in agriculture and in the harvesting of minerals and firewood.

Simple banks (accepting deposits, making loans, and charging interest) appeared in ancient Assyria, Babylonia, Greece, Rome, India, and China; banking enabled the easier stockpiling and transferring of larger amounts of wealth. More sophisticated banks emerged in northern Italy during the Middle Ages and Renaissance and proliferated throughout Europe. This proved a timely development, because European colonial adventures that began in the sixteenth century led to a substantial transfer of wealth from large swathes of the world to a few increasingly powerful maritime countries, for which plunder served as a pathway to economic growth.

At the same time, Europe came to be dominated by an economic theory and practice known as *mercantilism,* which promoted the establishment of national economic policies to encourage a positive balance of trade, especially in finished goods. These policies (forbidding colonies to trade with other nations, establishing high tariffs on imported manufactured goods, subsidizing domestic manufacturing and exports) both motivated and justified colonial expansion and coincided with over two centuries of trade wars and colonialist wars of conquest. Mercantilist theorists such as Thomas Mun (1571–1641) supposed that growth by these means could continue indefinitely.

By the late eighteenth century, mercantilism had found critics. An opposing doctrine, *physiocracy*, propounded by François Quesnay (1694–1774) and others, recognized the phenomenon of diminishing returns from successive applications of any input to the economic process (e.g., plowing the same field year after year), and predicted that humanity would eventually outgrow its resources. In effect, the physiocrats believed that there are inherent limits to economic growth. However, they failed to garner a significant or enduring following.

Greater success accrued to British economists Adam Smith (1723–90) and David Hume (1711–76), who criticized the mercantilists' notion that protectionist trade policies would lead to ever-increasing national wealth. Instead they advocated freer trade among nations— an idea that became part of the bedrock of what would come to be known as classical economics. Though some classical economists such as John Stuart Mill (1806–73) believed that economies would grow only until they achieved a satisfactory and sustainable size (a "stationary state"), and that unlimited growth would lead to the destruction of the environment and a reduced quality of life, others assumed that growth could and should continue indefinitely.[1]

In the nineteenth century, Karl Marx (1818–83) termed Western nations' emerging economic system *capitalism,* describing its fundamental motive and process as capital accumulation. He defined capital as money or goods devoted to the production of more money and goods (that is, not destined for direct consumption), and saw the accumulation of capital as an inherently self-reinforcing process based ultimately on capitalists' extraction of wealth from Earth's resources and from the surplus value of employed labor. Marx regarded capitalism as unsustainable primarily for social and political reasons, though he also recognized ecological limits to resource extraction.

Meanwhile, industrializing societies had begun using coal, then oil and natural gas, as energy sources. These fossil fuels provided energy that was cheaper, more concentrated, and (in the case of oil) more portable than previous energy resources. The effort that had to be put into mining or drilling for fossil fuels was trivial in comparison with the amount of energy unleashed. Even extremely inefficient uses of these fuels (such as in early steam engines) could do far more work than human or animal muscles.

Soon inventors realized that all this new energy represented an enormous opportunity, and they began finding ways to take advantage of it. In doing so, they built upon work that had already commenced: since the sixteenth century, physical scientists had been laying the theoretical groundwork for a rapid development of engineering principles and practices; new energy sources turbocharged the process, resulting in an explosion of inventions to make direct or indirect use of fossil fuels. Transport technologies (railroads and steamships, then motorcars and trucks) encouraged a further and now dramatic expansion of trade. In industrial nations, agricultural machinery freed a substantial portion of the population to move to cities and take up middle-class jobs in manufacturing, sales, engineering, marketing, customer service, advertising, banking, and on and on (initially two-thirds or more of the total populace worked at farming; eventually less than two percent could grow enough food for everyone). A burgeoning chemicals industry, using fossil fuels as feedstocks and heat sources, produced a cascading array of ingredients for industrial processes. Powered assembly lines sped up manufacturing, while powered mining equipment, saws, and fishing vessels drove an acceleration of resource extraction. Finally, electrification—based at first on coal and hydropower, and later also on nuclear energy and natural gas—supplied a highly versatile energy carrier to nearly every home and business, suitable for the operation of efficient motors as well as lighting and communications technologies.

The result was by far the fastest economic expansion in history in terms of both scope and intensity. As it occurred, national economies became more integrated, melding into a global economy—though a very lopsided one, with some nations seeing dramatic wealth expansion and others, especially in the Global South, suffering from colonial policies that exploited both human labor and natural resources.

In the early twentieth century, the economies of industrial nations diverged significantly in the degree to which they were planned and managed by government, and the degree to which government redistributed wealth among citizens. However, all accepted the idea that growth was good and that it could be maintained indefinitely. Each nation raced to grow the largest, the fastest, and indeed the historian J. R. McNeill has called growth the most important idea of the twentieth century.[2]

The history of economic growth can thus be described as a slow developmental period lasting many millennia, leading to roughly two centuries of rapid exponential increase in population, energy use per capita, GDP per capita, and total GDP.

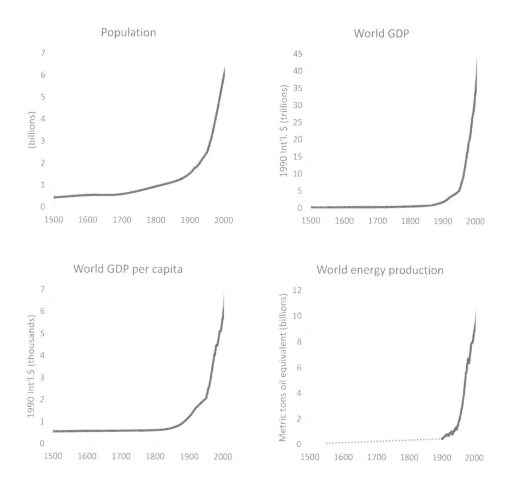

Figure 20.1 World population, GDP, GDP per capita and energy production, 1500–2000

Note: Int'l. $ = international (Geary–Khamis) dollars; see https://stats.oecd.org/glossary/detail.asp?ID=5528.

Sources: data from Angus Maddison, "Statistics on World Population, GDP and Per Capita GDP, 1–2008 AD," 2010, www.ggdc.net/maddison/historical_statistics/horizontal-file_02-2010.xls; The Shift Project, "Historical Energy Production Statistics," 2016, www.tsp-data-portal.org/Energy-Production-Statistics; see also Vaclav Smil, *Energy Transitions: History, Requirements, Prospects* (Santa Barbara, CA: ABC-CLIO, 2010). Images produced by author.

Ingredients for growth in industrial societies

The previous section retold a story that, in its general outline, nearly everyone already intuitively understands. We have, however, taken pains to highlight historical details that are helpful in revealing causal links. With this narrative in hand, it now may be helpful to tease apart some of the essential circumstances or ingredients that made industrial societies' recent, rapid economic growth possible. (These do not correspond exactly with the ingredients for growth listed in most macroeconomics textbooks; more about this below.) Doing so may give us helpful perspective on growth's vulnerabilities and potentials. The following are four of those ingredients; as we will see, it is possible to distill more from the historical data, but these are useful for the purposes of our discussion.

The capacity for capital formation

The existence of money, banking, debt, and interest appear to have been necessary preconditions for significant capital accumulation. These in turn presuppose an agrarian basis of production. (It seems safe to conclude that economic growth of the scale seen in the twentieth century could not happen in a hunting-and-gathering society.)

However, prior to the industrial era, money, debt, interest, and banking never provided a sure foundation for continuous economic growth. Agrarian economies were subject to cycles of expansion and decline. Further, economy-wrenching debt bubbles occurred as early as ancient Sumeria, requiring periodic debt jubilees to avert social collapse or conquest by neighbors. In their book *Secular Cycles* (2009), ecologist Peter Turchin and historian Sergey Nefedov document, describe, and attempt to account for the long-term (i.e. "secular") phases of expansion and contraction in several historically documented agrarian societies.[3] They see population growth and environmental carrying capacity as ultimately critical issues; however, sociopolitical factors such as the form of government can affect the specific trajectories that societies take. Turchin's and Nefedov's historical survey suggests that the average secular cycle plays out in about 300 years.

Clearly, while an agrarian base and the means for capital accumulation are necessary for a growth take-off, they are not sufficient by themselves.

Geopolitical opportunity

In his 1997 book *Guns, Germs and Steel,* geographer and physiologist Jared Diamond investigated why some societies have become rich and powerful while others stagnated.[4] He argued that geography largely determines national destiny. Eurasia gained an early advantage due to the greater availability of plant and animal species suitable for domestication; the continent's large landmass and long east-west axis increased these advantages, as did the microbial exchanges that built up resistance to disease among the diverse cultures of Eurasia. The plentiful supply of food and the dense populations that it supported ultimately made division of labor possible.

Further, there can be little doubt that the countries that succeeded first in colonizing others enjoyed a substantial advantage. For instance, in 1600, India was a far wealthier and more industrialized nation than England; by 1900, now a colony, India had seen most of its wealth appropriated, leaving England rich and India nearly destitute. However, colonization as a national wealth generator has proven to be subject to the law of diminishing returns, and the same will likely prove true of current forms of neocolonialism based on unequal trade rules. The growth of some nations at the expense of others, by whatever means, is a zero-sum game that, in its later stages, is likely to lead to revolution or rebellion.

Technology to apply energy (and thereby increase labor productivity)

Technologies to leverage the power of human labor have undoubtedly contributed greatly to the economic growth of the past century. Societies that devote effort and resources toward technological development therefore have a better chance at achieving growth. However, like colonization, technological development may also be subject to the law of diminishing returns; in any case, it depends upon the availability of appropriate forms of energy.

While a dizzying number of tools have been invented in recent decades, the most economically significant of them have been directed toward just a few universal human endeavors, including communicating; producing food, clothing, buildings, and other goods; and transporting people and goods. During recent decades, technologies have greatly sped up these endeavors and made them cheaper to accomplish. However, it is questionable whether further improvements of anywhere near the same magnitude are possible. A couple of centuries ago, if two people in distant cities wished to communicate, one of them traveled by foot, horse cart, or boat (or sent a messenger or private mail courier). It may have taken weeks. Now, using a smartphone, we communicate almost instantly. If we need to travel in person, we can arrive in a distant city in a matter of hours. Food, clothing, and houses are produced at minimal cost, using dedicated machinery at each stage of the process.

While we are never likely to reach zero in terms of time and cost in pursuing these essential human endeavors, we can be fairly certain that *the closer we get to zero time and cost, the higher the cost of the next improvement and the lower the value of the next improvement is likely to be.*[5] This means that, with regard to each basic technologically mediated pursuit (communication, transportation, food production, and so on), we will eventually reach a point where the cost of the next improvement is higher than its value. At that point, further economic "improvements" will be driven almost solely by aesthetic considerations identified by advertisers and marketers rather than by those achieved by engineers or inventors; and by the production of new kinds of goods for which humans have only marginal or peripheral needs. In the more highly industrialized societies, this stage has arguably already arrived.

The development of technologies to replace human labor in the commercial economy may reach a point of diminishing returns when machines replace laborers faster than new jobs appear; beyond this point, an ever-higher proportion of potential workers no longer has the wherewithal to purchase new goods, unless wealth can be redistributed via a program of taxes and government unemployment benefits (perhaps in the form of a guaranteed universal basic income).

In addition, as noted above, the further proliferation of technology depends on the availability of suitable energy; this constitutes the fourth of our ingredients.

High EROI energy sources

As highlighted in our brief history of growth, rapid economic take-off did not occur until the beginning of the fossil-fuel era. What was it about coal, oil, and natural gas that evidently made them catalytic to the growth process? We have cited some of the qualities of fossil fuels that enabled growth—their cheapness, abundance, portability, and energy concentration. One further quality demands attention: the high ratio of energy returned on the energy invested in accessing it (this is commonly measured in terms of energy return on investment, EROI; or energy returned on energy invested, EROEI). It takes energy to drill an oil well, build a nuclear power plant, or manufacture and install a wind turbine. It is the surplus energy left over after all energy costs in energy production, transformation, and delivery are accounted for that enables economic productivity and hence growth. Fossil fuels had a much higher EROI than firewood: both represent stored sunlight, but with fossil fuels nature did additional work in transforming and concentrating the energy resource over tens of millions of years at no cost to humans.

If coal, oil, and natural gas played a pivotal role in recent historic instances of rapid economic growth, this implies a vulnerability, since fossil fuels are non-renewable and therefore subject to depletion. Further, we extract these fuels based on the low-hanging fruit principle, so fuels with the highest EROI tend to be extracted first. Numerous scholars have found that, over time, the EROI of fossil fuels has tended to decline.[6] Some might argue that EROI declines could be offset simply by expanding gross production, i.e. by increasing both the EROI numerator and denominator. This would in theory increase the supply of net energy to society, but in practice the effort would be subject to limitations in available capital, skilled labor, economically recoverable reserves, and so on. Since all economic activity requires energy, declining EROI implies an ultimate limit to economic expansion, unless other high-EROI energy sources can be identified and substituted. But this must happen before the EROI of fossil fuels erodes so much that society is no longer able to organize and power the transition to these alternative sources. With oil, by one calculation, we are very close to this point.[7]

Most economic historians emphasize the first of these four ingredients for growth (the capacity for capital formation) and the third (technology); the second (geopolitical opportunity) is cited less frequently. Surprisingly, many economic historians omit the fourth (suitable energy resources) almost entirely, along with associated environmental considerations.

As we have seen, all four of these primary ingredients for growth are subject, in one way or another, to the law of diminishing returns. These ingredients, therefore, are also vulnerabilities: without them, growth stalls or reverses.

Other requisites for growth could be suggested. For example, in hindsight it would appear that one necessary ingredient for growth was a relatively stable global climate maintained for several millennia. Others might include an environment not already loaded with industrial waste products; or one that is not already depleted of freshwater, soil, or minerals.

Pondering these ingredients and vulnerabilities naturally leads to the question: In the face of diminishing returns, depletion, and increasing pollution loading, can economic growth nevertheless be maintained—for example, by transitioning from fossil fuels to alternative energy sources? We will discuss this prospect shortly. Before doing so, it may be helpful to recount how continual economic growth became normalized in economic theory and government policy. How could something so extraordinary become the universal basis of policy and planning?

How GDP normalized a boom

We humans are optimistic by nature: we easily get used to good times and often have unrealistically positive expectations about the future.[8] A recent example comes readily to mind. In the years after the 2008 financial crisis, petroleum prices were high and interest rates low; these circumstances set the stage for an oil boom in an unexpected place—North Dakota. Geologists previously had some general knowledge of the existence of significant oil resources there, but considered them uneconomic, as the oil was locked up in a formation (consisting mostly of shale) with low permeability. Very little oil could be recovered from a typical vertical well within the Bakken formation. But high prices appeared to justify using sophisticated technology (horizontal drilling and hydrofracturing, also known as fracking), and the drilling of more wells per unit of oil produced. Dozens of companies leased land and drill rigs, and company representatives promised decades of oil and high profits to potential investors. Tens of thousands of jobs soon appeared, and workers flooded into North Dakota from other states. An economic bonanza overwhelmed small towns surrounding the Bakken oilfields, with property values and hotel room rates skyrocketing. Oil production also shot up, offering tangible evidence to support the rising expectations of oil executives, investors, and local officials. Every responsible agency forecast long-term growth in production, profits, and economic benefits—and on this basis, towns in North Dakota began long-term investments in new schools, roads, and other infrastructure. Then, in mid-2014, the price of oil collapsed. Companies canceled drilling plans. Profits disappeared. Workers were laid off. Oil production rates fell. Suddenly former boom-towns faced falling real estate values, empty hotel rooms, and weak economies.

Dozens of similar examples from ancient to modern times could be cited. Booms feed on the natural human desire for an easier life, more wealth, and better days. And the bigger and longer the boom, the more likely it is that rosy but unrealistic expectations will be taken up and reinforced by society's most trusted leaders and institutions.

As it happened, the normalization and institutionalization of economic expansion in the twentieth century was eased by the development of a statistic, GDP, which came to be the very talisman of growth. By measuring economic activity in monetary terms that were further abstracted into a single number, whose derivation was poorly understood by the general public, the actual process of growth became mystified: people tended not to think of it so much in terms of resource extraction or population increase. If planners and policy makers had publicly said, "Let us build our economy upon the effort to perpetually increase the rate at which we extract Earth's resources and turn them to wastes, while we also encourage growth in the number of people consuming and wasting," many members of the public might have responded by saying, "Hang on, that's not likely to end well." (I am not arguing that the mystification of growth was a purpose of the adoption of GDP, only that it was a consequence.)

By most accounts, Simon Kuznets (1901–85), an economist at the National Bureau of Economic Research, deserves credit for inventing GDP, using it as the basis for his 1937 Report to the United States Congress, "National Income, 1929–32." His idea was to capture all economic production by individuals, companies, and the government in a single monetary measure, which should rise in good times and fall in bad. Following the 1944 Bretton Woods conference that established international financial institutions such as the World Bank and the International Monetary Fund, GDP became the standard tool for evaluating any and every country's economic success. However, Kuznets told Congress that GDP had inherent limits:

> The welfare of a nation can scarcely be inferred from a measure of national income. … Distinctions must be kept in mind between quantity and quality of growth, between costs and returns, and between the short and long run. Goals for more growth should specify more growth of what and for what.[9]

In 1962, Arthur Okun (1928–80), staff economist for US President John F. Kennedy's Council of Economic Advisers, formulated what came to be known as Okun's Law, which holds that for every three-point rise in GDP, unemployment will fall one percentage point. The theory thenceforth informed monetary policy, and the health of the body politic was assumed to depend on positive GDP numbers.

In 1978, economists Irving Kravis, Alan Heston, and Robert Summers compiled the first estimates of GDP per capita worldwide, listing figures for more than 100 countries. By this time, the usefulness of this statistical tool was almost universally praised, and Kuznets's caveats had been almost entirely forgotten; in 1999 the US Department of Commerce declared GDP "one of the great inventions of the 20th century."[10]

By now, it is surely fair to say the world is addicted to economic growth as measured by GDP. The fortunes of politicians and their parties rise or fall on their ability to deliver growth. Central banks manage interest rates to this end. Governments, businesses, and households borrow on the assumption that future growth will enable loan repayment. It is never enough merely to maintain the economy at its current size. Without growth, it is assumed that economic, political, and social upheaval will follow.

Criticisms of growth and GDP

During the twentieth century, as growth was becoming normalized, there were nevertheless those who regarded it as perilous and temporary, or who criticized GDP as a measure of economic health.

In 1959, economist Moses Abramovitz (1912–2000) became one of the first to question whether GDP accurately measures a society's overall well-being, cautioning that "we must be highly skeptical of the view that long-term changes in the rate of growth of welfare can be gauged even roughly from changes in the rate of growth of output."[11] This line of criticism was later taken up by other economists who pointed out that GDP measures economic activity without regard to its actual human benefit: a war or natural disaster can hike GDP numbers as readily as an improvement in labor productivity.

Mathematician and economist Nicholas Georgescu-Roegen (1906–94), in his *The Entropy Law and the Economic Process*, argued that all natural resources are irreversibly degraded when put to economic use, and that Earth's carrying capacity—that is, its capacity to sustain human populations and consumption levels—must eventually decrease as finite stocks of mineral resources (including fossil fuels) are extracted and used. Therefore, as long as the overall economy is growing, the world is accelerating toward an inevitable future collapse.[12]

In 1972, the book *Limits to Growth* by Donella Meadows, Dennis Meadows, Jorgen Randers, and William Behrens III (writing for the Club of Rome) caused a sensation, becoming the best-selling environmental book of all time. The authors examined the parameters of population, resource depletion, and pollution, using systems dynamics and computer modeling to develop various scenarios for future growth. Most scenarios showed a crash sometime in the 21st century, and recent research has legitimated their findings.[13]

In 1973, Herman E. Daly, a student of Georgescu-Roegen, published an anthology titled *Toward a Steady-State Economy,* arguing that rather than continual growth, policymakers should aim for a rate of material and energy usage that can be maintained more or less indefinitely.[14]

In his book *Overshoot* (1982), sociologist William Catton, Jr. (1926–2015) compared economic growth with biological phenomena. Catton observed that fossil fuels have enabled a temporary increase in Earth's carrying capacity for humans, both by increasing our food supply (via agricultural machinery and artificial fertilizers) and by decreasing our death rate (with better sanitation and medical care, including the use of fossil fuel-derived pharmaceuticals). With cheap transport fuel, we can bring needed resources (including food and medicines) from places of abundance to places of scarcity, again increasing overall carrying capacity. However, given the fact that fossil fuels are finite and depleting, human population and consumption may have overshot Earth's long-term carrying capacity, while also degrading it by depletion and pollution.[15]

In the 1990s, Mathis Wackernagel and William Rees developed *ecological footprint* analysis to quantify humanity's environmental impact, measured as the amount of natural resources consumed each year.[16] As of 2012, according to the Global Footprint Network, humans were consuming more than 1.6 Earth's worth of annual natural productivity, meaning that we are overshooting available carrying capacity by over 60 percent, doing so by drawing down future carrying capacity.[17]

Around the same time, Herman Daly and other ecological economists introduced Genuine Progress Indicator (GPI), which takes account of environmental and social factors, as an alternative to GDP.[18] In 2008, GPI was updated based on theoretical work and practical experience associated with government efforts to achieve and measure Gross National Happiness (GNH) in the nation of Bhutan.[19]

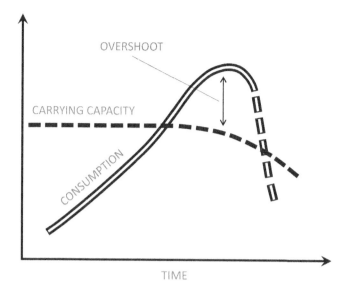

Figure 20.2   Earth's carrying capacity and humanity's resource consumption over time

Note: In this conceptual diagram, when total resource consumption increases beyond the ecological carrying capacity of the Earth ("overshoot"), carrying capacity diminishes and ultimately forces total consumption to decline. Total consumption decline may take the form of reduced per capita consumption, reduced population, or both.

Sources: The concept of overshoot was most famously developed in William Catton's *Overshoot* (Champaign, IL: University of Illinois Press, 1982); this simplified figure is based on Mathis Wackernagel and William Rees, *Our Ecological Footprint* (Gabriola Island: New Society Publishers, 1996). Image produced by author.

In 2001, the bursting of a tech bubble and the 9/11 attacks sent the United States economy into a brief but steep recession. During the subsequent recovery, between 2002 and 2006, GDP rose but personal incomes fell. Following the subsequent global financial crisis of 2007–9, GDP growth was restarted only through massive intervention by governments and central banks. Still, revived growth was sluggish and most income gains went to the financial elite. Economists including Robert Gordon and Lawrence Summers noted the trend toward ever-slower growth in already industrialized economies, describing this "secular stagnation" as a puzzling new normal.[20]

Growth by other means?

Not all critics of growth accept the inevitability of an eventual crash such as the one that Georgescu-Roegen forecasted. Some say a crash can be averted by resource substitution or by increasing the efficiency with which energy and materials are used.

Resource substitution is a strategy that might include, for example, replacing fossil fuels with renewable sources of energy like solar and wind, or with nuclear power. The EROI of solar, wind, and nuclear is lower than that shown by fossil fuels during the decades of highest world economic growth rates, but often better than that of biomass derived fuels.[21] Hydroelectricity often has a fairly high EROI, but its use is limited by geographies and its future is somewhat imperiled by expected changes to hydrological cycles. In any case, it is easy to forget how strong the energy returns once were on fossil fuels, and especially oil. Around 1900, oil often yielded 100 units of energy for every 1 unit of energy for extraction and refining—a ratio of 100:1. A hundred years later that number had fallen to 10:1, although the ratio can vary greatly depending on circumstance. Put differently, the economic growth of the twentieth century was fueled by access to cheap, energy-dense, and abundant fossil fuels, and it is unrealistic to think that renewables, which generally have a lower EROI, could magically replace depleted fossil fuels and continue to support the levels of growth and consumption that have come to define industrial society.

Further, high EROI is not the only criterion an energy source must satisfy. Concerns such as cost, infrastructural requirements, and environmental pollution need to be taken into consideration. Nuclear power is plagued by high costs and risks; for these reasons the nuclear industry is shrinking in most industrialized nations. With regard to solar and wind, there are doubts about whether there is sufficient time or investment capital for a robust transition, and about whether these alternative energy sources have the characteristics that would support economic growth in the way that fossil fuels have done. This author recently collaborated with David Fridley of the Energy Analysis Program at Lawrence Berkeley National Laboratory in assessing the prospects for a transition to 100 percent renewable energy.[22] We concluded that, while producing enough solar panels and wind turbines and integrating them into a grid system will be challenging enough, current energy-using infrastructure is largely designed to take advantage of specific characteristics of fossil fuels and will require very substantial overhaul or replacement in order to work with renewables. We conclude there is only a small likelihood of society having enough useful energy at the end of a rapid renewable energy transition (such as is being proposed to avert catastrophic climate change) in order to meet energy demand growth forecasts from official agencies such as the IEA.

In theory, growth can be achieved by using resources more efficiently, thereby decoupling growth in energy and materials use from GDP growth. With regard specifically to energy-GDP decoupling, there is evidence that some progress has occurred during recent decades, though recent analysis suggests that much of this apparent decoupling may be due to factors other than increased efficiency. For example, some decoupling in Western nations (notably the United States) is due simply to outsourcing of manufacturing: China is credited with the energy used for manufacturing goods ultimately consumed in the United States.[23] In addition, some observers claim that U.S. GDP growth statistics have been artificially pumped up in the last decade or so for political reasons by systematically underestimating inflation.[24] To the degree this is the case, the result would be an apparent but misleading decoupling. A recent study of the decoupling phenomenon by one group of scholars found that "achievements in decoupling in advanced economies are smaller than reported or even nonexistent."[25] Certainly a perspective that takes into consideration outsourcing and global trade tends to be less optimistic about absolute and even relative decoupling.

In the future, some energy–GDP decoupling could be achieved by increased efficiencies associated with switching energy sources. This is because processes that convert the energy from coal or gas to electricity are only about 40 percent efficient, at best, while solar panels and wind turbines produce electricity directly, without need for conversion; further, electric motors in cars are significantly more efficient than internal combustion engines at turning raw energy into motive force. However, the energy transition will also entail some new inefficiencies, such as the need for energy storage. Altogether, it is unclear how large, or how persistent, net efficiency gains from a transition to renewable energy will be.

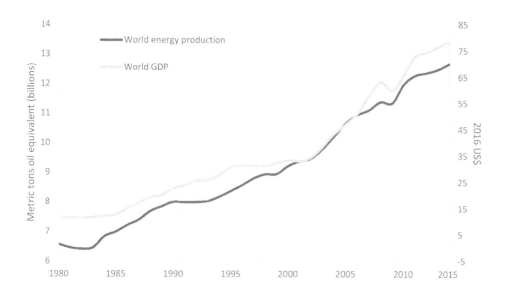

Figure 20.3 World energy production and world GDP, 1980–2015

Sources: The Shift Project, "Historical Energy Production Statistics," 2016, www.tsp-data-portal.org/Energy-Production-Statistics; The World Bank, "GDP (Current US$)," 2016, http://data.worldbank.org/indicator/NY.GDP.MKTP.CD. Image produced by author.

In the discussion above, of the ingredients for growth, we noted in passing that the list proposed here does not exactly correspond with the one recognized by most conventional macroeconomists. For the conventional theorist, economic output (and hence GDP growth) is typically seen as depending on only three factors: the amount of labor in the workforce, the amount of capital (e.g., factories, machines, computers, buildings), and "total factor productivity," or TFP—which is often defined as an economy's technological dynamism.[26]

TFP can be thought of as all of the reasons the economy grows not already accounted for by the quantity of labor and capital. In statistical terms this is called a "residual," or an amount unexplained by the factors in an equation. Macroeconomic analysis shows that a significant part of TFP can be attributed to the application of energy to economic processes (thereby increasing labor productivity) and by the increasing efficiency of converting energy to physical work.[27] Gains in TFP and labor productivity (economic output per hour of labor) have been relatively anemic in the last decade.[28]

Thus while standard economic theory indirectly accounts for the role of energy, it does not do so explicitly, and therefore fails to capture, for example, changes in energy quality over time. This may be a fatal flaw in conventional economics (as others have already noted).[29] Quite simply, conventional economic theory merely assumes the availability of energy, and does not inquire into its EROI let alone its impact on the natural environment. That could be a serious theoretical shortcoming in an era when a major societal shift in energy sources is necessary in response both to fossil fuel depletion and climate change.

Growth versus climate action

Theoretical problems aside, a much bigger difficulty with growth is that efforts to tackle climate change may require rethinking or abandoning it. In preparation for its fifth assessment report in 2014, the International Panel on Climate Change (IPCC), the United Nations agency charged with studying and responding to climate change, adopted Representative Concentration Pathway (RCP) scenarios for greenhouse gas concentrations during the next few decades.[30] All of the scenarios assume global economic growth, population growth, and growth in energy use. The scenarios model various tradeoffs between fossil fuel use, transitions (at differing rates) to nuclear or renewable energy sources, decoupling of energy use from GDP growth, and introduction of negative emissions technologies.[31]

While it is usually assumed that RCP8.5 (which features the highest emissions) is a "business as usual" scenario, the world's actual supply of economically producible fossil fuels may correspond more closely with RPC4.5 or even the RCP 2.6 scenarios.[32] But if that is truly the case (i.e., if total energy from fossil fuels peaks between 2020 and 2040) then further economic growth would be hard to achieve.

Replacing energy from coal, oil, and gas with energy from solar, hydro, wind, and nuclear would provide a way forward toward limiting warming nearer to 2 degrees Celsius, the officially agreed-upon maximum. But, as discussed above, enormous investments of both capital and energy (mostly from fossil fuels, at least in the early stages) would be required not only for

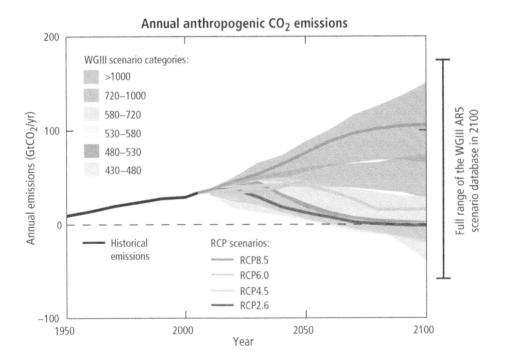

Figure 20.4 Possible carbon dioxide emission scenarios through 2100

Note: RCP = representative concentration pathway. WGIII = Working Group III (of the IPCC).

Source: IPCC, *Climate Change 2014: Synthesis Report*, Contribution of Working Groups I, II and III to the Fifth Assessment Report of the Intergovernmental Panel on Climate Change (Geneva: IPCC, 2014), 9.

the production of new energy supply technologies (solar panels, wind turbines, etc.) but also for constructing new energy-related infrastructure. This would surely rob capital and energy from other sectors of the economy, making business-as-usual economic growth problematic.

Most 2 degree Celsius scenarios rely heavily on negative emissions technologies, which aim to pull carbon out of the atmosphere. But many of these are unproven, while others are likely unscalable (one of the most promising, carbon sequestration in soils, would require massive changes to agricultural practices globally). There would be little incentive to develop these technologies absent a substantial price on carbon, which might itself inhibit economic growth.

In short, no combination of fossil fuel depletion, transition to alternative energy sources, or negative emissions technologies may realistically be up to the job of keeping the world below 2 degrees Celsius—as long as further economic growth is either assumed or required. The problem of how to increase energy supplies and reduce fossil fuels and their emissions, while still yielding economic growth in the meantime, is one that is stretching the keenest minds on the planet—some would say, to the point of breaking with reality.

Conclusion

Our exploration of the problem of economic growth leads us to four conclusions:

1   The rapid, multi-decade economic growth seen in recent decades is extraordinary and unprecedented in human history, and almost certainly cannot be maintained much longer.
2   We may have grown too much already, overshooting Earth's long-term carrying capacity.
3   There are signs that growth is indeed ending.
4   Insisting on further growth makes it harder to come to terms with the biggest environmental challenge in human history—climate change.

Political leaders' insistence on prioritizing further growth above all else forecloses certain options. Rather than voluntarily redesigning our economy so as to avert crisis, we are more likely to postpone adapting to planetary resource limits and pollution limits until a growth crisis forces us to do so. What will the crisis be? Most assume it will appear in the form of climate change; however, other critical limits are also converging—including global debt levels and a rapid decline in the EROI of oil. In any case, time for adaptation is limited, and policy makers who do become aware of encroaching limits should, in addition to efforts to reduce impacts, such as by limiting GHG emissions, also work to build societal resilience, so that as impacts arrive, crucial life-support systems can be maintained.

Notes

1   See Jeremy L. Caradonna, *Sustainability: A History* (New York: Oxford University Press, 2014).
2   John Robert McNeill, *Something New Under the Sun: An Environmental History of the Twentieth-Century World* (New York: W. W. Norton & Company, 2000), 236.
3   Peter Turchin and Sergey Nefedov, *Secular Cycles* (Princeton, NJ: Princeton University Press, 2009). Turchin and Nefedov describe secular cycles as "demographics–social–political oscillations of very long period (centuries long)" (ibid., 5).
4   Jared Diamond, *Guns, Germs, and Steel: The Fates of Human Societies* (New York: W. W. Norton & Company, 1999).
5   Mats Larsson, *The Limits of Business Development and Economic Growth: Why Business will Need to Invest Less in the Future* (New York: Palgrave Macmillan, 2004).
6   David J. Murphy and Charles A. S. Hall, "Year in Review—EROI or Energy Return on (Energy) Invested," *Annals of the New York Academy of Sciences* 1185, 1 (2010): 102–18; Charles A. S. Hall, Jessica

G. Lambert, and Stephen B. Balogh, "EROI of Different Fuels and the Implications for Society," *Energy Policy* 64 (2014): 141–52; Mikael Höök and Xu Tang, "Depletion of Fossil Fuels and Anthropogenic Climate Change—A Review," *Energy Policy* 52 (2013): 797–809.

7   Charles A. S. Hall, Stephen Balogh, and David J. R. Murphy, "What is the Minimum EROI that a Sustainable Society Must Have?" *Energies* 2, 1 (2009): 25–47. See also forthcoming work by Louis Arnoux and the Hill's Group on the same subject.

8   See Tali Sharot, Alison M. Riccardi, Candace M. Raio, and Elizabeth A. Phelps, "Neural Mechanisms Mediating Optimism Bias," *Nature* 450, 7166 (2007): 102–5; Neil D. Weinstein, "Unrealistic Optimism about Future Life Events," *Journal of Personality and Social Psychology* 39, 5 (1980): 806–20.

9   Simon Kuznets, "National Income, 1929–32, A Report to the US Senate," 73rd Congress, 2nd Session, document.

10 BEA, "GDP: One of the Great Inventions of the 20th Century," retrieved on June 10, 2017 from www.bea.gov/scb/account_articles/general/0100od/maintext.htm.

11 Moses Abramovitz, "The Welfare Interpretation of Secular Trends in National Income and Product," in *The Allocation of Economic Resources: Essays in Honor of Bernard Francis Haley* (Stanford, CA: Stanford University Press, 1959): 1–22.

12 Nicholas Georgescu-Roegen, *The Law of Entropy and the Economic Process* (Cambridge, MA: Harvard University Press, 1971).

13 Donella H. Meadows, Dennis L. Meadows, Jorgen Randers, and William W. Behrens III, *The Limits to Growth: A Report for the Club of Rome's Project on the Predicament of Mankind* (New York: Universe Books, 1972); see also Graham Turner, *A Comparison of the Limits to Growth With Thirty Years of Reality, Socio-Economics and the Environment in Discussion*, CSIRO Working Paper Series (Collingwood: CSIRO, 2008–9).

14 Herman E. Daly, ed., *Toward a Steady-State Economy* (New York: Freeman, 1973).

15 William R. Catton, *Overshoot: The Ecological Basis of Revolutionary Change* (Champaign, IL: University of Illinois Press, 1982).

16 Mathis Wackernagel and William Rees, *Our Ecological Footprint: Reducing Human Impact on the Earth* (Gabriola Island: New Society Publishers, 1996).

17 Global Footprint Network, "National Footprint Accounts, 2016 Edition," Global Footprint Network, retrieved on November 11, 2016, from www.footprintnetwork.org/en/index.php/GFN/page/public_data_package.

18 Herman E. Daly, John B. Cobb, and Clifford W. Cobb, *For the Common Good: Redirecting the Economy Toward Community, the Environment, and a Sustainable Future* (Boston, MA: Beacon Press, 1989).

19 Winston Bates, "Gross National Happiness," *Asian-Pacific Economic Literature* 23, 2 (2009): 1–16.

20 Robert J. Gordon, "Secular Stagnation: A Supply-Side View," *The American Economic Review* 105, 5 (2015): 54–9; Lawrence H. Summers, "Demand Side Secular Stagnation," *The American Economic Review* 105, 5 (2015): 60–65.

21 Murphy and Hall, "Year in Review."

22 Richard Heinberg and David Fridley, *Our Renewable Future*, Post Carbon Institute (Washington, DC: Island Press, 2016).

23 Marina Fischer-Kowalski and Mark Swilling, *Decoupling: Natural Resource Use and Environmental Impacts from Economic Growth* (Nairobi: United Nations Environment Programme, 2011).

24 John Williams, "Shadow Government Statistics," 2016, retrieved on November 11, 2016, from www.shadowstats.com.

25 Thomas O. Wiedmann et al., "The Material Footprint of Nations," *Proceedings of the National Academy of Sciences* 112, 20 (2015): 6271–6.

26 See Jesus Felipe and F. Gerard Adams, "A Theory of Production: The Estimation of the Cobb-Douglas Function: A Retrospective View," *Eastern Economic Journal* 31, 3 (2005): 427–45.

27 Maria Gabriela Ladu and Marta Meleddu, "Is There Any Relationship Between Energy and TFP (Total Factor Productivity)? A Panel Cointegration Approach for Italian Regions," *Energy* 75 (2014): 560–67; Can Tansel Tugcu and Aviral Kumar Tiwari, "Does Renewable and/or Non-Renewable Energy Consumption Matter for Total Factor Productivity (TFP) Growth? Evidence from the BRICS," *Renewable and Sustainable Energy Reviews* 65 (2016): 610–16.

28 John Fernald, *Productivity and Potential Output Before, During, and After the Great Recession* (Cambridge, MA: National Bureau of Economic Research, 2014).

29 Charles A. S. Hall and Kent A. Klitgaard, "Poverty, Wealth, and Human Aspirations," in *Energy and the Wealth of Nations: Understanding the Biophysical Economy* (New York: Springer, 2015), 5–8.

30 IPCC, *Climate Change 2013: The Physical Science Basis*, Contribution of Working Group I to the Fifth Assessment Report of the Intergovernmental Panel on Climate Change (Geneva: IPCC, 2013).
31 Note that the IPCC has been criticized for using, at times, unrealistic assumptions in creating their scenarios. See, for instance, Jeff Tollefson, "Is the 2 °C World a Fantasy?," *Nature*, retrieved on November 11, 2016, from www.nature.com/news/is-the-2-c-world-a-fantasy-1.18868.
32 Here I draw on forthcoming work by David Hughes.

References

Abramovitz, Moses. *The Allocation of Economic Resources: Essays in Honor of Bernard Francis Haley*. Stanford, CA: Stanford University Press, 1959.

Bates, Winston. "Gross National Happiness." *Asian-Pacific Economic Literature* 23, 2 (2009): 1–16.

BEA. "GDP: One of the Great Inventions of the 20th Century." Retrieved on June 10, 2017 from www.bea.gov/scb/account_articles/general/0100od/maintext.htm.

Caradonna, Jeremy L. *Sustainability: A History*. New York: Oxford University Press, 2014.

Catton, William R. *Overshoot: The Ecological Basis of Revolutionary Change*. Champaign, IL: University of Illinois Press, 1982.

Daly, Herman E., ed. *Toward a Steady-State Economy*. New York: Freeman, 1973.

Daly, Herman E., John B. Cobb, and Clifford W. Cobb. *For the Common Good: Redirecting the Economy Toward Community, the Environment, and a Sustainable Future*. Boston, MA: Beacon Press, 1989.

Diamond, Jared. *Guns, Germs, and Steel: The Fates of Human Societies*. New York: W. W. Norton & Company, 1999.

Felipe, Jesus, and F. Gerard Adams. "A Theory of Production: The Estimation of the Cobb-Douglas Function: A Retrospective View." *Eastern Economic Journal* 31, 3 (2005): 427–45.

Fernald, John. *Productivity and Potential Output Before, During, and After the Great Recession*. Cambridge, MA: National Bureau of Economic Research, 2014.

Fischer-Kowalski, Marina, and Mark Swilling. *Decoupling: Natural Resource Use and Environmental Impacts from Economic Growth*. Nairobi: United Nations Environment Programme, 2011.

Georgescu-Roegen, Nicholas. *The Law of Entropy and the Economic Process*. Cambridge, MA: Harvard University Press, 1971.

Global Footprint Network. "National Footprint Accounts, 2016 Edition." Global Footprint Network, retrieved on November 11, 2016, from www.footprintnetwork.org/en/index.php/GFN/page/public_data_package.

Gordon, Robert J. "Secular Stagnation: A Supply-Side View." *The American Economic Review* 105, 5 (2015): 54–9.

Hall, Charles A. S., and Kent A. Klitgaard. *Energy and the Wealth of Nations: Understanding the Biophysical Economy*. New York: Springer, 2015.

Hall, Charles A. S., Stephen Balogh, and David J. R. Murphy. "What is the Minimum EROI that a Sustainable Society Must Have?" *Energies* 2, 1 (2009): 25–47.

Hall, Charles A. S., Jessica G. Lambert, and Stephen B. Balogh. "EROI of Different Fuels and the Implications for Society." *Energy Policy* 64 (2014): 141–52.

Heinberg, Richard, and David Fridley. *Our Renewable Future*. Washington, DC: Island Press, 2016.

Höök, Mikael, and Xu Tang. "Depletion of Fossil Fuels and Anthropogenic Climate Change—A Review." *Energy Policy* 52 (2013): 797–809.

IPCC. *Climate Change 2013: The Physical Science Basis*. Contribution of Working Group I to the Fifth Assessment Report of the Intergovernmental Panel on Climate Change. Geneva: IPCC, 2013.

Kuznets, Simon. "National Income, 1929–32, A Report to the US Senate." 73rd Congress, 2nd Session, document.

Ladu, Maria Gabriela, and Marta Meleddu. "Is There Any Relationship Between Energy and TFP (Total Factor Productivity)? A Panel Cointegration Approach for Italian Regions." *Energy* 75 (2014): 560–67.

Larsson, Mats. *The Limits of Business Development and Economic Growth: Why Business will Need to Invest Less in the Future*. New York: Palgrave Macmillan, 2004.

McNeill, John Robert. *Something New Under the Sun: An Environmental History of the Twentieth-Century World*. New York: W. W. Norton & Company, 2000.

Meadows, Donella H., Dennis L. Meadows, Jorgen Randers, and William W. Behrens III. *The Limits to Growth: A Report for the Club of Rome's Project on the Predicament of Mankind*. New York: Universe Books, 1972.

Murphy, David J., and Charles A. S. Hall. "Year in Review—EROI or Energy Return on (Energy) Invested." *Annals of the New York Academy of Sciences* 1185, 1 (2010): 102–18.

Sharot, Tali, Alison M. Riccardi, Candace M. Raio, and Elizabeth A. Phelps. "Neural Mechanisms Mediating Optimism Bias." *Nature* 450, 7166 (2007): 102–5.

Summers, Lawrence H. "Demand Side Secular Stagnation." *The American Economic Review* 105, 5 (2015): 60–65.

Tollefson, Jeff. "Is the 2°C World a Fantasy?" *Nature*. Retrieved on November 11, 2016, from www.nature.com/news/is-the-2-c-world-a-fantasy-1.18868.

Tugcu, Can Tansel, and Aviral Kumar Tiwari. "Does Renewable and/or Non-Renewable Energy Consumption Matter for Total Factor Productivity (TFP) Growth? Evidence from the BRICS." *Renewable and Sustainable Energy Reviews* 65 (2016): 610–16.

Turchin, Peter, and Sergey Nefedov. *Secular Cycles*. Princeton, NJ: Princeton University Press, 2009.

Turner, Graham. *A Comparison of the Limits to Growth With Thirty Years of Reality, Socio-Economics and the Environment in Discussion*. CSIRO Working Paper Series. Collingwood: CSIRO, 2008–9.

Wackernagel, Mathis, and William Rees. *Our Ecological Footprint: Reducing Human Impact on the Earth*. Gabriola Island: New Society Publishers, 1996.

Weinstein, Neil D. "Unrealistic Optimism about Future Life Events." *Journal of Personality and Social Psychology* 39, 5 (1980): 806–20.

Wiedmann, Thomas O., et al. "The Material Footprint of Nations." *Proceedings of the National Academy of Sciences* 112, 20 (2015): 6271–6.

Williams, John. "Shadow Government Statistics." 2016, retrieved on November 11, 2016, from www.shadowstats.com.

21

FROM (STRONG) SUSTAINABILITY TO DEGROWTH

A philosophical and historical reconstruction

Barbara Muraca and Ralf Döring

Introduction: the story of symbols and their surplus of meaning

The history and destiny of the idea of sustainability, and of a sustainable development, can be symbolically captured by two images, both of which are logos. The first dates to 1992, when this idea found its first global success with the Rio Earth Summit; the second is from the more recent Rio+20 Summit that "reinvented" it in a new, allegedly more appealing fashion.[1] While the logo of the first Rio summit represents a hand/dove holding the planet and a branch of leaves, the logo from 2012 presents a more complex design that is supposed to summarize the core messages of what sustainable development had become. The first logo attempts to convey a message of peace and balance, holding and sustaining the planet in a rather literal sense. In the new logo, the planet is free-floating and constituted by the circular movement of three stylized icons, each representing one of the three dimensions of sustainable development identified during the first Rio summit: environmental (a green leaf), social (a red human figure), and economic (blue stairs). Not much fantasy is needed to read into the symbol. The green and the red icons flow into the blue of the central image that connects them and holds them together; the economic dimension is represented by a stairway going up from the bottom left low to the upper right-hand corner. It is the Western model of economic growth, from left to right, from low to high. Twenty years after the first Earth Summit it has now become manifest not only that growth is at the center of sustainability efforts, but also that it is supposed to be the core that allows for environmental and social policies to take place.

This reading of sustainable development represents—as we claim in this paper—a major shift with respect to the multilayered complexity of the original Rio Summit and even more with respect to the earlier sustainability-related meetings, documents, and declarations, such as the Stockholm Declaration and the Cocoyoc Declaration.[2] The key document on which the Rio Summit was based was the Brundtland report, which already represented a first step towards a reading of (sustainable) development in terms of growth.[3] As Erik Gómez-Baggethun and José Manuel Naredo have claimed, after the Brundtland report, ecological degradation was no longer correlated with economic growth, but with poverty.[4] They claim that the sustainability policy discourse shifted from "growth versus sustainability to growth *for* sustainability."[5] Early ecological economists such as Nicholas Georgescu-Roegen, when first faced with the concept of "sustainable development," rejected it by saying that it was an oxymoron, a *coup de genie*, a trick to revalorize economic growth (now translated into the more acceptable concept of "development" and framed according to the path followed by the Global North), while also considering the ecological and social basis that would render growth possible in the long run.[6] With his analysis, Georgescu-Roegen clearly identified the hidden meaning of sustainable development, anticipating its later re-definition and renegotiation during the years following the first Rio Summit. For him, given the temporal structure of life and the fundamental conditions for sustaining life on a finite planet, any idea of a sustainable development based on further economic growth was an impossible task. Georgescu-Roegen suggested instead a positive reading of what economists used to call the "declining state," later translated into *degrowth*, as a condition free from the diktat of competition, acceleration, the imperative of expansion, and savage exploitation.[7]

Is this the whole story? Are the concepts of sustainability or sustainable development as they have been developed in official documents to be abandoned altogether if we are to move on a path that can truly sustain life in the long run? But there is also a different story, one that ghosts behind the established narrative and needs to be unveiled. In following the master of utopian thinking, Ernst Bloch, we claim that such a search is led by what he called "militant optimism," that is, not just an optimism that interprets surface signs, but one that actively seizes on latent possibilities and illuminates them with radical imagination. This chapter is an exercise at reading the narrative of sustainability against the grain—not so much as a historical reconstruction, but as a philosophical interpretation—that unfolds along the line of strong sustainability and the more recent degrowth debate.[8]

When we ask students who are not familiar with the details of the history of sustainability and sustainable development to interpret the Rio+20 logo, it is illuminating to see how they read the symbols. Most of them read the blue stairs as a water flow that connects earth and humans. Others see the circularity and interdependence of all elements. Some read the human figure stylized at the bottom in terms of the Anthropocene and of human responsibility for the planet. No student ever reads the stair symbol as the economic dimension or as economic growth. Maybe the results would have been different if this exercise had taken place not in interdisciplinary classes on environmental values, resource management, and environmental economics, but in a classic, mainstream business class. Still, this informal exercise is instructive for two reasons. First, it says that people who are somehow curious about environmental issues in a very general and vague sense, do not think at a first glance that economic growth is the main path to sustainability. Second, that the symbol itself, in spite of the explicit meaning given to it by the Rio+20 conveners, bears what Bloch would have called a *surplus of meaning*, which can open up imagination towards alternative, deeper, and more radical interpretations of what sustainability and a sustainable development could and should be about.

In a time in which water scarcity, ocean acidification, and conflicts related to water are dramatically increasing, for example, it would indeed be interesting to read the blue stairs in terms of a waterfall, a flow that inevitably and intricately connects humans and earth. Or, even by remaining within the limit of economic thinking, one could read into the symbol the three *funds factors* that Georgescu-Roegen, the father of ecological economics, considered the agents of production that have to be maintained at a level of constant efficiency: land, labor, and capital. The stairs would not so much represent growth here, but the economy as a process, embedded in and dependent on the other two factors. In a bolder interpretation, one can reverse Western reading conventions and read the image from right to left, thus viewing the stairs in terms of degrowth. The twin sister of the Rio+20 logo, its mirror image, might entail thus a different message entirely: degrowth is needed to link the earth and humans. Giving up on the first Rio logo in this case would not be something to mourn. In fact, when talking about sustainability, it is not the planet as globe that is at risk (or not yet, at least).[9] It is the earth as the place where we dwell, the soil that sustains life, creative development, and flourishing both of humans and non-humans. From this point of view, the leaf is a much better symbol than the globe as it reminds us of the most fundamental process that allows for life on earth, photosynthesis. It represents the complexity of life-sustaining processes, not just discrete, monetized ecosystem services, nor the individual "natural" entities to be preserved in wilderness sanctuaries.[10] The economy, as a symbol of human creativity, industriousness, and flourishing, flows from the life-sustaining and life-creating processes of living beings, both in the figure of the leaf and of the human, and, indeed, links them in a process that sustains life instead of destroying it.[11]

Strong sustainability: a militant-optimistic reading

The concept of sustainable development is the offspring of two lineages that are often mixed together when trying to narrate its historical background. The English Wikipedia page, as with many other standard references, starts the history of sustainable development back in the seventeenth century with the introduction of the term "sustainability" in German forestry. With some gaps it then elegantly shifts to "sustainable development," introduced in the 1980s and established worldwide through the first Earth Summit in 1992. Not much is told on the same page about the other lineage—"development"—a concept still in the making (and unmaking), clearly framed after the Second World War to seal a specific model of societal structure based on industrialization, modernization, and ultimately economic growth, as it was embodied by the early industrialized countries of the so-called Global North.

By reconstructing the historical provenance of the sustainability concept, it becomes possible to retrace some of the core motivations that guided its original spirit, while also enabling a surplus of meaning that reaches beyond it. According to Ulrich Grober, when Hans Carl von Carlowitz introduced "sustainable" (from the German *nachhaltend*) he not only interpreted the term as meaning "cut only as much trees as you are able to replant" (in order to guarantee a yearly constant flow of timber) but also that the use of the resource should be as sufficient as possible (not wasting anything in the production process).[12] Carlowitz was not only driven by the urge to establish a new form of management of the key resource of the time for economic and military reasons. He was also concerned about future generations being deprived of what seemed to be the fundamental conditions of human survival and flourishing. In other words, from its very origin, the idea of sustainability was the bearer of a proto-normative claim for intergenerational justice.[13]

We do not want to reconstruct the history of the sustainability concept here, but rather to point out how a narrative has slowly emerged in which sustainable development and sustainability could be used interchangeably by social actors, policy-makers, and sometimes also scholars, thus delivering the whole idea of sustainability over to the dominant perspective of development.[14] The further story of the two Rio Summits confirms this line of interpretation. The idea of a sustainable development, thus conceived, reproduces the expectation that sustainability can be achieved as a byproduct of a somehow greener version of productivism and growth, that the world can grow itself out of ecological and social problems.

Under the guise of sustainable development, the idea of sustainability becomes a "plastic concept" that can be stretched beyond recognition, and which often conceals the status quo.[15] Following the first Rio Summit, and the Agenda 21 declaration that emerged out of it, governments in many parts of the world committed to sustainable development by creating indicator lists, dashboards, action plans, and ultimately monitoring committees. The Agenda

21 agreement combined some commitments to future generations with the three-pillar model by which economic, environmental, and societal objectives would somehow be balanced.[16] All three pillars are open to host a potentially infinite list of criteria and proxies for their assessment, depending on the perspectives of actors or institutions involved, thus indicating the troublesome plasticity of this concept. While at the local level, especially in Western Europe, municipalities implemented some important measures to reduce pollution, revivify neighborhoods, and increase livability (with the support of EU funding), at a larger scale the list of indicators also opened the path for a wide range of interpretations.[17] For example, the list of sustainability indicators for Germany, beside goals addressing land use, also includes development aid and open markets. The much-maligned metric of GDP remains the core indicator for wellbeing, although it is flanked by other proxies. In the 2015 version of the Sustainable Development Goals (SDGs), economic growth remains an explicit goal, albeit paired with what the document calls decent work.[18] This perspective is based on the assumption of a feasible path of absolute decoupling between GDP growth and the impact on resources and sinks. The list is inspiring in many ways, but it neither considers the internal contradictions and trade-offs among different goals, nor does it take a systemic perspective on how to address them. Moreover, the new "techniques of measuring, indexing, benchmarking and auditing are not themselves neutral but are rather deeply political ways of inscribing a particular view of the world—most frequently a neoliberal world of competitive states and entrepreneurial individuals amenable to rankings and zero-sum market exchanges."[19] That is, sustainable development still retains a strong belief in the social value and biophysical possibility of growth, without truly considering the ways in which growth has contributed to our unsustainable world.

In the following sections we will dedicate a few paragraphs to the analysis of the concept of sustainability from the point of view of its normative potential as well as from the perspective of the economic controversy between weak and strong sustainability. We will then show how the more recent degrowth discourse, which (re)emerged in new fashion in the late 1990s as an alternative narrative to the established path of a sustainable development stands not only for a reappropriation of the sustainability idea in terms of strong sustainability, but also for a re-politicization of the sustainability discourse altogether. It is a vision of sustainability that looks quite different than sustainable development.

Strong sustainability: a normative approach

In an attempt to systematize the concept of sustainability and articulate a normative theory that moves beyond the discursive consensus represented by indicator dashboards and international agreements, we draw on the work of an interdisciplinary group of scholars from the university of Greifswald, Germany who developed a framework that was later adopted by the German Advisory Council for the Environment (SRU) in its report to the government in 2002.[20]

The Greifswalder Approach addresses critically the very core of the sustainability idea, such as inter- and intragenerational justice and the diversified concept of natural capital, in order to shape a normative and comprehensive theory that offers a well-founded orientation to the societal and political process of decision-making.[21]

Through a normative reconstruction of the sustainability discourse, the Greifswalder approach proposes a multilayered theory, in which the different levels are not intended as a deductive hierarchy, but as a theoretical systematization. The first two levels, the core elements of the theory, consist of a theoretical reflection that aims at framing the concept of sustainability as a regulative ideal and articulates explicitly the surplus of meaning that it entails, and which goes beyond its actual implementation in policies and political agreements. The rules as bridging concepts lead to the last three levels that are supposed to facilitate a fruitful exchange with policy-making, praxis, and social movements:

- Idea (intergenerational justice)
- Concepts ("strong" or "weak" sustainability, intermediate concepts)
- Rules (constant natural capital rule, management rules)
- Dimensions of policy-making (climate change, fisheries, forestry, agriculture etc.)
- Objectives (targets, time frames, set of instruments, indicators)
- Implementation and monitoring.

In this chapter we focus only on the first two levels, which correspond to the theoretical foundations of sustainability. A normative reconstruction of the sustainability discourse does not simply make normative claims explicit; but also critically analyzes them from the point of view of conditions, implications, and immanent contradictions.

Sustainability as an ethical concept

Following the Greifswalder approach, we can say that the core of the idea of sustainability consists in the issue of intra- and intergenerational distributive justice and encompasses duties towards currently living generations and future generations regarding different goods, with a special focus on natural resources and processes.[22] Obligations to posterity are to be combined with an assessment of consequences and side effects of contemporary actions and institutions in order to identify how sustainable development should be established in policy-making. It is important here that the norms guiding political decisions be made explicit. Often times, in the socio-political discourse on sustainability, normative claims are left in the background, or buried beneath a veneer of objectivity. However, it is precisely at the level of normative claims, with respect to different understandings of justice, that different conceptions of sustainability policies collide and can be unmasked as reproducing unacceptable forms of injustice.

In terms of responsibility of justice towards future generations, the Greifswalder approach identifies the following questions as essential to the sustainability idea:

- Are there any obligations to future generations at all?
- Should responsibility for future generations be based on an egalitarian-comparative standard or on an absolute standard?
- What can be considered a "just" legacy?
- Are we permitted to discount future states of affairs?

While the idea of sustainability is often associated with *future* access to resources, and the ability of humans to live dignified, flourishing lives (this was already Carlowitz's concern), questions regarding intragenerational justice (i.e. justice among generations *currently* living on the planet) is equally important, if we want to avoid a situation in which inter- and intragenerational justice comes into mutual conflict. Moreover, a renewed focus on the present is imperative since climate change is no longer a matter for future states of affairs, but increasingly affects people today, and especially those in the poorer Global South. Finally, the sustainability discourse needs to address whether only human beings ought to be considered from a moral point of view as bearers of rights for their own sake or whether the "more-than-human-world" should count, too.

While from a common sense perspective it seems obvious to assume that we do have some kind of obligation towards our children and grandchildren in terms of leaving to them a livable world in which they can flourish, this perspective does not necessarily support a general argument for intergenerational justice. Common sense sometimes remains constrained within regional borders: for example, leaving a decent, livable world to the next generations of North Americans or Western Europeans, at current rates of consumption, would jeopardize the chances of livability for current and future generations in other parts of the world. When we take a philosophical perspective along the line of the Western tradition of thought, instead, we can argue that we have general obligations towards future generations by considering future people in their abstract status as right-bearers regardless of their individual characteristics, including geographical specificities.[23]

There are also many non-Western and indigenous ways of understanding intergenerational justice that we find compelling. In such traditions, individuals are considered as fundamentally relational, i.e. constituted by relations to other members of the community that share a common territory, history, and cultural narrative. Such a community does not only include ancestors and non-human animals, but also future generations. The underlying philosophical assumption that supports such a relational understanding of the self is not limited to the circle of nearby individuals, although it often takes the shape and the language of a specifically *located* community. Rather, it implies a more radical understanding of interdependence and relational entanglement among different forms of existence that are not linked only by spatial, temporal, or causal proximity.[24] This idea is very much alive in current perspectives critical of development and growth, such as the concept of *buen vivir* developed by the indigenous Quechua-speaking people of the Andes and which plays a major role in the degrowth movement.[25]

Once demonstrated *that* we have duties of justice towards future generations, the normative question arises of whether these duties should be based on an absolute standard (access to anything that is required for a life of human dignity) or a comparative one (no worse than current generations). The absolute standard ensures a "basic humane level," whereas the comparative standard raises the issue of an appropriate "equivalence" among generations. While the former allows current generations to bequeath less to future ones than they have inherited (provided that this would be sufficient to lead a decent or dignified human life), the latter requires that future persons be, at the very least, no worse off than current ones. A comparative standard might sound at a first glance like a better option, although this very much depends on the standard we are considering and the trade-offs between intra- and intergenerational justice that are implicated.

Following Ott and Döring, we hold that a strong and demanding absolute standard would offer a better platform for intergenerational justice than a comparative approach. As they suggest, an absolute standard does not have to be based on simply meeting "basic needs," but can shift attention to the material, cultural, and formal conditions to lead a dignified, flourishing life. Ott and Döring base their analysis on Martha Nussbaum's culturally interpretable and context-sensitive list of capabilities in her "thick vague theory of the good."[26] Nussbaum's capabilities approach does not focus on the *actual satisfaction* of needs or wants, but on the substantial conditions of possibility that enable the achievement of doings and beings that are considered an essential component of a life worthy of a human being, i.e. a good life. Such conditions do not only refer to personal abilities, but also and more specifically to material and institutional settings, including also cultural and intersubjective patterns, as well as to environmental conditions.[27]

Considering the capabilities approach as the basis for defining an absolute standard for intergenerational justice means that current living standards and ways of life must be scrutinized in terms of how they infringe upon some substantial conditions of possibilities for future generations to achieve a dignified and meaningful human life. By expanding Ulrich Brand's concept of the "imperial way of living"[28] to intergenerational justice, we could say that the life form of current generations in the Global North is imperialistic with respect to the chances of future generations to achieve a high minimum standard in the sense of the capabilities approach.[29] This is not so much an issue of individual guilt, but a systemic mode of organization of modern, capitalistic societies based on economic growth. Challenging the imperial way of living requires a radical transformation at the level of institutions, systemic power relations, and established mental infrastructures.[30]

Fair legacy: what do we owe to future generations? The economic controversy between weak and strong sustainability

The three production factors that traditionally were distinguished in economics—land, capital, and labor—were reduced to just two at the end of the nineteenth century. Land was seen as part of capital, that is, as an investment, and not as a production factor with specific characteristics. When, after the Second World War, economic growth became the core model in western societies, economists developed growth models using only labor and capital. Accordingly, economic growth was considered to depend on only the interaction between these two factors, without any connection to the biophysical world, and was represented as a closed cycle. Indeed, neoclassical economic thought in the middle and late twentieth century generally ignored the natural environment or considered it in terms of a substitutable factor.

This attitude toward the natural environment began to change in the late 1960s and 1970s with the emergence of systems theory and ecological economics. The Club of Rome's influential book *The Limits to Growth* (1972) anticipated the end of economic growth as a result of limited resources and contributed to a strong critique of the traditional view of production processes. Although in the subsequent years the book's detailed computer-modeled scenarios might have seemed entirely wrong, as both growth and an increasing use of resources have continued since the era of 1970s stagflation, its core message remains unchallenged.[31]

Right after its publication the report was heavily debated within economics on the ground of the assumption that shortages in certain resources would lead to—and even enhance—renewed growth, due to their substitution. In the 1970s, the economist Robert Solow introduced *resources* (instead of land) as a production factor into the growth model. He claimed that there was basically no problem in substituting between different production factors, including between resources and capital. In the mathematical representation of the production function, he stated that it would be the "best educated guess" to consider the elasticity of substitution between these two factors as 1. This means that *resources* as input in the production function can be indefinitely small, albeit never equal to zero, and that the production process and the economic activity can grow indefinitely. According to this model, currently living generations are entitled to "draw down the pool (optimally, of course!) so long as they add (optimally, of course!) to the stock of reproducible capital."[32] The optimal legacy that we ultimately owe to future generations is an overall portfolio—also termed *capital*—that would allow them to satisfy their needs and contingent preferences, which we do not know about and cannot judge. From this point of view, nature can be consumed provided that other capital stocks (man-made capital, human capital) are built up in its place.[33] The assumption is that the value of the natural environment can be converted into other forms of capital that can essentially perform the same economic functions.[34]

In hindsight, this argumentation formed the basis of what was later called "weak sustainability," according to which natural and man-made capital are considered as substitutes and future generations can be compensated for losses in one set by gains in the other one.[35] From this perspective it is not only possible and acceptable, but also necessary, to discount future benefits and costs and, therefore, to maximize net present value of future economic activities. According to the standard economic approach, discounting is justified because future generations are expected to be better off than current living people today, due to the expectation of exponential growth. The current generation is, therefore, allowed to devalue (or discount) future income streams and redistribute part of future higher incomes to the current generation.[36] While it might make sense to apply discounting when considering specific investment at the level of single enterprises, it turns out to be problematic when it is simply applied to every economic activity at a macro level. Especially in cases in which we can legitimately expect not to be better off in the future, applying a discount rate seems counterintuitive to basic justice assumptions.

Weak sustainability considers the different capital stocks of society in terms of an overall portfolio, in which natural capital is only one among many different stocks.[37] The ideal portfolio manager assesses possibilities of substitution by trying to maximize the net present value. From this point of view, the preservation of natural resources is a meaningful and feasible goal only if it is proved to be more efficient compared to other income types. This requires that, for the sake of comparability, natural resources be expressed in monetary terms.

Besides substitutability, there is a second very important assumption for weak sustainability: The idea that development follows an "Environmental Kuznets Curve" (EKC). Accordingly, with increasing income levels, societies would invest more and more in environmental programs. Negative "externalities," such as emissions, would decrease, while income would further increase. Again, the main assumption of this concept is the further growth of the economy. There are only very few, circumstantiated examples where the Kuznets curve could be measured in the real world (e.g. SO_2 emissions in Germany). The mathematical surface of the EKC conveys the illusion of an automatism driven by the market and neglects the role that social struggles, such as the green movement in Germany in the 1980s, political interventions, and the dislocation of heavily polluting industries to other countries, such as steel production to China or coal extraction to Columbia, have played for the occasional improvement of environmental conditions in the Global North.

Against weak sustainability, advocates of "strong sustainability" such as the economist Herman Daly, hold that the substitution between different capital stocks is highly limited. Accordingly, there is no sign that the elasticity of substitution between man-made and natural capital is near 1, but somewhat between 0 and 1 (substitution is possible but to a much lesser extent; where exactly, we do not know). If we look into statistics, although we may use fewer resources per single unit of production (relative decoupling), our economy requires more resource inputs than before.[38] As to discounting, in cases where we can legitimately expect to have less rather than more in the future (as is the case for example with biodiversity), we would have rather to invest into preservation and to discount negatively. The assumption of the EKC is not justified by real developments, is far too optimistic, and blind to the role of institutions.

The best, and sadly the most famous, example of an optimal weak sustainability path is found on the island of Nauru. After having extracted all of its phosphate in a relatively short timeframe, it is now totally reliant on the delivery of basic resources from elsewhere (including food for its population) and is dependent on the payment from Australia for refugee camps.[39] The mineral resources were transformed into financial capital, which proved short-lived, and which did not contribute to the long-term wellbeing and nutritional sustainability of the island's inhabitants.

From a weak sustainability perspective, our economic system is neither embedded into ecosystems nor dependent on them. Accordingly, the economy is seen as potentially limitless, and growth is encouraged in order to guarantee to future generations an increased capital stock (and possibly a higher flow of utility per person, as compared to today).

By contrast, ecological economists and advocates of strong sustainability draw on ecology, thermodynamics, energy studies, and systems theory to argue that our economic system is fundamentally embedded in natural (and social) systems and that growth (of populations, of economic activity, and of consumption) is necessarily limited by the biophysical capacities of the planet.[40] According to strong sustainability, since natural capital is now the scarce factor (and not man-made capital, as assumed in mainstream economics), it must be preserved on different scales (global, continental, national) out of moral respect for future generations. The loss of natural capital (trees, fisheries, minerals, biophysical processes), some of which can be irreversibly damaged, impoverishes the life prospects of future generations, their freedom of choice, and their ability to flourish. From his analysis of complementarity (i.e. non or highly limited substitutability) between different capital types, Daly derives specific management rules and policy suggestions, such as: not counting the depletion of natural capital in terms of income and the CNCR (Constant Natural Capital Rule), according to which natural capital should not decline over time.[41] For Daly, the assumption of complementarity is a sufficient argument to justify the normative framework of strong sustainability. In recent years, further arguments have been articulated to justify the CNCR. The point is not only whether or not and to what extent nature *can* be substituted in the production process, but also whether it is *morally acceptable* to do so, in terms of inter- and intragenerational justice.

The economic controversy between weak and strong sustainability reveals the different perspectives on what is considered a just legacy to future generations. On the one hand, we have an approach based on welfarism (i.e. the idea that what matters to people is the utility flow of goods, typically framed in terms of subjectively perceived happiness), according to which anything is in principle substitutable as a means of individual satisfaction.[42] Money in its abstraction is the best proxy to measure welfare thus intended. On the other hand, a strong sustainability approach relies on the idea that there are some specific characteristics of biological processes that cannot be reduced to a general utility function, due to the irreversibility of the qualitative transformation that affects them, and that values and political deliberation determine the framework for decisions about what to preserve or use, instead of relying on market-based instruments for "optimal" management paths.

According to the normative framework of the Greifswalder approach, several reasonable arguments support strong sustainability, such as a better representation of the multifunctionality and temporal structure of ecological systems; risk assessment based on the precautionary principle; greater freedom of choice for future generations in the face of likely irreversible change; better consideration of different value systems and languages of valuation with reference to human–nature relations.[43]

From sustainability to degrowth

The strong sustainability approach is rooted in the economic theory of Georgescu-Roegen, developed in *The Entropy Law and the Economic Process* (1971), which had a major influence on early ecological economists, including Daly. While Daly retained the concept of natural capital from mainstream economics, Georgescu-Roegen, his teacher, developed a different way of understanding the production function by introducing a *flow-fund model*.[44] He distinguished between two qualitatively different factors of productions, *funds* ("the agents of a process" of production that have to be kept at constant specific efficiency) and *flows* (the elements "which are used or acted upon by the agents" and enter *or* exit the process as inputs or outputs).[45] Capital, land, and labor are funds and cannot be "used up" during the process of production, whereas resources are inputs that enter the process and are transformed by it. Moreover, production is a process unfolding in (only one direction of) time. From this perspective we can infer that land refers to the conditions of generation and regeneration of natural processes, including resources, and *should* not (or anyway *cannot* in the long run) be consumed in the process, in the same way that labor cannot be an input or flow factor in the process either.[46]

Whereas Daly, who went on to work for the World Bank, adopted later the concept of sustainable development as a better version of his idea of a steady-state economy, Georgescu-Roegen rejected the term as a cynical attempt (masked by an alluring logo) to maintain the economic path of the early industrialized countries and their standard of living. Although the discourse of sustainable development acknowledged the needs of so-called developing countries, it did so without realizing the immanent contradiction of this very assumption. It held up countries in the Global North as the measuring stick of "development" and thus condemned poor countries to endless misery.[47] Daly, instead, tried to merge the principles of strong sustainability with sustainable development. His interpretation was based on the rather abstract distinction between material growth and non-material development. That is, societies can stop increasing the amount of stuff produced and energy employed but continue to develop spiritually, culturally, and socially.[48] Core elements for his understanding of a sustainable development would thus include population control and technological innovation, under the assumption that different types of capital are complementary, i.e. not readily substitutable.

Georgescu-Roegen, by contrast, argued that economists would soon have to rethink their entire profession and move from engaging with the issue of economic growth and toward identifying criteria to plan what he called with reference to John Stuart Mill the *declining state*. He proposed to use this term instead of sustainable development and used it in Mill's positive sense as a chance for a better social life. "Declining state" was then translated in the early 1970s—with Georgescu-Roegen's approval—into the French term "décroissance," and later back in English as "degrowth." For us, the idea of strong sustainability is best represented by Georgescu-Roegen's bioeconomic theory.

Georgescu-Roegen's provocative enunciation about the profession of economists plays a key role for the degrowth movement, with the difference that the planning of the declining state cannot simply be left to economists. Rather, it has to emanate from a process of deliberation and social activism that aims at creating a just, solidarity-based, and democratic path toward a radical transformation of society.[49] As we see it, degrowth best embodies the core message of a strong and viable conception of sustainability.

Georgescu-Roegen: strong sustainability as degrowth

According to Georgescu-Roegen, economic processes are similar to biological "open systems" and are therefore creative, metabolic, and qualitatively transformative. Like biological processes, they feed on the low entropy of their environment and occur along the line of *Time* as a continuum, which, just like historical time, cannot be reversed.[50] However, while biological organisms wisely feed on the single "infinite" source of low entropy available on the planet, i.e. the solar energy as it is captured and rendered available by the earth's surface (land), economic processes – starting with the Industrial Revolution – deplete the so-called terrestrial stocks of low entropy (fossil fuels, mineral resources), which are not infinite in size and whose regeneration takes geological epochs.

Whereas the flow rate of terrestrial stocks can be more or less adapted to society's needs, the flow of solar energy and of most renewable sources might be infinite in its amount, but not at our disposal with regard to the flow rate of its use. The shift from renewable to non-renewable (mainly fossil) resources enabled a remarkable acceleration and intensification in the production of new technologies, resulting in what we call the Industrial Revolution. As a result, humans disentangled themselves from the temporal limitation that the regeneration of renewable energies required. However, for Georgescu-Roegen, this is not an indefinite path, due to the limited availability of terrestrial stocks. In the end, the shift from renewable to non-renewable sources was based on the accelerated depletion of the terrestrial sources and is essentially parasitic.[51] We are caught in what Georgescu-Roegen calls an infinite regress, in which what is needed to keep an economic process ongoing is produced by another economic process and so on. Think, for example, of chemical fertilizers employed in conventional agriculture to enhance fertility beyond soil's regenerative capacity. Due to the temporal constraints of the regeneration of soil, water, air, and living organisms, the continuous intensification of productivity required a shift to limited fossil sources of energy and matter. In the long run, however, this acceleration spiral would lead back, according to Georgescu-Roegen, to the only unlimited source of low entropy, which is solar radiation captured by land and which is limited with respect to its flow rate.

While Daly, following Georgescu-Roegen's theory, rightly focused on the issue of scale as the first and most important dimension of sustainability, he missed the temporal perspective that his teacher had articulated. Daly famously claimed that the economic system is a "subsystem of a larger finite and non-growing system and consequently … the macroeconomy too has an optimal scale. A necessary requirement for this scale is that the economy's *throughput* … be within the regenerative and absorptive capacities of the ecosystem."[52] Accordingly, the priority order for sustainable development should be: scale – justice – efficiency, i.e. first and foremost the spatial limitation of ecological system in which the social system (justice) is embedded, and finally the economic logic of efficient allocation.[53]

Georgescu-Roegen's position presents a more sophisticated perspective that influenced the later degrowth debate. First, entropy for Georgescu-Roegen does not so much refer to absolute planetary limits (except in the sense of the spatial limitation of land as available surface for solar radiation and soil).[54] Rather, it is a theoretical symbol for the unidirectionality of time and the irreversibility of creative, qualitative, and cumulative processes that take place on the planet. Entropy and evolution—and development in any possible meaning—are thus the two sides of the very same coin. From this point of view, talking about planetary limits means to consider the *rate* of use of resources and the regeneration time of natural, living processes as the *limiting* as much as the *enabling* factor for creative production.[55] Entropy, moreover, is the physical representation of *Time* as continuum, and even if it is hard to model it in analytic-mathematical terms, it corresponds immediately to the unidirectionality and vectoriality of our own experience as living beings and as humans. As living beings endowed with reason, we "know" immediately about the irreversibility of living processes in our own experience, even if neoclassical economics had no tools to include it into its theory. Accordingly, entropy is a bridge-concept between the objective, analytical concept of time and the subjective and phenomenological experience of *Time* by conscious, living beings.

Recapitulating, Georgescu-Roegen allows us to frame the theory of strong sustainability in the following ways:

1   The relation between economy, ecology, and society is not simply a matter of concentric circles, but a dynamic and systemic interrelation. The "ecology" circle is not just an outer

house, within which social and economic processes are embedded, but it represents the temporal dimension of any creative transformation, including and intersecting with social and economic processes.

2   Economic processes are thus not just a subsystem of the biophysical realm, but are *qualitatively* different from mechanic, reversible processes. To adequately understand and analyze them, another economics is needed, modeled on biology and social sciences and able to frame *Time* as continuum. Social metabolism analysis (SMA) stems from this assumption.[56]

3   Given the complexity, temporality, and interrelatedness of dynamic systems, not only analytic tools, but also *dialectical* perspectives are needed to understand them. By this, Georgescu-Roegen means the grey zone of social and cultural concepts that do not have clear-cut boundaries, but which are constantly subject to societal and philosophical deliberation (such as justice). With a little stretch of imagination this could mean that, for example, in the sustainability debate, the very concept of scale is inextricably linked to justice and depends on the specific *societal relations to nature* that constitute society.[57] Accordingly, nature and society are the outcome of a continuous exchange. Without this dialectical perspective, there is a risk of reading the prioritization that Daly proposes in terms of a cynical setting of "naturally given" conditions that constrain the realm of political deliberation, as is the case for example with the US-based Carrying Capacity Network.[58]

4   Finally, and more concretely, following Georgescu-Roegen, it becomes clear, as degrowth scholars and activists claim, that the amazing development rendered possible by industrialization is fundamentally parasitic: it relies on and implies a continuous displacement of environmental impacts to other places (for example to the Global South), onto other social groups, and to future generations.[59] By abusing terrestrial stocks today and using renewable resources beyond their regeneration rate, we enact social injustice in the present day, and inevitably compromise the options of future generations to access both stocks and resources.[60] Growth, even so-called "green growth," relies accordingly on the exponential intensification and the accelerating spiral of infinite regress required to keep the economic process further going.

Degrowth: the repoliticization of strong sustainability

The term degrowth has a complex origin. On the one hand, it goes back to the French translation of Georgescu-Roegen's concept of "declining state." With the publication of some of his articles in a collection that appeared in France in 1979, the term degrowth clearly established itself as an alternative to zero-growth or steady-state economics.[61] This collection influenced not so much economic scholars, who for the most part kept ignoring his later work on entropy and bioeconomics, but intellectuals and political activists. André Gorz quoted him as an example of "ecological realism" in his critique of productivism, for instance.[62] Parallel to that, after the publication of the report by the Club of Rome, *Limits to Growth*, in 1972, and the subsequent reactions in Europe to the then-President of the European Commission Sicco Mansholt's explicit plea for a reorientation of the economy towards social utility instead of economic growth, a vivid and controversial discussion arose among French intellectuals.[63] In 1973, *Le Cahiers de la Nef* dedicated a whole issue to this topic under the headline "Les objecteurs de croissance" (the growth objectors). In a paper in this issue, André Amar introduced the term *décroissance*, which he intended in a rather unspecific way (not clearly distinguished from zero-growth). In a kind of "culturalist" analysis, Amar framed the paradigm of growth as rooted in the spirit of modern Western civilization and criticized the moral and anthropological aspects of it.[64]

Later on in France, sustainable development was translated in accordance with the mainstream weak sustainability paradigm into "développement durable" (durable or enduring development), in which development remained virtually synonymous with growth.

In Germany, the critique of growth was strongly present throughout the entire political spectrum (however with less of a specific distinction between zero growth and degrowth). It was an essential component of the early political demands of the Greens and was debated by the workers' unions. A vivid debate revolved around the best-selling series *Technologie und Politik: Das Magazin zur Wachstumskrise* (*Technology and Politics: The Journal of the Growth Crisis*), which was published between 1974 and 1985 by Rowohlt.[65] Flanking and contrasting the German model of ecological modernization that envisioned a technology and policy-led sustainable path compatible with growth, the critique of growth continued in the subsequent decades, albeit often out of the public eye.

As some scholars have pointed out, during the 1980s and 1990s, due to the end of the oil crisis (that sparked the debate around limits to growth in the 1970s), and due to the advent of neoliberalism, the debate on degrowth left the main stage.[66] The new in-vogue term "sustainable development" more adequately reflected the neoliberal paradigm: it not only embodied a narrative able to salvage economic growth as a top political goal and the Western model of development as a universal path, but was also more easily compatible with a managerial approach in the neoliberal mode of governmentality. "Sustainable development depoliticizes genuine political antagonisms about the kind of future one wants to inhabit; it renders environmental problems technical, promising win-win solutions and the (impossible) goal of perpetuating development without harming the environment."[67] As we have seen, sustainable development, once read through the lens of weak sustainability, classic development theory, and the neoliberal narrative of market regulation, technocratic approaches, and entrepreneurial holism, is removed from the realm of a true political deliberation among antagonistic perspectives and framed in terms of a technical issue of management.

It was not until the term development itself met with a renewed and harsh critique from a strange alliance between critics of globalization, such as the alternmondialistes,[68] in the Global North and the re-emerging voices of strong social movements in the Global South, including indigenous and peasants' movements, that degrowth re-emerged in a novel and radical form. According to Serge Latouche, one of the main proponents of the French *décroissance*, the date of birth of the degrowth movement was the colloquium "Défaire le développement, refaire le monde" (Unmake development and remake the world), which was held in spring 2002 at the UNESCO meetings in Paris, and which aimed at unmasking the destructive potential of the dominant model of economic development for Third World countries.[69]

Among its inspirational sources was not only Georgescu-Roegen's bioeconomic theory, but also the global environmental justice movement, post-development and alternmondialist theories, anti-productivist critique, and the French tradition of political ecology.[70] Especially in France, another "new" movement kindled interest in the *décroissance* discussion: the *casseurs de pub* (the French equivalent of the Canadian Adbusters), whose co-founder, Vincent Cheynet, introduced the term "sustainable degrowth" for the first time, explicitly against sustainable development (in French, *décroissance soutenable* instead of *développement durable*.)[71] From France it then conquered Italy and Spain with an even clearer link to global environmental justice and anti-globalization movements. In Germany it developed along a parallel line that unfolded for several years independently from the Southern European debate and was mainly sparked by the strong environmental and anti-nuclear movement as well as by anti-globalization groups, such as ATTAC.[72]

Finally, in the wake of the revival of the sustainable development paradigm under the narrative of a green economy, "degrowth emerged as a paradigm that emphasizes that there is a contradiction between sustainability and economic growth ... It argues that the pathway towards a sustainable future is to be found in a democratic and redistributive downscaling of the biophysical size of the global economy."[73] More specifically, this path should not stem from a top-down management decision, but result from bottom-up processes of societal deliberation, social experiments and innovation, alternative collective practices, and radical institutional reforms. Since 2010, degrowth has received greater attention and, in a sense, competes with other visions of sustainability and sustainable development in public and intellectual discourse.[74]

Degrowth: a project for a radical transformation of society

Since 2000, a "degrowth movement" has taken shape, especially in France, Germany, Spain, Italy and Greece, with some support in other parts of the world, that brings together scholars from different backgrounds and social activists concerned about the biophysical and social implications of growth. This movement is less visible and glamorous than the mainstream world of sustainable development, and its conferences are less well publicized than meetings such as the Earth Summit and Rio+20, but it is a movement that has gained increased notoriety in recent years, especially since the economic downturn of 2007–8.

Degrowth can be considered as a slogan for a social movement[75] or, as Paul Ariès suggested, a *mot obus,* a "missile word"[76] aimed at the re-politicization of the environmental and sustainability debate. As such, it hits at the core of modern, capitalistic societies not only by radically questioning their economic structure, but also the cultural infrastructure that legitimizes them.[77] In a more technical sense, degrowth relies on the recognition, based on scholarly research from ecological economics, philosophy, and sociology, that continuous economic growth is not only unsustainable but also undesirable.[78] That said, degrowth is not limited to advocating for a society with a smaller metabolism, but "more importantly, a society with a metabolism which has a different structure and serves new functions. Degrowth does not call for doing less of the same."[79] In this second sense, degrowth is not simply the opposite of growth that reproduces its logic by simply reversing it; that is, it is conceptually distinct from economic recession. In fact, degrowth challenges the very focus on growth as political goal and aims at reappropriating the political power to deliberate about the society we want to live in against any *pensée unique* or TINA[80] narrative enforced by the dominant neoliberal paradigm.

As Swyngedouw has shown, neoliberal capitalism, with its reduction of politics to a managerial set of interventions, tends to foreclose the space for genuine political debate in society as "the sphere of agonistic dispute and struggle over the environments we wish to inhabit and on how to produce them."[81] Accordingly, politics as public management is "hegemonically articulated around a naturalization of the need of economic growth and capitalism as the only reasonable and possible form of organization of socio-natural metabolism."[82] Growth appears against this background as no longer questionable or contestable.

The degrowth movement goes beyond the narrow paradigm of alternative environmental management and articulates a critique of growth not only from an economic and environmentalist perspective, but also from structural and cultural ones. Accordingly, the degrowth perspectives addresses "growth" in at least four different ways:

1 Monetary growth in terms of GDP, a highly inadequate measure of wellbeing. While the critique of GDP as a measure for wellbeing is as old as the introduction of the measurement itself (see for example the speech by Robert Kennedy in March of 1968), the

degrowth movement unmasks not only how GDP-growth is intimately related to material consequences directly (increase in the amount of matter and energy) or indirectly (for example the use of surplus money from the financialization of markets to increase land-grabbing), but also unmasks its structural contradiction (rebound effects), against the illusion of absolute decoupling.[83]

2   Material growth in terms of social metabolism (amount of stuff, matter, or energy) that is necessary for social reproduction.[84]

3   Growth as a structural driver for modern, capitalistic societies. Economic growth has played a structural role in stabilizing modern, industrialized societies of the Global North, by guaranteeing employment, social mobility, tax revenue, and social pacification. This was possible via ongoing expansion and intensification of productive activity and requires the constant occupation of new territories in a literal way (land-grabbing, shifting exploitation frontiers, and new extractivism) and in a metaphorical sense (acceleration of the pace of life, commodification of ecosystem services, deregulation of labor and money). The imminent crisis of this expansion logic might likely lead to destabilization, destitution, and increasing social conflicts.[85]

4   Growth as a cultural driver that has pervasively colonized the lifeworld and, by turning into a no-longer-questioned *mental infrastructure*, has molded our very self-understanding as social agents.[86] According to Latouche, we are growth-addicted, caught within a compulsive attitude, which is functional to the systemic drivers of modern societies.[87] As anticipated earlier, with Foucault we can say that the logic of growth is the hegemonic form of the individual subject in late-stage, neoliberal capitalism.[88] According to this analysis, in neoliberalism the main driver of economic growth is human capital, i.e. the innovative force of production, which drives economy and society, and which is ultimately the self-understanding of individuals as entrepreneurs of themselves, always concerned with themselves as an investment project oriented at constant improvement in competition with others.[89] Degrowth proponents struggle for a "decolonization of the imaginary" of growth at the level of individual and collective practices, institutional change, and social experiments.[90]

The structural analysis of growth and the re-politicization of the sustainability debate

Growth has not only served as dynamic stabilization for modern societies, but also as the output-legitimation ground for welfare state democracies in the Global North. With a widespread disaffection for political participation, the promise of increasing wellbeing, social mobility, and consumption-driven prosperity were triggered by the prospects of increased economic growth. Depoliticization easily followed the promise of increasing material prosperity, as a broad consensus on the values of economic growth replaced the collective deliberation about the conditions for a good life in society. George W. Bush's famous appeal to Americans in 2006 to go shopping as a civic duty in the face of the Iraq war was not a joke, but a perfect representation of this narrative: "As we work with Congress in the coming year to chart a new course in Iraq and strengthen our military to meet the challenges of the twenty-first century, we must also work together to achieve important goals for the American people here at home. This work begins with keeping our economy growing."[91] The active pursuit of economic growth has become a political goal that supposedly enables social cohesion.

As analysts observe, especially those within the degrowth discourse, we are now faced with a fundamental crisis of the dynamic of stabilization via growth, which turns out to have dysfunctional effects in terms of the socio-economic, political, and cultural reproduction of modern societies. Not only ecological constraints, such as resource scarcity and limits of sinks and absorption capacity, that increasingly reduce the margin of profitability of capitalist investments, or social limits (satiation in affluent societies and limits to the further intensification of individual performances), spell the approaching end of "easy" economic growth.[92] Industrialized countries seem to have reached a threshold at which the feasible growth rates no longer secure employment, social mobility, and welfare.

Under business-as-usual conditions the end of "easy" growth means recession, pauperization, and increasing social conflicts. Further, the growing awareness of climate change has forced a reconsideration of growth and its effects. Some economists, such as Nicholas Stern, have begun to call climate change a form of "market failure" that stems from the unsustainable expansion of the human economy and its associated pollutants.[93]

Moreover, the very legitimization of democracy is at risk. A re-feudalization path seems a rather plausible scenario.[94] Conservative critics of growth share this analysis and advocate for forms of structural and cultural adaptation to the "inevitable" end of the age of prosperity characterized by ongoing economic growth. The scenario they are depicting ranges from a rather neo-fascist representation of separated and ethnically homogeneous eco-communities[95] to a shifting towards spiritual and family values as coping strategies under a framework of social immobility, happy frugality, and high inequality.[96]

Against an adaptation model that remains inscribed in the neoliberal logic of late capitalism, the degrowth movement identifies in this crisis a chance for a radical transformation of society carried by new social movements, global alliances, and alternative social experiments.

The challenge is how to construct a society that is no longer dependent on economic growth for its stabilization and legitimation and in which real democracy, autonomy, and solidarity are strengthened within the collectively negotiated biophysical conditions for life on earth. Such a transformation cannot be the mere implementation of a blueprint dictated by political leadership or technocratic governance that would simply reproduce the managerial mode of neoliberal capitalism, but must emerge from a plurality of forces and their alliance, and emanate from self-organized, locally rooted and yet globally networked forms of resistance, subversion, and creative visions for alternative social experiments. In line with the tradition of political ecology, degrowth presents itself as a project to re-appropriate and enhance collective autonomy and economic democracy, i.e. the social deliberation on the conditions under which not only production, use, and consumption, but also social relations are shaped and organized in society.[97] Moreover, as a project of re-politicization, degrowth seeks to restore the sphere of the *political* and aims at a democratization of societal relations to nature against the dominant logic of neoliberal commodification.

A recent questionnaire survey about motives, attitudes, and practices of grassroots activists within the degrowth spectrum has shown two main points of consensus: 1) the profound conviction that sustainability and growth are incompatible and that growth is soon coming to an end in industrialized societies of the Global North, and 2) the idea that a positive social transformation "is critical of capitalism, pro-feminist, peaceful and bottom-up."[98] While the degrowth movement understands itself explicitly as a proposal for early industrialized countries of the Global North, in recent years interesting alliances with other movements worldwide have been built in a fruitful dialogue of common, yet different struggles.[99]

Given that basic social institutions in modern, capitalistic societies are fundamentally dependent on the prospect of economic growth for their stabilization, the transformation has to extend in the long term to institutions as well, such as the reorganization of work beyond the (gendered) division between paid and unpaid labor. Degrowth activists and scholars operate at the same time at the level of institutional transformation, through the experimentation of innovative collective, social practices, and towards a transformation of the social imaginary. Given the pervasiveness of the growth logic as mode of subjectivation, the role of alternative social experiments is far from being a niche of apolitical engagement. Rather, they are laboratories for liberation, protected spaces where subversive practices and alternative modes of subjectivation can be experimented. In such spaces alternative ways of conceiving needs, desires, and their satisfaction, are not only envisioned, but also experienced. By provisionally suspending the pervasive impact of dominant societal imaginaries, social experiments can crack open the established understanding of what is considered to be real and give room to alternative imaginaries, practices, and experiments of common living.[100]

Conclusion

In this chapter we have outlined, via a militant-optimistic reading of the sustainability discourse, a narrative that recovers some less visible tendencies in the long history of the concept, and span them from its origins in the nineteenth century to the current degrowth movement. While a clear legacy line can be identified that extends from the first articulations of a sustainable development in the 1980s to the current narrative of a green economy, this is not the whole story. By distinguishing between sustainability and sustainable development and looking back into the history of the sustainability concept, a counternarrative becomes manifest that links strong sustainability in a wider sense to degrowth and the degrowth movement. This analysis is not meant to be a historical rendering, but a normative reconstruction that focuses on embedded social critique, i.e. critique as it has been articulated by social movements, social actors and groups in the rather peripheral spaces flanking and challenging the mainstream narrative. In our reconstruction, an augmented theory of strong sustainability is sketched, which enriches the economic perspectives with social and philosophical analysis by disclosing the surplus of meaning embedded in the sustainability narrative and embodied in a renewed shape by the degrowth movement. Following recent scholarly research, we maintain that the degrowth movement does not only call for a repoliticization and reconceptualization of sustainability, but also envisions a radical social-ecological transformation that is driven by the democratization of societal relations to nature and characterized by a bottom-up approach of a multitude of diversified and networked social experiments worldwide.

Notes

1   See James Goodman and Ariel Salleh, "The 'Green Economy': Class Hegemony and Counter-Hegemony," *Globalizations* 10(3) (2013), n.p.
2   UN Stockholm Declaration, 1972, retrieved on February 13, 2017 from www.unep.org/documents.multilingual/default.asp?documentid=97&articleid=1503; UN Cocoyoc Declaration, retrieved on February 13, 2017 from www.mauricestrong.net/index.php/cocoyoc-declaration.
3   The document is called, in fact, *Our Common Future*, and touts growth at several spots. See, for instance: "What is needed now is a new era of economic growth—growth that is forceful and at the same time socially and environmentally sustainable." WCED, *Our Common Future* ("the Brundtland report") (Oxford: Oxford University Press, 1987), 7.
4   Eric Gómez-Baggethun and Jose Manuel Naredo, "In Search of Lost Time: The Rise and Fall of Limits to Growth in International Sustainability Policy," *Sustainability Science* 10(3) (2015), 385–95. The first chapter of the Brundtland report lists the symptoms and causes in the following order: (1) poverty; (2) growth; (3) survival; (4) the economic crisis (WCED, *Our Common Future*, 28).
5   Gómez-Baggethun and Naredo, "In Search of Lost Time," 391.
6   Nicolas Georgescu-Roegen, "Quo Vadis Homo Sapiens Sapiens?" in Mario Bonaiuti (ed.), *Bioeconomia* (Turin: Boleti Boringhierie, 2003), 219. See also the chapters in Iris Borowy and Matthias

Schmelzer (eds.), *History of the Future of Economic Growth: Historical Roots of Current Debates on Sustainable Degrowth* (London: Routledge, 2017), which demonstrate the problematic links between sustainable development and growth-oriented economics.

7   This does not mean free of all of this altogether of course. Rather, as we will show later in the text, free of the systemic addiction to any of this.

8   Degrowth can be explained as a gradual and equitable downscaling of production and consumption that aims at a quantitatively smaller and qualitatively different economy; see Francois Schneider, Giorgos Kallis and Joan Martinez-Alier, "Crisis or Opportunity? Economic Degrowth for Social Equity and Ecological Sustainability," *Journal of Cleaner Production* 18 (2010), 511–18. As we will articulate in this chapter, degrowth also calls for the re-politicization of the debate on socio-ecological transformations; see Federico Demaria, Francois Schneider, Filka Sekulova and Joan Martinez-Alier, "What is Degrowth? From an Activist Slogan to a Social Movement," *Environmental Values* 22(2) (2013), 191–215.

9   See Bruno Latour, "Facing Gaia: A New Enquiry Into Natural Religion," 2013, retrieved on June 3, 2017 from www.giffordlectures.org/lectures/facing-gaia-new-enquiry-natural-religion.

10 See Barbara Muraca, "Re-appropriating the Ecosystem Services Concept for a Decolonization of 'Nature'," in Bryan E. Bannon, *Nature and Experience* (Lanham, MD: Rowman & Littlefield 2016), 143–56.

11 Modern, industrial capitalism is read by feminist economist Amalia Orozco as life-destroyer in a most literal and systemic way: it is rooted in the direct and indirect exploitation of the so-called re-productive sphere beyond its regeneration capacity. See Amaia Perez Orozco, *Subversión feminista de la economía: Aportes para un debate sobre el conflicto capital-vida* (Madrid: Traficantes de Sueños, 2014).

12 Ulrich Grober, *Sustainability: A Cultural History* (Totnes: Green Books, 2012).

13 Konrad Ott and Ralf Döring, *Theorie und Praxis starker Nachhaltigkeit* (Marburg: Metropolis, 2008).

14 Both John A. Robinson and Jeremy Caradonna have argued that, as discourses, sustainability and sustainable development descend from different intellectual genealogies, or that sustainable development is a very particular, growth-oriented offshoot of a complex sustainability lineage. See John A. Robinson, "Squaring the Circle? Some Thoughts on the Idea of Sustainable Development," *Ecological Economics* 48(4) (2004), 369–84; Jeremy L. Caradonna, *Sustainability: A History* (Oxford: Oxford University Press, 2014).

15 Ott and Döring, *Nachhaltigkeit*.

16 Ibid.

17 See for example the network ICLEI (www.iclei.org) or the Sustainable Cities Network (www.sustainablecities.eu/the-aalborg-commitments).

18 UN, "Sustainable Development Goals," retrieved on February 13, 2017 from www.un.org/sustainabledevelopment/sustainable-development-goals.

19 Carl Death and Clive Gabay, "Doing Biopolitics Differently? Radical Potential in the Post-2015 MDG and SDG Debates," *Globalizations* 12(4) (2015), 597–612.

20 Sachverständigenrat für Umweltfragen (SRU), *Für eine neue Vorreiterrolle: Umweltgutachten* (Berlin: SRU, 2002). By following the so-called Greifswalder Approach, the SRU, somewhat surprisingly, departed from the assumptions about sustainable development associated with the Rio Summit. Konrad Ott was a member of the SRU, Ralf Döring a research associate.

21 Ott and Döring, *Nachhaltigkeit*. See also Konrad Ott and Ralf Döring, "Strong Sustainability and Environmental Policy: Justification and Implementation," in Colin L. Soskolne (ed.), *Sustaining Life on Earth: Environmental and Human Health through Global Governance* (Lanham: Lexington Books, 2009), 109–123.

22 Ott and Döring, *Nachhaltigkeit*.

23 For a comprehensive and detailed analysis of philosophical controversies regarding ethical obligations towards future generations, see Konrad Ott, "Essential Components of Future Ethics," in Ralf Döring and Michael Rühs (eds), *Ökonomische Rationalität und Praktische Vernunft. Gerechtigkeit, Ökologische Ökonomie und Naturschutz* (Würzburg: Königshausen & Neumann, 2004), 83–110.

24 See Barbara Muraca, *Denken im Grenzgebiet: Prozessphilosophische Grundlagen einer Theorie starker Nachhaltigkeit* (Freiburg: Alber, 2010).

25 Alberto Acosta and Esperanza Martínez (eds.), *El Buen Vivir: Una Vía Para El Desarrollo* (Santiago: Editorial Universidad Bolivariana, 2009).

26 See Martha Nussbaum, "Human Functioning and Social Justice. In Defense of Aristotelian Essentialism," *Political Theory* 20(2) (1992), 214; Ott and Döring, *Nachhaltigkeit*.

27 See Barbara Muraca, "Towards a Fair Degrowth-Society: Justice and the Right to a 'Good Life' Beyond Growth," *Futures* 44(6) (2012), 535–45.

28 See Ulrich Brand, "Green Economy, Green Capitalism and the Imperial Mode of Living: Limits to a Prominent Strategy, Contours of a Possible New Capitalist Formation," *Fudan Journal of Humanities and Social Sciences* 9(1) (2016), 107–21.

29 Ibid.

30 Harald Welzer, *Mental Infrastructures: How Growth Entered the World and Our Souls* (Berlin: Heinrich Böll Stiftung, 2012).

31 For recent analyses of the limits to growth that does not only consider environmental limits both in terms of resources and sinks, but also economic failures, see, for example, Richard Heinberg, *The End of Growth: Adapting to Our New Economic Reality* (Gabriola, BC: New Society Publishers, 2011).

32 Robert M. Solow, "The Economics of Resources or the Resources of Economics," *American Economic Review* 64(2) (1974), 1–14.

33 Konrad Ott, Barbara Muraca, and Christian Baatz, "Strong Sustainability as a Frame for Sustainability Communication," in Gerd Michelsen and Jasmin Godemann, *Sustainability Communication: Interdisciplinary Perspectives and Theoretical Foundations* (Berlin: Springer, 2011), 13–25.

34 Since at least the mid-nineteenth century what we call "nature" has been considered passive, inert, meaningless, exploitable. It is no longer an active factor of production and is relegated to a passive matter to be transformed by human labor. The theoretical basis of this understanding goes back to the roots of Western modernity. See Barbara Muraca, "Getting over 'Nature': Modern Bifurcations, Postmodern Possibilities," in Catherine Keller and Laurel Kearns (eds.), *Ecospirit: Religions and Philosophies for the Earth* (New York: Fordham University Press, 2007).

35 Eric Neumayer, *Weak versus Strong Sustainability* (Cheltenham: Edward Elgar, 1999).

36 Technically, discounting means using a discount rate (mostly comparably to long-term interest rates) and applying this rate to reduce, on an annual basis, the income by this percentage. Assuming an income of $1,000 in year 30, and a 3% discount rate, it would mean $970 in year 29, and so on.

37 Overall, six capital stocks are now distinguished: man-made, natural, social, cultivated natural, human, and knowledge capital.

38 Tim Jackson, *Prosperity without Growth: The Transition to a Sustainable Economy* (London: Sustainable Development Commission, 2009).

39 John Gowdy and Carl N. McDaniel, "The Physical Destruction of Nauru: An Example of Weak Sustainability," *Land Economics* 75(2) (1999), 333–8. See also Ilan Solomons, "Missed Opportunities," retrieved on February 14, 2017 from www.miningweekly.com/article/unsustainable-use-of-revenue-from-once-booming-phosphate-sector-haunting-nauru-2016-06-24.

40 Human activity based on economic growth over the past century has exceeded at least three of Earth's essential "planetary boundaries." See Johann Rockström et al., "A Safe Operating Space for Humanity," *Nature* 461 (2009), 472–5.

41 Herman E. Daly, *Beyond Growth: The Economics of Sustainable Development* (Boston, MA: Beacon Press, 1997).

42 For a critique of the happiness approach, see Amartya Sen, *The Idea of Justice* (Cambridge, MA: Harvard University Press, 2009), 277; Muraca, "Towards a Fair Degrowth-Society," 535–45.

43 Ott et al., "Strong Sustainability as a Frame for Sustainability Communication."

44 Nicholas Georgescu-Roegen, *The Entropy Law and the Economic Process* (Cambridge, MA: Harvard University Press, 1971).

45 Resources are inputs whereas products or waste are outputs. See Georgescu-Roegen, *The Entropy Law*, 230.

46 More precisely, whether a process is a flow or a fund is not a matter of its ontological structure, but rather depends on the perspective taken, the way in which the analytic boundary of the process is drawn, and how something is employed in and for the economic process. Consider for example a tomato sold on the market as a commodity (flow) or fallen on the ground and rotting while also regenerating the soil (fund). See Katharine N. Farrell and Kozo Mayumi, "Time Horizons and Electricity Futures: An Application of Nicholas Georgescu-Roegen's General Theory of Economic Production," *Energy* 34 (2009), 301–7.

47 Georgescu-Roegen, "Quo Vadis Homo Sapiens Sapiens?"; Barbara Muraca and Matthias Schmelzer, "Degrowth: Historicizing the Critique of Economic Growth and the Search for Alternatives in Three Regions," in Iris Borowy and Matthias Schmelzer (eds.), *History of the Future of Economic Growth: Historical Roots of Current Debates on Sustainable Degrowth* (forthcoming).

48 Daly, *Beyond Growth.*
49 Viviana Asara, Iago Otero, Federico Demaria, and Esteve Corbera, "Socially Sustainable Degrowth as a Social–Ecological Transformation: Repoliticizing Sustainability," *Sustainability Science* 10(3) (2015), 375–84.
50 Georgescu-Roegen, *The Entropy Law*, 125ff.
51 Muraca, *Denken im Grenzgebiet*; Panos Petridis, Barbara Muraca, and Giorgos Kallis, "Degrowth: Between a Scientific Concept and a Slogan for a Social Movement," in Joan Martinez-Alier and Roldan Muradian (eds.), *Handbook of Ecological Economics* (Cheltenham: Edward Elgar, 2015), 176–200.
52 Daly, *Beyond Growth:* 28–9.
53 Ibid., 51.
54 If we consider the current conflict between different uses of land (production of food or production of energy through biomass or solar panels), we can understand well the forward-looking perspective that Georgescu-Roegen had in the 1970s. The "scarcity" of land does not refer only to a measure of quantity (planetary surface), but also to quality (fertile soil, availability of water etc.), i.e. – mostly – to the incessant activity of living organisms in regenerating it.
55 Muraca, *Denken im Grenzgebiet.*
56 See, for example, Marina Fischer-Kowalski, "Society's Metabolism," *Journal of Industrial Ecology* 2(1) (1998), 61–78; Giorgos Kallis, "Societal Metabolism, Working Hours and Degrowth: A Comment on Sorman and Giampietro," *Journal of Cleaner Production* 38 (2013), 94–8.
57 Egon Becker, Diana Hummel, and Thomas Jahn, "Gesellschaftliche Naturverhältnisse als Rahmenkonzept," in Matthias Groß (ed.), *Handbuch Umweltsoziologie* (Wiesbaden: VS Verlag für Sozialwissenschaften, 2011), 75–96.
58 The CCN using the ecological concept of carrying capacity as the ground for immigration control, thus confusing political borders with ecosystem features. Daly served for a long time on the advisory board of the organization, but decided to step down recently, after a vivid and controversial discussion within the ecological economics community.
59 Discounting is the best way of masking this displacement. However, the very possibility of discounting future damages relies on the assumption of economic growth, i.e. of increasing wealth that would reduce the – merely monetary costs – of environmental damage. Moreover, it does not consider qualitative changes that might be irreversible and beyond any possible compensation. The very assumption of economic growth is thus an act of injustice towards future generations. See Roldan Muradian, Martin O Connor, and Joan Martinez-Alier, "Embodied Pollution in Trade: Estimating the 'Environmental Load Displacement' of Industrialised Countries," *Ecologial Economics* 41(1) (2002), 41–57.
60 Muraca, *Denken im Grenzgebiet.*
61 Nicholas Georgescu-Roegen, *La décroissance: Entropie—écologie—économie* (Nashville, TN: University of Tennessee, 1995); Barbara Muraca, "Decroissance: A Project for a Radical Transformation of Society," *Environmental Values* 22(2) (2013), 147–69.
62 Andre Gorz, *Ecology as Politics* (Montreal: Black Rose Books, 1980).
63 Thimotee Duverger, "De Meadows à Mansholt: L'invention du 'zégisme'," *Entropia* 10 (2011), 118ff.
64 André Amar, "La croissance et le problem morale," *Cahiers de la Nef* 52 (September–November 1973), retrieved on February 10, 2017 from www.decroissance.org/?chemin=textes/amar.
65 Muraca and Schmelzer, "Degrowth."
66 Giacomo D'Alisa, Federico Demaria, and Giorgos Kallis (eds.), *Degrowth: A Vocabulary for a New Era* (Abingdon: Routledge, 2014).
67 Ibid., 9. See also Jeremy L. Caradonna, "An Incompatible Couple: A Critical History of Economic Growth and Sustainable Development," in Iris Borowy and Matthias Schmelzer (eds.), *History of the Future of Economic Growth* (forthcoming).
68 Altermondialiste in French: the term is difficult to translate, but denotes an alternative path to global cooperation from below against the diktat of developmentalism.
69 Jóse Giornal, "Défaire le développement, Refaire le monde," *Réfractions* 9 (2002), 119–22.
70 See Muraca, *Decroissance.* See also Demaria et al., "What is Degrowth?"; D'Alisa et al., *Degrowth: A Vocabulary.*
71 D'Alisa et al., *Degrowth: A Vocabulary.*
72 Muraca and Schmelzer, "Degrowth."
73 Asara et al., "Socially Sustainable Degrowth as a Social–Ecological Transformation," 375.

74 See, for instance, Jeremy L. Caradonna et al., "A Degrowth Response to an Ecomodernist Manifesto," May 6, 2015, retrieved on June 3, 2017 from www.resilience.org/stories/2015–05–06/a-degrowth-response-to-an-ecomodernist-manifesto, which is a degrowth critique of ecomodernism.

75 Petridis et al., "Degrowth."

76 Paul Ariés, *Décroissance ou Barbarie* (Lyon: Golias, 2005).

77 Barbara Muraca, "Against the Insanity of Growth: Degrowth as a Concrete Utopia," in J. Heinzekehr and Philipp Clayton (eds.), *Socialism in Process: Ecology and Politics toward a Sustainable Future* (Anoka, MN: Process Century Press, forthcoming).

78 Petridis et al., "Degrowth."

79 D'Alisa et al., *Degrowth: A Vocabulary*, 4.

80 TINA comes famously from Margaret Thatcher's enunciation that "there is no alternative" to neoliberalism.

81 Eric Swyngedouw, "Depoliticization (the Political)," in Giacomo D'Alisa, Federico Demaria and Giorgos Kallis (eds.), *Degrowth: A Vocabulary for a New Era* (Abingdon: Routledge, 2014), 91.

82 Ibid., 92.

83 See Tilman Santarius, Hans Jacob Walnum, and Carlo Aall, *Rethinking Climate and Energy Politicies. New Perspectives on the Rebound Phenomenon* (Berlin: Springer, 2016).

84 Swyngedouw, "Depoliticization (the Political)," 92. See also Marina Fischer-Kowalski and Helmut Haberl, "Social Metabolism: A Metrics for Biophysical Growth and Degrowth," in Joan Martinez-Alier and Roldan Muradian (eds.), *Handbook of Ecological Economics* (Cheltenham: Edward Elgar, 2015).

85 Muraca, "Against the Insanity of Growth."

86 Welzer, *Mental Infrastructures*.

87 Serge Latouche, *Farewell to Growth* (Cambridge: Polity Press, 2009).

88 Michel Foucault, *The Birth of Biopolitics: Lectures at the Collège De France, 1978–79* (New York: Palgrave Macmillan, 2008).

89 Gary S. Becker, *Human Capital: A Theoretical and Empirical Analysis* (Chicago, IL: University of Chicago Press, 1993).

90 Muraca, "Against the Insanity of Growth."

91 See New York Times, "President Bush's News Conference," *New York Times* (December 20, 2006), retrieved on June 3, 2017 from www.nytimes.com/2006/12/20/washington/20text-bush.html.

92 Birgit Mahnkopf, *Peak Everything – Peak Capitalism? Folgen der sozialökologischen Krise für die Dynamik des historischen Kapitalismus* (Jena: DFG-KollegforscherInnengruppe Postwachstumsgesellschaften, 2013).

93 Nicholas Stern, *Stern Review on The Economics of Climate Change* (pre-publication edition) (London: HM Treasury, 2006).

94 Sieghard Neckel, "Refeudalisierung der Ökonomie. Zum Strukturwandel kapitalistischer Wirtschaft," *Neue Zeitschrift für Sozialforschung* 8(1) (2010), 117–28.

95 Alain de Benoist, *"Demain, La Décroissance!" Penser L'ecologie Jusqu'au Bout* (Paris: Édite, 2007).

96 Meinhard Miegel, *Exit: Wohlstand ohne Wachstum* (Berlin: Propyläen, 2010). To a certain extent, the recent developments in Europe and in the US with respect to Brexit and the election of Trump corroborate this analysis: the crisis of the promise of increasing wealth attached to growth for a relatively large number of people implies a delegitimization of democracy and nourishes the scape-goat narrative against immigrants. In the absence of creative visions for a radical transformation of society that reduces inequality, fosters participation, and restructures the economy, sealing-off the borders and backwards values are advocated as a bastion against the fear of loss and pauperization.

97 D'Alisa et al., *Degrowth: A Vocabulary.*

98 Dennis Eversberg and Matthias Schmelzer, "The Degrowth Spectrum: Convergence and Divergence within a Diverse and Conflictual Alliance," *Environmental Values* (forthcoming).

99 Ashish Kothari, Federico Demaria, and Alberto Acosta, "Buen Vivir, Degrowth and Ecological Swaraj: Alternatives to Sustainable Development and the Green Economy," *Development* 57(3) (2014), 362–75; Arturo Escobar, "Degrowth, Postdevelopment, and Transitions: A Preliminary Conversation," *Sustainability Science* 10(3) (2015), 451–62.

100 Barbara Muraca, *Gut Leben: Eine Gesellschaft jenseits des Wachstums* (Berlin: Wagenbach 2014).

References

Acosta, Alberto, and Esperanza Martínez (eds.), *El Buen Vivir: Una Vía Para El Desarrollo* (Santiago: Editorial Universidad Bolivariana, 2009).

Amar, André, "La croissance et le problem morale," *Cahiers de la Nef* 52 (September–November 1973), retrieved on February 10, 2017 from www.decroissance.org/?chemin=textes/amar.

Ariés, Paul, *Décroissance ou Barbarie* (Lyon: Golias, 2005).

Asara, Viviana, Iago Otero, Federico Demaria, and Esteve Corbera, "Socially Sustainable Degrowth as a Social–Ecological Transformation: Repoliticizing Sustainability," *Sustainability Science* 10(3) (2015), 375–84.

Becker, Egon, Diana Hummel, and Thomas Jahn, "Gesellschaftliche Naturverhältnisse als Rahmenkonzept," in Matthias Groß (ed.), *Handbuch Umweltsoziologie* (Wiesbaden: VS Verlag für Sozialwissenschaften, 2011), 75–96.

Becker, Gary S., *Human Capital: A Theoretical and Empirical Analysis* (Chicago, IL: University of Chicago Press, 1993).

Borowy, Iris, and Matthias Schmelzer (eds.), *History of the Future of Economic Growth: Historical Roots of Current Debates on Sustainable Degrowth* (London: Routledge, 2017).

Brand, Ulrich, "Green Economy, Green Capitalism and the Imperial Mode of Living: Limits to a Prominent Strategy, Contours of a Possible New Capitalist Formation," *Fudan Journal of Humanities and Social Sciences* 9(1) (2016), 107–21.

Caradonna, Jeremy L., *Sustainability: A History* (Oxford: Oxford University Press, 2014).

Caradonna, Jeremy L., "An Incompatible Couple: A Critical History of Economic Growth and Sustainable Development," in Iris Borowy and Matthias Schmelzer (eds.), *History of the Future of Economic Growth* (forthcoming).

Caradonna, Jeremy L., et al., "A Degrowth Response to an Ecomodernist Manifesto," May 6, 2015, retrieved on June 3, 2017 from www.resilience.org/stories/2015–05–06/a-degrowth-response-to-an-ecomodernist-manifesto.

D'Alisa, Giacomo, Federico Demaria, and Giorgos Kallis (eds.), *Degrowth: A Vocabulary for a New Era* (Abingdon: Routledge, 2014).

Daly, Herman E., *Beyond Growth: The Economics of Sustainable Development* (Boston, MA: Beacon Press, 1997).

De Benoist, Alain, *"Demain, La Décroissance!" Penser L'ecologie Jusqu'au Bout* (Paris: Édite, 2007).

Death, Carl, and Clive Gabay, "Doing Biopolitics Differently? Radical Potential in the Post-2015 MDG and SDG Debates," *Globalizations* 12(4) (2015), 597–612.

Demaria, Federico, Francois Schneider, Filka Sekulova and Joan Martinez-Alier, "What is Degrowth? From an Activist Slogan to a Social Movement," *Environmental Values* 22(2) (2013), 191–215.

Duverger, Thimottee, "De Meadows à Mansholt: L'invention du 'zégisme'," *Entropia* 10 (2011), 118ff.

Escobar, Arturo, "Degrowth, Postdevelopment, and Transitions: A Preliminary Conversation," *Sustainability Science* 10(3) (2015), 451–62.

Eversberg, Denis, and Matthias Schmelzer, "The Degrowth Spectrum: Convergence and Divergence within a Diverse and Conflictual Alliance," *Environmental Values* (forthcoming).

Farrell, Katharine N., and Kozo Mayumi, "Time Horizons and Electricity Futures: An Application of Nicholas Georgescu-Roegen's General Theory of Economic Production," *Energy* 34 (2009), 301–7.

Fischer-Kowalski, Marina, "Society's Metabolism," *Journal of Industrial Ecology* 2(1) (1998), 61–78.

Fischer-Kowalski, Marina, and Helmut Haberl, "Social Metabolism: A Metrics for Biophysical Growth and Degrowth," in Joan Martinez-Alier and Roldan Muradian (eds.), *Handbook of Ecological Economics* (Cheltenham: Edward Elgar, 2015).

Foucault, Michel, *The Birth of Biopolitics: Lectures at the Collège De France, 1978–79* (New York: Palgrave Macmillan, 2008).

Georgescu-Roegen, Nicholas, *The Entropy Law and the Economic Process* (Cambridge, MA: Harvard University Press, 1971).

Georgescu-Roegen, Nicholas, *La décroissance: Entropie—écologie—économie* (Nashville, TN: University of Tennessee, 1995).

Georgescu-Roegen Nicholas, "Quo Vadis Homo Sapiens Sapiens?" in Mario Bonaiuti (ed.), *Bioeconomia* (Turin: Boleti Boringhierie, 2003), 211–24.

Giornal, Jóse, "Défaire le développement, Refaire le monde," *Réfractions* 9 (2002), 119–22.

Gómez-Baggethun, Eric, and Jose Manuel Naredo, "In Search of Lost Time: The Rise and Fall of Limits to Growth in International Sustainability Policy," *Sustainability Science* 10(3) (2015), 385–95.

Goodman, James, and Ariel Salleh, "The 'Green Economy': Class Hegemony and Counter-Hegemony," *Globalizations* 10(3) (2013).

Gorz, Andre, *Ecology as Politics* (Montreal: Black Rose Books, 1980).

Gowdy, John, and Carl N. McDaniel, "The Physical Destruction of Nauru: An Example of Weak Sustainability," *Land Economics* 75(2) (1999), 333–8.

Grober, Ulrich, *Sustainability: A Cultural History* (Totnes: Green Books, 2012).

Heinberg, Richard, *The End of Growth: Adapting to Our New Economic Reality* (Gabriola, BC: New Society Publishers, 2011).

Jackson, Tim, *Prosperity without Growth: The Transition to a Sustainable Economy* (London: Sustainable Development Commission, 2009).

Kallis, Giorgos, "Societal Metabolism, Working Hours and Degrowth: A Comment on Sorman and Giampietro," *Journal of Cleaner Production* 38 (2013), 94–8.

Kothari, Ashish, Federico Demaria, and Alberto Acosta, "Buen Vivir, Degrowth and Ecological Swaraj: Alternatives to Sustainable Development and the Green Economy," *Development* 57(3) (2014), 362–75.

Latouche, Serge, *Farewell to Growth* (Cambridge: Polity Press, 2009).

Latour, Bruno, "Facing Gaia: A New Enquiry Into Natural Religion," 2013, retrieved on June 3, 2017 from www.giffordlectures.org/lectures/facing-gaia-new-enquiry-natural-religion.

Mahnkopf, Birgit, *Peak Everything—Peak Capitalism? Folgen der sozialökologischen Krise für die Dynamik des historischen Kapitalismus* (Jena: DFG-KollegforscherInnengruppe Postwachstumsgesellschaften, 2013).

Miegel, Meinhard, *Exit: Wohlstand ohne Wachstum* (Berlin: Propyläen, 2010).

Muraca, Barbara, "Getting over 'Nature': Modern Bifurcations, Postmodern Possibilities," in Catherine Keller and Laurel Kearns (eds.), *Ecospirit: Religions and Philosophies for the Earth* (New York: Fordham University Press, 2007).

Muraca, Barbara, *Denken im Grenzgebiet: Prozessphilosophische Grundlagen einer Theorie starker Nachhaltigkeit* (Freiburg: Alber, 2010).

Muraca, Barbara, "Towards a Fair Degrowth-Society: Justice and the Right to a 'Good Life' Beyond Growth," *Futures* 44(6) (2012), 535–45.

Muraca, Barbara, "Decroissance: A Project for a Radical Transformation of Society," *Environmental Values* 22(2) (2013), 147–69.

Muraca, Barbara, *Gut Leben: Eine Gesellschaft jenseits des Wachstums* (Berlin: Wagenbach 2014).

Muraca, Barbara, "Re-appropriating the Ecosystem Services Concept for a Decolonization of 'Nature'," in Bryan E. Bannon, *Nature and Experience* (Lanham, MD: Rowman & Littlefield 2016), 143–56.

Muraca, Barbara, "Against the Insanity of Growth: Degrowth as a Concrete Utopia," in J. Heinzekehr and Philipp Clayton (eds.), *Socialism in Process: Ecology and Politics toward a Sustainable Future* (Anoka, MN: Process Century Press, forthcoming).

Muraca, Barbara, and Matthias Schmelzer, "Degrowth: Historicizing the Critique of Economic Growth and the Search for Alternatives in Three Regions," in Iris Borowy and Matthias Schmelzer (eds.), *History of the Future of Economic Growth: Historical Roots of Current Debates on Sustainable Degrowth* (forthcoming).

Muradian, Roldan, Martin O'Connor, and Joan Martinez-Alier, "Embodied Pollution in Trade: Estimating the 'Environmental Load Displacement' of Industrialised Countries," *Ecologial Economics* 41(1) (2002), 41–57.

Neckel, Sieghard, "Refeudalisierung der Ökonomie. Zum Strukturwandel kapitalistischer Wirtschaft," *Neue Zeitschrift für Sozialforschung* 8(1) (2010), 117–28.

Neumayer, Eric, *Weak versus Strong Sustainability* (Cheltenham: Edward Elgar, 1999).

New York Times, "President Bush's News Conference," *New York Times* (December 20, 2006), retrieved on June 3, 2017 from www.nytimes.com/2006/12/20/washington/20text-bush.html.

Nussbaum, Martha, "Human Functioning and Social Justice. In Defense of Aristotelian Essentialism," *Political Theory* 20(2) (1992), 214.

Ott, Konrad, "Essential Components of Future Ethics," in Ralf Döring and Michael Rühs (eds), *Ökonomische Rationalität und Praktische Vernunft. Gerechtigkeit, Ökologische Ökonomie und Naturschutz* (Würzburg: Königshausen & Neumann, 2004), 83–110.

Ott, Konrad, and Ralf Döring, *Theorie und Praxis starker Nachhaltigkeit* (Marburg: Metropolis, 2008).

Ott, Konrad, and Ralf Döring, "Strong Sustainability and Environmental Policy: Justification and Implementation," in Colin L. Soskolne (ed.), *Sustaining Life on Earth: Environmental and Human Health through Global Governance* (Lanham: Lexington Books, 2009), 109–123.

Ott, Konrad, Barbara Muraca, and Christian Baatz, "Strong Sustainability as a Frame for Sustainability Communication," in Gerd Michelsen and Jasmin Godemann, *Sustainability Communication: Interdisciplinary Perspectives and Theoretical Foundations* (Berlin: Springer, 2011), 13–25.

Perez Orozco, Amaia, *Subversión feminista de la economía: Aportes para un debate sobre el conflicto capital-vida* (Madrid: Traficantes de Sueños, 2014).

Petridis, Panos, Barbara Muraca, and Giorgos Kallis, "Degrowth: Between a Scientific Concept and a Slogan for a Social Movement," in Joan Martinez-Alier and Roldan Muradian (eds.), *Handbook of Ecological Economics* (Cheltenham: Edward Elgar, 2015), 176–200.

Robinson, John A., "Squaring the Circle? Some Thoughts on the Idea of Sustainable Development," *Ecological Economics* 48(4) (2004), 369–84.

Rockström, Johann, et al., "A Safe Operating Space for Humanity," *Nature* 461 (2009), 472–5.

Sachverständigenrat für Umweltfragen (SRU), *Für eine neue Vorreiterrolle: Umweltgutachten* (Berlin: SRU, 2002).

Santarius, Tilman, Hans Jacob Walnum, and Carlo Aall, *Rethinking Climate and Energy Politicies. New Perspectives on the Rebound Phenomenon* (Berlin: Springer, 2016).

Schneider, Francois, Giorgos Kallis and Joan Martinez-Alier, "Crisis or Opportunity? Economic Degrowth for Social Equity and Ecological Sustainability," *Journal of Cleaner Production* 18 (2010), 511–18.

Sen, Amartya, *The Idea of Justice* (Cambridge, MA: Harvard University Press, 2009).

Solomons, Ilan, "Missed Opportunities," retrieved on February 14, 2017 from www.miningweekly.com/article/unsustainable-use-of-revenue-from-once-booming-phosphate-sector-haunting-nauru-2016-06-24.

Solow, Robert M., "The Economics of Resources or the Resources of Economics," *American Economic Review* 64(2) (1974), 1–14.

Stern, Nicholas, *Stern Review on The Economics of Climate Change* (pre-publication edition) (London: HM Treasury, 2006).

Swyngedouw, Eric, "Depoliticization (the Political)," in Giacomo D'Alisa, Federico Demaria and Giorgos Kallis (eds.), *Degrowth: A Vocabulary for a New Era* (Abingdon: Routledge, 2014), 90–93.

UN Stockholm Declaration, 1972, retrieved on February 13, 2017 from www.unep.org/documents.multilingual/default.asp?documentid=97&articleid=1503; UN Cocoyoc Declaration, retrieved on February 13, 2017 from www.mauricestrong.net/index.php/cocoyoc-declaration.

WCED, *Our Common Future* ("the Brundtland report") (Oxford: Oxford University Press, 1987).

Welzer, Harald, *Mental Infrastructures: How Growth Entered the World and Our Souls* (Berlin: Heinrich Böll Stiftung, 2012).

22

SUSTAINABILITY BEYOND GROWTH

Toward an ethics of flourishing

John R. Ehrenfeld

If names be not correct, language is not in accordance with the truth of things. If language be not in accordance with the truth of things, affairs cannot be carried on to success. When affairs cannot be carried on to success, proprieties and music do not flourish. When proprieties and music do not flourish, punishments will not be properly awarded.

When punishments are not properly awarded, the people do not know how to move hand or foot.

(*Confucius,* Analects)

Global concerns about the constellation of threats to humanity and the Planet that have been labeled unsustainable have changed gradually over the past three to four decades, like the proverbial frog in hot water. If you do not already know this tale about the effects of small changes, it goes like this. If you drop a frog into a pot of boiling water, it will immediately try to jump out. But if you put it in the pot and slowly raise the temperature, it will remain there until it is cooked. The story may not be biologically accurate, but it does metaphorically describe the failure of people to observe very slow changes until it is too late to respond. It is particularly apt in the case of climate change and other challenges to sustainability. To follow this history, we have to examine the origins of both sustainability and its negative twin, unsustainability.

Our public concerns about the deteriorating conditions of the Planet go back at least centuries to the times of soft coal burning in crowded London in the seventeenth century. Maybe the cave dwellers had similar concerns about their sooty abodes. John Evelyn, in what is generally accepted as the first published critique of environmental conditions, *Fumifugium, or, The inconveniencie of the aer and smoak of London dissipated together with some remedies humbly proposed by J. E. esq. to His Sacred Majestie, and to the Parliament now assembled*, complained about the terrible state of the atmosphere:

It is this horrid Smoake which obscures our Churches, and makes our Palaces look old, which fouls our Clothes, and corrupts the waters, so as the very Rain, and refreshing Dews which fall in the several Seasons, precipitate this impure vapour, which, with its black and tenacious quality, spots and contaminates whatsoever is expos'd to it.[1]

He proposed a surprisingly modern solution: to move the most serious production facilities away from the city, and a not-so-modern one: to ring the city with gardens planted with sweet-smelling plants and shrubs. One huge can of Glade:

> That these Palisad's be elegantly planted, diligently kept and supply'd, with such Shrubs, as yield the most fragrant and odoriferous Flowers, and are aptest to tinge the Aer upon every gentle emission at a great distance: Such as are (for instance amongst many others) the Sweet-briar, all the Periclymena's and Woodbinds; the Common white and yellow Jessamine, both the Syringa's or Pipe trees; the Guelder-rose, the Musk, and all other Roses; Genista Hispanica: To these may be added the Rubus odoratus, Bayes, Juniper, Lignum-vitae, Lavander: but above all, Rosemary, the Flowers whereof are credibly reported to give their scent above thirty Leagues off at Sea, upon the coasts of Spain; and at some distance towards the Meadow side, Vines, yea Hops.[2]

Now, fast forward to the twentieth century when the same problems were arising in the United States. Not much happened until the situation became intolerable in places near power plants and steel mills, and rivers started to catch on fire. Rachel Carson published *Silent Spring* in 1962, connecting the loss of many raptor species to a thinning and subsequent destruction of eggshells from the effects of ingesting DDT.[3] The chemical industry vigorously attacked the book and the author, but the scientific community largely supported her findings. Ten years later, in 1972, with great public support and little opposition from the chemical industry, Congress banned the use of DDT, as a part of the first round of environmental statutes enacted in the 1970s in the United States. The frog had jumped out of the environmental pot and pushed the Congress to pass a series of federal anti-pollution statutes that still remain the centerpiece of our environmental concerns. The early statutes mainly reflected environmental impacts that were tangible and visible: soot, smog, polluted rivers and lakes.

Starting in the 1980s and 1990s, many people began to recognize more subtle deteriorating trends, for example the impact of toxic substances on health, but had no conventional way of generalizing them. The early environmental laws were organized around discrete media: air, water, solid waste, and so on. You could talk about ozone depletion, global warming, fisheries collapse, inequality, and so on as isolated problems, but it was difficult to convey the presence and enormity of the situation without a simple, neat word. "Unsustainability" arose to fill this "semantic void," a phrase used by Leo Marx in describing the evolution of the usage of the word "technology." Marx, in writing about the emergence of technology into the cultural vernacular from its technical roots, noted that "[social] changes, whatever they were, created a semantic—indeed, a conceptual—void … an awareness of certain novel developments in society and culture for which no adequate name had yet become available." Technology became "a keyword in the lexicon we rely on to chart the changing character of society and culture."[4] This collapsing of a constellation of problems and characteristics into a single word facilitated its diffusion and the subsequent high level of attention paid to it.

As public consciousness of unsustainability became widespread, another frog jumped out and gave us the concept of *sustainable development*, a broad program presumed to continue modernity's economic progress without further damaging a world already showing signs of severe stress. It entered our vocabulary as a remedy to the serious social inequality between the Northern and Southern Hemispheres, and to slow down the damages human society was wreaking on the earth. The original form of this notion was defined in the 1987 Brundtland Report (*Our Common Future*) as "[economic] development that meets the needs of the present without compromising the ability of future generations to meet their own needs."[5] The intent of those who wrote the report was to place environment solidly into the hegemonic, economically driven political agendas of governments. The concept of sustainable development usually assumes that we can grow our way out of the mess we are in if we are very, very careful, but ignores the fact that growth had played a major role in creating the mess.[6] Perhaps unsurprisingly, it mimics the same argument that economic growth is the universal solution for almost all other societal ills. As a policy instrument, it has been largely impotent to stop either continuing environmental predation and degradation and unfair economic development. Despite a series of global summits designed to promote sustainable development, little or nothing has been done to explicitly alleviate the economic disparities among and within nations.

The positive reaction to unsustainability was a plethora of remedial programs, largely carried out by businesses. Initially, activities with the label of sustainable development were scattered and, without a name to describe them, the institution of business could not say much about what was going on. A new word, "greening," was coined, to encompass all the activities aimed at reducing unsustainability. Being green showed concern for the (green) environment. This new word sufficed for about a decade until it was clear that greening was inadequate to counteract growing unsustainability. Greening could not encompass the enormity of social problems as well as those of the environment. A new phrase was needed to describe the early activities aimed at the larger set of issues. This semantic void was filled with "sustainability," but the meaning of sustainability was never clear. It quickly became a euphemism for business-as-usual or, in some scattered cases, business-almost-as-usual. This drift toward meaninglessness is due, partly, to the nature of the word, itself, and, partly, to the desire of businesses to keep doing whatever they had been doing.

The broad, but fuzzy concept of sustainability was reduced in the everyday vernacular to catchphrases like the triad of environment, economy and equity/ethics/social, or people, profits, and planet (the triple bottom line; see Figure 22.1).

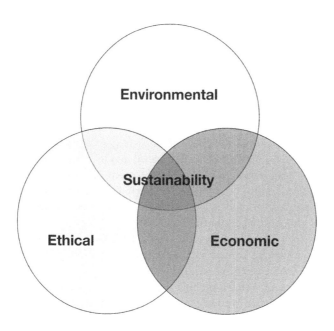

Figure 22.1 The three Es of sustainability

Businesses found that the new term had a nice ring to it with more market power than greening. They could designate the firm as a "sustainable business" and what they market as sustainable luxury (*sic*), sustainable furniture, sustainable T-shirts, etc. This usage escaped from the business world to become much more generally used as sustainable consumption, sustainable cities, sustainable transport, and so on. Virtually all who are working on sustainable [you supply the noun] have neglected to notice that "sustainable" is an adjective. The primary attention is focused on the noun it modifies; this use of sustainable has little or nothing to do with the state of the world. Sustainable business is about business, not the world. The original intent of the Brundtland Report, which focused on the global environmental and social system, became diffused and limited to the interests of individual enterprises.

Anticipating the agenda of the first United Nations Conference on Environment and Development (UNCED) held in Rio de Janeiro in 1992, global businesses formed the World Business Council for Sustainable Development (WBCSD). Its response to the challenges of sustainable development was elaborated in a book, *Changing Course*, in which the Council members unveiled their principal strategy, eco-efficiency, defined as producing more value with less environmental (and social) impact.[7] The concept of eco-efficiency was first described by Schaltegger and Sturm, but remained mostly as an academic idea until the WBSCD adopted it.[8] Since then it has been accepted as the key strategic theme for global business in relation to commitments and activities directed at sustainability or sustainable development. The WBCSD describes eco-efficiency as "being achieved by the delivery of competitively priced goods and services that satisfy human needs and bring quality of life, while progressively reducing ecological impacts and resource intensity throughout the life cycle, to a level at least in line with the Earth's estimated carrying capacity."[9]

The concept's practical and theoretical importance lies in its ability to combine performance along two of the three axes of sustainable development: environment and economics. Issues concerning equity and other social properties are only weakly included in the concept of eco-efficiency. The notion that increasing economic development would have to be correlated with lowering of environmental impact was not a new concept. Even before people started talking explicitly about eco-efficiency, the tight relationship between it, growth, and impact was shown in the so-called IPAT identity ($I = PAT$), formulated by Paul Ehrlich and John P. Holdren.[10] Global impacts (I) equal population (P) times a measure of affluence (or economic output) per capita (A) times a technical efficiency term (T). The latter term is equivalent to what has now become called eco-efficiency, a ratio of impact per unit of economic output. Given increasing population, the product of A and T must decrease to hold constant or to reduce impact. Some have argued that the IPAT identity as written above is too simplistic to describe the interrelationships between affluence (A) and technical efficiency (T), but, even with a more intricate relationship, the fact remains that eco-efficiency has historically lagged behind economic growth and shows no signs of catching up. Later, environmental scholars spoke of Factor 4, Factor 10, or more as estimates of just how much "dematerialization" would be needed to offset economic and population growth.[11]

But a more important question always lurks in the background. Eco-efficiency is a notion that is meaningful only in the context of the economic model of sustainable development. Although the WBCSD's statements indicate that production output should be kept "in line with the Earth's carrying capacity," there is nothing in the analytic representation of eco-efficiency that provides a clue to this. Standard economic theory assumes that limitless resources will always be available as scarcity incentivizes sufficient innovation to produce substitutes. Carrying capacity simply does not enter the economic calculus. The present, clearly unsustainable state of the world trumps this theory, rendering eco-efficiency only a partially useful concept.[12]

Sustainability has become almost universally used in business, education, or government to refer to whatever they are doing to reduce their impacts on the natural world (many, many activities) and on inequality (hardly anything at all). I suspect few would be able to clearly enunciate what they are trying to sustain, other than their term in office or their competitive position. The key question of what was to be sustained was rarely explicitly asked. Nor were questions raised about the source of the causative factors of the deteriorating conditions of the Earth. If there was any image behind its use, it was the continuation of societal life based on an economic system of benign growth.

In practice, most sustainability initiatives have involved only the first two parts of the tripartite representation illustrated in Figure 22.1: environment and economy. Fewer programs have been directed at the third dimension, improving the lot of the humans on the Planet. Under the rubric of corporate social responsibility (CSR), companies divide their activities into three categories: philanthropy, employee ethics, and, curiously, environment, assuming that doing good things for the Earth is good for people as well. True, but a long way from the intent of the term, which arose from a sense that institutions needed to have distinct, but perhaps overlapping, strategies for the social and natural dimensions of sustainability. Whether explicit or not, CSR is overwhelmed by the primacy of profit. Milton Friedman's famous cry, made in 1970, that "the social responsibility of business is to increase its profits" still reverberates in the halls of CEOs' offices.[13]

I do not have well-documented numbers related to the proportion of efforts going to eco-efficiency versus CSR, but my own experience and research indicate that the first concern far outweighs the latter. Eco-efficiency is generally defined as providing equal or more value with less environmental impact. Efficiency is always a ratio of an output to an input. Whether couched in material, energy, or economic terms, higher efficiency means producing a fixed amount of goods or value, while using up less of the world's resources. Taken out of the context of the whole system, efficiency increases appear, by themselves, to be a good strategy because more resources are left over for subsequent use. However, increased efficiency in the economic system of a nation leaves capital on the table that can be reinvested in more production capacity, thus, enabling growth. This is the goal of policy-oriented neoclassical economists and politicians but is incompatible with the finite capacity of the Earth's resources, whether they are fuels, agricultural land, or waste assimilation. In the case of eco-efficiency, the benefit of a given amount of improvement can be and has been overwhelmed by economic growth exceeding the amount of efficiency gain—the so-called rebound effect. Whatever evidence we have on eco-efficiency gains made since 1991 is small with respect to the economic growth, and will become even less effective in reducing impact as the new BRIC (Brazil, Russia, India, and China) giants continue to boom. Climate change is, perhaps, the most important example of this intimate tie between global environmental impact and economic growth, even with marginal gains in eco-efficiency. Indices of both environmental and social unsustainability have continued to increase.

Let us say that 45 years have passed and the last surviving frog is still sitting in the pot. The metaphor has taken on more real meaning as the Earth is slowly warming up in real, not metaphorical, terms. This frog has been calling out to us to get moving on lowering the thermostat, but not too fast, arguing that we might be able to keep things cool by staying on the track called sustainable development. "I like it in this pot," the frog has been saying, "so do something to keep the powers that be from turning up the heat further." The business community might have responded with, "sustainable development is being done by the UN and world governments; our company will do its part by continuing to make us sustainable." Governments have continued to do very little. Successive global summits after the first in Rio have made some progress toward meaningful action plans. The Paris Agreement on Climate Change, signed in 2015 by most countries, was the first sustainability initiative to gain a global consensus, but its impact is ambiguous.

Greening and then sustainability struggled to solve the problems they were designed to counter. Producing even more green laundry soap was no answer to the problem. The producers of goods and services were deaf to the word "more." They were being eco-efficient, using up less of the Earth's capabilities while producing the same value for their customers, but always looking to grow. Consumers did the same, buying more green products, using totes made of recycled plastic to carry the groceries home, but also buying bottled water shipped the very long distance from Fiji, advertised as "Earth's finest water." Sustainability had the same history as greening. Each actor, both producers and consumers, gave themselves a pat on the back for being green or sustainable, all the time failing to notice that nothing was happening to slow down the rise of unsustainability. The frog's bathwater was now, perhaps, heating up more slowly, but the frog remained happy. The use of "sustainability" grew and grew until just about everyone was using the word and believed the new day was at hand. The frog stopped taking the temperature of the pot and instead started listening to those who said, look at what we are doing; don't worry about the hot water.

This is where we are today, sitting in a pot of ever hotter water, but oblivious of the fact that it is getting to the point where we may not be able to jump out. The misuse of "sustainability" lulls us into neglecting to take the "temperature" and observe how it is continuing to rise. Even as those who are doing "sustainability" are being recognized for their work—for example, membership in the Sustainability Hall of Fame or building larger market share—the temperature continues to rise and will continue to as long as "sustainability" is misunderstood and misused.

What might it take to change this state of affairs? The first step is to take a critical look at the situation and grasp the reasons that so little progress has been made. The full analysis of the underlying causes is far beyond the scope of this chapter, but it is reasonable to highlight several major concerns. They are:

1   The lack of a proper normative end point (vision) to attain and then sustain.
2   The continuing use of outmoded beliefs as the foundation of the key institutions of our modern, industrialized societies.
3   The dominance of reductionist knowledge and rationality, and failure to operate within a systems framework. Addressing the symptoms rather than causes of unsustainability.
4   Related to items 2 and 3, excessive reliance on technological solutions for all kinds of problems.

The word sustainability has little or no power in producing intentional actions in the absence of an explicit reference to what quality or material thing is to be sustained. Further, it leaves the choice up to the individual actors, making collective, aligned action problematic. Sustainability, per se, has a basic semantic problem associated with its use. It is an empty word, lacking any practical sense until whatever it refers to is explicit or tacitly understood by all the actors involved. Its use is especially cloudy when the adjective sustainable is used to modify some noun, like business or luxury. No matter what was intended, this usage will always point to the noun it modifies. The wide-spread use of the adjectival form, sustainable X, adds further confusion as it points to the noun being modified.

The tie between sustainability and its predecessor, sustainable development, is development, and particularly economic development, which points to growth as what is to be sustained.

This interpretation conforms to the primary normative goal of Western, neo-liberal polities that run their economies primarily to increase GDP. Well-being, as represented in public policy, business strategies, and most individual norms, is to be measured by some indicator of material wealth. This aspect of sustainability is self-defeating and is one reason for the disappointing results so far. The Earth is mostly a closed system, except for the energy that comes from the Sun. The material equivalent of GDP has grown to a point that aspects of that system are being stressed beyond some point that allows the system to remain within a stable, health-providing regime. Some earth scientists are arguing that we have entered a new geologic epoch, the Anthropocene, the first in the 4.7 billion-year history of the Earth where the great planetary systems are being perturbed by human action.[14] More specifically, a group of scientists has argued that we have already exceeded safe "operating boundaries" of the present long-lived Holocene era and are moving into a new regime with unwanted, destabilizing environmental impacts. This team presents data that suggest we have already exceeded such limits in three of nine categories: biodiversity loss, climate change, and changes in the global nitrogen cycle.[15] These effects are unintentional consequences of the materiality of the global economy.

Growth as the objective of sustainability, even if tempered by eco-efficiency, is a self-defeating norm. It has a negative social as well as environmental impact; inequality has been increasing with rising GDP in industrialized countries, such as the United States. Thomas Piketty has posited that inequality is positively coupled to output: more growth creates more disparity of wealth between those at the bottom and the top levels.[16] If the ideals behind sustainability are to be realized, a new objective other than growth must be used as the norma-tive target. I have argued for flourishing as this substitute.[17] Flourishing is the realization of the potential of human and non-human living organisms. It is a qualitative result of social and environmental systems that are operating such that flourishing is able to emerge. Flourishing is closer to representing well-being than proxy metrics such as wealth because it is a direct indicator of the existential health of living creatures. Human flourishing has an additional dimension related to the manner in which individuals express societal norms.

Flourishing is beginning to enter what have been conversations about sustainability with more frequency.[18] It is closer to what it means to be human than any term based on some refer-ence to a fixed human essence. Well-being and flourishing are linguistic relatives, but not when well-being takes on a metrical sense, as it does in economics. Flourishing is inherently verbal and active. It reflects the outcome of some kind of existential process, that is, having to do with how life is being played out. If humans were nothing but animals whose nature is determined largely, if not only, by their genes, flourishing would be a metaphor for life itself, living out the potential provided by one's genes.

Flourishing, or other observable qualities of life, can enable the idea of sustainability to be used as a policy and strategy guide, where growth cannot. At this moment in time, however, sustainability appears to be a poor choice because much of the globe's life forms and natural environment are not flourishing. The related word, attainability, would be more appropriate. Before we can sustain anything, we have to create it. Attainability, semantically, is more power-ful in forcing a critical examination on the situation than sustainability, which suggests that the system is already performing as desired.

The second barrier to those who hold sustainability as a worthy normative goal is the fail-ure to identify and address the root causes of the very problems these actors aim to mitigate or eliminate. Ironically, the norm of growth is one of the proximate causes of unsustainability, and can be traced back to the model of human behavior on which modern economic institutions are constituted and have developed. The concept of the human being as an autonomous, isolated, insatiable, self-interest-serving individual, using her rational powers to obtain the maximal quantity of material goods and services, leads inevitably to the conclusion that a future of more or different is always better than whatever is present-at-hand. *Homo economicus*, a name to describe this form of the human species, has become embedded in all Western political economies as a given, but has is origins in an unscientific presumption put forth a long time ago by Adam Smith and other Enlightenment thinkers.

This belief, while effective in building social institutions that clearly created highly productive economies and gains in the conditions of human life, appears to have run its course. In any case, another model or belief about the nature of human beings stands ready to be used to redesign and operate societal systems that are aligned with flourishing as the end point. That model might be called *Homo caritas*, representing the caring nature of human beings. Care is the fundamental way of interacting with the human and non-human world tied to concerns that arise out of a consciousness of connection to whatever worldly object or situation rests in the focal plane of an actor. For much of human evolution, care was the primary motivation for intentional action. As social beings, early humans slowly built cultural institutions to concretize their experience of caring.

Care can also be related to the phenomenon of human consciousness and to the meaningful actions that humans perform. The eminent neuroscientist Antonio Damasio's way of defining consciousness ties care to the biology of human life:

> Consciousness is, in effect, the key to a life examined, for better or worse, our beginner's permit into knowing about the hunger, the thirst, the sex, the tears, the laughter, the kicks, the punches, the flow of images we call thought, the feelings, the words, the stories, the beliefs, the music and the poetry, the happiness and the ecstasy. At its simplest and most basic level, consciousness lets us recognize an irresistible urge to stay alive and develop a concern for the self. At its most complex and elaborate level, consciousness helps us develop a concern for other selves and improve the art of life.[19]

The first sentence speaks about a "life examined." Of all living creatures, human beings are the only species that may ask the question, "What does it mean to be?" The self-awareness necessary to ask this question requires consciousness. This use of "concern" is important, as it is a prerequisite to care. He adds that concern of self and others is critical to "the art of life," another way to point to the necessity to care for self and others to survive well. The use of "art" suggests that humans are capable of adding meaning to their perceptions and are able to learn from their experience of living. This way of describing human existence does not rest on the presumption of some set of inherent interests, other than those necessary to maintain life itself, food, for example.

The idea of the world as a machine whose secrets could be revealed by scientific methods is another belief behind the rise of unsustainability. Since the time of Descartes, we have come to believe that humans could both come to know all the parts of and consequently control the cosmic world. We, now, certainly know much more about the world, but are slow to recognize the limits of this form of knowledge. As the Earth has become more populous and technically developed, this model increasingly fails to produce the results our intentional individual and collective actions are supposed to. A better model or belief about the cosmos and parts of it, like the Earth with its myriads of life forms, is that of complexity, a word used to describe systems where the interconnections are as important as the parts, and lead to behaviors that cannot be described by summing up the knowledge of all the parts.

Using the mechanistic, scientistic model to design and operate complex physical or social systems leads to the possibility of unintended consequences because that kind of model cannot fully describe the system. Inequality arises out of the failure of macro-economic models and the institutional structures and rules based on them to fully account for the way the system plays out in the real world. Climate change arises because we fail to connect how we produce and use energy with the global meteorological system. Almost all the undesirable aspects of unsustainability can be attributed to failures to account for the complexity of the real world, as contrasted with the partial, abstracted worlds of designers and governors.

To avoid or minimize unintended consequences, those who design and implement the systems and rules to attain whatever endpoints are to be used as the normative objectives of societies should augment these hegemonic, abstract, analytically derived models with forms of thinking and analysis based on complexity and interconnectedness. Complexity-based planning, design, and operations are consistent with qualitative norms like flourishing and other similar normative qualities, like trust or solidarity. Such qualitative norms are emergent properties, a property peculiar to complex systems. They remain as possibilities until the state of the system comes to a state where they start to show up as if by magic.

Space does not permit for a fuller exploration of the root causes of unsustainability. This choice is buttressed by well-accepted sociological models that hold beliefs as the basic building blocks of institutions ranging from individuals to families to entire societies.[20] Beliefs are the filters used to make sense out of the intrinsically meaningless phenomenal world. The ability to attribute meaning to our perceptions and, consequently, to act intentionally underpins our unique form of social life. A few other social creatures also create structured worlds, but do not know why.

Complexity requires planners, designers, and actors to think systemically, as opposed to the reductionist framework associated with the mechanistic model. Analytic frames focusing on the parts tend to focus attention on symptoms of the problems, and stop at the proximate cause level in seeking a solution. Rather than dig deeper, the overwhelming tendency in almost all situations is to seek a quick fix, usually of technological or technocratic origin. This common methodology is behind the persistence of social and environmental problems existing at the global level and also within most institutions. However one frames sustainability, systems thinking and related methods are critical. Technical solutions are likely to be invoked as a part of the solution, but other parts of the system below the proximate layer must be addressed. The deeper into the system the examining body goes, the more likely they will discover other causal elements, for which proposed remedies are more likely to have the desired effect.

The Dutch, some time ago, recognized the systemic nature of unsustainability and allocated pollution reduction goals to industrial sectors, which then passed along targets to individual entities, using models of "environmental utilization space," a concept that "reflects that at any given point in time, there are limits to the amount of environmental pressure that the Earth's ecosystems can handle without irreversible damage to these systems or to the life support processes that they enable. ... The 'society' for which the biosphere provides services is of course global."[21] Environmental space means the "space" available to humanity as a whole as a source of materials and a sink for their wastes. Their definition applied to sources that are globally tradable, and sinks that are global in extent, but they also point out that the recognition of global limits forces us to face the issue of how environmental space is to be allocated between nations and regions. While no such general agreement exists, this basic concept has underpinned the Global Agreement on Climate Change, The Montreal Protocol on Ozone Depletion, and the general strategy of emissions trading.

The failure to frame the issues around sustainability or, as I would propose, flourishing, in a systems context is a critical part of the failure to alleviate or eliminate the set of problems and issues that created the call for action in the first place. Action-intending words, such as sustainability, tend to fail to produce the desired results unless the context and goals are explicit, as legal and linguistic contextualist scholars tend to note. Today, when companies, particularly, or governments speak of sustainability, they look only at their own narrow contexts, not the wider, systemic whole. They presume that what they do will have positive impacts on the whole system, but they have the wrong system and wrong properties, just as the original Brundtland Report did. It had the world system right, but sought to sustain economic growth, both then and now, as the system property of primary and singular concern. Those invoking sustainability, also then and now, tend to make a categorical error. They mistake an internal process, growth, as the important system property, instead of emergent systemic properties like flourishing or justice, or other possibilities.

Almost all efforts carrying the name sustainability or sustainable in them look only at the system of negatives being produced as unintended consequences of our normal, dominant belief in growth as the purveyor of human happiness. Most sustainability efforts are some form of eco-efficiency or remediation, neither of which affects the entire system. They focus on a mere piece of the puzzle, for example, a company's explicit contributions to unsustainability. These efforts are self-referential. Such efforts are to be welcomed for making, hopefully, the harms less worse, but also tend to allow the actors to become even less mindful of the systemic nature of the problems. This is why all the sustainability reports in the world cannot bring forth justice or flourishing. They are useful for discriminating among individual efforts, but not about the effectiveness of these efforts as a whole. Reducing unsustainability cannot create flourishing.

One reason for the continuing approach to collapse is that sustainability, as it is being used and practiced, is not lowering the flames under our pot—the world we inhabit. While we are putting out some fires, maybe, the critical source of heat goes merrily along unquenched. The fire that is heating up our pot is not stoked by sustainability's prime target–unsustainability. It is fueled by our faulty cultural beliefs. We must begin, quickly, to point our fire extinguishers towards these causes, or we will discover that we can no longer jump out of the hot pot into a safe place. The fibrillating efforts to cope with unsustainability demand that we quickly eliminate the linguistic dither and find a suitable substitute to motivate collective action. It is critical to get the words right, not just the intent behind them. Language is the medium of human action and coordination. Words guide our actions. If someone says, "Dinner is ready" to me, I do not start the bath water. If someone says sustainability to me, I start thinking about flourishing and the need to address the cultural roots of unsustainability, but all those Chief Sustainability Officers (CSOs) and Sustainability Hall of Famers start thinking about eco-efficiency or CSR or something similar.

No matter how many folks get named to the Sustainability Hall of Fame, the temperature is going to keep rising. Words always precede coordinated action. The wrong words may produce actions that seem to do something good, but also can and do produce unintended consequences. It is those unintended consequences called unsustainability that are turning up the heat. First, let us get the words right. Then and only then can we begin to collectively act to cool down the planetary pot enough to live comfortably within it.

Notes

1   J. Evelyn, *Fumifugium, or, The inconveniencie of the aer and smoak of London dissipated together with some remedies humbly proposed by J. E. esq. to His Sacred Majestie, and to the Parliament now assembled* (London, 1661; reprinted 1976), 6.
2   Ibid., 14.
3   Rachel Carson, *Silent Spring* (Boston, MA: Houghton Mifflin, 1962).

4   L. Marx, "Technology, the Emergence of a Hazardous Concept," *Technology and Culture* 51, 3 (2010): 563.

5   G. H. Brundtland, *Our Common Future: Report of the World Commission on Environment and Development* (New York: United Nations General Assembly, 1987), 42.

6   See the essays in Iris Borowy and Matthias Schmelzer, eds., *History of the Future of Economic Growth: Historical Roots of Current Debates on Sustaianble Degrowth*, Routledge Studies in Ecological Economics (New York: Routledge, 2017).

7   See S. Schmidheiny, *Changing Course: A Global Business Perspective on Development and the Environment* (Cambridge, MA: MIT Press, 1992).

8   See S. Schaltegger and A. Sturm, *Okologie-Induzierte entscheidungsprobleme des Managements: Ansatzpunkte zur Ausgestaltung von In-strumenten*, WWZ Discussion Paper No. 8914 (Basel: WWZ, 1989).

9   See C. Holliday and J. Pepper, *Sustainability Through the Market: Seven Keys to Success* (Geneva: World Business Council for Sustainable Development, 2001).

10   See P. R. Ehrlich and J. P. Holdren, "Impact of Population Growth," *Science* 171 (1971): 1212–17.

11   See L. Reijnders, "The Factor 'X' Debate: Setting Targets for Eco-Efficiency," *Journal of Industrial Ecology* 2, 1 (1998): 13–22.

12   See, for instance, Thomas O. Wiedmann et al., "The Material Footprint of Nations," *PNAS* 112, 20 (2015): 6271–6.

13   M. Friedman, "The Social Responsibility of Business is to Increase its Profits," *The New York Times Magazine*, 32 (1970): 173–8.

14   See W. Steffen, P. J. Crutzen, and J. R. McNeill, "The Anthropocene: Are Humans Now Overwhelming the Great Forces of Nature?," *Ambio* 36, 8 (2007): 614–21.

15   See J. Rockstrom et.al., "A Safe Operating Space for Humanity," *Ecology and Society* 14, 2 (2009): article 32.

16   See T. Piketty, *Capital in the Twenty-First Century* (Cambridge, MA: Belknap Press of the Harvard University Press, 2014).

17   See J. R. Ehrenfeld, *Sustainability by Design: A Subversive Strategy for Transforming Our Consumer Culture* (New Haven, CT: Yale University Press, 2008); J. R. Ehrenfeld and A. Hoffman, *Flourishing: A Frank Conversation about Sustainability* (Palo Alto, CA: Stanford University Press, 2013).

18   See C. Laszlo et al., *Flourishing Enterprise: The New Spirit of Business* (Stanford, CA: Stanford University Press, 2014).

19   See A. R. Damasio, *The Feeling of What Happens: Body and Emotion in the Making of Consciousness* (Boston, MA: Mariner Books, 2000).

20   See A. Giddens, *The Constitution of Society* (Berkeley, CA: University of California Press, 1984).

21   See R. Weterings and J. B. Opschoor, *Towards Environmental Performance Indicators Based on the Notion of Environmental Space* (Rijswijk: Advisory Council for Research on Nature and Environment, the Netherlands, 1994).

References

Borowy, Iris and Matthias Schmelzer, eds. *History of the Future of Economic Growth: Historical Roots of Current Debates on Sustaianble Degrowth*. Routledge Studies in Ecological Economics. New York: Routledge, 2017.

Brundtland, G. H. *Our Common Future: Report of the World Commission on Environment and Development*. New York: United Nations General Assembly, 1987.

Carson, Rachel. *Silent Spring*. Boston, MA: Houghton Mifflin, 1962.

Damasio, A. R. *The Feeling of What Happens: Body and Emotion in the Making of Consciousness*. Boston, MA: Mariner Books, 2000.

Ehrenfeld, J. R. *Sustainability by Design: A Subversive Strategy for Transforming Our Consumer Culture*. New Haven, CT: Yale University Press, 2008.

Ehrenfeld, J. R. and A. Hoffman. *Flourishing: A Frank Conversation about Sustainability*. Palo Alto, CA: Stanford University Press, 2013.

Ehrlich, P. R. and J. P. Holdren. "Impact of Population Growth." *Science* 171 (1971): 1212–17.

Evelyn, J. *Fumifugium, or, the Inconveniencie of the Aer and Smoak of London Dissipated Together with Some Remedies Humbly Proposed by J.E. Esq. To His Sacred Majestie, and to the Parliament Now Assembled*." London, 1661. Reprinted Exeter: University of Exeter, 1976.

Friedman, M. "The Social Responsibility of Business is to Increase its Profits." *The New York Times Magazine*, 32 (1970): 173–8.

Giddens, A. *The Constitution of Society*. Berkeley, CA: University of California Press, 1984.

Holliday, C. and J. Pepper. *Sustainability Through the Market: Seven Keys to Success*. Geneva: World Business Council for Sustainable Development, 2001.

Laszlo, C., J. S. Brown, J. R. Ehrenfeld, M. Gorham, I. B. Pose, R. Saillant, D. Sherman and P. Werder. *Flourishing Enterprise: The New Spirit of Business*. Stanford, CA: Stanford University Press, 2014.

Marx, L. "Technology, the Emergence of a Hazardous Concept." *Technology and Culture* 51, 3 (2010): 561–77.

Piketty, T. *Capital in the Twenty-First Century*. Cambridge, MA: Belknap Press of the Harvard University Press, 2014.

Reijnders, L. "The Factor 'X' Debate: Setting Targets for Eco-Efficiency." *Journal of Industrial Ecology* 2, 1 (1998): 13–22.

Rockstrom, J., et.al. "A Safe Operating Space for Humanity." *Ecology and Society* 14, 2 (2009): article 32.

Schaltegger, S. and A. Sturm. *Okologie-Induzierte entscheidungsprobleme des Managements: Ansatzpunkte zur Ausgestaltung von In-strumenten* [*Ecology Induced Management Decision Support: Starting Points for Instrument Formation*]. WWZ Discussion Paper No. 8914. Basel: WWZ, 1989.

Schmidheiny, S. *Changing Course: A Global Business Perspective on Development and the Environment*. Cambridge, MA: MIT Press, 1992.

Steffen, W., P. J. Crutzen, and J. R. McNeill. "The Anthropocene: Are Humans Now Overwhelming the Great Forces of Nature?" *Ambio* 36, 8 (2007): 614–21.

Weterings, R. and J. B. Opschoor. *Towards Environmental Performance Indicators Based on the Notion of Environmental Space*. Rijswijk: Advisory Council for Research on Nature and Environment, the Netherlands, 1994.

Wiedmann, Thomas O., Heinz Schandl, Manfred Lenzen, Daniel Moran, Sangwon Suh, James West, and Keiichiro Kanemoto. "The Material Footprint of Nations." *PNAS* 112, 20 (2015): 6271–6.

23

BUSINESS, SUSTAINABILITY, AND THE BOTTOM OF THE PYRAMID

Ana Maria Peredo

The 1972 publication of *Limits to Growth* sent a shock wave through those sectors of the industrialized world where people were concerned about human society and its environment (Meadows et al. 1972). It attracted a variety of reactions, but prominent among them was a mushrooming interest in what came to be called "sustainability," an interest that was amplified by the report of the Brundtland Commission in 1987 (World Commission on Environment and Development 1987). That interest brought with it a sense that widespread and fundamental changes in societal arrangements were necessary if a durable place for human habitation was to be recovered on earth. Those sharing that conviction warned that the kinds of changes needed were likely to be resisted by many with vested interests, and those interests were especially strong in the profit-driven corporate and business sector that was doing quite well in the setting of existing arrangements (see, e.g., Brown 1981: 318–26). Sustainability and business interests were seen as sources of potential, and even likely, conflict.

It is hard not to think that the warnings were well founded. A 2002 report from the UN Environment Programme observed that "the majority of companies are still doing business as usual" (UN News Centre 2002). Twelve years later, Naomi Klein (2014) maintained that in fact big business not only continues to exploit the natural environment, it has also managed to co-opt environmental movements in its continuing resistance to reforms that would protect the environment from destructive externalities.

It would be a mistake, however, to overlook significant attempts that began as early as the 1980s and continued into the current century to incorporate principles of sustainability into the practice as well as the study of business. This chapter centers on one of those attempts: an approach focusing on what is known as the "Bottom" or "Base of the Pyramid" (BoP). I will outline the program as it has developed as a sustainability response from the world of business and business scholarship. Before undertaking an assessment of BoP from several perspectives, considering its applicability as a sustainability initiative, I will begin by locating the BoP program in the business response to the call for sustainability.

The business response to sustainability

First responses from the business community to the call for sustainability were clearly focused on the environment. A key figure in the early business recognition of a need for sustainability was American activist, environmentalist, and entrepreneur Paul Hawken. His background as civil rights activist, promoter of small business, ecological business entrepreneur and author led to a 1993 book aptly titled *The Ecology Of Commerce: How Business Can Save The Planet*. Ray Anderson, celebrated innovator in sustainable business modeling, credited that book with inspiring the radical shift that took place in his multi-national flooring business, Interface. In 1999, Hawken partnered with environmentalists Amory and Hunter Lovins to publish *Natural Capitalism: Creating the Next Industrial Revolution,* a book championing the recognition of "natural capital" and full cost accounting for services from the natural environment. The book's message that business could harmonize decreased environmental impact with profitability made a great impression on national leaders as well as practitioners and business educators.

A parallel movement developed in the 1990s under the title of "the triple bottom line" (TBL), a term coined in 1994 by John Elkington, founder of British consultancy SustainAbility. His contention was that the standard "bottom line" of economic gain ("profit") must be accompanied by equally careful calculation of social ("people") and environmental impacts ("planet") if a business is really determining its net impact. Elkington's publication of *Cannibals with Forks: The Triple Bottom Line of Twenty First Century Business* (1997) helped popularize the approach with businesses, consulting groups and management studies. The TBL concept echoed the "natural capital" idea in maintaining that paying attention to environmental and social effects could actually increase profitability. Both conceptions of sustainability as a business imperative had an impact on the broad notion of "corporate social responsibility" (CSR) movement that entered the vocabulary of business scholarship in the 1960s, but flourished in the 1990s and into the current century. At least some consideration of social and environmental impacts were increasingly considered essential by both participants in and observers of business practice, though it must be admitted that such considerations could sometimes be seen as much as "greenwashing" and marketing as genuine commitments to the underlying values.

These responses dominated business responses to the call for sustainability that multiplied in the late twentieth and early twenty-first centuries, along with programs with titles such as "natural step," "eco-efficiency" and "green entrepreneurship." Although this family of initiatives included consideration of social sustainability, it is fair to say that the main concentration was on environmental impact. The prevailing concept was that commerce could not continue to disregard the ecological costs of business as usual, but—and this was central—taking seriously impacts on the environment could actually create opportunity for additional profit. There was a concern that business should take into account harms to people and the social fabric as well as the environment (e.g. Hawken 1993: 1). But a market-based system, with profit incentives driving innovation, was firmly embraced and enthusiastically endorsed as the best mechanism for achieving environmental sustainability (Caradonna 2014: 168–72).

A ground-breaking business initiative was launched in 2002 that was equally rooted in market capitalist assumptions but shifted its focus to the social dimension of sustainability. A publication appeared in the influential *Harvard Business Review* titled "Serving the Poor, Profitably" (Prahalad and Hammond 2002). The publication seized on a term coined by Franklin Roosevelt in a 1932 radio address, where he referred to "the forgotten man at the bottom of the economic pyramid" (Roosevelt 1938: 624). Rather than viewing those at the bottom of the pyramid (BoP) as in need of aid, however, business scholar C. K. Prahalad and others, such as sustainability scholar Stuart Hart and entrepreneur Al Hammond, saw those in the BoP as potential beneficiaries from a process that would at the same time generate significant profit for corporations. The idea immediately struck a responsive chord. C. K. Prahalad filled out the argument in a 2004 book the title of which summarized its thesis: *The Fortune at the Bottom of the Pyramid: Eradicating Poverty through Profits.* This launched a sustainability response from business on which I focus in this chapter.

Business and the bottom of the pyramid

The BoP program shift from environmental degradation to entrenched poverty retained the fundamental commitment to a market solution: the poor, it contends, can be lifted from hardship by inclusion in capitalist markets.

The central proposition of Prahalad's BoP approach is that businesses can tap large potential markets, and simultaneously help alleviate poverty, by seeing things "through a new lens of inclusive capitalism" (Prahalad and Hart 2002: 1). An essential element in this proposal is the principle that poverty is deeply rooted in a lack of access to the market system that brings such benefits to the richer parts of the world. A further conviction is that the poor would welcome the opportunity to enlarge their market participation, and with it their welfare. Rather than being viewed as helpless victims in need of charity, the poor should be seen—Prahalad and his associates insisted—as eager potential participants in globalized markets, first as consumers currently frustrated by lack of access to goods they need and want, but also as entrepreneurs unsatisfied with their existing access to markets. This way of seeing things, it was argued, opens the door not only to profitability for companies doing business with the poor but improved welfare for the poor themselves.

Prahalad identified the BoP as those earning $2 per day or less (purchasing power parity rates for 1990), a population he reckoned to number more than four billion. It is the sheer size of this potential market that Prahalad appealed to as creating its great potential. Markets in the richer parts of the world are close to saturation with many goods and services, but the BoP are severely under-served by the corporations that ignore the vast numbers of prospective customers in the poor parts of the world.

Prahalad clearly regarded multi-national corporations (MNCs) as the principal prospects for engaging with the BoP market. (It must be noticed that many of his examples of innovation actually came from traditional development sectors such as cooperatives, NGOs, microcredit schemes, and local businesses.) For MNCs to penetrate as they should, he argued, would require a change in mindset on several topics; but with that change MNCs could expect to do very well in those markets, at the same time benefitting their customers. Those changes in outlook, he insisted, must be dictated by the circumstances of those in the BoP, whose location and habits as well as their meagre resources differ markedly from the richer clientele normally served by MNCs. Accordingly, technology must be blended with attention to locale and customs to offer products that are a match for what markets in those circumstances will provide a demand for. Smaller package sizes, products formulated to accord with differences in such things as access to water, and other adaptations to the circumstances of the BoP are essential. Research will be vital in discovering both what will be attractive in those circumstances and how to market it in the evolving circumstances of its potential customers. Alliances and decentralized supply chains will need to be developed out of recognition that many of the BoP are in comparatively remote areas, and markets are involved in dynamic social systems. MNCs will also have to adjust their thinking about profitability. Lower margins per unit will have to be accepted, and profit looked for in volume of business. Prahalad included a number of case studies he suggested illustrated his thesis.

Publication of the idea brought an enthusiastic reaction from the public press—*The Economist* listed *The Fortune at the Bottom of the Pyramid* among the best business and economics books of 2004. Public figures and government officials were equally warm in their reception. Bill Gates and US Secretary of State Madeleine Albright were among those who praised the book and endorsed its program.

The concept of the "bottom" or "base of the pyramid" quickly became a staple of business discussions concerning poverty. In business circles it became the standard way of referring to the population at the extreme of poverty, as well as representing a way of dealing with their plight. The paradigm also had an immediate effect on international as well as national policy. The highly influential UNDP Commission on the Private Sector and Development immediately included Prahalad in its membership, and his influence in the commission was evident. The introduction to the commission's report to the Secretary General of the United Nations included the statement, "[t]his report highlights many instances of large companies that have targeted bottom-of-the-pyramid markets and developed products and processes to serve the poor profitably or to operate sustainably in very challenging environments" (Commission on the Private Sector and Development 2004: 9). The report was built on an advocacy of private sector, market solutions for poverty, noting quite candidly the opportunities for profit taken to exist there for multi-national companies. "As today's advanced economies become a shrinking part of the world economy, the accompanying shifts in spending could provide significant opportunities for global companies" (ibid.: 30). At the same time, as BoP discourse insists, the commercial activity of these large corporations is seen as a device for bringing economic growth as well as added product choices for the poor and lower prices (ibid.: 8).

The BoP paradigm simultaneously came to dominate academic discussions of the relations between business and conditions of poverty; in fact, the word "poverty" in management circles was largely replaced by "BoP." This central impact was soon evident in the curriculum and courses offered by business schools. The University of Michigan created a unit called "The Base of the Pyramid Initiative," and Harvard quickly led many business schools in offering courses in "Business at the Base of the Pyramid." From its inception in 2009, "Oikos," an important international student organization which hosts an annual "Young Scholars Development Academy" for PhD students, stated in its call for applications: "Research on inclusive business models, market development and sustainability at the Base of the Pyramid (BoP) is a promising and challenging field for young researchers and PhD students."

Early BoP literature included some debate about the appropriate income figure to be used in identifying the population and the actual numbers and potential market to be found there (Karnani 2007c). This led some who saw promise in the approach to suggest that attempting to fix a precise figure as the definition was a mistake, especially given the multi-faceted nature of poverty. Instead, they proposed that the BoP population be recognized simply as that sector that lives primarily in the informal economy without access to a "Westernized" economic environment (London 2007).

As the idea spread and in response to critics, some advocates of the BoP paradigm suggested that formulating it as basically a business strategy left it vague on actual poverty alleviation implications, and called for a more developed idea as to how the approach would in fact aid the very poor. London (ibid.) stated that more attention needed to be paid to forming collaborative alliances with those in the BoP, "co-creating" with them as producers and not just consumers, and building on market structures already in place. This approach was echoed by Simanis and Hart (2008), who argued that "most 'first generation' corporate BoP strategies … failed to hit the mark" (ibid.: 1). Suggesting that those initial strategies were too easily seen as simply marketing to the poor on the assumption that this would alleviate poverty and promote sustainable development, they outlined an approach they labeled "BoP 2.0." "BoP 1.0" was tagged as "selling to the poor," while BoP 2.0 they labeled "business co-venturing." The emphasis in their revision was on the need for companies to "work in equal partnership with BoP communities to imagine, launch, and grow a sustainable business" (Simanis and Hart 2008: 3). "Co-development," they said, "catalyzes business imagination and ensures the business model is culturally appropriate and environmentally sustainable by building off of local resources and capabilities. Importantly, it also expands the base of local entrepreneurial capacity" (ibid.).

The BoP approach has continued to evolve in other respects. Rangan et al. (2011) argued that three different categories need to be distinguished in the BoP: "low income," "subsistence" and "extreme poverty." Each of these segments offers different challenges, and different opportunities. Different constellations of consumers, co-producers and clients will be found at different levels of resources. To be successful, programs must therefore take into account the significant differences in these populations.

The BoP paradigm has continued to attract attention from corporations. An online slide presentation (Fox 2013) vividly illustrates the step-by-step, BoP approach taken by corporate giant Unilever in marketing a soap product to the BoP in Brazil. International bodies such as the UN interested in alleviating poverty continue to use the BoP model, often in cooperation with MNCs, as a framework for bringing capitalist interests to bear on the problems of destitution.

Assessing the BoP paradigm

Challenging the basics

In among the enthusiastic responses that greeted the development by Prahalad and others of the BoP paradigm there were nevertheless criticisms to be found. One of the earliest and most trenchant critics was a colleague of Prahalad at the University of Michigan, A. Karnani. In a flurry of papers published between 2007 and 2009, Karnani contested many of the underpinnings of the BoP perspective (Karnani 2007a, 2007d, 2007b, 2007c, 2008, 2009).

Karnani began by questioning whether the BoP market, even assuming Prahalad's criterion of $2 PPP or less per day, numbered anything like the four or five billion Prahalad foresaw. Karnani pointed to figures provided by the World Bank and other researchers to suggest that the number may be as small as 600 million. Even working with the World Bank figure of 2.7 billion, his suggestion was that there is no fortune to be found there. Prahalad's submitted potential of $13 billion, he contended, should be amended to something like $0.3 trillion, given the meager purchasing power of the very poor and the fact that profits would be repatriated at market exchange rates and not the levels suggested by parity purchasing power. The geographic dispersal, cultural diversity and price sensitivity of BoP populations could be expected, Karnani argued, to offer challenges to distribution and marketing that would seriously undermine potential profitability. He went on to challenge in detail the case examples offered by Prahalad of successes and prospects at the BoP. Karnani also contested the idea that MNCs were the place to look for increasing economic opportunity in the BoP, suggesting that smaller, local businesses were better placed to take the necessary initiatives.

Karnani went on to attack the idea that the poor are, as portrayed by Prahalad, "resilient and creative entrepreneurs and value-conscious consumers." The poor, Karnani suggests, are like the rich in being subject to buying things that do not help them, even hurt them, diverting purchasing power from things that would be benefits. The evidence is that that imprudent expenditures on things like tobacco and alcohol reflect no greater savvy on the part of BoP than richer consumers. The problem is that for very poor people, the problems created by those purchasing preferences are more serious and disadvantageous. As for entrepreneurship, Karnani called for a distinction between entrepreneurship defined merely as forced self-employment, and entrepreneurship understood as creative, skilled and self-motivated embarking on new ventures. There is no evidence, he insisted, that very poor people are more inclined to the latter outlook than are populations elsewhere; but they are forced into the former by their impoverished circumstances.

This links directly to Karnani's contention that the second part of the BoP proposal—that seeking profits in the BoP will at the same time help the poor—is unrealistic. MNCs can be expected, he contends, to market to the poor what they will find meets or attracts a demand and promises a profit, regardless of the benefit to the poor. A favourite example was the fashion product "Fair & Lovely," a skin-whitening compound cited by Prahalad as an example of BoP entrepreneurship. The product is marketed quite profitably by Unilever, an active BoP participant, to women of India. Karnani notes the widespread objection to the racial and gender implications of "Fair & Lovely," and contests Prahalad's assertion that freedom to buy this product amounts to empowerment. Unilever should, of course, be entitled to sell the item, he agreed, and consumers free to buy it. But, he submitted, the idea that companies empower women and help eradicate poverty with products like this is "morally problematic" (Karnani 2007a: 22).

The solution, Karnani concluded, lies not in increasing consumer or entrepreneurial opportunity for the poor, but in creating opportunities for regular employment at reasonable wages. Governments, he insisted, and not just "free" markets, should be principal agents in bringing that about.

Prahalad replied with a response to several points in Karnani's critique, but suggested that the main point of contention was ideological. Prahalad maintained that the principle underlying the BoP approach was respecting freedom of choice: freedom to purchase and consume what one wishes, and freedom to engage in the trading one wishes. The debate attracted attention in business literature circles, but did not seem to detract from the public influence of the BoP paradigm. As *The Guardian* newspaper observed in 2014, "Ten years later, businesses big and small continue to pursue profits at the bottom of the pyramid" (Gunther 2014).

But does it work?

Perhaps the best way of weighing critiques like Karnani's is simply to ask whether the program delivers. Does the BoP enterprise produce the results its advocates have projected from marketing to the very poor? The short answer seems to be that it is difficult to know.

There are two sorts of outcome sought in BoP markets: the first is profits, and the second is benefits to the very poor in terms of increased access to products and services at reasonable prices but also the opportunity to capitalize on their entrepreneurial talents.

Begin with profits. Proponents of the BoP approach have certainly pointed to examples of what they see as success. As noted, Prahalad launched the initiative with instances where he saw profits being made as well benefits bestowed. Early reviewers of Prahalad's proposal noted, however, that his most striking examples were not of private enterprise targeting a market but not-for-profits operating as social enterprises or MNCs partnering with governments and international agencies such as the World Health Organization (Walsh et al. 2005). There may have been innovative approaches to delivery and partnership and resourceful approaches to generating income, but standard calculations of "profit" did not seem to apply. The private company most celebrated by Prahalad and others in the BoP movement was Hindustan Unilever Limited (HUL), the Indian subsidiary of Unilever, which Prahalad and Hart (2002: 5) referred to as "a pioneer among MNCs exploring markets at the bottom of the pyramid."

Prahalad and associates cited two products in particular: candy and salt. Candy, however, was abandoned by HUL in 2005 because of unprofitability, and salt is sold mainly to somewhat better off customers where it is still at a severe cost disadvantage to locally produced product (Jaiswal 2008). Unilever remains a prime reference for BoP supporters. It is said that Unilever generates more than half of its sales from "developing" markets (Gunther 2014); but it is by no means clear how profitable this is or whether those sales are to the poorest, who occupy the BoP, or those who are somewhat better off.

It is generally recognized that in practice, trade at the BoP has been less profitable than hoped. Procter & Gamble could not generate a competitive return on its Pur water-purification powder after launching the product on a large scale in 2001, so gave up on the project in 2005. DuPont subsidiary Solae marketed a soy-based food supplement to the poor that would enrich their protein intake, but abandoned the venture when it generated inadequate sales (Simanis 2012).

Promoters of the BoP perspective have worked to identify the challenges that led to these disappointing results and to develop practical responses to them; but it seems clear the profit objective has been demanding. The most focused consideration of evidence as to profitability in the BoP was undertaken by Kolk et al. (2014). Their judgement was that a majority of the relevant research reported a somewhat positive impact for firms investing in BoP initiatives, but a wide disparity in how outcomes were evaluated (e.g. does learning from the experience count as a success?) made it hard to draw any conclusions. Their conclusion was the judgements as to profitability need a much clearer set of criteria, and much clearer reporting, than anyone has provided. The question about profitability at the BoP appears moot.

What about benefits to the BoP population? Again, it appears that we do not really know. In their 2014 review of the academic literature on BoP, Kolk et al. noted that the "scarcity of objective assessments" of social impact seemed remarkable given claims about the potential of the concept for poverty reduction. Several articles did claim improvements in areas such as education, health care, and water quality, and some suggested improvements in such impressionistic variables as "quality of life," "empowerment" and "reduced exploitation." But the overall impression is that there has not been much attempt to identify precise indicators of the kinds of improvement that might be hoped for, and even less attempt to use them to gauge outcomes. The BoP program's outcomes for poverty relief are, according to reviews such as these, unclear.

A revealing perspective on this question appears in a pair of publications by BoP advocate Simanis (2012, 2013). Simanis admitted that the BoP concept has proven far less profitable than hoped, and suggested that in fact the low-margin, high-volume approach at its centre simply does not work. Costs are much higher, and the purchasing customs of the target population create much higher barriers than traditional markets. The solution, Simanis proposed, is to be forthright in pursuing higher profits, and to be clear that business at the BoP is business. BoP enterprises, he asserted, are not social enterprises. Shareholder demand must be satisfied. The idea that corporations can use a "blended value" approach in working at the BoP is, Simanis says, "a myth" (Simanis 2013: 224). Simanis is quite clear that profits can be made at the BoP and offers concrete suggestions as to how this is possible. But he is equally clear that poverty relief is a hoped-for spin-off. It is not another line in the ledger that is balanced with profit (and environmental?) concerns. Profit must prevail.

This line of reasoning certainly suggests that the link between the BoP program and sustainability program is questionable at best.

BoP and "co-development"

As noted earlier, there were advocates for the BoP approach that in response to critics recognized its potential for actually assisting the poor was obscured when it was advanced mainly as a program for increasing their role as consumers. London, Simanis and Hart were foremost among many who acknowledged that the BoP must benefit not just by selling more things to them, but by engaging them in the productive activities. They held that "co-venturing" and "co-development" were vital parts of an approach that could be expected to bring benefits to the BoP.

However, Kolk et al. identified very few examples of BoP engagement where the poor are actually involved as co-operators in innovation, production or marketing. "Most BOP models reported in the literature," they observe, "view the poor as consumers of existing or adapted products" (Kolk et al. 2014: 356). A review by Goyal et al. (2014) comes to a similar conclusion. To the extent that BoP initiatives are supposed to deliver benefits to the poor by engaging them not just in consumption but production, the record is not encouraging.

What about the environment?

Practitioners and educators situate the BoP paradigm in the field of business sustainability. Of course that field is not, as sometimes supposed, concerned exclusively with environmental considerations; but those are clearly central to the program. How does the BoP concept align with that concern?

The results have not been impressive. There have certainly been those among BoP advocates who have included an environmental focus. In their 2008 report aimed at moving the BoP concept beyond a "selling to the poor" strategy—published, incidentally, by Cornell University's Center for Sustainable Global Enterprise—Simanis and Hart included as part of their "co-creation" approach a concern for "environmental management" and "environmentally sustainable" business models (Simanis and Hart 2008: 3). Kolk et al. noted in their survey, however, that the environmental aspect of sustainability receives very little attention in the literature concerning the BoP paradigm, a situation they found surprising given the concern that entry into BoP markets might simply reproduce and extend the exploitative behavior of commerce operating in the industrialized world. Brix-Asala et al. (2016) echo that concern, and offer a case study of BoP enterprise in Africa that they suggest illustrates the requirement for unacknowledged trade-offs among the three pillars of sustainability when projects are implemented. Environmental considerations, it seems, may often be in real tension with the alleviation of poverty and with profitability.

BoP discourse and market capitalist hegemony

A more radical assessment of the BoP program has emerged in the last decade. Recent scholars (e.g. Bonsu and Polsa 2011) have suggested that the BoP paradigm in fact embodies an ideology and a world-view that needs to be examined critically instead of simply being applied without question. From this viewpoint, the BoP perspective is just an outcropping of the hegemonic status of market capitalism. As an hegemonic outlook, market capitalism—especially in the prevailing era of neoliberalism—establishes itself as the normal, natural way of seeing things (Gramsci 1971; Barrett 1991). The BoP framework simply adopts the "taken-for-grantedness" of that viewpoint.

A discourse analysis of the literature advocating the BoP program seems to reveal the ways in which those in the BoP are in fact constructed as eager consumers and entrepreneurs by consistently locating them in the framework of market transactions as liberating (Montgomery et al. 2012). A 2007 publication of the World Resources Institute and the International Finance Corporation, for instance, introduces its case for a market-based approach to poverty as follows: "The starting point for this argument is not the BOP's poverty. Instead, it is the fact that BOP population segments for the most part are not integrated into the global market economy and do not benefit from it" (Hammond et al. 2007: 4). Acosta et al. articulate clearly the assumption behind the BoP perspective: "[S]uccessful participation as consumers or producers in markets (or 'market inclusiveness') is *sine qua non* in permanently exiting poverty" (Acosta et al. 2011: 50). There is no argument for this assertion. It is assumed to be obvious.

The BoP attitude to other forms of aid can be seen as demonstrating its commitment to a way of seeing things that should take the place of other, unsatisfactory perspectives:

- other responses to poverty are articulated as charity- or state-based;
- charity- and state-based solutions to poverty are said to represent poor people as victims and dependents;
- these solutions to poverty are historical failures; and finally
- the BoP perspective is a different and preferable approach to poverty alleviation both because it is based on something that works (i.e. market-based transaction) and because it does not dismiss the poor as victims but sees them as open to empowerment by means of that transaction.

In Prahalad's words: "If we stop thinking of the poor as victims or as a burden and start recognizing them as resilient and creative entrepreneurs and value-conscious consumers, a whole new world of opportunity will open up" (Prahalad 2004: 1). In this way, the argument goes, poor people are made intelligible as further examples of *Homo economicus*: self-interested individuals aspiring for more consumer choice and capitalist markets.

Shiva (2010) is eloquent on the way in which the industrialized, market-centered outlook actually creates poverty, in two senses. First, it perceives poverty in circumstances where the subjects are well sustained (as in Sahlins's "original affluent society"; see Sahlins 1972) but not engaged in the cycle of resource exploitation and over-consumption on which industrial societies depend. Second, in its attempts to remedy, by inclusion in the market economy, what it sees as poverty, the market-based program of industrialization frequently destroys the means of sustenance upon which populations have depended.

The problem with accepting the market capitalist hegemony as the BoP paradigm does, the argument continues, is its blinkering effect: it obscures important factors that are relevant to effectively addressing poverty (Montgomery et al. 2012).

One thing arguably hidden by the BoP perspective is the many ways in which poverty may actually be experienced. Stipulating a daily income as the marker for poverty, as BoP does, discounts the diversity convincingly outlined by Amartya Sen in his argument against using income as a standard for identifying the poor (Sen 1999: 87ff). It also ignores the nuanced understanding of poverty from the inside represented, for instance, in the eloquent *Voices of the Poor* trilogy (Narayan and Petesch 2002; Narayan et al. 2000; Narayan-Parker 2000).

Closely related to this is the way that the BoP framework disregards the manifold roots of poverty. The forces that create and sustain genuine poverty clearly include historical, social and especially political factors (see, e.g., Ferguson and Lohmann 1994). Sen's account of famine, perhaps the starkest expression of poverty, underlines the way in which political forces combine with other elements in a way that must be taken into account in attempts to relieve and prevent famine (Sen 1999). It appears that approaching poverty as basically a problem of access to markets simply ignores factors vital to understanding and responding to poverty.

Equally significant is the way that the BoP paradigm hides from view the diverse economies available to those living in hardship, including some employed by the poor to produce and improve their livelihood. Gibson-Graham, for instance, have engaged people who might well fit the simplistic income criterion for inclusion in the BoP as active participants in "discursive destabilization," where "poverty" and the hegemonic image of poor victim (or its supposed solution, the free entrepreneur) could be rethought in collective terms (Gibson-Graham 2006: 127–63). Co-operatives, as well as what have come to be known as "community-based enterprises" (Peredo and Chrisman 2006), and barter markets (Argumedo and Pimbert 2010) are among many examples of collective economic activity that are engaged with markets, if at all, in ways quite different from the standard market-capitalist assumptions implicit in the BoP program.

Conclusion

I have tried to outline, in some detail, the evolved BoP paradigm as a business response to the call for social sustainability, and outlined some criticisms that have emerged in the literature. At least two conclusions can be drawn from this survey.

The first is that the BoP, as a candidate for inclusion among the tools for improved sustainability, faces formidable challenges. Its promise of profitability stands in need of support it has not received, and its assurance of assistance to the poor is likewise open to question. The uncritical acceptance of a market-capitalist paradigm as the preferred instrument of alleviating poverty seems definitely to invite thoughtful consideration of the factors that assumption ignores. Further, the BoP paradigm does not seem to offer significant ecological benefits, and may even contribute to environmental degradation.

A second conclusion extends the concerns just mentioned to a general worry about business-based approaches to sustainability. As noted earlier, the BoP paradigm shares with other sustainability proposals from the business sector the assumption that standard, market-based activity can be harnessed to address sustainability at the same time it supplies the profits that businesses seek.

The idea that environmental sustainability can be achieved by market means has been subjected to searching criticism (Prudham 2009; Harris 2013; Smith 2011). We have seen that the BoP commitment to the market mechanism has drawn its share of disapproval. It seems fair to conclude from this that the business approach to sustainability, insofar as it rests on an assumption of pairing sustainability outcomes with standard market profitability, faces some formidable challenges. The warning cited at the beginning of this chapter about the resistance to sustainability initiatives that might be expected from business interests seems well placed. It may well be that if sustainability is our goal, approaches to economic livelihood that do not rely on standard, individualistic and profit-based assumptions will have to receive serious consideration as viable "business" models.

References

Acosta, P., N. Kim, I. Melzer, R. U. Mendoza, and N. Thelen. 2011. "Business and Human Development in the Base of the Pyramid: Exploring Challenges and Opportunities with Market Heat Maps." *Journal of World Business* 46 (1): 50–60. doi: 0.1016/j.jwb.2010.05.017.

Argumedo, Alejandro, and Michel Pimbert. 2010. "Bypassing Globalization: Barter Markets as a New Indigenous Economy in Peru." *Development* 53 (3): 343–49.

Barrett, M. 1991. *The Politics of Truth: From Marx to Foucault*. Stanford, CA: Stanford University Press.

Bonsu, S. K., and P. Polsa. 2011. "Governmentality at the Base-of-the-Pyramid." *Journal of Macromarketing* 31 (3): 236–44. doi: 10.1177/0276146711407506.

Brix-Asala, Carolin, Rüdiger Hahn, and Stefan Seuring. 2016. "Reverse Logistics and Informal Valorisation at the Base of the Pyramid: A Case Study on Sustainability Synergies and Trade-Offs." *European Management Journal* 34 (4): 414–23.

Brown, Lester R. 1981. *Building a Sustainable Society*. New York: W. W. Norton & Company.

Caradonna, Jeremy L. 2014. *Sustainability: A History*. Oxford: Oxford University Press.

Commission on the Private Sector and Development. 2004. *Unleashing Entrepreneurship: Making Business Work for the Poor*. New York: United Nations.

Elkington, John. 1997. *Cannibals with Forks: The Triple Bottom Line of Twenty First Century Business*. Mankato, MN: Capstone.

Ferguson, James, and Larry Lohmann. 1994. "The Anti-Politics Machine: 'Development' and Bureaucratic Power in Lesotho." *The Ecologist* 24 (5): 176–81.

Fox, David. 2013. "Unilever Presentation." Retrieved October 24, 2016 from www.slideshare.net/DavidFox7/unilever-presentation.

Gibson-Graham, Julie Katherine. 2006. *A Postcapitalist Politics*. Minneapolis, MN: University of Minnesota Press.

Goyal, Sandeep, Mark Esposito, Amit Kapoor, M. P. Jaiswal, and Bruno S. Sergi. 2014. "Understanding Base of the Pyramid Literature—a Thematic, Methodological and Paradigmatic Review." *International Journal of Business and Globalisation*.

Gramsci, A. 1971. *Selections from the Prison Notebooks*. London: Lawrence & Wishart.

Gunther, Marc. 2014. "The Base of the Pyramid: Will Selling to the Poor Pay Off?" *The Guardian*, May 22. Retrieved October 26, 2016 from www.theguardian.com/sustainable-business/prahalad-base-bottom-pyramid-profit-poor.

Hammond, A. L., W. J. Kramer, R. S. Katz, J. T. Tran, and C. Walker. 2007. *The Next 4 Billion: Market Size and Business Strategy at the Base of the Pyramid*. Washington, DC: World Resources Institute, International Finance Corporation.

Harris, Jerry. 2013. "Can Green Capitalism Build a Sustainable Society?" *International Critical Thought* 3 (4): 468–79.

Hawken, Paul. 1993. *The Ecology of Commerce: How Business Can Save the Planet*. London: Weidenfeld & Nicolson.

Hawken, Paul, Amory B. Lovins, and L. Hunter Lovins. 1999. *Natural Capitalism: Creating the Next Industrial Revolution*. 1st ed. Boston, MA: Little, Brown & Co.

Jaiswal, Anand Kumar. 2008. "The Fortune at the Bottom or the Middle of the Pyramid?" *Innovations* 3 (1): 85–100.

Karnani, Aneel. 2007a. *Fortune at the Bottom of the Pyramid: A Mirage*. Working paper. Ann Arbor, MI: Ross School of Business, University of Michigan.

Karnani, Aneel. 2007b. "The Mirage of Marketing to the Bottom of the Pyramid: How the Private Sector Can Help Alleviate Poverty." *California Management Review* 49 (4): 90–111.

Karnani, Aneel. 2007c. "Misfortune at the Bottom of the Pyramid." *Greener Management International* 51: 99–111.

Karnani, Aneel. 2007d. *Romanticizing the Poor Harms the Poor*. Working paper. Ann Arbor, MI: Ross School of Business, University of Michigan.

Karnani, Aneel. 2008. "Help, Don't Romanticize, the Poor." *Business Strategy Review* 19 (2): 48–53.

Karnani, Aneel. 2009. *The Bottom of the Pyramid Strategy for Reducing Poverty: A Failed Promise*. Working Paper # 80. DESA.

Klein, Naomi. 2014. *This Changes Everything: Capitalism vs. the Climate*. New York: Simon & Schuster.

Kolk, Ans, Miguel Rivera-Santos, and Carlos Rufin. 2014. "Reviewing a Decade of Research on the "Base/Bottom of the Pyramid" (Bop) Concept." *Business and Society* 53 (13): 338–77. doi: 10.1177/0007650312474928.

London, T. 2007. *A Base-of-the-Pyramid Perspective on Poverty Alleviation*. Working Paper. Ann Arbor, MI: William Davidson Institute, University of Michigan.

Meadows, D. H., D. L. Meadows, J. Randers, and W. W. Behrens III. 1972. *The Limits to Growth*. New York: Universe.

Montgomery, N., A.M. Peredo, and E. Carlson. 2012. "The Bop Discourse as Capitalist Hegemony." *Academy of Management Annual Meeting Proceedings.*

Narayan, Deepa, and Patti Petesch. 2002. *Voices of the Poor: From Many Lands*: Washington, DC: World Bank and Oxford University Press.

Narayan, Deepa, Meera K. Shah, Patti Petesch, and Robert Chambers. 2000. *Voices of the Poor: Crying Out for Change*: New York: Oxford University Press for the World Bank.

Narayan-Parker, Deepa. 2000. *Voices of the Poor: Can Anyone Hear Us?* New York: Oxford University Press for the World Bank.

Peredo, Ana Maria, and James J. Chrisman. 2006. "Toward a Theory of Community-Based Enterprise." *Academy of Management Review* 31 (2): 309–28.

Prahalad, C. K. 2004. *The Fortune at the Bottom of the Pyramid: Eradicating Poverty through Profits.* Philadelphia, PA: Wharton School Publishing.

Prahalad, C. K., and A. L. Hammond. 2002. "Serving the Poor, Profitably." *Harvard Business Review* 80 (9): 48–59.

Prahalad, C. K., and S. L. Hart. 2002. "The Fortune at the Bottom of the Pyramid." *Strategy+Business* 26 (54–67): 1–14.

Prudham, Scott. 2009. "Pimping Climate Change: Richard Branson, Global Warming, and the Performance of Green Capitalism." *Environment and Planning* 41 (7): 1594–1613.

Rangan, V. Kasturi, Michael Chu, and Djorjiji Petkoski. 2011. "Segmenting the Base of the Pyramid." *Harvard Business Review* 89 (6).

Roosevelt, Franklin D. 1938. *The Public Papers and Addresses of Franklin D. Roosevelt, Vol. 1, 1928–32.* New York: Random House.

Sahlins, Marshall David. 1972. *Stone Age Economics.* Chicago, IL: Aldine-Atherton.

Sen, Amartya. 1999. *Development as Freedom.* New York: Alfred A. Knopf.

Shiva, Vandana. 2010. "Resources." In *The Development Dictionary: A Guide to Knowledge and Power*, edited by Wolfgang Sachs, 228–42. London: Zed Books.

Simanis, Erik. 2012. "Reality Check at the Bottom of the Pyramid." *Harvard Business Review* (July).

Simanis, Erik. 2013. "Bringing Bottom of the Pyramid into Business Focus." In *A Planet for Life: Reducing Inequalities: A Sustainable Development Challenge*, edited by Rémi Genevey, Rajendra K. Pachauri and Laurence Tubiana, 217–28. New Delhi: The Energy and Resources Institute (TERI).

Simanis, E., and S. Hart. 2008. *The Base of the Pyramid Protocol: Toward Next Generation BoP Strategy.* Ithaca NY: Center for Sustainable Global Enterprise.

Smith, Richard. 2011. "Green Capitalism: The God That Failed." *Real-World Economics Review* 56: 112–44.

UN News Centre. 2002. "Efforts to Reduce Industrial Effect on Environment 'Uneven', UN Agency Reports." Retrieved on October 15, 2016 from www.un.org/apps/news/story.asp?NewsID=3677&Cr=sustainable&Cr1#.WTrXPRQ9GQI.

Walsh, James P., Jeremy C. Kress, and Kurt W. Beyerchen. 2005. "Book Review Essay: Promises and Perils at the Bottom of the Pyramid." *Administrative Science Quarterly* 50 (3): 473–82.

World Commission on Environment and Development. *Our Common Future* ("the Brundtland Report"). Oxford: Oxford University Press, 1987.

24

AT THE CROSSROADS

Sustainability and the twilight of the modern world

John B. Robinson and David Maggs

Introduction

As the "Anthropocene" solidifies into a common diagnosis of our age, the underlying challenges of environmental sustainability arrive at a crossroads: Can they be addressed from within the dominant paradigm of modernist rationality or do they demand a departure from the paradigm altogether? In other words, does the Anthropocene indicate a difference of degree (i.e. more and better quantification of problems at a global scale, and informing/convincing of publics), or a difference of kind (reimagining ourselves, our worlds, and the nature of the problem itself)?

While not new, the dilemma is distinct from many of the pressing sustainability debates that marked the turn of the last century. Take, for example, the tension between the language of "sustainability" and that of "sustainable development," which is so often conflated.[1] In this case, both options leave the larger assumptions of modernist rationality relatively unchallenged: positivist standards of truth; dualist instincts separating fact from value, object from subject, nature from culture; or linear, cause-effect, command-and-control approaches rooted in a belief in the describability of systems and the predictability of interventions.

Over the past few decades, however, a growing number of fields have become restless within such a paradigm. Health, mental health, obesity, addictions, poverty reduction, and disease control are just some areas whose "wicked problems"—problems where complex, emergent interactions make predictable interventions elusive—suggest they may not be problems *for* modernist rationality, but problems *of* modernist rationality.[2]

When it comes to sustainability this restlessness appears to manifest in a curious ambivalence. On one hand, approaches to research and action on sustainability challenges are expanding to include increasing public engagement on a diversity of issues, expanding relevant communities of knowledge and practice (traditional ecological knowledge, citizen science, sporting, cultural, and religious organizations, etc.) and incorporating perspectives beyond natural sciences and quantitative social sciences that include the arts, humanities, and qualitative social science.

And yet this expansion has encountered a surprising tenacity of the modernist paradigm and the scientistic hegemony that supports it. Even as interest in diverse approaches grows, a genuine shift in how we understand sustainability challenges, and what counts as useful knowledge, remains unrealized. Marginal approaches appear to be introduced and re-marginalized, or forced to impoverish their own offerings to fit modernist idioms. Why the ambivalence? What is it about the relationship between sustainability and its encompassing worldview that presents such simultaneous restlessness and dependence?

Exploring this question might benefit from something of an overview of the development of the modernist worldview itself, viz. the relationship between humans and their environment. Worldviews are deeply relevant to how we conceive of ourselves, how we think and act, what we consider true, valuable, or possible for our worlds. And yet these significant forces behind our realities often remain invisible. In anticipation of exploring the evolution of the modernist worldview and its relevance to sustainability, the following points are worth considering:

1   Powerful, but largely concealed and unexamined assumptions about human nature and "external" nature guide the formation of worldviews, ideologies, and beliefs.
2   The assumptions have changed significantly over time, usually through slow and subtle developments but sometimes dramatically and radically.
3   The development of modern science and technology, economics and industrialism, even education and institutions, has been guided by a more-or-less consistent set of assumptions about human and external nature which are qualitatively different from pre-modern conceptions of the natural world.
4   A variety of philosophical, social, spiritual, and political criticisms of the assumptions underlying modern industrial society emerged with the rise of modernist thought in the seventeenth and eighteenth centuries, and continue to be expressed today.
5   In recent years, building on those critiques, many thinkers have argued that current sustainability problems pose a significant challenge to the attitudes, beliefs, values, and practices characteristic of modern industrial society.
6   It is useful to examine these assumptions and the critiques made of them, in order (i) to determine to what extent current sustainability problems are rooted in the basic assumptions of industrial society, and (ii) to develop appropriate analyses, critiques, and proposed solutions that do not simply treat the symptoms and thus perpetuate the diseases of modern society.

With these points in mind, the somewhat cavalier investigation that follows is not seeking the definitive origins or destiny of "the modernist worldview." Rather, we hope to ground an examination of sustainability in these often hidden assumptions about reality, identify their influence over our collective sense of truth and value today, and, if possible, find useful leverage in their exposure.

Our preoccupation here is with tracing a Western, and primarily European narrative regarding human–nature relationships. There is a rich, varied, and essential tradition of indigenous thought and practice on this theme which is outside the field of our immediate concern, and more importantly, outside our expertise as well. Certainly a comparative discussion would be valuable.

In the beginning?

Wondering about the origins of our industrial society understandably leads to curiosity about pre-historic humankind. Given a lack of hard evidence concerning the character of early hominid cultures, however, the exercise has typically evolved (or devolved) into more of an image of the observers than the observed. Precisely such projective musings established a formative dialectic for Western modernist thought, the famous "Hobbes versus Rousseau" debate. Was pre-civilized life "solitary, poor, nasty, brutish, and short" (Hobbes)[3], or was it a pastoral Arcadia of noble savagery (Rousseau)[4]? Ultimately, the debate is less descriptive of pre-historical humanity and more indicative of contemporary societal norms: Is the state of nature miserable or glorifying? Does civilization dignify or corrupt? Is an admirable society one that exists within "natural" parameters or one that transcends them? Is human history a rise towards a civilization or a fall from natural grace? These questions persist and modulate throughout the development of modernism, manifesting in a tension between Enlightenment rationality and Romanticism, and in many ways, marking the very crossroads at which we currently stand.

The Mythical Age

From the shadows of our pre-historical ancestors emerges what is sometimes known as the Mythical Age. Norse, Egyptian, or early Greek mythologies paint a common picture of the gods—plural—as immanent in nature (i.e. the divine emerges from within nature). When we encounter nature, we encounter the gods, and vice versa. Secondly, these gods are highly anthropomorphic and deeply entangled with human affairs.[5] As Karen Armstrong points out, truth and accuracy in the Mythical Age was about revealing the interdependence of human, divine, and natural forces.

A notable feature of the age is its ambivalence towards technology and change. Change is often represented by mischievous trickster gods who defy rules and incur punishment for thwarting the order of things. Technology is often represented with lameness or mutation, for instance, as Hephastus/Vulcan, the Greek/Roman god of technology who was born lame, or the smiths of Norse mythology, who were dwarves, stunted, evil, and greedy. At the root of this ambivalence is a powerful notion of a lost "Golden Age," a sense that present society is a pale echo of lost past harmony between social, natural, and divine agencies. As we will see, the notions of the divine as immanent in nature, and of the mutual inter-penetration of the human and the divine would be jettisoned in the dominant traditions of Western culture, while the idea of a past Golden Age is transmuted into a particular religious vision of the original state of innocence.

The Hebraic and Christian traditions

The Hebraic tradition that eventually distinguished itself from the pagan mystery cults and other polytheistic traditions of surrounding empires initiated a slow transition away from an integrative animism and toward a binaried view of man and nature. The long struggle for monotheism would ultimately turn immanent gods into a singular, and increasingly abstract, remote, transcendent God. As a result, the natural world was reimagined as a creation of—rather than the source of—divinity. The sanctity of God is no longer to be found *in* the world, but rather beyond it. What remains is a disenchanted world of mere matter, which, in later Christian cosmology, is slated either for abandonment or destruction.[6]

As the being for whom the world was made, man was the highest creation in the terrestrial realm. As ecofeminists have argued, this dominant position extended beyond non-human creation to include women, that subservient afterthought, "a helper suitable for [Adam]" (Genesis 2:18). This helpfulness, however, is immediately undermined as Eve is soon blamed for precipitating Adam's (and therefore humanity's) fall from innocence. Here the earlier notion of a lost "Golden Age" solidifies into the Garden of Eden, from which, as a result of our disobedience, we are banished and plunged into an adversarial relationship with the natural world:

"The ground is cursed because of you. All your life you will struggle to scratch a living from it" (Genesis 3:17).

Ultimately, this interpretation of ancient, monotheistic cosmologies, which dates back to Lynn White Jr.'s famous work of 1967, helps us see the lasting legacy of monotheism on our dominant perspective of the natural world—a disenchanted, recalcitrant Earth, an object to be dominated and appropriated, under the adversarial dominion of the masculine half of humanity.[7] Men sit at the apex of creation and the natural world is considered the "devil's playground," or a "howling wilderness." Here the seeds of unsustainability appear well planted.

Greek rationalism

However influential in Western cultural history, the Hebraic and Christian traditions were not especially oriented towards understanding the world on its own terms, leaving little inclination to figure out how the world worked in order to dominate it better. Left alone, this worldview might have scratched away at the Earth rather sustainably for millennia. However, contemporary with the development of the teachings that became encoded into the Christian Old Testament, and only 1,200 kilometers away, the evolving tradition of Greek rationalism would turn the world into precisely such a knowable entity: a question to be answered through reasoned argument and observation. To a certain extent, this transformation began with the ancient Hebraic tradition, whose written script was passed from the Hebrews to the Greeks, eventually becoming the Greek alphabet. David Abram has argued that writing played an enormous role in creating a mental universe that came to see human culture as "in here" and the natural world "out there" as, over time, letters were no longer associated with specific natural phenomena and eventually became total abstractions.[8]

Under towering thinkers of this tradition, such as Parmenides, Heraclitus, Democritus, Socrates, Plato, and Aristotle, the pursuit of wisdom became the supreme virtue and the exercise of reason the highest good. Two crucial ideas emerged: First, that humans can ennoble themselves through reason instead of existing perpetually in a fallen state. History need not be seen as a fall from prior glory, indeed, such a thing as progress can emerge. Second, while the Greeks did not have a strong experimental tradition, it is hard to overemphasize the importance, or the novelty, of the view that humans can figure out how the world works through the use of language and reason, and further that the world works according to natural laws.

The last lost world: the medieval perspective

As Roman-era and medieval Christianity slowly conquered the Greek and pagan cultural traditions, ancient secular wisdom fell by the wayside for nearly a millennium. By the eleventh century, however, with the rediscovery—through translations from the Arabic—of Greek texts that had been lost for centuries, the two worldviews came into dynamic tension through the brilliant fusion of St. Thomas Aquinas. This fusion gave rise to a framework of meaning, value, and understanding that would come to dominate Western culture for several centuries.

This framework established the cradle of ideas from which the modernist worldview would emerge. Initially formed within the medieval framework, these new ideas would eventually destroy it. In this way, the Middle Ages offers us a glimpse of Western civilization fundamentally distinct from the one we know and inhabit today, offering a sense of how sealed and self-generating a worldview is, and how such a thing might be so all encompassing and yet all but invisible too.

Conceptually speaking, the world of the Middle Ages was finite, ordered, and meaningful. The terrestrial realm contained the mutable, corrupt Earth and all of creation, with humanity at the top. The heavenly realm, perfect and unchanging, contained God, the *primum movens*, or unmoved mover, and the heavenly bodies. Binding it together in hierarchical and meaningful fashion was the Great Chain of Being. Time had an absolute beginning, and an absolute end. The physical world was explained using Aristotelian physics, with four elements, four humors, and four qualities. Scholarship dedicated itself to aligning religion with Aristotelian science. Religion was not merely a set of beliefs or opinions, but as self-evident as geometry. Knowledge and representation relied on symbolism and meaning. Ultimately, truth was metaphorical, not literal; accuracy was proven against the question "why" not "what" or "how."

Brief interlude: the decline of magic in natural philosophy

All of these characteristics of the medieval worldview were overthrown from the sixteenth to the eighteenth centuries—the period of the Scientific Revolution and the Enlightenment. While the initial motives of this revolution were strongly religious, over time, and especially in the flowering of the new science in the eighteenth century, the religious and mystical under-pinnings of the new worldview were stripped away. By the early nineteenth century, God and religious understanding had been thoroughly edged out of scientific thought. In 1802, when asked by Napoleon what place God had in his system of celestial mechanics, the celebrated astronomer Laplace was able to say "Sire, I have no need of that hypothesis."

As it turns out, the initial cracks in the medieval synthesis came about through the devel-opment of a perspective that later became anathema to the modernist mind. This was a magical or occult understanding of the world that still flourished in the sixteenth and seventeenth centuries. Its origins lay in a deep-rooted spiritual and folkloric healing culture, but it was popularized via a second wave of translations of Greek texts during the Renaissance—the so-called Hermetic writings.

Attributed to Hermes Trismegistus, a powerful sage said to have been a contemporary of Moses, the Hermetic texts presented a view of the world, and the role of humanity within it, that was in key respects fundamentally challenging to the dominant Christian tradition. While Renaissance scholars such as Marsilio Ficino and Pico della Mirandolla (author of *Oration on the Dignity of Man*, 1486) tried to remain within a Christian framework, their Hermetically inspired writings contradicted the passive role for humanity within the medieval framework. The new occult framework promoted a highly active role for humanity in completing God's work, something that strongly influenced early natural philosophers such as Kepler and Newton. As Frances Yates has pointed out, this shift from a passive to an active relationship with the mysteries of creation was a crucial turning point in the seventeenth century.[9]

As a symbol of this increasingly active role in the puzzles of creation, the laboratory was first developed during this period. While early examples from mystics and apothecaries were secre-tive and hidden, eventually the laboratory would become a highly public entity, a tool for demonstrating to the general public the veracity of the new scientific methods.[10]

While the magical perspective represented an important step for the role of humanity and the nature of the cosmos, its mystical elements would ultimately lose out to a purely mechan-ical picture of the world.[11] This framework of thought adopted the activist role for humanity of the magical worldview but saw the world in very different terms—as an essentially mechan-ical system of forces in motion.

René Descartes argued in this period that all that was needed was matter and motion to "build a world." Although Isaac Newton—the most influential early scientist—was an alchemist and a mystic, crucially, he kept those practices quite secret, since they were increasingly frowned upon by the late seventeenth-century scientific culture associated with the Royal Society. His published works, combined with his legacy, which papered over his non-scientific interests, helped to bring about a new view of science as rational, objective, secular (or passively Protestant), masculine, and reductionistic.

Newton's work laid the groundwork for the fusion of empirical and rationalist thinking that was to give rise to a cosmology and physics that reigned triumphant from the later seventeenth to the late nineteenth century.[12] Indeed, the predictive power of this new mechanical understanding was such that it was widely held for centuries that Newton had figured out the universe:

> Nature and nature's laws lay hid in night
> God said, let Newton be, and all was light[13]

Arguably, however, the magical perspective played an indispensable role in bringing about the mechanical perspective. As Table 24.1 makes clear, that perspective shared a number of views with that of the organic medieval worldview, while also containing views characteristic of the mechanical worldview that supplanted it, providing some suggestive evidence of the process of fundamental paradigm change.

Like clockwork: the mechanical world

By the time the transition had solidified, this mechanistic view of the world has splintered reality into primary and secondary properties, where primary properties were those that existed "out there" in the world, and had mathematical characteristics (solidity, extension, motion, number, and figure) and the secondary qualities belong only to the subjective world of perceived characteristics (color, smell, taste, etc.). For the Enlightenment rationalists, only primary qualities were true. Evidence of the senses, if it could not be expressed mathematically, was treated as subjective and untrustworthy.

This basic distinction underwrites the primacy of facts over values, natural laws over cultural practices, and objective truth over subjective experience.[14] In this a sense of modernist dislocation takes root since subjective experience is fundamental to the meaning we attach to existence. Asked to subvert such experience to the primacy of truths independent of it, we grow vulnerable to a sense of meaninglessness. Compelled to live either in the truth of facts

Table 24.1 A comparison of worldviews

Medieval	Magical	Mechanical
Nature purposive ⟷	Nature purposive	Nature purposeless
Imbued with meaning ⟷	Imbued with meaning	Meaningless, except for utility
Truth allegorical ⟷	Truth allegorical	Truth literal
Inherently religious ⟷	Inherently religious	Potentially secular
Non-mathematical	Mathematical (qualitative) ⟷	Mathematical (quantitative)
Non-empirical	Empirical ⟷	Empirical
Passive role for humanity	Active role for humanity ⟷	Active role for humanity

that have no human meaning, or in the experience of values we cannot trust (as they are contested human beliefs), the rational subject grows "infinitely remote from the world."[15] Irrelevant to our realities, we experience a growing sense of detachment, disenchantment, alienation, and perhaps even futility.[16]

Situate this perspective inside a cosmos that has transformed from symbolical and semiotic security to infinite meaninglessness and Galileo's statement about the world he helped engender is especially poignant:

> You wonder that there are so few followers of the Pythagorean opinion [that the Earth moves] while I am astonished that there have been any up to this day who have embraced and followed it. I cannot find any bounds for my admiration, how that reason was able in Aristarchus and Copernicus to commit such a rape on their senses, as in despite thereof to make herself mistress of their credulity.[17]

Thus the modern condition is to inhabit a rather debilitating paradox, and one that reaches all the way to our present dilemma of sustainability.

The science of society

Importantly, Enlightenment rationality did not stop at descriptions of the universe. The beauty of Newtonian laws for the physical world was too tempting for the social world to resist. Invisible truths like the forces of gravity were sought in social domains, as well. Auguste Compte invented the new discipline of sociology explicitly in order to provide a science of human progress based on natural laws. Adam Smith, a devoted follower of Newton's conception of a rule-bound natural order, crafted the "invisible hand," for example, which supposedly (and naturally) balanced production and consumption, and relied on an essential self-interest to generate a beneficent outcome (wealth).[18]

A similar crossing of science into society occurred in the late nineteenth century with social Darwinism, where evolutionary theory was appropriated to justify social norms, such as laissez-faire governance, letting the weak die out, or reading failure and success as just desserts. In other words, a "systematic apologetics of liberal capitalism and conservative politics."[19] Even without social Darwinism, evolutionary theory would have a significant social and psychological impact. Not long after Copernicus moved humanity from the center of the Cosmos to a peripheral planet, Darwin moved humanity from the top of the Great Chain of Being to a branch off the Ape's family tree.

Beyond these two demotions, modernism had one last blow for the status of humanity, this from Freud and the "discovery" of the subconscious mind:

> Man's craving for grandiosity is now suffering the third and most bitter blow from present-day psychological research which is endeavouring to prove to the "ego" of each one of us that he is not even master in his own house, but that he must remain content with the veriest scraps of information about what is going on unconsciously in his own mind.[20]

The full paradox of modernism might be felt in the contradictory dimensions of this statement, mocking humanity's grandiose cravings with humiliation at the hands of our own dazzling capacity to strip the universe of its veils and stare into once hidden secrets. Unfortunately, modernist dualism failed us in the end, leaving us inescapably part of that now denuded world.

Richard Tarnas summarizes the paradox:

> Thus Western man enacted an extraordinary dialectic in the course of the modern era—moving from a near boundless confidence in his own powers, his spiritual potential, his capacity for certain knowledge, his mastery over nature, and his progressive destiny, to what often appeared to be a sharply opposite condition: a debilitating sense of metaphysical insignificance and personal futility, a spiritual loss of faith, uncertainty in knowledge, a mutually destructive relationship with nature, and an intense insecurity concerning the human future. In four centuries of modern man's existence, Bacon and Descartes had become Kafka and Beckett.[21]

It is a remarkable arrival, when one compares the stunning technological achievements of modern industrial society with the underlying, and arguably worsening sense of malaise that hangs over the modern world.[22]

Centuries of second thoughts: romanticism

By the end of the eighteenth century, with the full paradox of Enlightenment rationality and its attendant Scientific and Industrial Revolutions at hand, romanticism was brewing in revolt. Its basic values were the exaltation of imagination, and its correlates sensation, genius, intuition, emotion, creativity, passion, sensibility—everything, that is, dismissed by the coldness of a Newtonian Universe. Consider the following lines of Keats's *Lamia*:

> Do not all charms fly
> At the mere touch of cold philosophy?
> There was an awful rainbow once in heaven:
> We know her woof, her texture; she is given
> In the dull catalogue of common things.
> Philosophy will clip an Angel's wings,
> Conquer all mysteries by rule and line,
> Empty the haunted air, and gnomèd mine—
> Unweave a rainbow as it erewhile made
> The tender-person'd Lamia melt into a shade.

The Romantic agenda was to re-weave the rainbow and rescue the world from the clutches of Enlightenment rationality, a desire to abdicate the modernist agenda outright. Comparing the competing worldviews, in a much simplified form, reveals just how antithetical they often were (Table 24.2). Perhaps the only substantial common ground was agreement on the basic tenets of humanism, a belief that it is within this world and within human consciousness that meaning and purpose are to be found.

Beyond the contrasts and commonality, however, is a strange coexistence. Romanticism did not replace Enlightenment rationality but travelled alongside the growing momentum of industrialism, scientific advance, and rationalization of social structures, eventually transforming itself into multiple forms of resistance across time. Thus, the current crossroads of sustainability can be found in this initial tension between Enlightenment rationality and Romanticism.

Table 24.2 Different worlds

Romantics	Enlightenment
World as unitary organism	World as atomistic machine
Exalted ineffability of inspiration	Exalted enlightenment of reason
Emotion, imagination, creativity	Reason
Nature as live vessel of spirit and source of inspiration	Nature as object for observation and experimentation
Truth as transfiguring and sublime	Truth as testable and objective
Goals: new person, new world	Goal: understand and control world
World as construct of mind	World as independent reality

Social criticism

As the modernist worldview crept down from the cosmos and into social life, disruption was abundant, as was a burgeoning movement of social criticism. Across numerous industries technology was wiping out traditional work practices. In early nineteenth century England, General Ludd led the Army of Redressers attacking mills and destroying machinery, ultimately leading to the passage of the Frame Breaking Act, awarding the death penalty for machine-breaking. Feminism began to emerge in the late eighteenth century, with Mary Wollstonecraft pushing for educational equality.[23] Wollstonecraft's husband, William Godwin, refuted class structures with anarchist arguments: "The good things of the world are a common stock, upon which one man has as valid a title as another to draw for what he wants."[24] Cotton entrepreneur Robert Owen created an inspiring example of socialism in 1800 with his model industrial town of New Lanark, Scotland. Here an economically thriving community offered sanitation and housing, non-profit stores, good schools and working conditions, proving that oppression and profit need not go hand in hand.

The most influential social critic of the nineteenth century was Karl Marx. Like much of the social critique taking aim at modernism, Marx drew on the methods and systems of analyses that emerged out of Enlightenment rationality as a means to dismantle the social outcomes they themselves produced. Marx saw capitalism as a robbery of surplus value from the workers by the capitalists and capitalist rule as an inevitable but temporary phase en route to revolution and, ultimately, a communist utopia. While the political outcomes he envisaged proved fairly unsuccessful, the lens of critique he fashioned proved essential to understanding the modernist political economy.

Contrary to Romanticism, much nineteenth-century social criticism worked from within the structures of Enlightenment rationalism, attempting to turn the tools of modernist thought on its own social outcomes rather than rejecting the encompassing worldview outright. In this way, these two streams of nineteenth-century critique generated a complex pattern of acceptance and rebellion towards modernist ideas.

Environmental critique

This same ambivalence would emerge in the awakening of the environmental movement. In late nineteenth-century America, in particular, a tension emerged almost immediately between preservationist and conservationist approaches to the issue of America's dwindling wilderness. The preservationists stemmed from Romanticism, seeing wilderness as a spiritual font essential to human character, in contrast to the decadent, urban bourgeoisie,[25] whereas the conservationist argument prioritized the value of nature to society. "Man has a right to the use, not the abuse, of the products of nature."[26]

The anthropocentrism of conservationist arguments allowed greater receptivity to the social dimensions of environmental concern, whereas the wilderness preoccupation of preservationist positions was often attacked for its elitism. So while the environmentalism that managed to drive policy was often the former, the emerging world of environmental NGOs and environmental activism was built on the latter. This paradox, along with the anxiety of "running out" of wilderness, and a relative neglect of social dimensions, are several key features of early environmentalism that cast a long shadow through the movement's proliferation.

From the 1960s onwards, modern environmentalism expanded on these basic dimensions. "Deep Ecology," built from Arne Naess, shares the preservationist or Romantic dismissal of the modernist worldview.[27] Murray Bookchin's "Social Ecology" descends from nineteenth-century anarchism, aligning the domination of nature with domination embedded in social structures.[28] "Ecofeminism" argues a similar line, connecting the patriarchal domination of women to the domination of nature.[29] "Green politics" developed as an amalgamation of environmental, peace, social justice, and women's movements, challenging the anthropocentric, technocratic, patriarchal, colonialist, militarist perspectives that damage both the natural world and social relations. "Ecological Economics" focuses on carrying capacity, resource depletion, and environmental impacts, asserting human systems as part of larger ecological systems with limits and constraints.[30]

Since the 1980s, "sustainable development" has linked social and ecological concerns with poverty and development issues, advocating, in the original formulation of the Brundtland Commission,[31] a ten-fold increase in gross world economic output. That is, the assumption within this branch of the sustainability movement is that economies can grow their way out of social and environmental problems. While appealing to government and industry, many question whether environmental redress is possible amidst significant economic growth. Another controversial presence is the "business and sustainability" field, including Corporate Social Responsibility (CSR), where the private sector seeks efficiency gains along with a larger gloss of social responsibility. Critics decry the "greenwashing" dimensions to this, but the innovative capacity of businesses has advanced sustainability nonetheless.

Despite this diversity of responses to environmental and social problems, scientific languages remain almost hegemonic in defining and calibrating environmental challenges. At the same time, so much environmental action is driven not by rational comprehension of scientific data but by Romantic inclinations, themes of identity, emotion, sense of place and belonging, pursuit of meaning, and value beyond cost-benefit analyses. It would appear that not only is sustainability at a crossroads of sorts, but it may be stuck there, dependent on both paths and unable to choose one for fear of abandoning the other.

Post-modernism

Importantly, this sense of being stuck presumes that the distinction between Romanticism and Enlightenment rationality is clear and eternal, as sturdy now as it was in 1827; it presumes that philosophy is still entitled to its coldness, that rainbows have been successfully unwoven, and that Newton's light illuminates our world as well as it did his. In other words, we stand at the crossroads assuming that the two paths ahead still lead away from each other.

Fortunately, new and innovative fields, such as science and technology studies (STS), the sociology of scientific knowledge, complexity studies, Mode 2 science, post-normal science and others have created new methods and forms of knowledge that strongly suggest that Enlightenment "objectivity" and rationality are losing their grip on the increasingly complex world of the Anthropocene. As these approaches show, the founding dichotomies of modernism break down, with neither side living up to their purified ideals.[32] Facts involve heavy doses of "contingency," "negotiation," and "judgement"[33] supporting Thomas Kuhn's famous argument that facts remain artefacts of their paradigms.[34] Knowledge feels like knowledge thanks to its participation in larger frameworks of value, and utility (i.e. institutional programs of knowledge-discovery and validation). Politics, economics, language, history, and belief are not simply the context in which knowledge happens, but constituents from which it is made.

But beyond a Kuhnian critique of knowledge paradigms and the tenacity of worldviews, modernist objectivity must now weather a deeper challenge. As the emerging field of complexity studies illustrates, the pressing "objects" of the Anthropocene—climate change, oceans, resources, pathogens, etc.—are no longer simply natural objects playing out their fates according to their natural laws, but instead are such entanglements of social, technological, and natural agencies that the interactions of these differing agencies produce emergent dynamics sufficient to render them inherently unpredictable. In the words of Allenby and Sarewitz: "The systems we seek to understand are created by the queries we pose."[35] In contexts where human systems (institutions, technologies, cultures, etc.) are constitutive of the object in question, old-fashioned objectivity is an increasingly futile hope.

Instead, an adequate engagement with the Anthropocene may require that we innovate not only with method, but with ontology. That is, not just with how we pursue facts, but what sort of things we think they are once we have found them and, crucially, what capacity we assume they offer in response to challenges such as sustainability.

Philosophy's last hurrah

The work of the late American philosopher Richard Rorty addresses this issue with his historical notion of redemptive truth:

> I shall use the term "redemptive truth" for a set of beliefs that would end, once and for all, the process of reflection on what to do with ourselves. Redemptive truth ... would fulfill the need that religion and philosophy have attempted to satisfy ... To believe in redemptive truth *is to believe that there is something that stands to human life as elementary physical particles stand to the four elements*—something that is the reality behind the appearance, the one true description of what is going on, the final secret.[36]

Rorty suggests Western civilization has drawn redemptive truth from three successive sources: God, then philosophy, and finally literature.

The age of philosophy, Rorty argues, culminated in two intellectual currents: German Idealism (now dead) and materialistic metaphysics (i.e. science).

> Materialist metaphysics, however, is, in fact, pretty much the only version of redemptive truth presently on offer. It is philosophy's last hurrah, its last attempt to provide redemptive truth and thereby avoid being demoted to the status of a literary genre.[37]

However, as we may be seeing in the stalled progress of environmental action, materialist metaphysics has no redemptive value, argues Rorty.

> Modern science is a gloriously imaginative way of describing things, brilliantly successful for the purpose for which it was developed—namely, predicting and controlling phenomena.[38] But it should not pretend to have the sort of redemptive power claimed by its defeated rival, idealist metaphysics.[39]

In other words, according to Rorty, science cannot serve the scientistic role it has so often been assigned in modernist culture.

An inevitable truce?

If the crossroads of sustainability are conceived as representing "modernism vs. something new" (or, more crudely, as "more science vs. no more science"), it is little wonder that we feel stuck. An outright rejection of modernist epistemic advances and technological capacity would seem suicidal now that the fate of the planet is in our hands like never before. Yet given the levels of complexity generated by this new circumstance, science no longer fashions the kind of clarion certainty that once made it such an appealing vision to live by.[40] As Rorty's argument suggests, fundamental questions of how to live on this Earth cannot be answered by science even at the best of times.

Following thinkers such as Richard Rorty, Jerome Ravetz, Silvio Functowicz, Naomi Oreskes, Bruno Latour, Lorraine Daston, Brian Wynne, Sheila Jasanoff, and many others, it seems essential to free scientific description from unreasonable expectations to offer traditional truth-telling (objective, value-independent, yet morally compelling) in a very different world. What may be useful is a shift away from the way we conceive of, practice, and wield science in its wide range of discursive contexts, in order to release it from increasingly contradictory positions brought about by deepening conditions of complexity.[41] The agenda here is neither to depart from science or continue serving its hegemony, but to free it from its social role of redemptive truth and release it from its epistemic task of objectivity. That is to say, rather than choosing one direction or another at the crossroads, it seems best we leave the intersection altogether and take another route.

In this we might hope to rehabilitate ourselves from a kind of metaphysical schizophrenia that forced modernism and its rebellions into its fractured realities— subject *or* object, fact *or* value, science *or* sentiment, truth *or* experience, nature *or* culture. The Anthropocene is a world in which such disjunctions are not only incoherent but costly. Eliminating subjective elements of meaning and value in order to determine what is objectively real is perilous when such elements are inextricable at not only symbolic but also material levels as well. The hope for a world whose reality exists in objective and absolute terms rests on a faith in false dichotomies, an encircling moat that keeps "the real" safe from "the imaginary" while—as the crises of sustainability attest—we splash around in the inextricable mixtures of the two, whether we notice it or not. Taking full stock of a hybrid world is the conceptual, perceptual, and linguistic challenge of the late modernism.[42]

Recent history: incoherence and the
Intergovernmental Panel on Climate Change

We are all, to a degree, Trojan horses to our own transformative visions. Old paradigms remain hidden in our languages, institutions, value-structures, and the instincts and imaginative resources we deploy with every effort to map and make sense of experience. Indeed, there may be precious little in our minds that can inhabit a world in which facts and values, objectivity and subjectivity, flow together in any kind of intuitive, useful, or acceptable way. Thus arguments such as these often smack of philosophical indulgence peripheral to the real work of saving the real world. To try and hold out beyond the crossroads long enough to feel at home elsewhere, we should remind ourselves that this is no theoretical extravagance, that the context of practice has already departed the intersection, and that it is our theoretical, conceptual, imaginative resources that are lagging behind. Consider the activities of the Intergovernmental Panel on Climate Change (IPCC), for example.

The IPCC was created to provide a global assessment, by the community of researchers studying climate change, of the state of the field with respect to climate science, impacts and adaptation, and mitigation, and to communicate this assessment to policy-makers.[43] To accomplish this goal, the IPCC developed in the 1990s and early 2000s several approaches to the science–policy interface, such as the publishing a periodic *Summary for Policymakers*, holding plenary sessions between lead authors and government representatives, and creating contact groups at these plenary meetings to discuss controversial texts. These approaches were intended to facilitate interaction between science and policy communities, and thus contribute to situating the IPCC scientific assessment process within an intergovernmental framework. In many ways, these processes represent compelling examples of moving beyond fact–value dichotomies in order to contend with Anthropocenic complexity in full force, as scientists and politicians wrestle with data, meaning, and expression in developing what would become authoritative representations of "the climate."[44]

However, any hope that such texts would be received as welcome examples of navigating the Anthropocene would prove naïve. IPCC processes have sparked significant controversy and criticism with regard to the methodological rigor and credibility of their interpretations and products, with critics decrying the "politicization of science" of such processes.[45] In other words, the hybrid practices of the IPCC, which arguably demonstrate a keen sense of the inapplicability of modernist dichotomies in this complex arena are then contradicted by official IPCC rhetoric which returns to, and celebrates, such distinctions.[46]

In this regard, we might understand the failure here not as one of science nor of politics, but, rather, of ontology; a failure of our fundamental imaginative instincts to make sense of an entity—the climate—that is giving us little choice *but* to move beyond that familiar crossroads: a failure to clear sufficient conceptual, imaginative terrain needed to openly display such an entity within the public sphere without wondering which path it belongs to.[47]

Conclusion

As this historical sweep has attempted to show, the emergence of the modernist view of human–nature relations has generated a deep ambivalence that has marked environmental thought and action since the inception of environmental concern. This ambivalence is rooted in the founding dichotomies of the Modernist epistemic project, (e.g. primary/secondary properties, subject/object, fact/value, etc.) and produces a deep dilemma regarding the best way to understand and act within contexts of strained human–nature relations. This dilemma has either polarized environmental concern, (e.g. science vs. sentiment, objective descriptions vs. subjective meanings), thereby impoverishing action by pressuring engagement to choose one path or the other; or it has left environmental thought and action stalled, knowing any genuinely transformative progress needs to have it both ways, while nonetheless accepting the dichotomy that insists they are mutually exclusive.

Several ideas from the present discussion might be understood as rendering this dilemma inert. First, the rise of sociological and cultural understandings of science practices, where the activities of science itself fail to maintain their purified ideals of objectivity. In this, we see the active involvement of human values, human subjects, human cultures in debating the nature of inevitably human worlds, yet in ways in which we might take increasing comfort, even reassurance, freeing ourselves from the belief that human subjectivity has no place in the epistemic (and therefore metaphysical, moral, existential) dimensions of society. In this, the alienation of modernism might soften a little.

Second, now that we have entered the age of the Anthropocene we find ourselves living on a planet that operates not according to eternal natural laws, but according to a deep integration of natural tendencies with the volition of human and technological systems. Thus, even if science could free itself from the "subjective" influences of paradigms, beliefs, values, interests, perspectives, capacities, etc. and look objectively upon the objects of "nature," just as the IPCC example makes clear, it would have little choice but to contend with an increasingly complex, "post-normal," socio-technical-natural world.

Lastly, this brief and rather schematic historical review not only uncovers some of the roots of our current deeply-held assumptions about the world, but also reveals that these assumptions are mutable, and have indeed changed radically in the past. This is a hopeful finding, because it suggests that the entangled challenges of the Anthropocene may not only give rise to the dilemmas we noted above, but also call out new ways of thinking, new approaches based on forms of ontological hybridity that may offer us an escape from these dilemmas, and new ways forward that offer hope in a time of apocalyptic pessimism.

Notes

1   John B. Robinson, "Squaring the Circle? Some Thoughts on the Idea of Sustainable Development," *Ecological Economics* 48, no. 4 (2004): 369–84.
2   Bruno Latour, *Politics of Nature* (Cambridge, MA: Harvard University Press, 2004).
3   Thomas Hobbes, "Of the Naturall Condition of Mankind as Concerning their Felicity, and Misery," in *Leviathan*, R. Tuck, ed. (Cambridge: Cambridge University Press, 1991). Originally published 1651.
4   Jean-Jacques Rousseau, *A Discourse on Inequality*, M. Cranston, trans. (London: Penguin, 1984). Originally published 1754.
5   Karen Armstrong, *A Short History of Myth*, vol. 1 (London: Canongate Books, 2004).
6   In Christian cosmology, this material world becomes subordinate to a linear sense of time. Created as an artefact that becomes the home of man, it will be destroyed as an integral step in how the relationship between God and man unfolds. This destruction of the present Earth was central to medieval Christianity in Europe.
7   Lynn White, Jr., "The Historical Roots of our Ecologic Crisis," *Science* 155, no. 3767 (1967): 1203–7. White's interpretation of the legacy of Christianity has been contested by many subsequent authors who have identified stewardship traditions within Christianity (cf. White's own identification of St. Francis of Assisi as the patron saint of environmentalism). It seems clear, however, that stewardship was not the dominant instinct during medieval times and that, in any case, the unique status and role of humanity in the world was common to the stewardship tradition as well and highly influential in Western cultural history.
8   David Abram, *The Spell of the Sensuous: Perception and Language in a More-Than-Human World* (London: Vintage, 2012).
9   Frances A. Yates, *Giordano Bruno and the Hermetic Tradition* (Abingdon: Routledge, 2014).
10   Steven Shapin and Simon Schaffer, *Leviathan and the Air-Pump: Hobbes, Boyle, and the Experimental Life* (Princeton, NJ: Princeton University Press, 2005).
11   This mechanical perspective also had its roots in Greek rationalism, as famously stated by Democritus in the fourth century BCE: "According to convention there is the bitter and the sweet, the hot and the cold. And according to convention there is colour. In reality, there are atoms and the void."

12 Alexandre Koyré, *From the Closed World to the Infinite Universe*, vol. 1 (Library of Alexandria, 1957).

13 Alexander Pope, "Epitaph on Sir Isaac Newton" (1727).

14 It is no coincidence that mathematics became *the* language of science in this period. Newton's famous work of 1687 translates (from Latin) as the *Mathematical Principles of Natural Philosophy*.

15 Bruno Latour, *We Have Never Been Modern* (Cambridge, MA: Harvard University Press, 2012), 56.

16 Marshall Berman, *All That is Solid Melts into Air: The Experience of Modernity* (London: Verso, 1983).

17 Galileo Galilei, *Dialogue Concerning the Two Chief World Systems* (Berkeley, CA: University of California Press, 1967). Originally published 1632.

18 Adam Smith. *The Wealth of Nations* (Chicago, IL: University of Chicago Bookstore, 2005). Originally published 1776.

19 Floyd Matson, *The Broken Image: Man, Science, and Society* (New York: George Braziller, 1964).

20 Sigmund Freud, *A General Introduction to Psychoanalysis* (New York: Boni and Liveright Publishers, 1920). Delivered as lectures 1915–17, and first translated into English in 1920.

21 Richard Tarnas, *The Passion of the Western Mind: Understanding the Ideas that Have Shaped our World View* (New York: Ballantine, 1991), 393–4.

22 Charles Taylor, *The Malaise of Modernity* (Toronto: House of Anansi, 1991).

23 "How many women thus waste life away the prey of discontent, who might have practised as physicians, regulated a farm, managed a shop, and stood erect, supported by their own industry, instead of hanging their heads surcharged with the dew of sensibility, that consumes the beauty to which it at first gave lustre." Mary Wollstonecraft, *Vindication of the Rights of Woman*, vol. 29 (Calgary: Broadview Press, 1982), ch. 9. Originally published 1792.

24 William Godwin, *Enquiry Concerning Political Justice* (Kitchener: Batoche, 2001). Originally published 1793.

25 Henry David Thoreau: "The story of Romulus and Remus being suckled by a wolf is not a meaningless fable. The founders of every state which has risen to eminence have drawn their nourishment and vigour from a similar wild source. It is because the children of the Empire were not suckled by the wolf that they were conquered and displaced by the children of the northern forest who were." Henry David Thoreau, "Walking," in *Excursions: The Writings of Henry David Thoreau* (Boston, MA: Houghton Mifflin, 1893).

26 George Perkins Marsh, *Man and Nature,* David Lowenthal, ed. (Seattle, WA: University of Washington Press, 2015). Originally published 1864.

27 Arne Naess and Satish Kumar, *Deep Ecology* (Phil Shepherd Production, 1992).

28 Murray Bookchin, *The Ecology of Freedom* (Naperville, IL: New Dimensions Foundation, 1982).

29 Maria Mies and Vandana Shiva, *Ecofeminism* (London: Zed Books, 1993).

30 David Pearce, "An Intellectual History of Environmental Economics," *Annual Review of Energy and Environment*, 27 (2002): 57–81.

31 World Commission on Environment and Development (WCED), *Our Common Future* (Oxford: Oxford University Press, 1987). The commission was chaired by Gro Harlem Brundtland.

32 Steve Woolgar and Bruno Latour, *Laboratory of Life: The Construction of Scientific Facts* (Princeton, NJ: Princeton University, 1986); Lorraine Daston, "Science Studies and the History of Science," *Critical Inquiry* 35, no. 4 (2009): 798–813.

33 Karin Knorr-Cetina, *The Fabrication of Facts: Toward a Microsociology of Scientific Knowledge* (Konstanz: Bibliothek der Universität Konstanz, 1984), 115–40.

34 Thomas Kuhn, *The Structure of Scientific Revolutions* (Chicago, IL: University of Chicago Press, 1962).

35 Braden R. Allenby and Daniel Sarewitz, *The Techno-Human Condition* (Cambridge, MA: MIT Press, 2011), 186.

36 Richard Rorty, "Philosophy as a Transitional Genre," in his *Philosophy as Cultural Politics*, Philosophical Papers, vol. 4. (Cambridge: Cambridge University Press, 2007), 7.

37 Ibid., 14.

38 Given the recent discussion on complexity and the Anthropocene, the key term here is "phenomena." Rorty is addressing purified constituent elements subject to prediction and control and not the complex objects of the Anthropocene. We can confidently predict the motion of a specified gas in a vacuum, but not its nature and actual behavior in the world, let alone its human health effects. Predicting and controlling these simple elements does not allow us to predict and control the imbroglios into which they fall in complex contexts.

39 Rorty, "Philosophy as a Transitional Genre," 20.

40　See Naomi Oreskes: "But the idea that science ever could provide proof upon which to base policy is a misunderstanding (or misrepresentation) of science, and therefore of the role that science ever could play in policy. In all but the most trivial cases, science does not produce logically indisputable proofs about the natural world. At best it produces a robust consensus based on a process of inquiry that allows for continued scrutiny, re-examination, and revision." Naomi Oreskes, "Science and Public Policy: What's Proof Got to Do With It?," *Environmental Science and Policy* 7 (2004): 369–83.

41　J. R. Ravetz, "'Climategate' and the Maturing of Post-Normal Science," *Futures* 43 (2011): 149–57.

42　Perhaps one of the more detailed accounts of how to proceed with such a challenge is offered in Latour's *Politics of Nature*. For an extended discussion of the implications of the approach argued for here for science and technology policy, see Brian Wynne et al., *Taking European Knowledge Society Seriously*, Report of the Expert Group on Science and Governance to the Science, Economy and Society Directorate, Directorate-General for Research, EUR2270 (Brussels: European Commission, 2007).

43　For more, see www.ipcc.ch.

44　Alison Shaw and John Robinson, "Relevant but not Prescriptive? Science Policy Models within the IPCC," *Philosophy Today*, 48 (Supplement) (2004): 84–95.

45　See, for instance, S. Boehmer-Christiansen, "Global Climate Protection Policy: The Limits of Scientific Advice (Part I)," *Global Environmental Change* 4, no. 2 (1994): 140–59.

46　Shaw and Robinson, "Relevant but not Prescriptive?," 84–5.

47　One attempt to think through some of the practical consequences for sustainability of the arguments we have made here can be found in David Maggs and John B. Robinson, "Racalibrating the Anthropocene," *Environmental Philosophy* 13, no. 2 (2016): 175–94.

References

Abram, David. *The Spell of the Sensuous: Perception and Language in a More-Than-Human World*. London: Vintage, 2012.

Allenby, Braden R. and Daniel Sarewitz. *The Techno-Human Condition*. Cambridge, MA: MIT Press, 2011.

Armstrong, Karen. *A Short History of Myth*, vol. 1. London: Canongate Books, 2004.

Berman, Marshall. *All That is Solid Melts into Air: The Experience of Modernity*. London: Verso, 1983.

Boehmer-Christiansen, S. "Global Climate Protection Policy: The Limits of Scientific Advice (Part I)." *Global Environmental Change* 4, no. 2 (1994).

Bookchin, Murray. *The Ecology of Freedom*. Naperville, IL: New Dimensions Foundation, 1982.

Daston, Lorraine. "Science Studies and the History of Science." *Critical Inquiry* 35, no. 4 (2009): 798–813.

Freud, Sigmund. *A General Introduction to Psychoanalysis*. New York: Boni and Liveright Publishers, 1920.

Galilei, Galileo. *Dialogue Concerning the Two Chief World Systems*. Berkeley, CA: University of California Press, 1967.

Godwin, William. *Enquiry Concerning Political Justice*. Kitchener: Batoche, 2001.

Hobbes, Thomas. "Of the Naturall Condition of Mankind as Concerning their Felicity, and Misery." In his *Leviathan*, R. Tuck, ed. Cambridge: Cambridge University Press, 1991.

Knorr-Cetina, Karin. *The Fabrication of Facts: Toward a Microsociology of Scientific Knowledge*. Konstanz: Bibliothek der Universität Konstanz, 1984.

Koyré, Alexandre. *From the Closed World to the Infinite Universe*, vol. 1. Library of Alexandria, 1957.

Kuhn, Thomas. *The Structure of Scientific Revolutions*. Chicago, IL: University of Chicago Press, 1962.

Latour, Bruno. *We Have Never Been Modern*. Cambridge, MA: Harvard University Press, 2012.

Latour, Bruno. *Politics of Nature*. Cambridge, MA: Harvard University Press, 2004.

Maggs, David and John B. Robinson, "Recalibrating the Anthropocene." *Environmental Philosophy* 13, no. 2 (2016): 175–94.

Marsh, George Perkins. *Man and Nature*, David Lowenthal, ed. Seattle, WA: University of Washington Press, 2015.

Matson, Floyd. *The Broken Image: Man, Science, and Society*. New York: George Braziller, 1964.

Mies, Maria and Vandana Shiva. *Ecofeminism*. London: Zed Books, 1993.

Naess, Arne and Satish Kumar. *Deep Ecology*. Phil Shepherd Production, 1992.

Oreskes, Naomi. "Science and Public Policy: What's Proof Got to Do With It?" *Environmental Science and Policy* 7 (2004): 369–83.

Pearce, David. "An Intellectual History of Environmental Economics." *Annual Review of Energy and Environment* 27 (2002): 57–81.

Pope, Alexander. "Epitaph on Sir Isaac Newton" (1727).

Ravetz, J. R. "'Climategate' and the Maturing of Post-Normal Science." *Futures* 43 (2011): 149–57.

Robinson, John B. "Squaring the Circle? Some Thoughts on the Idea of Sustainable Development." *Ecological Economics* 48, no. 4 (2004): 369–84.

Rorty, Richard. "Philosophy as a Transitional Genre." In his *Philosophy as Cultural Politics*, Philosophical Papers, vol. 4. Cambridge: Cambridge University Press, 2007.

Rousseau, Jean-Jacques. *A Discourse on Inequality*. M. Cranston, trans. London: Penguin, 1984.

Shapin, Steven and Simon Schaffer. *Leviathan and the Air-Pump: Hobbes, Boyle, and the Experimental Life.* Princeton, NJ: Princeton University Press, 2005.

Shaw, Alison and John Robinson. "Relevant but not Prescriptive? Science Policy Models within the IPCC." *Philosophy Today*, 48 (Supplement) (2004).

Smith, Adam. *The Wealth of Nations.* Chicago, IL: University of Chicago Bookstore, 2005.

Tarnas, Richard. *The Passion of the Western Mind: Understanding the Ideas that Have Shaped our World View.* New York: Ballantine, 1991.

Taylor, Charles. *The Malaise of Modernity.* Toronto: House of Anansi, 1991.

Thoreau, Henry David. "Walking." In *Excursions: The Writings of Henry David Thoreau.* Boston, MA: Houghton Mifflin, 1893.

White, Jr., Lynn. "The Historical Roots of our Ecologic Crisis." *Science* 155, no. 3767 (1967): 1203–7.

Wollstonecraft, Mary. *Vindication of the Rights of Woman*, vol. 29. Calgary: Broadview Press, 1982.

Woolgar, Steve and Bruno Latour. *Laboratory of Life: The Construction of Scientific Facts.* Princeton, NJ: Princeton University, 1986.

World Commission on Environment and Development. *Our Common Future.* Oxford: Oxford University Press, 1987.

Wynne, Brian, et al. *Taking European Knowledge Society Seriously.* Report of the Expert Group on Science and Governance to the Science, Economy and Society Directorate, Directorate-General for Research, EUR 2270. Brussels: European Commission, 2007.

Yates, Frances A. *Giordano Bruno and the Hermetic Tradition.* Abingdon: Routledge, 2014.

25

RETHINKING THE HISTORY OF AGRICULTURE WITH SUSTAINABILITY IN MIND

Jeremy L. Caradonna

The history of agriculture forms the backbone of the origin story that our culture has been telling itself for ages. The story goes something like this. Agriculture was founded in Mesopotamia, along the fertile Tigris and Euphrates Rivers, both of which originate in modern-day Turkey and meander through the Middle East, before spilling into the Persian Gulf. The decision to domesticate grasses and legumes was an ingenious response to the climatic fluctuations at the end of the last Ice Age. The main changes took place between 13,000 and 9,500 years ago, as nomadic scavengers slowly became more sedentary, less reliant on hunted foods, and more inclined to domesticate animals on a new invention called the farm.

As agriculture developed, so too did our species, as we shifted gradually from simple hunter-gatherers to more complex and civilized herders and farmers. Populations grew, fed by the huge surplus of cereal grains. Cities appeared, and with them, fortifications. Informal militias became well-organized armies. Written languages were invented, as were specialized crafts and professions. Religion and a clerical class appeared on the scene. Complex social hierarchies and art forms ended the ageless and artless state of anarchy. Although it is fundamentally a story of Progress, it is also littered with moral and ecological lessons. As agriculture spread to Europe and Asia, the Middle East was left in the dust—literally. The irrigation that watered the crops had brought salt deposits to the soil, which eventually lowered the region's fertility and output. By 2000 BC, yields had fallen by half. The Middle East became desertified, dusty, infertile. The area that is today Iraq was once a fertile breadbasket, yet became a man-made desert through a drawn-out experiment in depletion. Likewise North Africa, today a desert, was once the granary of the ancient world. In both areas, soils were not properly rebuilt, and the newly invented plow led to over-tillage and the loss of topsoils. Although agriculture brought civilization, it also furnished an early lesson in unsustainable living.

The problem with this story is not so much that it is wrong, as that it leaves out quite a lot, ignores the agricultural developments in other parts of the world, and equates agriculture with merely one style of farming—and one that proved rather unsustainable in the long run.[1] For context, it turns out that the indigenous Aborigines of Australia have been practicing agriculture for at least 30,000 years—much earlier than any food-growing culture that existed in Mesopotamia—and this agriculture thrived for some tens of thousands of years. In *Dark Emu: Black Seeds—Agriculture or Accident?*, Bruce Pascoe shows that grindstones found at Cuddie Springs in New South Wales were grinding flour, at this early date, from seeds harvested from grasses that were, at the very least, tended by human hands.[2] Yet when Europeans eventually came to Australia some thousands of years later, all they saw were "boundless grassy plains"— not ecosystems that had been carefully cultivated by native peoples.[3]

Europeans eventually adopted Mesopotamian-style agriculture, affecting not only very impactful events in world history, but also the way in which agriculture is conceived. Eurocentrism pervades the history of agriculture precisely because what counts as "agriculture" is defined narrowly and with European habits in mind—the most prominent of which is the growing of annuals or semi-annuals, such as cereal grains (wheat, barley) or leguminous plants (lentils, peas), on clearly demarcated socio-agro-ecological plots of land called "farms." The European-style farm is a culturally specific form that was assumed to be the norm everywhere, and lack of a "farm" therefore meant a lack of agriculture.

The problem is that agriculture can exist in countless forms, many of which would look quite foreign to eyes accustomed to seeing vast monocultures of soil-depleting annuals. There are many ways to farm; many ways to be a farmer. If agriculture is defined as any kind of edible-food cultivation—which is how it should be defined—then agriculture is indeed both very ancient and very plainly evident all over the globe. If we look beyond the fields of wheat and barley, we see whole ecosystems cultivated through the strategic use of fire to fertilize soils and create the conditions for beneficial species to grow; we see diverse, resilient, and highly complex polycultures that weave together symbiotic plant communities; we see a whole range of perennial and tree-grown crops that look quite different than fast-growing, short-lived annuals. Many extant "food forests" across the Americas are anthropogenic in origin, not uncultivated wilderness, and these too are farms (or orchards) of a sort.[4]

Indeed, as historians have begun to rethink the narrow focus on Mesopotamia and Europe, the diverse global traditions of food cultivation have forced a wholesale reappraisal of the history and meaning of agriculture. Some of the key terms that now appear in the historiography of food growing include "agroforestry" (the growing of tree-derived crops), "food forests" (human-made forests that center on food-producing species), "agrosilviculture" (combining food-yielding trees with timber species), "silvopastoralism" (non-crop trees and domesticated animals, for instance the cork-yielding oak forests and acorn-eating pigs of Iberia), "agrosilvopastoralism" (combining trees that yield fruits, nuts, and vegetables with timber crops, domesticated animals, and potentially annuals, biennials, and biannuals). One frequently sees the terms "closed-loop agriculture," which refers to farming systems that circulate nutrients and are self-sustaining, and "polyculture," which is any kind of growing system that brings together diverse species and varieties for systemic resilience. These terms and concepts are largely absent from classic agricultural history, and derive largely from agroecology, permaculture, and indigenous studies.

Rethinking the history of agriculture also has profound impacts on how we think about sustainability, both its past and its future. There is a very clear set of stepping-stones that link together Mesopotamian agriculture, European and North American industrial (or "conventional") agriculture, the Green Revolution of the 1960s, and the Gene Revolution of the 1990s. There is no single, universally agreed upon method for measuring sustainable agriculture, but a system's ability to produce a consistent quantity of food, the wellbeing of those who grow or tend it, the system's overall diversity and resilience, the wise use of water resources, and the health of soils and micro-fauna are all key considerations.[5] By any measure, our current agricultural system, which clearly has roots in Mesopotamia, is not particularly sustainable. It depletes soils at an alarming rate, drains aquifers, introduces enormous quantities of chemical fertilizers and pesticides into ecosystems (and human bodies), simplifies ecosystems by reducing genetic diversity, and concentrates tremendous wealth and power in the hands of an ever-dwindling number of corporations.[6] Even more disconcerting is the fact that, in many places, yields are declining, just as they did in the ancient Middle East. Consequently, those interested in the history and practice of sustainable agriculture should probably look elsewhere for inspiration, towards societies that actually *increased* their agro-biological diversity and soil fertility over time.

Luckily, the history of food cultivation and ethnobotany (the human use of plants) is becoming ever more interesting. In the works of Nancy J. Turner, we learn that Coast Salish people of southern Vancouver Island—a people who Charles Mann incorrectly labels as "nomads"—cultivated camas bulbs and other root vegetables, such as silverweed and riceroot, and used fire to encourage the growth of Garry oaks, which furnished edible acorns.[7] The Salish and their neighbors also created clam beds and other forms of "aquaculture"—a subject that has been sorely under-studied by historians of food-making. Indeed, only recently have historians and ethnobotanists come to realize that the "nomads" of Coastal British Columbia actively farmed their own food. Take, for example, the cultivation of springbank clover (*Trifolium wormskjoldii*):

> Springbank Clover rhizomes were a highly important vegetable for a number of coastal aboriginal groups in British Columbia, including Straits Salish, Nuu-chah-nulth, Kwakwaka'wakw, Nuxalk, Haisla and Haida. Just as the growing and harvesting of Blue Camas bulbs by the Straits Salish around Victoria can be termed semi-agricultural, the harvesting of Springbank Clover roots by the Kwakwaka'wakw, Nuu-chah-nulth, and Haida had characteristics akin to agriculture. They divided extensive patches of Clover growing along river flats into rectangular beds that were owned by families or individuals in a village group and passed from generation to generation.[8]

Comparable forms of agriculture or ecosystem cultivation existed all over the Americas and, indeed, all over the world. In the work of Charles Mann, we learn that maize has been cultivated by Mesoamericans for possibly 10,000 years, and that it forms part of a highly sustainable and polycultural system called the *milpa*, in which beans, corn, squash, peppers, tomatoes, avocadoes, and many other species were (and are) grown together, as part of a sustainable closed-loop.[9] Jared Diamond teaches us that the Highlanders of New Guinea have been farming continuously for 7,000 years, although the practice of agroforestry could extend back much further.[10] In the works of William Balée, William Denevan, Susannah Hecht, and many other archaeologists, we learn that the native peoples of the Amazon cultivated over a hundred species of trees and plants, and that agroforestry has dominated the agricultural landscape throughout the region, where palms, mango, acai, guava, papaya, avocado, and many other fruit, vegetable, and nut trees have been grown for thousands of years. We also learn that Amazonians have been building fertile, man-made soils called "Terra Preta" for millennia.[11] Increasingly, we also learn that highly sustainable forms of agriculture arose in Northern China, too, along with other parts of Asia and the South Pacific.

Not all of these farming systems antedate those of Mesopotamia, but they are examples, nonetheless, of very ancient and sustainable forms of food and ecosystem cultivation that deserve a place in the grand narrative of agriculture. They also tend to share a common, unhappy trait, which is that Europeans, when they eventually stumbled upon these inventive agricultural splendors, *always* misjudged what they saw. Armed with a very narrow understanding of what constituted agriculture, European invaders had difficulty identifying the biologically diverse and very un-farm-like forms of agriculture that existed elsewhere. Thus, for instance, when the Spaniard Francisco de Orellana floated down the Amazon River in the

1540s—he and his men were the first European infiltrators in the area—his untrained eye found lots of people, but, confusingly, no farms. He saw "jungle," not a dense and productive form of polycultural agroforestry.[12]

In a sense, many historians still see only "jungle"; still see only "boundless grassy plains," rather than the diverse, non-monocultural farming systems of pre-industrial peoples. We repeat the mistakes of the colonial powers by imposing narrow definitions of agriculture onto the past, and by reducing complex and resilient societies to the level of hapless scavengers. (Not that there is anything wrong with being a hunter-gatherer.) But by viewing agriculture through European eyes, we miss out on surprising truths. One example comes from Amazonia, where new research suggests that the region had huge pre-industrial cities and dense populations, and was able to achieve as much without European-style wheat fields. The archaeologist Betty Meggers spent her entire career arguing that this was impossible, and that "the level to which a culture can develop is dependent upon the agricultural potentiality of the environment it occupies."[13] Her "law of environmental limitation of culture" could well be correct, in a general sense, but she certainly misjudged the "agricultural potentiality" of Amazonia.

In the end, the Mesopotamian-agriculture-as-Progress story is problematic for many reasons, not the least of which is that this form of agriculture ultimately proved hopelessly unsustainable, and further, that post-Mesopotamian-style agriculture has now conquered the globe and is reproducing a soil-depleting, unsustainable system on a much larger scale, which means we live in a destructive global food system with low resilience and high levels of associated risk. But beyond this sad fact is the simple reality that agriculture has many homelands, many inventors, many forms. As we rethink the history of agriculture, we need, first and foremost, to write indigenous peoples and ecosystem cultivation back into the story of agriculture. We also need to cast a much wider net when we employ the culturally specific concepts of "farm," "farming," and "agriculture." Indeed, those who study the history of agriculture, along with those who practice agriculture in the present, could stand to learn quite a lot from pre-industrial, aboriginal, and non-Western modes of food growing and ecosystem cultivation, and especially from the wonders of agroforestry, aquaculture, perennial grains, and the many colorful and resilient forms of polyculture that nourished humble peoples long before chemical fertilizers and pesticides came on the scene.

Acknowledgement

A different version of this essay first appeared on the History News Network (http://historynewsnetwork.org) in 2016.

Notes

1   David R. Montgomery, *Dirt: The Erosion of Civilizations* (Berkeley, CA: University of California Press, 2007).
2   Bruce Pascoe, *Dark Emu: Black Seeds—Agriculture or Accident?* (Broome, Australia: Magabala, 2015). It helped that Australia, in the Southern Hemisphere, was mostly free of ice during the last Ice Age.
3   Larissa Behrendt, "Indigenous Australians Know We're the Oldest Living Culture—It's in our Dreamtime," *The Guardian* (September 22, 2016).
4   A. Gómez-Pompa, J. Salvador Flores, and V. Sosa, "The 'Pet Kot': A Man-Made Tropical Forest of the Maya," *Interciencia* 12 (1987): 10–15; William Cronon, "The Trouble with Wilderness; or, Getting Back to the Wrong Nature," in *Uncommon Ground: Rethinking the Human Place in Nature*, W. Cronon, ed. (New York: W. W. Norton & Co., 1995), 69–90.
5   Jonathan R. B. Fisher et al., "How Do We Know an Agricultural System is Sustainable?," 2013, retrieved from www.nature.org/science-in-action/science-features/ag-sustainability-metrics.pdf.

6   Marcel Mazoyer and Laurence Roudart, *Histoire des agricultures du monde: Du néolithique à la crise contemporaine* (Paris: Seuil, 2002); Thomas R. Sinclair and C. J. Sinclair, *Bread, Beer and the Seeds of Change: Agriculture's Imprint on World History* (Wallingford: CABI, 2010); Vandana Shiva, *The Violence of Green Revolution: Third World Agriculture, Ecology, and Politics* (London: Zed Books, 1992); P. Oosterveer and D. A. Sonnenfeld, *Food, Globalization and Sustainability* (New York: Routledge, 2012).

7   Nancy J. Turner, *Ancient Pathways, Ancestral Knowledge: Ethnobotany and Ecological Wisdom of Peoples of Northwestern North America* (Montreal: McGill-Queen's University Press, 2014).

8   Nancy J. Turner, *Food Plants of Coastal First Peoples* (Victoria: Royal BC Museum, 1995).

9   Charles A. Mann, *1491: New Revelations of the Americas Before Columbus* (London: Vintage, 2006).

10   Jared Diamond, *Collapse: How Societies Choose to Fail or Succeed* (London: Penguin, 2005).

11   William Denevan, *Cultivated Landscapes of Native Amazonia and the Andes* (Oxford: Oxford University Press, 2001); William Balée, *Cultural Forests of the Amazon: A Historical Ecology of People and Their Landscapes* (Tuscaloosa, AL: University of Alabama Press, 2015); Susannah B. Hecht, "Indigenous Soil Management and the Creation of Terra Mulata and Terra Preta in the Amazon Basin," in *Amazonian Dark Earths: Origins, Properties and Management*, eds. J. Lehmann et al. (New York: Springer, 2003), 1–21; see also Charles A. Mann, *1493: Uncovering the New World Columbus Created* (London: Vintage, 2011).

12   Robert Tindall, Frédérique Apffel-Marglin, and David Shearer, *Sacred Soil: Biochar and the Regeneration of the Earth* (Berkeley, CA: North Atlantic Books, 2017).

13   Mann, *1491*, 328.

References

Balée, William. *Cultural Forests of the Amazon: A Historical Ecology of People and Their Landscapes*. Tuscaloosa, AL: University of Alabama Press, 2015.

Behrendt, Larissa. "Indigenous Australians Know We're the Oldest Living Culture—It's in our Dreamtime," *The Guardian* (September 22, 2016).

Cronon, William. "The Trouble with Wilderness; or, Getting Back to the Wrong Nature." In *Uncommon Ground: Rethinking the Human Place in Nature*, W. Cronon, ed., 69–90. New York: W. W. Norton & Co., 1995.

Denevan, William. *Cultivated Landscapes of Native Amazonia and the Andes*. Oxford: Oxford University Press, 2001.

Diamond, Jared. *Collapse: How Societies Choose to Fail or Succeed*. London: Penguin, 2005.

Fisher, Jonathan R. B., et al. "How Do We Know an Agricultural System is Sustainable?" 2013. Retrieved from www.nature.org/science-in-action/science-features/ag-sustainability-metrics.pdf.

Gómez-Pompa, A., J. Salvador Flores, and V. Sosa. "The 'Pet Kot': A Man-Made Tropical Forest of the Maya." *Interciencia* 12 (1987): 10–15.

Hecht, Susannah B. "Indigenous Soil Management and the Creation of Terra Mulata and Terra Preta in the Amazon Basin." In *Amazonian Dark Earths: Origins, Properties and Management*, J. Lehmann et al., eds., 1–21. New York: Springer, 2003.

Mann, Charles A. *1491: New Revelations of the Americas Before Columbus*. London: Vintage, 2006.

Mann Charles A. *1493: Uncovering the New World Columbus Created*. London: Vintage, 2011.

Mazoyer, Marcel, and Laurence Roudart. *Histoire des agricultures du monde: Du néolithique à la crise contemporaine*. Paris: Seuil, 2002.

Montgomery, David R. *Dirt: The Erosion of Civilizations*. Berkeley, CA: University of California Press, 2007.

Oosterveer, P., and D. A. Sonnenfeld. *Food, Globalization and Sustainability*. New York: Routledge, 2012.

Pascoe, Bruce. *Dark Emu: Black Seeds—Agriculture or Accident?* Broome, Australia: Magabala, 2015.

Shiva, Vandana. *The Violence of Green Revolution: Third World Agriculture, Ecology, and Politics*. London: Zed Books, 1992.

Sinclair, Thomas R., and C. J. Sinclair. *Bread, Beer and the Seeds of Change: Agriculture's Imprint on World History*. Wallingford: CABI, 2010.

Tindall, Robert, Frédérique Apffel-Marglin, and David Shearer. *Sacred Soil: Biochar and the Regeneration of the Earth*. Berkeley, CA: North Atlantic Books, 2017.

Turner Nancy J. *Food Plants of Coastal First Peoples*. Victoria: Royal BC Museum, 1995.

Turner, Nancy J. *Ancient Pathways, Ancestral Knowledge: Ethnobotany and Ecological Wisdom of Peoples of Northwestern North America*. Montreal: McGill-Queen's University Press, 2014.

26

THE GENE REVOLUTION AND THE FUTURE OF AGRICULTURE

Thierry Vrain

Biology and its many sub fields of research have seen a recent boon with the emergence of molecular biology. Much of the gigantic progress of biochemistry and physiology over the past three decades can be ascribed to this new field, and in particular to genetic engineering. We cannot watch life's processes under a microscope—the molecular realm is out of reach of all our most sophisticated instruments—but the ability to transfer DNA code for proteins and their RNA master molecules from one organism to another has allowed scientists to peer into and manipulate cellular processes at the molecular level.

This chapter will provide an overview on and summarize the major developments of the new science of genetic engineering, and consider its impacts with sustainability in mind. The Gene Revolution, as I call it, has had a profound impact on agriculture, and has, in effect, extended the reach of the Green Revolution of the 1960s and 1970s.

The Green Revolution was fundamentally about growing crops that would increase their yield, as the specially hybridized seeds could withstand and utilize higher amounts of synthetic nitrogen and other fertilizers. It also brought about a dramatic increase in the amount of chemical pesticides and insecticides that were used in agriculture. Indeed, the Green Revolution was a boon to the chemical industry, which became a powerful player in agriculture beginning in the 1960s. Recent research has shown that the Green Revolution, especially in Mexico and India, which became the poster children of this movement, greatly increased fossil fuel consumption, soil degradation, chemical usage, and water consumption (since Green Revolution plants tended to have high water needs), and was a major driver of climate change.[1] The Revolution increased food production in some areas—although it failed disastrously in others. It did not solve the poverty issues in countries such as India, and has led to a decrease in social wellbeing among Indian farmers, who are now reliant on seed and chemical companies to function as farmers.

Then, in the 1990s, came the Gene Revolution, which has been a huge commercial success in agriculture and one more boon to the chemical industry. And yet this Revolution started as a green and social enterprise. The Gene Revolution goes back to the early 1970s, in fact. In 1972, Paul Berg at Stanford University combined the DNA of two viruses, and in 1973 Stanley Cohen at Stanford and Herbert Boyer at the University of California in San Francisco combined bacterial DNA. Shortly thereafter, Paul Berg and others organized the International Congress on Recombinant DNA molecules—at Asilomar, near San Francisco, in February

1975. One hundred and forty scientists and lawyers came to evaluate the potential risks and benefits and establish safety and containment rules. Paul Berg stated in 2008: "In the 33 years since Asilomar, researchers around the world have carried out countless experiments with recombinant DNA without reported incident. Many of these experiments were inconceivable in 1975, yet as far as we know, none has been a hazard to public health."[2]

In the 1970s, most of the scientists engaged in recombinant DNA research were working in universities and government labs and were able to work and voice opinions without having to look over their shoulders. This is no longer the case, as much if not most molecular biology research (and universities in general) is now funded by chemical companies, and the research is driven by commercial considerations.

While the first efforts were mostly in the medical field, with human proteins made in bacteria and on antibodies, the technology has been most successful in agriculture. Indeed, agriculture has been transformed in the past 20 years. Genetically modified crops are modified to perform different functions, and thus there are many different kinds of GMOs. For example, many crops have been modified—developed mostly with public funding—to become resistant to insect pests or pathogens. The best known case is the modified papaya fruit in Hawaii. When the papaya industry in Hawaii lost significant numbers of trees to a new virus disease, researchers found a way to vaccinate the trees by genetic modification. The trees are true GMOs and they saved the Hawaiian papaya industry—with no pesticides involved.[3]

A second category of GMO crops has been genetically modified for an improved nutrient profile. The best example of this is Golden Rice, a variety of rice modified with two genes to produce and accumulate provitamin A (that is, beta carotene) that the human body converts into vitamin A. Golden Rice is a true GMO, but while it is intended to improve lives in tropical countries, the rice has never been shown to eradicate vitamin A deficiency. It is a "commercial" bust, and it has been thoroughly rejected by prospective recipients.[4]

The third category of GMOs is intended to reduce spoilage. The best example of this is the Arctic brand of modified apples that were created in the department of Agriculture Canada in which I worked, at the Summerland Research Station. The apple represses one of its own genes to interrupt the normal maturing process and thereby considerably increase shelf life. There are no foreign genes and no pesticides involved. It is a modified apple, a true GMO, and though there has been significant debate on the impacts and value of this apple, it has not had a significant impact on the apple industry.[5]

There are still other crops that have been modified to tolerate drought, salt, high temperatures, or which are able to produce particular chemicals. We might consider these various crop experiments as a fourth, more speculative category of GMO, since none of these modified crops have had an extensive impact and are grown on a very small acreage, a few thousand acres perhaps, and many remain on the shelf or in the lab. For one reason or another they are very minor or non-commercial successes. For instance, despite extensive laboratory tests on genetically modified wheat varieties, there is no GMO wheat that is currently on the market.

The fifth and most important category of GMO—what represents the overwhelming success of the Gene Revolution—is the RoundUp Ready crops from Monsanto, sold mostly in North and South America. RoundUp Ready crops are crops that have been modified to withstand a chemical called glyphosate, which is the main ingredient in the herbicide RoundUp. Once a crop has been modified, it can be doused with RoundUp, which will kill all the surrounding plants, while sparing the modified organism. This still is a revolutionary technology that simplifies weed management programs. In 2017, there are well over 500 million acres of genetically modified crops, and practically all of it is planted with RoundUp

Ready soy and corn, canola, cotton, sugar beet, and alfalfa. Those six RoundUp Ready crops dominate all the other "minor" developments of the Gene Revolution.

Monsanto released RoundUp Ready soy and corn in 1996, and the other crops in the years following. Instead of spraying weeds early in the season before seeding the crop, the RoundUp Ready technology allows farmers to spray the weeds much later, after emergence of the crop. A huge convenience and a more efficient weed management help to explain the corporate success of these crops in the last 20 years. The practice of post-emergent weed control, as well as the corporate ownership of the modified seeds, introduced a new paradigm in agriculture. Most of the decisions regarding how food and feed are grown and harvested are now in the hands of a very small and shrinking number of chemical corporations. The chemical corporation Monsanto with a line of products including Agent Orange with Dioxins, PCBs, polystyrene, saccharin and aspartame, recently morphed into a "Life" company. They own the genetically modified seeds and they sell the chemical necessary to activate them. A farmer who chooses to buy the RoundUp herbicide and the Roundup Ready seeds buys only the right to plant the seeds once. The farmers have become sharecroppers on their own land, and the ancient practice of seed-saving has quickly eroded. Farmers are forced to sign a contract that illegalizes the replanting of seeds from their own crops. In effect, family farms have turned into corporate farms growing corporate crops—the corporation does not even need to own the land. Farmers plant and pocket only what they are given.

As of 2016, the herbicide RoundUp is the most successful agrochemical ever. Glyphosate is the active molecule of the formulated RoundUp mix. Glyphosate was created as a glycine analog in 1950 by a chemist working for Cilag, a Swiss chemical company. It was then shelved and reappeared many years later at the US Patent Bureau in 1964 as a chelator capable of cleaning up (dissolving) mineral plaques in industrial pipes and boilers. Glyphosate was a "descaling agent" that essentially broke down the mineral buildups that occurred on the inside of industrial equipment.

The next public appearance of glyphosate was again at the US Bureau of Patents in 1969, but this time it was listed as a herbicide. Indeed, it was (and is) a very powerful non-selective herbicide that kills all plants that it touches. It was promptly tested for mammalian toxicity by Monsanto, the chemical company owner of the herbicide patent. The Environmental Protection Agency (EPA) in the United States deemed it safe and Monsanto formulated it into the herbicide RoundUp, containing 40 percent glyphosate and a few other ingredients, some of which are now known to be toxic.

RoundUp became a fast success as the herbicide of choice for all situations. Farmers had never seen a herbicide so lethal and non-selective. From 1974, when it was first commercialized, until 1996, RoundUp became a leading herbicide, alongside other heavy-hitters, such as Atrazine and 2,4-D.

Since 1996, RoundUp has become even more widely used, since, as of that year, it was now in an intimate relationship with RoundUp Ready crops, which only made sense when doused with this herbicide. It is now the weed killer of choice, sprayed all over the place to control vegetation in forestry and agriculture, and therefore its residues are found everywhere. It is in the soil, in the water (in drinking water in most places), in the forests, in gardens and in the fields, and of course in food and feed and animals. The most successful agrochemical ever has accumulated and bioaccumulated throughout our ecosystems.

Since 1996, more and more RoundUp Ready crops have hit the market, thus increasing the use of RoundUp itself. First came soy and corn, which quickly gobbled up the market share. Today, over 90 percent of the corn and soybeans that are grown in the US and Canada have been genetically modified to withstand RoundUp. Then came a cavalcade of new RoundUp

Ready crops: cotton, canola, sugar beet, and alfalfa. The farmers adopted the crops instantly because the technology works, although a few stopped to consider the health and ecological effects of these new technologies. Today, many of these new GMOs, including sugar beet and canola, possess nearly the entire market share.

As an important aside, canola was developed over many years of intense breeding by a team of scientists working for the Department of Agriculture in Canada—hence funded by Canadian tax dollars. The amazing success of the RoundUp technology makes for a troubling and historically unprecedented situation in which a chemical corporation literally owns 100 percent of the crop, because the GMO technology is considered proprietary. This is the situation that India and many other countries are trying to avoid—what is now called "biopiracy," in which genetic material is privatized and copyrighted, drastically undermining the independence and liberties of farmers who now farm, literally, someone else's crops.[6]

The total acreage of GMOs worldwide is half a billion acres, and the total use of RoundUp on GMOs and many other crops and areas as stated above, is several quarts per acre. The current estimates stand at 8.6 billion kilograms of glyphosate sprayed over the last 20 years, with a sizable portion of the usage found in the United States and Canada. Since 1996, there has been nearly a 15-fold increase in the use of glyphosate.[7]

Surprisingly, many other crops are also sprayed with RoundUp that *are not* RoundUp Ready. Why would farmers want to kill their own plants? In short, to simplify the harvesting of the plants. In the last 15 years or so, it has become routine and normal to spray cereal grains and other seed crops, such as peanut and sunflower, at harvest time. Since these grains and plants are not engineered, the chemical spray kills the weeds and the crops and dries them out. So-called "chemical drying" is now widely practiced, and glyphosate turns out to be an effective desiccant.[8]

The increased use of glyphosate has raised serious questions about the effects that this chemical has had on ecosystems and human health. But given the complexity of the debate over GMOs and glyphosate, it is important to draw some clear distinctions before moving on. First, it is unclear whether the practice of genetic modification itself has any adverse health effects on mammals. The famous and controversial Séralini study, which studied the effects of GMOs and glyphosate on rats, found that rats that ate GMO feed without RoundUp developed tumors and suffered from many health defects.[9] But in general, the research on GMOs shows that genetically modified crops are, on their own, quite harmless. The modified papaya might make for a case in point. Second, however, and broadening the scope, GMOs are generally farmed as part of a system of agriculture—called conventional or industrial agriculture—that has depleted soils worldwide and has toxified the environment through the increased usage of chemical fertilizers, which cause eutrophication in waterways, as well as pesticides, herbicides, insecticides, and fungicides, which damage ecosystems and have poisoned many communities and farm workers.[10] Further, GMOs have contributed to a decline in biological diversity and a lack of agricultural resilience, and thus create vulnerabilities in the food supply. Ecologists agree that agroecosystem resilience requires biological diversity, since diverse species and polycultures—the opposite of monocultures—are better equipped to deal with pests and climatic stresses.[11] That is, GMOs now play a central role in an unsustainable agricultural system. Third, and more to the point of this chapter, although GMOs themselves appear to be, on the whole, benign for human consumption, a growing body of peer-reviewed scientific literature suggests that the herbicide glyphosate is anything but benign.

Glyphosate is a synthetic amino acid analog. Amino acid analogs can easily be mistaken for the real thing by the protein metabolism of the cells. Genes code for proteins that make all the other molecules in the cells. A gene is read by a protein complex (RNA polymerase) that transcribes the DNA message into an RNA molecule, called a "messenger." The messenger RNA molecule—pulled by unknown proteins—travels outside of the cell nucleus into the cytoplasm and is delivered to another RNA and protein complex called a ribosome. Ribosomes translate RNA messages written with four letters into proteins written with twenty letters. The messenger RNA is read three letters at a time by a group of twenty or so RNA molecules, called transfer RNAs. Each transfer RNA molecule carries one amino acid. It appears that the transfer RNA that carries the amino acid glycine is easily fooled into carrying the analog glyphosate instead. Glyphosate is glycine methyl phosphonate, the amino acid glycine with a small tail. Glyphosate misincorporated into a large number of proteins changes the structure of the proteins, and of course their function. Protein dysfunction means cell death and eventually organ failure—that is, chronic illness.[12]

Indeed, a spate of articles on glyphosate's dangers have appeared over the past ten years or so in the most reputable journals in microbiology, toxicology, and chemistry, including *Current Microbiology, Ecotoxicology and Environmental Safety, Chemical Research in Toxicology, International Journal of Toxicology, Toxicology in Vitro*, and many others. Although the research on glyphosate can be contradictory, what is striking is that studies not funded by industry tend to find that, in fact, this chemical is a highly potent antibiotic, an endocrine disruptor, and an agent that causes oxidative stress—that is, it is a carcinogen.

In fact, 15 years ago Monsanto filed glyphosate as an antibiotic with the US Bureau of Patent. The patent was granted in 2010. The claims of the patent are very clear that all organisms making their own proteins, such as plants and bacteria and most microscopic animals, are killed by this chemical. Specifically one part per million obliterates the malaria agent and all bacteria studied. A German group confirmed in 2013 that 1 ppm of glyphosate kill the bacteria of the poultry microbiome. Glyphosate is a powerful antibiotic masquerading as a herbicide.[13]

There is currently no data on the effects of glyphosate on the human microbiome since there has been a noticeable lack of follow-up on the safety of glyphosate from the regulatory agencies. However, there have been legal limits, established in 1996, of the ppm of glyphosate that is allowed to exist in foodstuffs. Curiously, the US and Canada recently raised these limits in 2013 to much higher levels, well above 20 ppm for all GMO food ingredients and chemically dried crops, despite the fact that the recent science suggests that even low exposure to glyphosate is toxic. The only reasonable explanation for the increase in limits is that industry lobbied for higher allowances, given that glyphosate is now very commonly found in human bodies.

The research on glyphosate's effects on the microbiome are particularly disconcerting. The microbiome is a superorgan that all animals have, usually connected to the digestive tube. Our intestine is a 26 foot long bioreactor where thousands of species of bacteria—about 100 trillion of them, easily the size and weight of a human brain—are bathing in and feeding from the nutrient solution we provide. In exchange for being fed, these bacteria finish the digestion of our food and give us many of the molecules we require for proper functioning at the cellular level. We call them vitamins and such. A good, basic explanation of the microbiome is found in the popular best-seller by Giulia Enders, called *Gut*.[14]

The last five years have seen an explosion of microbiome research. We are learning very quickly how much influence and control the microbiome has over most of our organs, particularly the brain, the immune system, and of course all the digestive system organs. Chronic disturbances in the equilibrium of these bacteria result in inadequate controls and dysfunctional organs and chronic illness.

We have no epidemiological data about glyphosate damaging the human microbiome. However, we do know that antibiotics in general do damage to the gut flora in our intestines, and we also have a growing body of science that shows that glyphosate has adverse effects on the microbiome and immune system of other mammals, including those of rats and cows.[15] Further, we know that the legal residue limits of glyphosate in grains and seeds crops are more than powerful enough to kill the microbiome bacteria of birds and other animals. Levels of glyphosate as low as 1 ppm can do severe damage, let alone the 20 ppm or more that are, as of 2013, allowed to exist on human foodstuffs. In addition to the research on glyphosate's effects on the microbiome, there is also a growing realization that this chemical affects mammalian hormones, too, which mean it is an endocrine disruptor. One study has found that "glyphosate based herbicides are toxic and endocrine disruptors in human cell lines."[16]

Finally, there is the question of the carcinogenicity of glyphosate. The International Agency for Research on Cancer (IARC), the authority that identifies carcinogens to the World Health Organization, classified glyphosate as a probable carcinogen in 2015—"probable" because there is no human study. This designation was based on the many peer-reviewed studies that have been published since 2000, and indeed the literature is full of reports identifying glyphosate as causing oxidative stress.[17] All of the criteria that point to carcinogenicity are met in these studies, which is why the IARC decided to act. The WHO originally accepted the recommendation of the IARC, although after heavy pressure from the industries that create and profit from these chemicals and crops, the WHO suddenly decided to reverse its decision.[18] These sorts of about-faces have become commonplace, given the tension that exists between self-interested industry and independent scientific researchers. Indeed, the EPA had listed glyphosate as a carcinogen as early as 1985, until Monsanto successfully pressured to have that designation dropped in 1991. Although the WHO changed its mind about the carcinogenicity of glyphosate, as of 2016, the State of California still lists it as a carcinogen.

If it is true that glyphosate is a carcinogen, this would have profound implications for the GMO industry, our agricultural system, and indeed our entire society. It would mean that we have poured 10 million metric tons of a carcinogen onto edible and textile crops. There would be a backlash against the chemical industry, and much soul-searching, along the lines of what we saw in the 1960s, when Rachel Carson exposed DDT as ecosystem hazard.[19] The same goes for the research suggesting that glyphosate is an endocrine disruptor, and likewise for its status as an antibiotic, which is even less in question, given that it is now patented as an antibiotic as of 2010.

In closing, it would be hard to say that glyphosate and the bulk of GMOs have, on the whole, contributed much to sustainable agriculture. They have undermined the self-sufficiency and liberties of farmers worldwide, decreased biological diversity in agriculture, and have introduced unprecedented levels of a harmful chemical into the ecosystem. And the concerns about GMOs and glyphosate will not go away. Sixty-four countries now require GMO labeling and the European Union (EU) tightly regulates the use of glyphosate. In fact, as of 2016 many EU states have banned over the counter sales of RoundUp. Polls in North America show that over 90 percent of Americans want GMOs labeled, over health concerns, although thus far industry has succeeded in preventing such labeling.[20] Further, there is now a large and growing body of scientific literature that suggests that glyphosate, even at low levels of exposure, is a powerful antibiotic, an endocrine disruptor, and a carcinogen. Even though the literature is still divided on the matter, the precautionary principle would suggest that we should listen more carefully to the non-industry-funded scientists who have discovered the adverse health effects caused by glyphosate. This issue is all the more important because the chemical industry is fast developing new and even more powerful herbicides to supplant glyphosate. Glyphosate was never going to last very long because weeds quickly adapt to herbicides, and in fact over 20 commonly found weeds are now resistant to glyphosate.[21] As glyphosate becomes an ineffective herbicide, and as concerns for its toxicity grow, and finally, as more and more countries, states, and municipalities restrict or ban the usage of glyphosate, the time is now to reconsider the Gene Revolution and figure out a more sustainable approach to agriculture.

Notes

1   Vandana Shiva, *The Violence of Green Revolution: Third World Agriculture, Ecology, and Politics* (London: Zed, 1992); Pingali L. Prabhu, "Green Revolution: Impacts, Limits, and the Path Ahead," *PNAS* 109, 31 (2012): 12,302–8; R. B. Singh, "Environmental Consequences of Agricultural Development: A Case Study from the Green Revolution State of Haryana, India," *Agriculture, Ecosystems and Environment*, 82, 1–3 (2000): 97–103.
2   Cited in Paul Berg, "Meetings that Changed the World: Asilomar 1975: DNA Modification Secured," *Nature* 455, 290–291 (2008): published online.
3   M. M. M. Fitch, R. M. Manshardt, D. Gonsalves, J. L. Slightom, and J. C. Sanford, "Virus Resistant Papaya Derived from Tissues Bombarded with the Coat Protein Gene of Papaya Ringspot Virus," *BioTechnology* 10 (1992): 1466–72.
4   X. Ye et al., "Engineering the Provitamin A (Beta-Carotene) Biosynthetic Pathway into (Carotenoid-Free) Rice Endosperm," *Science* 287, 5451 (2000): 303–5.
5   N. Carter, "Petition for Determination of Nonregulated Status: Arctic Apple (*Malus × domestica*), Events GD743 and Gs784," 2012, retrieved on June 10, 2017 from www.aphis.usda.gov/brs/aphis-docs/10_16101p.pdf.
6   See Vandana Shiva, *Biopiracy: The Plunder of Nature and Knowledge* (Brooklyn, NY: South End Press, 1999).
7   Charles M. Benbrook, "Trends in Glyphosate Herbicide Use in the United States and Globally," *Environmental Sciences Europe* 28, 3 (2016), doi:10.1186/s12302-016-0070-0.
8   See Jeremy Caradonna and Thierry Vrain, "The 'Non-GMO' Label Doesn't Go Far Enough: Taking Stock of GMOs and Glyphosate," retrieved from www.resilience.org/stories/2016-09-29/the-non-gmo-label-doesn-t-go-far-enough-taking-stock-of-gmos-and-glyphosate.
9   G.-E. Séralini et al., "Republished Study: Long-Term Toxicity of a Roundup Herbicide and a Roundup-Tolerant Genetically Modified Maize," *Environmental Sciences Europe* 26, 14 (2014), doi:10.1186/s12302-014-0014-5. This controversial paper, which linked GM maize to the development of tumors and other diseases in rats, first appeared in 2012, but was retracted in 2013, after heavy pressure from the chemical industry. An independent panel reviewed the findings and the paper was republished, in a different journal, in 2014.
10 Michael Pollan, *Omnivore's Dilemma: A Natural History of Four Meals* (London: Penguin, 2006); P. Oosterveer and D. A. Sonnenfeld, *Food, Globalization and Sustainability* (New York: Routledge, 2012); Samuel Fromartz, *Organic, Inc.: Natural Foods and How They Grew* (New York: Harcourt, 2006); Johan Rockström et al., "Planetary Boundaries: Exploring the Safe Operating Space for Humanity," *Ecology and Society* 14, 2 (2009): article 32, retrieved from www.ecologyandsociety.org/vol14/iss2/art32; Regis Magauzi et al., "Health Effects of Agrochemicals Among Farm Workers in Commercial Farms of Kwekwe District, Zimbabwe," *Pan African Medical Journal* 9, 26 (2011), retrieved from www.ncbi.nlm.nih.gov/pmc/articles/PMC3215548.
11 See Stephen R. Gliessman, *Agroecology: The Ecology of Sustainable Food Systems, Third Edition* (London: CRC Press, 2014).
12 A. Samsel and S. Seneff, "Glyphosate Pathways to Modern Diseases V: Amino Acid Analogue of Glycine in Diverse Proteins," *Journal of Biological Physics and Chemistry* 16 (2016): 9–46.
13 A. Shehata, W. Schrödl, A. Aldin, H. Hafez, M. Krüger, "The Effect of Glyphosate on Potential Pathogens and Beneficial Members of Poultry Microbiota In Vitro," *Current Microbiology* 66, 4 (2013): 350–58.
14 Giulia Enders, *Gut: The Inside Story of our Body's Most Underrated Organ* (Vancouver: Greystone, 2015).
15 A. Samsel and S. Seneff, "Glyphoste's Suppression of Cytochrome P450 Enzymes and Amino Acid Biosynthesis by the Gut Microbiome: Pathways to Modern Diseases," *Entropy* 15, 4 (2013): 1416–63; W. Ackermann, M. Coenen, W. Schrödl, A. A. Shehata, and M. Krüger, "The Influence of Glyphosate on the Microbiota and Production of Botulinum Neurotoxin During Ruminal Fermentation," *Current Microbiology* 70, 3 (2015): 374–82, doi:10.1007/s00284-014-0732-3; W. Schrödl et al., "Possible Effects of Glyphosate on Mucorales Abundance in the Rumen of Dairy Cows in Germany," *Current Microbiology* 69, 6 (2014): 817–23.

16 C. Gasnier et al., "Glyphosate-Based Herbicides are Toxic and Endocrine Disruptors in Human Cell Lines," *Toxicology* 262, 3 (2009): 184–91.

17 Two examples are: S. Thongprakaisang et al., "Glyphosate Induces Human Breast Cancer Cells Growth via Estrogen Receptors," *Food and Chemical Toxicology* 59 (2013): 129–36; J. George, S. Prasad, Z. Mahmood, and Y. Shukla, "Studies on Glyphosate-Induced Carcinogenicity in Mouse Skin. A Proteomic Approach," *Journal of Proteomics* 73, 5 (2010): 951–64.

18 Daniel Cressey, "Widely Used Herbicide Linked to Cancer," *Scientific American*, March 25, 2015; Arthur Neslen, "Glyphosate Unlikely to Pose Risk to Humans, UN/WHO Study Says," *The Guardian*, May 16, 2016.

19 Rachel Carson, *Silent Spring* (New York: Houghton Mifflin, 1962).

20 See, for instance, ABC News, "Poll: Skepticism of Genetically Modified Foods," retrieved from http://abcnews.go.com/Technology/story?id=97567.

21 In a sense, pesticides and herbicides have backfired greatly in the medium term. Here is Samuel Fromartz: "Cornell University entomologist David Pimentel estimates that despite a more than tenfold increase in pesticide use since 1945, crop losses due to pests have nearly doubled. The NAS counted over 440 pesticide-resistant insects in 1986 and found the number rising. An international survey identified 289 herbicide-resistant weeds in 2004, with the biggest number in the United States. The adoption of GE crops has allowed freer use of herbicides, but this has also led to a rise in herbicide-resistant weeds" (Fromartz, *Organic, Inc.*, 51–2).

References

ABC News. "Poll: Skepticism of Genetically Modified Foods." Retrieved from http://abcnews.go.com/Technology/story?id=97567.

Ackermann, W., M. Coenen, W. Schrödl, A. A. Shehata, and M. Krüger. "The Influence of Glyphosate on the Microbiota and Production of Botulinum Neurotoxin During Ruminal Fermentation." *Current Microbiology* 70, 3 (2015): 374–82, doi:10.1007/s00284-014-0732-3.

Benbrook, Charles M. "Trends in Glyphosate Herbicide Use in the United States and Globally." *Environmental Sciences Europe* 28, 3 (2016), doi:10.1186/s12302-016-0070-0.

Berg, Paul, "Meetings that Changed the World: Asilomar 1975: DNA Modification Secured." *Nature* 455, 290–291 (2008): published online.

Caradonna, Jeremy, and Thierry Vrain. "The 'Non-GMO' Label Doesn't Go Far Enough: Taking Stock of GMOs and Glyphosate." Retrieved from www.resilience.org/stories/2016-09-29/the-non-gmo-label-doesn-t-go-far-enough-taking-stock-of-gmos-and-glyphosate.

Carson, Rachel. *Silent Spring*. New York: Houghton Mifflin, 1962.

Carter, N. "Petition for Determination of Nonregulated Status: Arctic Apple (Malus × domestica), Events GD743 and Gs784." 2012. Retrieved on June 10, 2017 from www.aphis.usda.gov/brs/aphisdocs/10_16101p.pdf.

Cressey, Daniel. "Widely Used Herbicide Linked to Cancer." *Scientific American*, March 25, 2015.

Enders, Giulia. *Gut: The Inside Story of our Body's Most Underrated Organ*. Vancouver: Greystone, 2015.

Fitch, M. M. M., R. M. Manshardt, D. Gonsalves, J. L. Slightom, and J. C. Sanford, "Virus Resistant Papaya Derived from Tissues Bombarded with the Coat Protein Gene of Papaya Ringspot Virus." *BioTechnology* 10 (1992): 1466–72.

Fromartz, Samuel. *Organic, Inc.: Natural Foods and How They Grew*. New York: Harcourt, 2006.

Gasnier, C., et al. "Glyphosate-Based Herbicides are Toxic and Endocrine Disruptors in Human Cell Lines." *Toxicology* 262, 3 (2009): 184–91.

George, J., S. Prasad, Z. Mahmood, and Y. Shukla. "Studies on Glyphosate-Induced Carcinogenicity in Mouse Skin. A Proteomic Approach." *Journal of Proteomics* 73, 5 (2010): 951–64.

Gliessman, Stephen R. *Agroecology: The Ecology of Sustainable Food Systems*, third edition. London: CRC Press, 2014.

Magauzi, Regis, et al. "Health Effects of Agrochemicals Among Farm Workers in Commercial Farms of Kwekwe District, Zimbabwe." *Pan African Medical Journal* 9, 26 (2011), retrieved from www.ncbi.nlm.nih.gov/pmc/articles/PMC3215548.

Neslen, Arthur. "Glyphosate Unlikely to Pose Risk to Humans, UN/WHO Study Says." *The Guardian*, May 16, 2016.

Oosterveer, P., and D. A. Sonnenfeld. *Food, Globalization and Sustainability*. New York: Routledge, 2012.

Pollan, Michael. *Omnivore's Dilemma: A Natural History of Four Meals*. London: Penguin, 2006.

Prabhu, Pingali L. "Green Revolution: Impacts, Limits, and the Path Ahead." *PNAS* 109, 31 (2012): 12,302–8.

Rockström, Johan, et al. "Planetary Boundaries: Exploring the Safe Operating Space for Humanity." *Ecology and Society* 14, 2 (2009): article 32, retrieved from www.ecologyandsociety.org/vol14/iss2/art32.

Samsel, A., and S. Seneff. "Glyphosate Pathways to Modern Diseases V: Amino Acid Analogue of Glycine in Diverse Proteins." *Journal of Biological Physics and Chemistry* 16 (2016): 9–46.

Samsel, A., and S. Seneff. "Glyphoste's Suppression of Cytochrome P450 Enzymes and Amino Acid Biosynthesis by the Gut Microbiome: Pathways to Modern Diseases." *Entropy* 15, 4 (2013): 1416–63.

Schrödl, W., et al. "Possible Effects of Glyphosate on Mucorales Abundance in the Rumen of Dairy Cows in Germany." *Current Microbiology* 69, 6 (2014): 817–23.

Séralini, G.-E., et al. "Republished Study: Long-Term Toxicity of a Roundup Herbicide and a Roundup-Tolerant Genetically Modified Maize." *Environmental Sciences Europe* 26, 14 (2014), doi:10.1186/s12302-014-0014-5.

Shehata, A., W. Schrödl, A. Aldin, H. Hafez, and M. Krüger. "The Effect of Glyphosate on Potential Pathogens and Beneficial Members of Poultry Microbiota In Vitro." *Current Microbiology* 66, 4 (2013): 350–58.

Shiva, Vandana. *Biopiracy: The Plunder of Nature and Knowledge*. Brooklyn, NY: South End Press, 1999.

Shiva, Vandana. *The Violence of Green Revolution: Third World Agriculture, Ecology, and Politics*. London: Zed, 1992.

Singh, R. B. "Environmental Consequences of Agricultural Development: A Case Study from the Green Revolution State of Haryana, India." *Agriculture, Ecosystems and Environment* 82, 1–3 (2000): 97–103.

Thongprakaisang, S.. et al. "Glyphosate Induces Human Breast Cancer Cells Growth via Estrogen Receptors." *Food and Chemical Toxicology* 59 (2013): 129–36.

Ye, X., et al. "Engineering the Provitamin A (Beta-Carotene) Biosynthetic Pathway into (Carotenoid-Free) Rice Endosperm." *Science* 287, 5451 (2000): 303–5.

INDEX

For Product Safety Concerns and Information please contact our EU
representative GPSR@taylorandfrancis.com Taylor & Francis Verlag GmbH,
Kaufingerstraße 24, 80331 München, Germany

Printed and bound by CPI Group (UK) Ltd, Croydon, CR0 4YY
08/05/2025
01864359-0004